Household Equipment

LOUISE JENISON PEET, Ph.D.

Professor of Family Environment
Iowa State University

MARY S. PICKETT, Ph.D.

Professor of Family Environment
Iowa State University

MILDRED G. ARNOLD, B.S.

Associate Professor, Consumer Sciences and Housing
Colorado State University

With Chapter 19, Maximizing Satisfaction in
Work in the Home by

ILSE H. WOLF, Ed.D.

Professor Emeritus Home Management
Texas Tech University

NEW YORK/LONDON/SYDNEY/TORONTO

Household Equipment

SEVENTH EDITION

JOHN WILEY & SONS, INC.

Library of Congress Cataloging in Publication Data:

Main entry under title:

Household equipment.

First-5th ed. by L. J. Peet and L. E. S. Thye.
1. Home economics—Equipment and supplies.
2. Domestic engineering. I. Peet, Louise Jenison, 1885- Household equipment.
TX298.H68 1975 640′.28 74-13454
ISBN 0-471-67786-8

Printed in the United States of America

10-9 8 7 6 5 4 3 2

THE EQUIPMENT DOES THE WORK, BUT THE HAND GETS THE CREDIT.

—Persian Proverb

Foreword

Changes take place in equipment from month to month, sometimes almost from week to week. Undoubtedly the present emphasis on conserving energy will initiate changes to less energy-demanding equipment than in the past, but modern space-age living has already invaded the kitchen. A growing number of appliances now have the advantage of solid state controls. In many instances emphasis is on reliability and longevity, qualities developed as a result of up-to-date methods of testing before the appliances are put on the market. Consequently, service calls have been reduced during the last few years.

No one can keep house without a certain amount of equipment. Satisfactory use of appliances includes their correct selection, arrangement, operation, and care, so that the homemaker may accomplish the greatest amount of work in the shortest possible time with a minimum of effort. The equipment should give continued service for years, preventing irritation and expenditures for repairs and replacement.

At the same time, greater emphasis has been placed on safety. In a recent report sponsored by Underwriters' Laboratories, Inc., it was noted that out of 100 percent of accidental deaths and 100 percent of deaths from diseases, based on vital statistics released by the Department of Health, Education, and Welfare in 1967, only 1 percent of the deaths was caused by electrical shock, compared to 45.8 percent of deaths from auto accidents and 54.7 percent from heart disease. This was true in spite of the fact that in the last 15 or 20 years there has been an increase of 72 percent in the number of electrical appliances bought annually and there are more than 2½ times as many persons using these appliances.

Lighting plays an important part in any safety program. Adequate lighting makes it possible to work more rapidly, more easily, more safely, and with less fatigue.

The introductory chapter calls attention to the wide and greatly increased use of household appliances during the last 30 or 40 years, with the attendant growth in the equipment industry. Also discussed are family investments in equipment, the need to investigate the comparative desirability of buying a top-of-the-line range, refrigerator, or washer, or a less expensive model. At the same time the homemaker is urged to learn about the reliability of manufacturer and dealer, the availability of servicing, guarantees and what they stand for, and the relative cost of buying on time and paying cash. Finally, illustrations of the Approval Seals for gas and electric appliances are shown and their worth evaluated.

Other chapters in this textbook supply basic information. The chapter on materials used in manufacturing household equipment aids in the wise selection and care of utensils. The review of fundamental facts about electricity and gas is intended to give an intelligent understanding of the efficient operation of appliances. In this chapter is a section on household electronics, including solid state controls, already referred to. A chapter on heat completes the background information.

Within the last few years, personal care appliances have appeared on the market in increasing numbers and varieties. A separate chapter discusses their uses and advantages. In other chapters a new approach has been followed in an effort to make the material more meaningful and helpful. Suggested experiments have been added at the end of most of the chapters. To meet the needs of senior students doing advanced laboratory experiments on ranges and refrigeration a few experiments requiring the use of measuring and testing instruments have been included. In other chapters certain experiments are also designed for advanced work.

At the suggestion of a teacher who uses this textbook a list of terminologies has been added at the end of several chapters. They serve as partial summaries of the material covered and should aid the student in reviewing the subject matter of these chapters.

In general the terminology used in the book has followed that listed in the Handbook of Household Equipment Terminology, published by the American Home Economics Association.[1] It is assumed that all of the terms used will be easily understood by the students and other readers, without the necessity of consulting a dictionary.

This type of book, which gives information on the selection, operation, and care of the appliances in your home, should prove as useful

[1]Handbook of Household Equipment Terminology. American Home Economics Association. 1600 Twentieth Street, N.W. Washington, D.C. 20009. Price $2.00.

as your cookbooks. Indeed, cooking operations and kitchen equipment are quite inseparable, each depending on the other for the success of the finished product.

At the end of the book there is a partial list, including addresses, of manufacturers of household appliances and of associations connected with interests in this field. To these lists has been added a list of periodicals containing articles and advertisements featuring new equipment.

For complete information on current appliances, study of the textbook should be supplemented with selected articles from these magazines, and also available leaflets, specification sheets, and other advertising material from the manufacturers.

Louise J. Peet
Mary S. Pickett
Mildred G. Arnold

Contents

Household Equipment

CHAPTER 1
Introduction

Studies show that the average family owns 15 to 18 different pieces of household equipment with the number increasing as new and more enticing labor savers appear. The number of models of each piece of equipment available in each manufacturer's line has resulted in more choice for consumers and, at the same time, has imposed higher-than-ever requirements of knowledge for selection, care, and use of all types of household equipment.

Although many of the household chores that have required large amounts of physical energy on the part of the homemaker are being replaced by mechanical servants, a recent study[1] shows that the total time Ms. Homemaker used "for the family" in 1967–1968 was not less, on the average, than it was 40 years earlier. However, an analysis reveals that less time is spent on some routine chores, and more time is spent on marketing, record keeping, and management. Homemakers with children continue to have time-demanding household work (Fig. 1.1), but the physical drains of many household chores have been radically altered by improved household equipment. An automatic clothes washer and dryer, for example, have removed the onerous quality from doing the family laundry, have eliminated "wash day," and have liberated the homemaker from the uncertainties imposed by the weather. At the same time, to produce satisfactory results, the homemaker needs more knowledge of new fibers and finishes, water temperatures, detergents, agitation and spin speeds, and drying techniques.

[1]U.S.D.A. and College of Human Ecology, Cornell University.

Wired homes with: number	%	Appliance	Wired homes without: %	number
33,960,000	48.9	Room air conditioners	51.1	35,488,000
37,433,000	53.9	Bed coverings (elec.)	46.1	32,015,000
29,793,000	42.9	Blenders	57.1	39,655,000
35,210,000	50.7	Can openers	49.3	34,238,000
65,837,000	94.8	Coffeemakers	5.2	3,611,000
23,821,000	34.3	Dishwashers	65.7	45,627,000
37,432,000	53.9	Dryers, clothes (elec. & gas)	46.1	32,016,000
24,515,000	35.3	Disposers, food waste	64.7	44,933,000
26,321,000	37.9	Freezers, home	62.1	43,127,000
42,085,000	60.6	Frypans	39.4	27,363,000
17,431,000	25.1	Hotplates & buffet ranges	74.9	52,017,000
69,379,000	99.9	Irons (total)	0.1	69,000
64,795,000	93.3	Irons (steam & steam/spray)	6.7	4,653,000
60,628,000	87.3	Mixers	12.7	8,820,000
69,379,000	99.9	Radios	0.1	69,000
32,224,000	46.4	Ranges, free–standing (elec.)	53.6	37,224,000
12,015,000	17.3	Ranges, built–in (elec.)	82.7	57,433,000
69,379,000	99.9	Refrigerators	0.1	69,000
69,379,000	99.9	Television, B & W	0.1	69,000
46,600,000	67.1	Television, color	32.9	22,848,000
67,017,000	96.5	Toasters	3.5	2,431,000
67,712,000	97.5	Vacuum cleaners	2.5	1,736,000
67,900,000	97.8	*Washers, clothes	2.2	1,548,000
26,460,000	38.1	Water heaters, (elec.)	61.9	42,988,000

***Fig. 1.1** Routine household chores are less burdensome with the use of well-chosen equipment. ("Trudy" by Jerry Marcus.)*

A NEW CHALLENGE TO CONSUMERS AND TO HOUSEHOLD EQUIPMENT MANUFACTURERS

With the new, worldwide awareness of the unities of dynamic balance called ecosystems and the widespread acceptance of the fact that environmental pollution is an inescapable by-product of industrial development, one of the manifestations of which is the voracious appetite for energy, both manufacturers and consumers must take a new look at the wide array of household equipment now on the market.

The increase in the use of mechanical equipment in homes in the United States (Fig. 1.2 and Fig. 1.3) has greatly increased the energy consumption at the same time that energy demands are increasing in other sectors of the economy and throughout the world. The use of electricity per household increased by 84% between 1960 and 1970. The consumption of energy, generally, has risen at a rate five times greater than the rate of population increase. Almost 20% of the energy consumption of the United States is in our homes. Water heating, cooking, clothes drying and refrigeration account for 8.2% of home consumption of energy. Space heating uses 11.0% and cooling 0.7%[2]

Needs for energy in the United States appear to be outstripping supplies. The impact of energy shortages has been felt in most sections of the United States by "brownouts"[3] and some "blackouts"[4] during periods of peak demand, in closing of schools during periods of cold weather because of unavailability of gas and heating oil in sufficient quantities, in factories having to shut down because of shortages of natural gas, in refusal of utilities to add new gas customers, and in threatened rationing of gasoline.

In terms of specific pollutants, electric power is responsible for 50% of the U.S. total for sulfur oxides which are harmful to human, animal, and plant life and to paint, metals, building materials, fibers, and automobile tires; 20% of the total for nitrogen oxides, forming smog and ozones; and 20% of the total for particulate matter—fly ash or soot. Other undesirable by-products of producing electric power include strip-mined land, noise from generating facilities, and landscapes cluttered with transmission lines.

[2]The Potential For Energy Conservation, A Staff Study, Executive Office of the President, Office of Emergency Preparedness, Oct. 1972.

[3]A brownout is a planned reduction in electrical voltage to a given area for a specified time.

[4]A blackout is the cutting off of all power to some areas within the utility distribution system for a period of time.

HOW PRODUCT OWNERSHIP GREW AND CHANGED IN THE PAST DECADE (as of December 31, 1972)

PRODUCTS	1962	1963	1964	1965	1966	1967	1968	1969	1970	1971
AIR CONDITIONERS, ROOM	18.8	19.4	20.2	24.2	27.9	30.7	33.5	36.7	40.6	44.5
BED COVERINGS, ELECTRIC	28.6	29.6	32.4	34.7	38.7	42.3	45.6	47.5	49.5	51.1
BLENDERS	9.4	9.9	11.0	13.0	16.0	20.0	25.9	31.7	36.5	40.0
CAN OPENERS	11.2	15.0	19.7	24.7	29.8	34.5	39.4	43.2	45.5	48.1
COFFEEMAKERS	64.1	65.5	68.5	71.7	76.0	79.6	82.9	86.4	88.6	91.0
DISHWASHERS	8.9	9.0	11.8	13.5	15.7	18.1	20.8	23.7	26.5	29.6
DRYERS, CLOTHES, ELEC. & GAS	22.9	23.5	24.2	26.4	30.5	34.6	38.8	40.3	44.6	47.6
DISPOSERS, FOOD WASTE	12.5	13.4	13.5	13.6	15.9	18.0	20.5	22.9	25.5	28.4
FREEZERS, HOME	25.6	26.4	26.7	27.2	27.5	27.7	28.5	29.6	31.2	32.7
FRYPANS	47.8	48.9	49.0	49.2	50.3	51.8	53.4	55.2	56.2	58.0
HOTPLATES & BUFFET RANGES	24.6	22.4	22.5	22.7	23.0	23.4	23.7	24.1	24.5	24.8
IRONS, TOTAL	87.9	97.4	98.3	99.1	99.3	99.3	99.5	99.5	99.7	99.8
IRONS, STEAM & STEAM/SPRAY	67.3	70.6	73.9	77.5	81.2	83.3	85.8	87.1	88.2	90.1
MIXERS, FOOD	61.2	68.3	70.4	72.8	76.0	78.5	80.5	81.7	82.4	84.4
RADIOS	94.8	97.9	99.3	99.5	99.5	99.5	99.7	99.7	99.8	99.8
RANGES, ELEC. FREE-STANDING	31.4	31.7	31.9	32.1	32.4	34.1	36.2	38.3	40.5	42.6
RANGES ELEC. BUILT-IN	7.6	8.4	9.5	10.3	12.2	12.9	13.7	14.4	15.0	15.7
REFRIGERATORS	99.0	99.1	99.3	99.5	99.6	99.7	99.8	99.8	99.8	99.8
TELEVISION, B&W	92.7	93.4	94.1	97.1	97.8	98.1	98.5	98.7	98.7	99.8
TELEVISION, COLOR	–	–	5.1	9.5	15.0	26.2	35.7	38.2	42.5	51.1
TOASTERS	76.6	79.9	81.1	83.6	86.3	87.6	89.3	91.0	92.6	94.2
VACUUM CLEANERS	78.2	79.5	81.2	83.5	85.6	87.0	89.1	90.7	92.0	94.4
WASHERS, CLOTHES	86.2	86.5	86.9	87.4	88.2	89.3	90.8	91.9	92.1	94.3
WATER HEATERS, ELECTRIC	21.8	22.9	23.2	23.4	24.7	26.1	27.8	29.6	31.6	33.8
(BASE) TOTAL NUMBER WIRED HOMES	53,683,000	54,890,000	56,410,000	57,580,000	58,845,000	60,062,000	61,296,500	62,699,000	64,025,000	65,550,000

38 MERCHANDISING WEEK

Fig. 1.2 Saturation index of key household equipment products. 1974 Statistical and Marketing Report, **Merchandising Week.**

The President has warned that the costs of fuel and electricity will rise considerably in the months and years ahead. The price of fuel in the future will include environmental costs which have not previously been included in the price consumers paid for energy.

Studies of the material and mineral resources remind all of us that the supplies of these are not infinite and that present supplies, at present rates of consumption, are threatened.[5]

[5]U.S. Geological Survey, 1973; **Limits to Growth,** Massachusett Institute of Technology.

Fig. 1.3 How product ownership of key household equipment products grew and changed in the past decade. **Merchandising Week.**

The challenge to the household equipment industry is to produce equipment which will perform the functions of our present equipment but with more efficient use of energy, and to produce equipment which will have an increased life expectancy which will relieve the strain on the finite supply of materials.

The challenge to the consumer is to select the most efficient equipment available, to use and care for the equipment to conserve energy, and to prolong its life.

The study of household equipment now takes on a new dimension requiring more and better knowledge in the selection, performance, and care of all types of household equipment.

THE HOUSEHOLD EQUIPMENT INDUSTRY IN THE UNITED STATES

The $10 billion appliance industry produces more than 30 million[6] major appliances annually. Demographic studies make it certain that the population of the world will double from its present number before the end of the century. The growth of the industry is assured to supply new homes and to provide replacements of equipment in older homes.

[6]Guenther Baumgart, President of AHAM in a speech at Chattanooga Tenn. April 16, 1973.

Fig. 1.4 The complete module is placed on a slab basement. (Westinghouse)

The future of the household equipment industry will depend on the application of new technologies to the entire home package. At the present time, the industry has utilized only a few. To keep pace with changing living patterns, a greatly increased population, and a more mobile society will require a new approach to housing and equipment needs. Housing construction methods are largely the same today as they have been for hundreds of years. The shell of the house is still put together piece by piece with expensive labor. Then a heating system, a cooling system, lighting fixtures, laundry equipment, ranges, and refrigerators are added. As the wage scale of skilled builders goes up and the number of skilled craftsmen goes down, the time must come when all of these separate systems become an integral part of the home package produced on an assembly line. The feasibility and the economy of a single mechanical core have been demonstrated in a number of instances in the past 30 years. A compact core containing the heating unit, water heater, plumbing, wiring, and other utility fittings, and around which are complete kitchen, laundry, and bath appliances and fixtures, can be assembled at the factory, trucked to the site of the house, and quickly installed and placed in operation (Fig. 1.4).

Home packages need not be as regimented and unimaginative in design as are the present housing developments in every community suburb. They should be flexible (Fig. 1.5) and able to be quickly

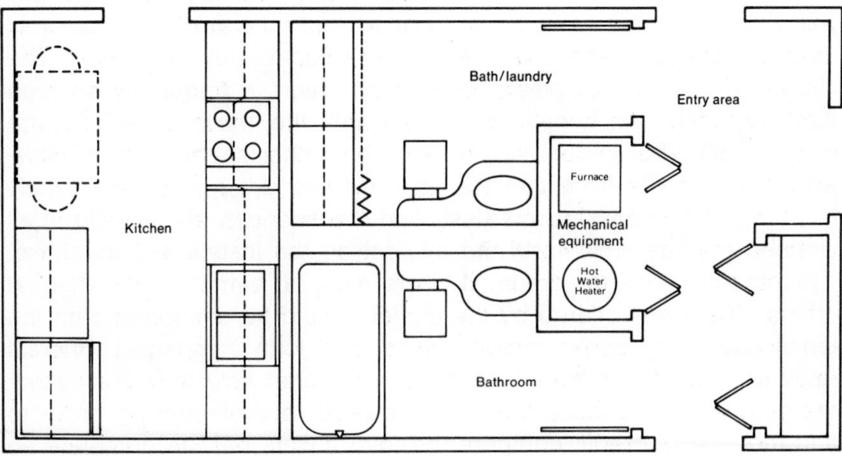

***Fig. 1.5** The floor plan is flexible. Doors and windows can be placed where desired. (Westinghouse)*

erected and dismantled, easily expanded or contracted, and easily moved from one site to another. High prices and scarcity of land will force housing into more and higher multiple dwellings. Architects and builders must be trained to produce more imaginative and practical solutions for family needs in these compact quarters.

Predictions point to computers for the home which will perform a variety of functions from keeping stock of household supplies to paying bills and making deductions automatically from checking accounts.

FAMILY INVESTMENT IN HOUSEHOLD EQUIPMENT

The investment of an average family in household equipment has been estimated at from $2000 to $5000 and upward. Certainly the major pieces of household equipment account for the largest family expenditures after the purchase of a home and a car. The investment increases gradually over the years as new appliances appear on the market and as family income increases.

New families, with limited incomes and the largest requirements for all types of household equipment at one time, need knowledge, which they seldom have, not only of the types of household equipment available, but of ways to evaluate their functions in order to make priority judgments that will give the most satisfaction.

The homemaker must plan, first of all, for the initial purchase of the equipment. This may involve techniques such as planning to stagger major purchases—a refrigerator this year and a range two years later. She must decide whether to wait until the family can pay cash for the equipment desired or whether to buy on credit. The increased cost of equipment bought on credit is frequently so high that the family can benefit by waiting until they have saved enough to pay cash. The amount saved might buy other needed items. However, the benefits from the use of a clothes dryer when the family is young, the needs are greatest, and the finances are usually most strained may be well worth the interest on the installment purchase.

It pays to shop for credit. It costs more at some places than at others. The terms offered by the dealer should be compared with the terms offered by banks, credit unions, and loan companies. Interest rates are stated in different ways but, no matter how they are stated, the best terms are those that cost the least total amount. In order to find the cost of credit, compare the total dollars paid to complete an installment contract (total of all installments plus down payment, if any) with the cost of the cash purchase.

Expenditures for household equipment, like expenditures for automobiles, are continuous. A washing machine is not bought this year and the problem settled for life. The water heater, the refrigerator, and the washing machine wear, must sometimes be repaired and, after a time, must be replaced, as must all mechanical equipment.

A study by the United States Department of Agriculture (U.S.D.A.) and the Census Bureau in 1960 shows the life expectancy of major pieces of household equipment to be from 9 to 25 years.[7] A study published in 1970 by the Whirlpool Corporation had similar findings, as seen in Table 1.1.

Table 1.1
Life Expectancy of Appliances
(years of use by one owner)

Appliances	Department of Agriculture	Whirlpool
Washing machines	10–11 years	10 years
Clothes dryers	14	12
Refrigerators	16	15
Freezers	15	18
Ranges	16	16
Vacuum cleaners		
Upright	18	15
Tank	15	
Sewing machines	24	
Toasters (automatic)	15	
TV sets (b&w, not portable)	11	
Dishwashers		10
Room air conditioners		12

In recent years, the household equipment industry, in its effort to capture a larger share of the consumer dollar, has apparently taken lessons from the automobile industry and placed more emphasis on styling, color, and change of appearance of each year's model. If a study were undertaken covering replacements of household equipment for the next 10 years, we might find homemakers replacing household equipment more often than they formerly have in order to keep up with the latest look, color, or convenience features currently available.

[7]Jean L. Pennock and Carol M. Jaeger. "Household Service Life of Durable Goods," **Journal of Home Economics,** January 1964, **56,** (1), pp. 22–26.

Planning ahead for replacements is one of the most difficult areas of family budgeting. Few homes have yet adopted the common business practice of setting aside a regular amount each year for replacement of equipment. How much to set aside varies for each family. Some heavily used pieces of equipment will have to be replaced sooner than those that are used infrequently. It is estimated that in an average home an automatic washer is in operation 250 hours a year—about the same amount of hours as the average family automobile (10,000 miles at 40 mph). It is not yet a part of the thinking of most families that a washing machine may need to be replaced as often as the family car. Suggested percentages which should be saved for replacement of equipment vary from 5% to 15% each year.

The care which is given the equipment will also determine how long it will last.

FACTORS TO BE CONSIDERED IN THE SELECTION OF HOUSEHOLD EQUIPMENT

Considerations which apply to specific equipment will be treated as the individual pieces are studied. Only the general factors that may be applied to the selection of any type of household equipment will be considered here.

Saving of Time and Energy

The addition of a piece of household equipment may save time and energy, or it may do a better job than could be done without the use of the equipment. The use of an upright vacuum cleaner to clean a large rug will, no doubt, save time and energy and do the job better as well. However, there are some pieces of equipment that might do the job itself well but the time and energy expended in preparation for their use, putting them together, disassembling them, and cleaning up afterward would nullify any advantage that might be realized in time saved in doing the job.

Cost

In today's market there are a number of models of appliances in each manufacturer's line which perform the same function. These represent a wide price range, from the least expensive models, usually referred to as low-end-of-the-line, to the most expensive models, the top-of-the-line, with middle-of-the-line models at prices in between. The least expensive models usually have only the essential characteristics to perform the operation for which they were designed.

More convenience features and sometimes different materials are found on the middle models and more automaticity and higher styling are usually found on the top-of-the-line models. Differences in initial cost of top-of-the-line and low-end-of-the-line equipment in six appliances of one manufacturer are shown in Table 1.2.

Table 1.2
Comparison of Initial Cost of Top-of-the-Line and Low-End-of-the-Line Models of One Manufacturer[a]

Equipment	Top of the Line	Low End of the Line
Range, 30 in.	$1060.00+	$ 168.00++
Refrigerator-freezer	890.00 (24′)+	399.00 (21′)++
Dishwasher, portable		
(15 place settings)	340.00+	118.00++
Built in	290.00	180.00
Waste disposer	99.95+	39.95++
Washer (automatic),		
family size	290.00+	195.00++
Compact		200.00
Clothes dryer	230.00+	140.00++
Compact		120.00
Total	$2909.95+	$1059.95++

[a]Rocky Mountain Region suggested retail prices. Figures from Operation Electric, Fort Collins, Colorado, May 1973.

Sales taxes add to the cost of major pieces of equipment and are usually not quoted in the price of equipment. A 5% sales tax would increase the cost of top-of-the-line items in Table 1.1 by $145.50.

The difference in possible initial investment on the different pieces of equipment shows the necessity of wise decisions in their selection. Each family and homemaker will have different values and the features which are important to one may seem to be luxuries to another. Features should be evaluated in terms of those which will meet the needs of the family now and in the forseeable future. The money spent for unused features could be invested in other, more needed items. Frequently, homemakers are persuaded to buy a number of special features which they never use, either because they never take the time to learn how to use them or they are not suited to the activities of the family.

In addition to the investment in the equipment itself, costs for operation and repair must be added.

The cost of both gas and electricity varies in different parts of the country. Utility rates have increased in the last few years in most parts of the country, and, in all probability, more increases may be expected. Since utility expenses are affected by the design of the appliance model and by the use of the equipment by the consumer, it is more important than ever that energy requirements be considered when buying an appliance and that consumers learn to use equipment to keep current consumption to the minimum. For example, a fully frostless model refrigerator costs almost twice as much to operate as one which is frost free only in the refrigerator section. A quick-recovery electric water heater will, on the average, use 600 more kilowatt-hours of electricity in a year than a standard water heater of the same size.

Energy requirements of equipment are also increased by thoughtless use of an appliance. For example, in a refrigerator, excessive door openings, storing hot foods, failure to defrost, and allowing dust to accumulate on the condenser coils all increase current use. In a hot-water heater, leaking faucets, even one that leaks one drop a second, adds up to 700 gallons of hot water a year.

Installation costs must also be added to the cost of the equipment. In some homes, for example, additional electrical capacity will be needed if an electric clothes dryer is purchased, or a considerable change in the plumbing may be required to install a food-waste disposer and a dishwasher.

Methods of retailing have changed considerably in the last 10 years. Many manufacturers no longer publish list prices or even suggested retail prices. The identical pieces of equipment are usually available in the same community at different prices. Therefore, it is advisable to shop different outlets to become acquainted with the offering prices and then evaluate the delivery and repair services, the reliability of the dealer, and the cost of credit, if the purchase is to be made on the installment plan.

Ease of Cleaning

Many pieces of equipment are seldom used because of the difficulty in cleaning them. The design, the material, and the finish influence the ease of cleaning. A purchaser should look for a design without crevices where dirt can collect and be difficult to remove; the placement of controls on cooking equipment where they are out of the area of spattering grease from cooking; a material which will not stain and can be cleaned easily with a damp cloth and detergent; a finish that does not need continuous polishing or the use of special cleaners; and the availability of self-cleaning features.

Safety

The Underwriters' Laboratories Approval Seal (UL) on an electric appliance, a cord, or a plug indicates that the article has passed certain tests for fire casualty and electrical safety. The Underwriters' Laboratories is a private organization to which a manufacturer may submit his electrical product for examination and test. Costs of the test are charged to the manufacturer. Tested and approved equipment may bear the Underwriters' Seal (Fig. 1.6). On an appliance the seal is usually stamped on the name plate. Since July 1, 1968, no UL label has been placed on the cord of a new appliance. This prevents the consumer from being misled into assuming the appliance itself is UL investigated, when this may not be the case. UL-inspected appliances carry their own marker. Since July 1, 1968, only replacement power supply cords for appliances have carried a brown UL label, clearly identified as "Replacement Power Supply Cord" (Fig. 1.7).

Underwriters' Laboratories follows up its initial tests with further inspections, sometimes at the factory where the item is manufactured or in its own laboratories after obtaining the article in the open market. The letter **R** on the Underwriter's label means that a product has been retested under this service. The Laboratories publish lists annually of equipment they have checked and approved. The UL label indicates that the article has passed requirements for safety; it makes

Fig. 1.6 The Underwriters' Laboratories seal.

***Fig. 1.7** New label effective 1968: "Replacement Power Supply Cord" (left); former label shown at right.*

no claims with regard to performance. UL recognizes colored appliance cords, giving its listing to the synthetic rubber process developed by DuPont. With this listing the colorful cords are expected to appear on more and more portable appliances. A heater cord, called Hypalon, is the first of its kind to offer a full range of decorator colors.

The Laboratories Certification Seal of the American Gas Association, Inc. (a trade association) is a blue star emblem (Fig. 1.8). Its display is the manufacturer's representation that the design and construction of the gas appliance bearing it complies with national safety standards for both construction and performance. In addition, at the manufacturer's option, the appliance design may be tested and certified for compliance with national standards for efficiency and convenience requirements in construction and performance. The designs of accessories such as valves, thermostats, automatic burner ignition, and safety shutoff devices may also be certified for compliance with national standards in which case the accessory may display the monogram Laboratories' Certification Symbol (Fig. 1.9). The Directory of Certified Appliances and Accessories is published semiannually in January and July with Supplemental Directories published on the first of each intervening month. The directory lists the specific models

Fig. 1.8 The A.G.A. Laboratories certification seal for appliances.

of appliances and accessories, the designs of which are certified as of the date of publication.

Minimum standards of performance and safety have been set for many types of household equipment by trade organizations, the American National Standards Institute (ANSI) (formerly the American Standards Association), and the U.S. Department of Commerce through its Simplified Practice Recommendations and Commercial

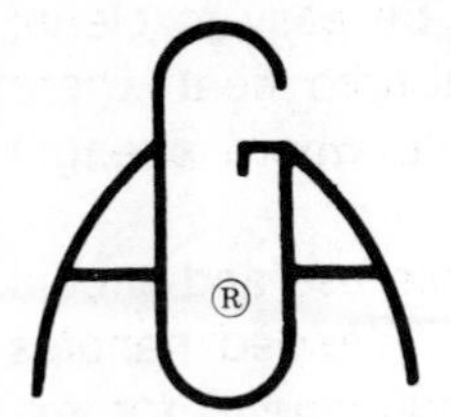

Fig. 1.9 The A.G.A. certification symbol for accessories.

Standards Division, now the Commodity Standards Division of the Office of Technical Services. Since no labels or other means of identifying products meeting these standards exist, they are of little value to the consumer at the point of purchase except in the case of air conditioning units which do carry the trade association tag of A.H.A.M. (Association of Home Appliance Manufacturers) and the A.R.I. (Air Conditioning and Refrigeration Institute), and refrigerator-freezers, and humidifiers and dehumidifiers which carry the A.H.A.M. tag.

Name Plate

Each appliance made by a reputable manufacturer will carry a name plate, either stamped into the metal of the equipment or inscribed on a separately affixed plate. The name plate specifies the conditions under which the equipment is constructed to operate. These specifications include: type of current (AC or DC), frequency of current, voltage and either wattage or amperage, and sometimes horsepower for motor-driven appliances. The manufacturer's name is also given, together with his address and the model number of the appliance. This information will be needed in case of repairs, especially if no local dealer is available, since the homemaker will then have to write to the manufacturer. Occasionally, special directions for satisfactory operation are found, as for example: "Do not immerse in water" or "Connect to wall outlet only."

Mobility

Families living in rented apartments or homes should select movable equipment that does not need to be permanently installed.

All purchasers of equipment should be sure that its installation and use are permitted by local plumbing or electrical codes.

Materials and Construction

No one material is suitable for every use, and most household equipment is made of more than one material. The material should be suitable to the use to which it is applied; it should be easy to clean; it should be judged on its ability to conduct or absorb heat where heat is involved; and it should be rigid and durable where strength is needed.

Most household equipment will receive considerable use and should be durably constructed to withstand wear and tear. Welded handles on pans and welded frames on large pieces of equipment, for example, are strong methods of joining two pieces of material.

Appliances are made by assembling and joining preformed parts.

Parts are stamped, drawn, or forged and are moved, usually by conveyor belts, from one point of assembly to another. Parts are joined by welding, brazing, soldering, riveting, or bolting, and finishes are applied.

Electrical resistance welding is usually used to join parts of major appliances. Heat and pressure cause the steel parts to fuse together making a strong joint. Pressure welding is used to join aluminum parts. Bolting or screwing is used to join parts which might be removed for replacement if damaged. Soldering may be used to join materials when the end product will not be exposed to high heat or rough handling. Riveting, brazing, and spot welding are used frequently in joining parts of small utensils. These are discussed in Chapter 3.

It is sometimes difficult to judge either material or construction when examining equipment on a sales floor. Frequently the salesperson is not well informed. Manufacturer's specification sheets which are supplied to retailers will give the information on size, material, and construction. A purchaser should ask to see specification sheets when specific information is needed.

Warranties or Guarantees

Although there may be a technical or legal difference between a warranty and a guarantee as they are now used by most manufacturers, from the standpoint of the consumer and from some court decisions, in essence they are the same.

Guarantees and warranties should be carefully read by the purchaser. Some guarantee the appliance against defective parts only and for a limited time. Some include labor in replacing defective parts. Some are for a specified length of time on a full guarantee basis and a limited guarantee after some designated time has elapsed.

MACAP (Major Appliance Consumer Action Panel sponsored by the Association of Home Appliance Manufacturers, the Gas Appliance Manufacturers Association, and the National Retail Merchants Association), which is composed of professional specialists in consumer education and communications and represents consumers in their dealings with the appliance manufacturing, marketing and service industries suggests consumers ask:

1. Does the warranty or guarantee cover the entire product? Only certain parts? Is labor included?
2. Who is responsible for repairing the product? The dealer? A service agency? The manufacturer?
3. Who pays for repairs? Parts? Labor? Shipping charges?

4. How long does the warranty or guarantee last on the entire product? On individual parts or assemblies?
5. If the product is out of use because of a service problem, or if it has to be removed from the home for repair, will a substitute product or a service be provided? By whom?

Availability of Service

A combination of a reliable manufacturer and a reliable dealer are the best guarantees of service on a piece of equipment. The manufacturer of the equipment is responsible for the design and workmanship incorporated into the equipment. The guarantee or warranty which accompanies a piece of equipment is assumed by the manufacturer. The user's instruction booklet as well as the instructions to the serviceman are also a responsibility of the manufacturer. A reliable manufacturer will be known for his previous record of responsibility in each of these areas.

A reliable dealer will present the product honestly to the purchaser and will acquaint the new owner with the kinds of service he is equipped to provide and their cost.

The user also has responsibility to read the instruction book carefully, to use the equipment as it was intended, and to care for it in the manner advised by the manufacturer. This will in itself reduce the servicing required.

Many service calls would be unnecessary if the consumer consulted the checklist in the manufacturer's instruction booklet. The problem of an inoperative appliance may merely be a blown fuse, a clogged filter, a stopped-up lint tray. If a service call **is** necessary, the model number of the appliance, which is on the name plate, is a very important piece of information to the service company.

If the problem cannot be solved locally, the consumer should write or call the manufacturer giving details. If this does not provide satisfaction, the consumer may write to Macap Complaint Exchange, 20 North Wacker Drive, Chicago, Illinois 60606.

Experiments and Projects

1. Examine name plates on electrical and gas equipment in the laboratory. Tabulate the items of information given on name plates. How can this information aid you when buying equipment?
2. From the name plates on equipment in the laboratory and at home, list the wattages (power requirements) of the following items:

electric range, electric refrigerator, electric clothes dryer, gas clothes dryer, automatic washing machine, food-waste disposer, a waste compactor, a dishwasher, an electric mixer, an electric skillet, a blender, a portable oven-broiler, an iron, a toaster, an electric coffee maker, an electric grill, an electric clock, a vacuum cleaner, an electric floor polisher, an electric fan, a room air conditioner, a sewing machine, color TV, black and white TV, an electric shaver, an electric blanket, and an electric heating pad. Classify appliances into three groups:
   1. Those using over 1000 watts.
   2. Those using between 200 and 1000 watts.
   3. Those using less than 200 watts.
3. From examination of wattages on names plates, what generalization can you make regarding the power requirements of heating and motor-driven equipment?
   What generalization can you make about energy costs of cooking one dish (i.e., a stew) on a unit on top of the range and the same dish cooked in the oven?
   How can the energy costs of using an oven be more efficiently utilized?
4. Visit an appliance store. Compile prices on top-of-the-line, middle-of-the-line, and low-end-of-the-line models of gas and electric ranges, refrigerators, washing machines, and dryers.
5. Visit an appliance store and ask the dealer to show you the specification sheets for ranges, refrigerators, washers, and dryers. How can the examination of specification sheets help you when purchasing equipment.
6. Pick a major appliance that costs $300 or more at a retail store. Find out how much more it would cost to buy the appliance on a two-year contract.

CHAPTER 2

Materials Used in Household Equipment

The materials and finishes used in household equipment are important to its life and durability and to consumer satisfaction with performance and maintenance. Because many new materials and finishes are being introduced each year, and many old materials and finishes are being used in new ways, consumers are finding it more difficult to make intelligent evaluations of the materials themselves and the suitability of the applications being made in household equipment. In addition, the recent awareness that shortages of natural resources may actually threaten modern civilization[1] make it important that our supply of materials be used to produce goods which will last longer and thereby conserve our dwindling mineral resources.

A knowledge of the characteristics of the base materials and finishes will help in evaluating the practicability of the material and finishes of household equipment.

Aluminum

Pure aluminum exists in three distinct physical forms: cast, hard-worked, and annealed. In the cast form pure aluminum is soft and ductile and finds little practical use. Cast aluminum cooking utensils usually contain 6 to 8% copper. One manufacturer adds a small percentage of magnesium and calls its product magnalite.

[1]U.S. Geological Survey, 1973; also **Limits to Growth,** Massachusetts Institute of Technology.

Pure cast aluminum, however, is the first stage in the production of aluminum sheets and wire. The pure aluminum pigs are melted and recast into rectangular slabs; the slabs are heated, and while hot they are rolled under enormous pressure into sheets ½ to ¼ inch thick; and the sheets are then cold-rolled into plates of any desired thickness. During the cold-rolling process the hardness and hence the tensile strength of the sheet metal increase and ductility decreases. The manufacturer usually adjusts the tensile strength of the sheet by varying the thickness or by removing some of the hardness, due to cold-rolling, by annealing.

Aluminum is light in weight and is easily fabricated. It has a low melting point (659°C) compared to iron (1535°C). Aluminum is a good conductor of heat and electricity (Table 3.1). Aluminum resists corrosion in air. It is darkened by alkalies found in some foods and the hard water of certain regions; strong alkaline soaps and dishwasher detergents should not be used for washing these utensils. A special aluminum cleaner or acid foods such as rhubarb and tomato or lemon and vinegar water may be used to remove the discoloration.

Pitting may occasionally occur when very saline water is used or when moist food is allowed to stand in a kettle of aluminum, but more frequently it results from the molecular structure of the sheet from which the utensil is stamped.

Experiments at Johns Hopkins and other universities entirely discredit the theory that cancer or other diseases may be caused by foods cooked in aluminum utensils.

Aluminum is inert when in contact with oils, fats, and acids—one of the reasons food manufacturers are packaging convenience and frozen foods in aluminum foil containers.

Aluminum is widely used in household equipment for such articles as pots and pans, baking utensils, refrigerator shelves, freezer compartments, range drip pans, ice cube trays, and for trim.

Iron

Iron is obtainable in two forms: cast and sheet. Cast iron is only a fair conductor of heat. Therefore, it heats slowly and, because of its thermal mass, retains heat well. Because of its dark color and rough surface it is a good absorber of radiant heat. Cast iron is heavy and rather brittle and will corrode unless cared for. The rust formed on damp iron is red iron oxide. In contrast a dark blue or black iron oxide which protects the surface may be obtained by heating the iron in oil. This is known as "seasoning." Cast ironware may be preseasoned by the manufacturer, or the consumer may be required to season a utensil before use by heating an unsalted oil in the utensil in the oven for two hours. Seasoning will be removed if the utensil is washed in strong detergents or scoured, and the utensil must be reseasoned to prevent rusting.

New iron alloys that impart greater strength permit the casting of thinner items than formerly was possible.

Cast iron currently is used for skillets, roasters, Dutch ovens, broilers, griddles, and muffin pans.

Very little sheet iron is used in kitchen utensils. With present fuels and high input units in modern ranges, sheet iron warps and becomes unusable.

Steel

Steel is produced by refining pig iron. Steel is lighter in weight than iron and has replaced iron for many uses in household equipment. Steel can be forged, rolled, drawn, or stamped and can vary in properties with chemical formulation and treatment. Special properties are required by different pieces of equipment, and it is essential that the manufacturer obtain the steels suitable for the intended end use.

Sheet steel can be welded. It is strong and durable, and when it contains at least 1% carbon it will take and hold a sharp edge. However, it will rust if it is exposed to dampness and, therefore, may require a protective finish.

New developments in the steel industry are leading to new applications which will affect many areas of household equipment and furnishings. A process of continuous casting is being used by some steel mills in the United States which allows molten steel to be cast directly into basic products. One steel corporation announced it will continuously cast cold-rolled sheet steel, an important basic product used in household appliances and automobiles.

Sheet steel is now being coated at the plant with colorful paints and plastics which can be shaped into parts for household appliances.

A new "soft" steel can be bent by hand for use of water pipes, rainspouts, and gutters. Extremely fine steel wires are being woven into carpets and draperies to eliminate static electricity.

"Weathering" steels that do not need painting are being produced. When exposed to the elements they form their own protective coating.

A new type of architecture, utilizing superhigh-strength steel cables as supports for floors and roofs, eliminates beams in the center of large buildings.

According to the industry, more than half of the steel products made today were not available 10 years ago.

Stainless Steels

In stainless steel some of the carbon has been replaced by chromium and nickel unless it is designated as high-carbon stainless steel. A common stainless steel is known as 18-8, and contains 18 parts of chromium and 8 of nickel.

Although not good conductors, stainless steels have high tensile strength, are not easily dented, are resistant to corrosion, take and retain a high polish, do not scratch with scouring, and are not affected by most food acids and alkalies.

Because of its smooth, fine-grained surface, stainless steel is easy to clean with detergent and water. It should be dried promptly to prevent water spotting.

Strong bleaches will corrode stainless steel and, therefore, should not be allowed to remain in contact with stainless steel sinks or utensils for more than 30 minutes. Such foods as mustard, mayonnaise, lemon juice, vinegar, salt, or dressings containing these will also corrode stainless steel if allowed to remain in contact with a stainless steel surface for a period of time.

If overheated, some stainless steels develop a blue-gray or brownish tinge from surface oxidation. Minor darkening may be removed with a stainless steel cleaner. If excessively overheated, the darkening cannot be removed by any home method.

Vanadium and Titanium Steels

Vanadium is added to steel to be made into cutlery; it increases the steel's ability to take and maintain a sharp edge. Vanadium stainless steel will not tarnish, stain, or otherwise discolor, even from excessive heat.

Titanium steel, known by the trade name Ti-Namel, has a titanium content five times the carbon content. This type of steel resists sagging under heat, eliminates the need of using a ground coat of porcelain enamel, and makes possible the use of thinner coats of enamel.

Tin

Tin was early used as a protective covering for metals. It was mined in England in very early times and about 1670 was employed in the manufacture of utensils for the home. The extensive production in the United States dates from 1892. Tin, as used in the home, is an applied finish.

Copper

Copper was one of the earliest metals used. It is separated from its ores and cast into ingots, from which it is made into sheets, wire, and castings. Pure copper is one of the best-known conductors of electricity and of heat (Table 3.1). It is malleable and ductile and lends itself readily to various forms of manufacture. Copper tarnishes easily, however, and requires constant burnishing to be kept bright. If cooked foods are left in contact with copper, a toxic substance may form. For this reason copper cooking surfaces are always lined with tin or are coated with a nonstick finish. Copper is also plated on the outside of some stainless steel utensils to increase the conductivity of the utensil, but the surface exposed to the food is stainless steel.

Monel

Monel is a trade name for a nickel-copper alloy of high nickel content. Monel metal has a silverlike luster and takes and retains a good polish. It has high tensile strength, is very resistant to denting, scratching, and staining (but does show water-spotting), does not corrode, is unaffected by food acids and alkalies, and is a comparatively poor conductor of heat. It is used for sinks, laundry equipment, oven linings, and for fittings and trimmings, although in a number of applications it has been replaced by stainless steel.

Earthenware and Heatproof China

Earthenware and heatproof china are clay products. Various kinds are manufactured under different trade names. These products vary chiefly in the quality of the clay, the method of making, and the glaze that is used. Earthenware and the china are opaque, comparatively coarse in texture (the earthenware more than the china), and porous when fractured. They are little affected by food acids and alkalies or by sudden changes in temperature.

Heat-Resistant Glass

Heat-resisting glass does not rust and is not affected by food alkalies and acids. It can be scratched by coarse abrasives, it may break if subjected to sudden and extreme changes of temperature (especially if moisture is present), and it may shatter under impact.

There are two types of heat-resisting glass: glass that can be used for surface cooking and glass that can be used for oven cooking. Glass to be used for surface cooking and over direct heat is one with a low coefficient of expansion. It is recommended that a metal grid be used on a surface unit when this glass is used on an electric unit. Glass utensils for use on top of the range may be used in the oven. However, **oven** type glassware cannot be used on top of the range units. Some heat-resistant glass can go directly from the freezer to a preheated oven.

Glass is a very poor conductor of heat (Table 3.1), but is one of the best absorbers of radiant heat.

Glass Ceramics may be glazed or unglazed. Unglazed glass ceramics withstand extreme changes in temperature because of a low coefficient of expansion. They can go directly from freezer to a hot surface unit or hot broiler without damage. Unglazed glass ceramics are also used as a cooktop with heating elements mounted beneath.

Glazed glass ceramics are used for plates, cups, saucers, bowls and platters. They may be used in the oven but cannot be used on the range top or in microwave appliances.

FINISHES

There are two methods of finishing the surface of materials—applied and mechanical—depending on the corrosive properties of the base metal.

Applied Finishes

The noncorroding applied finish put over base metals that corrode may be metallic or nonmetallic in nature.

Metallic Applied Finish

Common metallic finishes are copper, nickel, chromium, tin, and zinc. Copper, nickel, and chromium are electroplated onto the base. The base material is made the cathode in an electric bath and a film of the metal is deposited over it.

Nickel

Nickel was once widely used as a finish on small electric appliances. Nickel has a yellowish tinge with age, and stains are difficult to remove from its surface. Nickel is added to stainless steel and is used to add strength and heat resistance in heating elements in many appliances.

Chromium

Most small electric appliances are finished in chromium plate. Chromium has an attractive silvery color and is kept in good condition simply by wiping with a damp cloth and polishing with a dry one. To provide a durable finish, a layer of copper is plated on a steel base, then a layer of nickel is added, and then the chromium plating is applied. Coarse abrasives should not be used on chromium. Whiting will remove burned-on grease.

Tinned Steel

Tin is pliable and can be applied before the sheet is made into a utensil. Tinnedware is light in weight and is graded according to the thickness and quality of the base plate of steel and of the tin coat. "Block" tin is a term applied to the most heavily coated sheets, and refers to the quality of the tin. This quality largely determines the life of tinnedware; the presence of tiny pinholes in the tin exposes the steel base, which corrodes rapidly when reached by moisture. Tin is affected by food acids. It darkens with use; hence its absorption of radiant heat changes rapidly. This surface tarnish furnishes a protective coating, and the utensil should not be scoured simply for the sake of making it bright. Moreover, scraping tends to scratch the surface and may expose the steel or iron base which will then rust.

Galvanized Ware

Zinc-coated material is known as galvanized iron or steel, depending on the base. Zinc is not pliable and ductile and, hence, does not lend itself to most forming operations; heavy coatings of zinc are likely to crack and peel off under severe strain. It is best adapted to flatware and utensils in which the strains are slight, and is used for pails, washboards, lids for fruit jars, and other articles that come into contact with moisture. These are dipped after forming.

Nonmetallic Applied Finishes

Porcelain enamel, synthetic enamels, Teflon, and polyimides are nonmetallic finishes which are frequently used in household equipment.

Porcelain Enamel

Porcelain enamel is a glasslike substance, inorganic in nature, which is **fused** to the surface of metal. The metal substrate that is used most often is sheet steel; however, a good deal of porcelain enamel is also applied to aluminum and cast iron, and a lesser amount may be found on copper, stainless steel, and even bronze and sterling silver.

The porcelain enamel coating material itself is made by smelting together a number of carefully formulated mineral ingredients at very high temperatures to form a molten alumina-boro-silicate glass. The molten glass is drawn from the furnace and quenched rapidly, either in cold water or by running it through chilled rollers. This sudden cooling causes the glass to shatter into fine particles known as "frit." It is in this form that the raw porcelain enamel goes to the factory or point of application. Here the frit is pulverized and mixed with water, clay, and other additives to form a thick cream known as "slip."

The base metal to which the porcelain coating is to be applied normally is drawn or fabricated to the desired shape first, since the glasslike characteristics of the fused enamel limit its pliability. Flanges, handles, or joining seams are welded, and the formed product or part goes through a "pickling" process in solutions of acids and cleaners—partly to remove impurities from the metal and partly to apply a slight etch to the surface. During the same process the part may also receive a "flash" coating of nickel to promote good adherence between the enamel and the base metal.

The product or component is then ready for the slip, or liquid enamel, application—usually either by spraying or dipping. Most porcelain enamels are applied in two coats. The so-called "ground coat," usually dark blue, contains a certain percentage of cobalt and nickel oxides which have a strong affinity for iron. It is nearly always applied by dipping or "flow-coating." The part is then dried and placed in an enameling furnace, where temperatures of 1400°F to 1600°F cause the minute frit particles to melt and fuse into the pores of the base metal. One or more "cover" coats are then applied to the ground coat and fused by essentially the same process to form the final finish.

The cover coat on most porcelain enameled appliances is very similar in composition to the ground coat, except that titanium oxide is used in place of cobalt and nickel. Titanium greatly increases the opaqueness of the enamel so that normally only a single thin coat is needed to cover the dark cobalt. Titanium also gives the enamel a very pure white color which is extremely uniform from one sheet to another. Other inorganic materials are generally added to the porcelain enamel to give it an alkali or acid-resistant glaze.

Colored porcelain enamels are produced by the addition of various metal oxides that fuse into the enamel; hence, colored enamels will never fade or wear off.

An enamelware product has practically all the properties of a glass product, plus the strength and stability of the metal substrate. It is nonsoluble, will not rust or discolor, and is not affected by atmospheric conditions. Because it is processed at very high temperatures, it is also immune to any normal heat attack. Today's porcelain enamels are also increasingly resistant to thermal shock (sudden extreme changes in temperature) because of thinner coatings and improved processing techniques. However, if subjected to undue torsion or impact, the finish may crack or chip; acids allowed to remain on the enamel will tend to etch the glaze and make cleaning more difficult. It can be marked by metal spoons or beaters, but such marks can be safely removed by use of a mild cleanser.

In most cases, warm water and detergent is all that is required to clean a porcelain surface. If scouring is necessary, use baking soda, whiting, or a mild, nonabrasive cleanser; coarse scouring powder used too vigorously or too often will damage the glaze.

Porcelain enameled appliances, or critical components of appliances, find extensive use in every home: ranges, refrigerator food liners and crisper pans, washers (including tops, tubs, and some cabinets), dryer tops and tubs, dishwasher interiors, water heater linings, sinks, bathtubs, and cookware.

Porcelain enamelware, like glass, is a poor conductor of heat and a good absorber of radiant heat.

SYNTHETIC ENAMELS

Epoxy Enamel

Epoxy enamel is an organic resin coating applied to metal surfaces and cured at about 450°F. It is available in a variety of colors. Epoxy enamel is chip and scuff resistant, stain resistant, and dishwasher safe.

Acrylic Enamel

Acrylic enamel is an organic coating, made from synthetic resins, baked on to a metal surface at approximately 450°F. It is used largely for exterior surfaces of cabinets of laundry equipment and refrigerators and freezers. Acrylic enamel is available in a variety of colors. It has good resistance to stains and can be cleaned fairly readily. The finish can be permanently damaged by ammonia, strong solvents, or abrasive cleaners and is susceptible to heat attack above 180°.

Alkydenamel

Alkydenamel is an organic resin coating baked on metal surfaces at about 350°F. Alkydenamel is used for cabinet, refrigerator, and freezer liners. It is less durable than acrylic enamel and more susceptible to alkali attack. It is available is a variety of colors and is chip and scuff resistant. This finish is sometimes referred to as "baked enamel," although technically any baked-on organic finish could be called baked enamel.

Polyurethane Enamel

Polyurethane enamel is an organic resin coating applied to exterior metal and plastic surfaces and cured at 350°F. It is chip and scuff resistant, stain resistant, and dishwasher safe. Polyurethane enamel is available in a variety of colors.

Teflon

Teflon is Du Pont's registered trademark for its family of fluorocarbon resins and materials based on these resins. It does not react chemically with food, water, or detergents. It is applied as a coating to provide a surface so smooth that other materials will not adhere to it, making the surface easy to clean. There are at present 20 different colors available in Teflon coatings.

Teflon is widely used on inside surfaces of skillets, saucepans, baking utensils, and many small electrical appliances. It is being used by some range manufacturers as a coating on oven linings, range griddles, and on the inside of range hoods. Spatulas, stirring spoons, measuring spoons, blades of electric and hand beaters, rolling pins, and other food preparation tools have been coated with Teflon. Telflon coatings are being used on the soleplates of some steam irons. Coatings have also been applied to snow shovels and hand saws and as insulation on underground electrical cable.

The first Teflon coatings were soft and easily damaged with metal forks, knives, or spatulas, and discolored from heat and food stains. Hard-base finish systems to apply Teflon have been developed recently which make it scratch resistant. This hard-base finish (Teflon II) is approved by Du Pont only for top-of-the-range cookware and some electrical appliances.

In the original, standard, two-coat process the surface of the utensil is first roughened and cleaned by a sandblasting treatment. This forms tiny teeth and pits and provide a foothold for the nonstick finish. A liquid coating of the nonstick finish is then sprayed onto the metal so it flows into the small pits and crevices. It is baked on at about 450°F. A second, final coat of the nonstick finish, containing pigments that add color, is sprayed on and baked at 750°F.

For the newer, more durable finish, a triple-coat process is used. First, the surface is roughened and cleaned as in the two-coat

Fig. 2.1 Du Pont certification marks for housewares coated with Teflon non-stick finishes. (Du Pont)

process. Then a special material of either ceramic frit or molten metal is sprayed on at 1000°F to form an extra-hard surface, assuring permanent adhesion to the utensil. This hard surface emerges with many microscopic peaks and valleys which are then coated with the standard two coats of nonstick finish.

The Du Pont Company has set up quality standards for application of the finish for each of these types. Companies meeting these standards may use the Du Pont certification marks (Fig. 21.).

Polyimide

Polyimide is a synthetic material which is chemically related to nylon. It is applied as a shiny coating on the outside but not the bottoms of cookware. It is unaffected by detergents, food, and water.

Before applying polyimide, the utensil shell is drawn and cleaned and then the polyimide is sprayed and baked on at about 600°F. The finish looks like porcelain and is available in a wide variety of colors. Polyimide coatings have flexibility which makes them resistant to damage from impact or from sudden changes in temperature. They can be scratched by harder materials such as a knife edge, fork tine, or a sink faucet. However, scratches will not open up, and the polyimide will not peel.

Polyimide does not offer the no-stick, no-scour, no-scratch advantages of hard-base Teflon and so is not used on the interior of utensils. The Food and Drug Administration has approved its use for cover interiors.

Heavy grease deposits or badly burned food can be removed with car cleaner—one containing a fine abrasive material which cleans without scratching. The polyimide finish should never be cleaned with scouring powders or scouring pads. Polyimide-coated utensils can be washed in a dishwasher.

Polyimide coatings act as a heat-absorbing finish and are suitable for those bakeware utensils on which crisp crusts are desired.

Because of its lower application temperature, polyimide can be applied on medium-weight aluminum. This places polyimide-coated utensils in a moderately priced category.

Paint

Paint also is a nonmetallic finish. It changes color and will chip, rub off, and become marred; it needs, therefore, fairly frequent renewal.

Mechanical Finishes

A mechanical finish is the finish of the metal itself, not something applied on the outside. It is usually a polish, made with a burnishing tool, or a satin-type, and occasionally a pebbled or hammered finish (Fig. 2.2).

Anodized Finish

This finish may be considered in a class by itself. It is an oxide deposited on the surface of aluminum by making the utensil or tray the anode in an electric bath, after the aluminum has been polished or given a satin finish. The anodized finish then takes the nature of this original finish, which may be shiny or dull. Anodized finishes last indefinitely; they form a hard, wear-resisting coating that eliminates the smudging tendency of ordinary aluminum. The oxide film is porous and absorbs color readily. Since high temperatures used in surface cookery, in the oven, or in a mechanical dishwasher may remove

***Fig. 2.2** The surface of the aluminum saucepan has been given a distinctive hammered or pebbled finish. (Club Aluminum)*

these colors, only the covers of saucepans are colored by this process and it is not recommended that these be washed in the dishwasher.

An anodized finish on bakeware increases the absorption of radiant heat in the oven.

Another anodic coating, known under one trade name as Gem Coat, produces a very hard and dark finish on aluminum. The electrochemical process takes place at a temperature below freezing. The finish is claimed not to crack, peel, chip, or flake.

Plastics

To a greater or lesser degree, plastics are a part of all up-to-date household equipment. They are light in weight but strong, colorful, resistant to moisture and deterioration, good insulators of both heat and electricity, easy to clean, and some types are extremely durable. They have become an indispensable aid in saving the homemaker's time and energy.

Plastics are man-made materials and are manufactured from everyday substances—air, water, petroleum, wood, coal, natural gas, lime, and salt—by the action of chemicals, heat, and pressure. Many different kinds are available, of which perhaps 15 are basic and, of these, five or six lead in production.

Plastics are divided into two principal groups, depending on their reaction to heat. **Thermoplastics** are softened by heat but harden again when cooled, and this change may be repeated a number of times without alteration in the physical properties. Polyethylene is an example of a thermoplastic. The **thermosetting** plastics, by contrast, undergo a chemical change when heated. They harden into permanent shape and cannot be remelted. Phenolic is a prime example of a thermosetting material.

Thermoplastic Products

Representative plastics in this group include polyethylene, polystyrene, nylon, vinyl, acrylics, and fluoroplastics. There are many grades, each tailored to the requirements of the application.

Polyethylene is used in bottles for milk, bleaches, and detergents, squeeze bottles for cosmetics, semirigid mixing and refrigerator bowls, juice containers, coffee can lids, dishpans, trash containers, food storage bags, and in sheet form, for economical table cloths and food wraps.

The **polystyrene** family of plastics is a broad one and many types are produced. They are all rigid but have different levels of toughness, chemical and heat resistance, transparency, and cost.

Opaque, colored polystyrene is used in many applications requiring toughness and durability. Refrigerator door liners, food bins, table and portable transistor radios, air conditioner and other appliance parts are examples. A fast growing new use is for wood-grained furniture parts such as stereo cabinet fronts. Cake dish covers, shoe storage boxes and clock faces are made of a transparent grade and should be handled with care. Transparent, heavy-weight tumblers and blender bowls are made from a copolymer resin which is both tough and more heat resistant. Like polyethylene, large quantities of polystyrene are being used in packaging applications. Disposable drink cups, cottage cheese containers, and foamed meat trays are commonplace.

Nylon is used for gears in sewing machines, for bearings in egg beaters, for the roller parts in drawers and sliding shelves in refrigerators and cabinets, and for other parts of pumps and equipment where its resilience and resistance to abrasion and fatigue are beneficial.

Vinyl comes in both flexible and rigid varieties. Refrigerator gaskets, floor tile, placemats, upholstery material, and shower curtains are familiar uses of the flexible types. Rigid vinyl bottles provide transparent, shatter-resistant containers for dishwasher detergent, hair preparations, and certain foods.

ABS Plastics are used for refrigerator inner door liners and food liners, housings for vacuum cleaners, food mixers, floor washers and hair dryers; for soap and bleach dispensers and ventilating panels in washing machines, fans, wheels and cleaning tools in vacuum cleaners; and for grilles and deflectors, fan blades, drip pans and knobs in air conditioners; and for sewer and drain pipes.

Acrylics are used in lighting fixtures because of their high transparency, good diffusion characteristics, and light stability.

Fluoroplastics are finding increasing utility as antistick surfaces for cooking pots and pans and other kitchen utensils.

All of these plastics are basically nontoxic and essentially free of odor and taste. They are not affected by water or by normal concentrations of household chemicals.

Polycarbonates are used for coffee maker sight glass, for refrigerator food liners and for housings for electrical and electronic equipment.

Thermosetting Products

In this group fall melamine, phenolics, polyesters, urea, and polyimides.

Melamine, obtainable in a wide variety of translucent and opaque colors, is made into dinnerware and mixing bowls, and is used for laminated counter tops.

Phenolics are opaque and dark in color, usually brown or black; they make ideal utensil and hand-iron handles. Other important uses include light plugs and switches, appliance bases, washing machine agitators, and telephones.

Polyester resins are combined with reinforcing fibers to make corrugated panels for wall dividers and awnings, appliance housings, and light-weight laundry tubs. Polyester film is used to package boil-in-bag foods.

Polyimides are used as exterior coatings on cooking utensils (see Finishes).

Urea is manufactured into buttons, cosmetic jar tops, electric plugs, clock cases, and picnic ware.

Members of this group are resistant to heat, moisture, and scratches.

Silicones

Silicones, a group that includes over a hundred different products, may be considered a distinctive group by themselves. Intermediate between inorganic and organic substances, they have some of the characteristics of glass. In equipment they find numerous applications: to treat the surfaces of waffle grids, ice cube trays, and baking pans in order to prevent the food from sticking; as silicone rubber, which is both odorless and stainless; in the gaskets of steam irons to prevent leakage; in the form of an oil to increase the accuracy of toaster timers and the clocks of range ovens. Certain silicone products form the insulation on electric wires and in motors subject to high temperatures. In the discussion of thermostats it was noted that a silicone fluid is used in the probe tube of some small electric appliances.

Bakeware utensils with silicone linings have been less costly than the same items with Teflon linings. There is no appreciable difference in performance. At present, silicone linings are not used on skillets and saucepans. The silicone nonstick finishes require the same care as other nonstick finishes.

Care of Plastics

Plastics are easily cleaned with a damp cloth, or by washing in lukewarm water with mild soap or detergent. Abrasive household cleaners, steel wool, or sandpaper will scratch the surface and should not be used. Most plastics are softened and even charred by high heat such as oven heat. Certain ones cannot be exposed to high dishwasher temperatures. Wraps or bags should never be placed in an oven because they will melt. On the other hand, phenolics and the fluoroplastics are designed for exposure to heat.

Most plastics are strong and will not break with normal handling. However, they are not indestructible and some will be damaged if they are subjected to hard blows or dropped on the floor.

Other Uses of Plastics

Plastics are used in certain polishes that need no rubbing, in resins mixed with house paints (plastics flow readily, dry fast, resist mars, and prevent white paints from yellowing), in oils and varnishes, and for coatings on bread wrappers and milk cartons. Used on fabrics, plastics render them stain and water repellent. They are also used to bind other materials together to form laminates for counter tops, knife handles, etc.

Plastics are now being molded into parts for household equipment which were formerly made of metal castings. Plastic pipes are being used in some areas in household plumbing and in some communities in city water delivery systems.

MATERIALS USED FOR INSULATION

Since heat always flows from a warmer to a cooler area, insulating materials are commonly used in home equipment to keep the heat where it belongs, out of the food chambers of freezers and refrigerators and inside range and roaster ovens. Insulation in the walls of houses also has this double role, depending upon the location of the home and the season of the year (Chapter 18). Other insulating materials like plastics prevent the heat of many small appliances from marring the surfaces on which they are placed and from burning the fingers. Still others keep electricity in its proper path and lessen the possibility of a short circuit. In kitchen cabinet doors and air conditioners, insulation decreases transfer of sound and minimizes noise. Insulating materials are recognized, therefore, as important aids in the successful operation of the home.

To be most effective insulating materials should be impervious to moisture and odors, be light in weight, be unaffected by both low and high temperatures, be of open structure to allow breathing (they should not sag), be proof against charring or burning, and be in a form that can be easily cut.

Mica

Chemically a silicate, mica is one of the most commonly occurring minerals. In mining, the mica crystals usually break up into plates an inch or more thick. The plates are passed through a cleaning and trimming process and are then split. Owing to the almost perfect cleavage of the crystals, the split pieces are exceedingly thin, 0.002 inch thick; they are built into sheets by the use of binding materials. In the best grades, the mica sheets are highly transparent, and also soft, flexible and elastic, but strong. Mica is a fairly good conductor of heat but a poor conductor of electricity, two properties that make it adaptable as an insulator in electric heating appliances. For example, in the toaster mica sheets are usually used as the base around which the Nichrome heating element is wound.

Asbestos

Asbestos, mined chiefly in Canada and Rhodesia, occurs in slender, flexible fibers that are separated easily from the rock. The fibers are then graded according to length and fabricated into different types of material; the long ones spun into yarn and woven into cloth, the short ones pressed into sheets. Asbestos is light in weight, has high heat-resisting qualities, and is noncombustible, all of which make it a good insulating material. It is not moistureproof, however, unless specially treated. In the equipment field it is used primarily as a layer beneath the outer fabric of appliance cords for heating appliances.

Rock and Mineral Wool

Rock wool, a mineral product made from calcium magnesium silicate, is heated until melted, then blown into cooling chambers. The resulting product is a fine white fibrous material of low thermal conductivity which is manufactured into felts, flexible blankets, insulating bricks and blocks, acoustic plaster, pipe coverings, and insulating cement.

Mineral wool will not burn, deteriorate, or attract vermin or mice. It is durable and chemically stable but not moistureproof unless specially treated. Certain manufacturers treat the wool in the molten state so that the fibers are annealed as blown, a process that tends to toughen the fibers and render them resistant to moisture.

Although Fiberglas has largely replaced mineral wool as an insulation in household appliances, it is still used for wall and floor insulation in homes (Chapter 18).

Fiberglas

Fiberglas is the registered trademark for a variety of inorganic, man-made fibers manufactured by the Owens-Corning Fiberglas Corporation. The method of manufacture is similar to that used for mineral wool. By varying the ingredients and their proportions, regulating the temperature, and controlling the diameter of the fibers, products of different characteristics are obtained.

Fiberglas is divided into two basic types: (1) fibers that are a resilient fleecelike mass, fabricated into bats, blankets, boards, and other forms and are used extensively for thermal and sound insulation (Fig. 2.3); (2) fibers that are twisted into yarns and woven into braids, tapes, and cloth. The fiberglas bats are used for insulation in ranges, roasters, and water heaters; rigid boards, often treated with a binder, are more commonly used in refrigerators and freezers.

Regardless of the type, Fiberglas has certain basic characteristics. It is flexible, noncombustible, does not absorb moisture, has high tensile strength, and will not mildew, rust, or rot.

Plastic Foams used in refrigerators, freezers and in insulation in buildings

Polystyrene foam is used as a core between a sheet of the polystyrene and Fiberglas cloth. It forms a rigid panel that may be cut to any desired shape and size, and is light in weight, noncombustible, and moisture-resistant.

Polyurethane foam uses Freon 12. The Freon is mixed with other chemicals which react to generate heat and to change the Freon liquid to a gas. The final mixing of the ingredients in the cavity between the refrigerator walls immediately starts the formation of the foam, which expands to fill every bit of the space, and, as it does so, entraps the Freon in the millions of tiny cells. The foam becomes rigid and bonds to all the surfaces, blocking any possible path for heat leakage. It is moisture-resistant, and 1¼ inches of it is equivalent to 3 inches of conventional insulation.

This insulation provides a strong, rigid structure that will make possible the use of lighter-weight materials in the frame, while at the same time preventing even slight distortion in alignment of the walls during manufacture and shipping. This is a definite asset, since a slight change in shape may affect the sealing of the door gasket.

Polyurethane foam and polystyrene foam have been used as insula-

Fig. 2.3 Types of Fiberglas insulation to control heat, cold, or sound. (Owens-Corning)

tion in houses and have also been used in furnishings. However, because these have recently been declared by F.T.C. as serious fire hazards, many local building codes are forbiddding their use (Chapter 18).

Reflective Insulation

Reflective types of insulation are usually a metal foil or foil-surfaced material. They operate on the principle that polished metallic surfaces will reflect infrared or radiant heat rays. The number of reflecting surfaces, not the thickness of the material, determines its insulating value. Each reflecting surface is equivalent to ½ inch of insulating board.

FLOOR, WALL, AND COUNTER COVERINGS

Floor Coverings

The Building Research Institute groups the many types of floor coverings on the market into five general groups: (1) linoleum, which has been used for nearly 100 years and is still a favorite with many homemakers, (2) rubber, now usually a synthetic product, (3) asphalt tile, the first resilient flooring that could be laid on concrete at ground level or below, (4) cork, and (5) vinyl plastic, available in three forms: homogeneous, asbestos tiles, and backed vinyl, in both tile form and sheets.

A perfect smooth-surface, resilient flooring should have certain desirable characteristics. It should be soft under foot, be resistant to moisture, stains, household acids, alkalies, oils, and other chemicals, be proof against permanent indentations from heavy furniture, be able to withstand penetration and scuffing from dirt, and it should require cleaning only at long intervals by the simplest method possible. None of the floor materials under consideration possesses all these advantages, but each one has several.

Linoleum Sheet and Tile

Linoleum is made from linseed oil to which gums, natural and synthetic resins, and pigments like those used in paint are added. The filler was once cork but is now usually wood flour.

This mixture is sifted over a burlap or saturated felt backing and is pressed into a solid sheet by a heated roller, a process that welds the mix to the burlap or felt; in other words, it is a thermoset material. The heat curing tends to make the linoleum more resistant to shrinkage and opening seams and also more resistant to indentations.

Most manufacturers make at least five kinds: plain, Jaspé, marbleized, inlaid, and printed. They may be purchased in a wide choice of color and design and in several thicknesses to provide for different floor requirements.

Plain linoleum is of solid color without design. The thicker gauges are known as battleship linoleum because of their use in United States battleships. Jaspé has a two-tone striated appearance instead of one color. Marbleized linoleum has an overall pattern that resembles a sheet of marble; it comes in various colors and is very practical since it does not show soil readily. In both the Jaspé and marbleized types the graining goes through to the burlap.

Inlaid linoleum is made by laying sections of plain, Jaspé, or marbleized linoleum in a pattern on a burlap back. Raised or felt-embossed designs give a pleasing effect but are not as easily cleaned as those with a smooth surface. Printed linoleum, as the name implies, has the pattern printed on the surface of the product in oil paints; since the design does not extend through to the backing, it will wear off with constant use.

The felt base type is less expensive than the burlap back and may be cemented directly to the floor, eliminating the expense of the underlay. Most linoleums, however, give better service if laid over a felt or sponge-rubber underlay which in turn is cemented to the floor. The underlay takes up any expansion and contraction of the wood floor and prevents the linoleum from buckling and splitting. It also makes the floor warmer and more resilient. The foam-rubber lining increases quietness and comfort and reduces rate of soiling, perhaps because the foot does not exert so much pressure on the softer surface. Any seams should be joined with waterproof cement.

Linoleum is resilient, does not crack or mar easily, is highly resistant to grease and oils, and is comparatively inexpensive. It is not recommended for floors in direct contact with the ground because of dampness. The backing of burlap or felt can be damaged by fungus. With the exception of the printed variety, linoleum should not be varnished, shellacked, or lacquered.

If properly laid and cared for, linoleum is durable and easy to keep clean. Some linoleums are given a wax finish in the factory, while others require waxing as soon as laid. Wax seals the pores, preventing dirt from being ground in, and also helps preserve the resiliency. Paste or liquid wax may be used, and it is well to apply a second coat after the first one has thoroughly dried. Then, under normal conditions, going over the floor daily with a dry mop and wiping it once a week with a damp mop will be sufficient. For more thorough cleaning, use warm water and a mild soap solution, and then rinse with clear water—rinsing is always advised for all smooth floors. Hot water, strong soaps, and cleaning agents should never be used since they tend to destroy the natural oil in the linoleum and make it dry and brittle. In fact manufacturers of linoleum warn against frequent washing of any kind; they claim that more linoleum is washed out than is worn out.

Rubber

Since 1946 most rubber flooring, both sheet and tile, has been made almost exclusively from synthetic styrene-butadiene rubber. It also contains reinforcing clay and whiting, resin, oil, waxes, color pigments, and sulfurs. The sulfurs vulcanize the rubber, causing it to retain its shape under stress and heating and cooling. Practically without exception, marbleized patterns are used—in all colors, including white and black.

This synthetic rubber material is resilient, flexible, quiet, and resistant to wear and indentation under load. The floor may be washed with a mild soap or synthetic detergent and between washings damp-mopped with clear water. Water emulsion wax is recommended, but never waxes or cleaners containing a solvent. Varnishes, shellacs, and lacquers are detrimental also, as are oily dust mops.

Asphalt

Marketed in 9- and 12-inch squares, asphalt is made from hydrocarbon resin builders and asbestos fibers, with some inert filler and color pigment added and then pressed into shape while hot. The colors, extensive in variety, go all the way through to the back.

Asphalt was the first resilient flooring that could be used over concrete in contact with the ground. It is still one of the most economical in cost for any installation, and it is durable, resistant to moisture and fire, but it may be dented by heavy furniture. Asphalt is easy to install. Regular asphalt tile is affected by oil and grease, but a greaseproof variety is now available. Asphalt tile can be softened by solvents used in some paste waxes. Harsh cleansers can harm the colors.

Cork

Granules made from the bark of a species of oak tree are the basic material for cork tiles. An early process was to bond the granules together under high heat and pressure with only the natural resins of the cork for a binder. Modern methods, however, employ lower heat and vinyl or other phenolic binders to produce a tile of more uniform color, with greater tensile strength, less porosity, and more resiliency. The vinyl forms a protective layer over the cork. Since cork is a wood product, the amount of moisture allowed to come into contact with the cork should be kept always to a minimum; excessive use of water will cause the granules to swell. It is necessary to pre-condition the tiles to the temperature of the area in which they are to be laid by stacking them there for at least 24 hours before installation.

Vinyl Tile and Sheet

Vinyl asbestos and homogeneous tile are composed of vinyl resins, plasticizers, fillers, and pigments, and, in the one case, of asbestos fibers. They are formed under pressure while hot. In the backed vinyl, a wearing layer of the same components is overlaid on a backing which may or may not be of alkaline-resistant material.

In general vinyl products are flexible, resilient, and quiet. They are resistant to acids, alkalies, oil, and grease, and will withstand wear and indentation. The asbestos variety, in addition, is extremely resistant to wide variations in temperature and to excessive moisture. Colors remain clear and bright over long periods of use. The surface sheen may become dull in common traffic lanes.

Vinyl floor tiles are made to resemble many other materials, such as brick, marble, leather, and random-width wood. Solid vinyl tile is made of polyvinylchloride resin. It has a tough, dense surface which is highly stain and grease resistant. It will withstand pressure up to 250 pounds per square inch without damage—more than any other resilient floor covering.[2] A solid inlaid vinyl tile has a pattern running throughout the tile. Another type of all-vinyl tile is one in which the design comes from embedded rotogravure printing. Its surface is textured.

Vinyl floor coverings are also available in sheets in 6-foot widths. They can be cut to fit around pipes and corners and come in a number of qualities and thicknesses.

Cushion vinyl is sheet vinyl composed of a thick layer of foam vinyl heavily coated with layers of tougher, long-wearing vinyl. Cushion vinyl muffles sound and cushions footsteps. Cushion vinyl products are made in a number of different types and grades, in sheets and in tiles. Some have waterproof backings which allow them to be used on floors that are below ground, such as basement floors. Some may be installed by new methods to provide sealed seams. Some need not be cemented down and can be cut with scissors and laid in place.

One type of vinyl flooring is poured on the subfloor. The liquid is covered with vinyl chips scattered randomly. After the first coat has dried, four or more additional coats are applied on top.

A recent development is the introduction of vinyl floor coverings, either cushioned or noncushioned, which do not require waxing. These coverings have a high or moderate initial gloss. If these coverings retain their gloss under continuous foot traffic conditions and are at the same time surfaces which are easy to keep clean, they should find wide acceptance. Preliminary wear tests indicate a wide

[2]B. R. Stewart, O. R. Kunze, and Price Hobgood, **Indentation and Recovery Tests of Common Resilient Floor Coverings, Texas Agr. Ex. Sta. Bul.** 961, 1960.

difference in the no-wax floor coverings, as in other floor coverings, in retention of original gloss, in the amount of cleaning necessary to keep a dirt-free appearance, in resistance to permanent indentation, and in resistance to staining by household chemicals.

Carpeting

Colorful and practical carpeting is now available for the kitchen. It is mainly available in polypropylene (Olefin) and nylon fibers with a backing of high density foam rubber or waffle-design rubber. Nylon has good abrasion resistance and is resilient. Mat-nylon yarns resist soil to a greater degree than the lustrous types. Because nylon does not absorb water readily, water soluble stains are easily removed. Nylon is also resistant to household acids and organic solvents. Nylon tends to generate static electricity and may pill and develop a fuzzy appearance.

Polypropylene is strong, moisture, stain, and soil resistant. Its nap crushes readily.

Investigations into the bacterial retention in carpeting are somewhat contradictory. In two hospital studies, fungi and bacteria survived for months even after the floor covering had been scrubbed with a germicide. A Swedish study reported that a damp carpet was a good breeding ground for bacteria but that bacteria and fungi died within 24 hours in a dry, lighted habitat. The recommendation of the Swedish group was to clean carpets by a dry method rather than with use of a detergent-water solution.

Carpeting can be bought from a roll and scissor cut to fit. The rubber backing creates a suction so the carpeting cannot skid. It can be removed when soiled and taken outside to be cleaned.

Kitchen carpeting is also available in squares with self-adhesive backings. The advantages of these is that they are easy to install, and, if individual tiles become soiled, they can be removed and cleaned and replaced. Some difficulties have been experienced by consumers in the even adherence of these "tiles" to the floor.

Ceramic Tile

Three basic types of ceramic tile are available in over 100 different shapes and sizes, 500 different designs, and more than 1000 different shades and color combinations.

Glazed wall-type comes in large sizes, usually four inches square, and is primarily used for wall surfaces. It is available in decorative designs that can be used over an entire area or as an accent with coordinating solid-color tiles.

Ceramic mosaic tile usually is less than two inches square and is used for such decorative purposes as inlaying a coffee table. Mosaic tiles are usually unglazed, with the color throughout the tile body. Colors range from deep blue to subdued earth shades. For easy installation, mosaic tile is sold mounted on large sheets.

Quarry tile is usually in six inch squares, although both smaller and larger sizes are available. It is thicker than other types of tile and is designed for heavy wear. It is used primarily on floors. The most familiar is earth red, but quarry tile also can be glazed.

A new, nonporous, extremely hard Mediterranean ceramic floor tile which withstands fire and frost, rain and sun, is now being used in kitchens as well as in bathrooms, porches, and patios. Like the ceramic tile floors which have been used for a number of years in kitchens and bathrooms, ceramic tile can be very attractive in appearance and very practical in areas where water is frequently spilled or dripped on the surface of the floor. However, since ceramic tile is not resilient, it is not a comfortable surface on which to stand for any period of time.

Installation Requirements

It has been pointed out by manufacturers that a smooth-surface resilient floor is part of a system, which includes the subfloor, underlay, adhesives, mechanical fasteners, and even environmental factors. Since each factor in this system is important, it is probably better to have a skilled professional do the job, in spite of the present emphasis on the "do-it-yourself" home mechanic, unless the mechanic is well versed in basic information and knows the various factors that must be taken into consideration.[3]

Self-adhesive vinyl asbestos tiles adhere to any floor in good condition. These come with a paper-covered adhesive backing. The paper is peeled off and the tiles adhere to the floor.

Tiles, other than the self-adhesive type, require less time to install than they formerly did. One company recommends that the adhesive be applied only to the perimeter and seam areas of each tile.

Maintenance Procedures

Accelerated wear tests on six common floor-covering materials were made by the Agricultural Engineering Department of the Texas Agri-

[3]Valuable help may be obtained from **Publication** 597, Installation and Maintenance of Resilient Smooth-Surface Flooring, Building Research Institute, National Academy of Sciences—National Research Council, 1958. Is probably in many public libraries.

cultural Experiment Station.[4] The investigators found that rubber tile and sheet vinyls showed significantly less wear than did vinyl asbestos tile, cork, linoleum, and asphalt tile. Asphalt tile showed most wear. In general, wet abrasives caused more wear than dry abrasives. Only on linoleum and cork did waxing seem to increase the resistance to wear.

All resilient floor coverings tend to be somewhat porous and softer than the dirt tracked in on shoes, which, therefore, causes abrasive action that scratches the material and may even fill the pores and is not always removed by sweeping. Wet cleaning is often necessary.

Excessive use of water or cleaner should be avoided. Hot water may penetrate seams and reach the adhesive, causing warping of the flooring. Too much cleaner means more rinsing and consequently overuse of liquid. Avoid abrasives, solvents, and strong alkalies, because they scratch and dissolve the natural oils in the products and may cause colors to fade.

The water emulsion waxes may be used on all floors except natural cork, which is treated like a wood floor. Treatment of other cork flooring varies with the type selected, and specific information should be obtained if needed. The solvent waxes are not used on asphalt and rubber tiles, but may be used on other resilient floor coverings. The water-emulsion resins, used on all except natural and prewaxed cork, produce tough, hard coatings with high luster, but do scratch and should be avoided in sandy and cindery areas. They have otherwise proved very satisfactory.

Since some dirt will penetrate into the lower coating of wax and the wax tends to build up in nontraffic zones, behind doors, for example, the floor must be thoroughly cleaned at intervals and all the wax removed; the procedure is then repeated.

Cleaning materials should be those specifically manufactured for the purpose. The Rubber Manufacturers Association and the Asphalt and Vinyl Asbestos Institute have set up standards, and cleaner manufacturers have usually followed their recommendations. These associations publish lists of approved cleaners, making correct selection easy.

Many falls in the home are the result of slippery floors. Application of wax to floors heightens the possibility of slipping. Since impartial tests show that only on linoleum and cork did waxing seem

[4]B. R. Stewart, O. R. Kunze, and Price Hobgood, Accelerated Wear Tests on Common Floor-covering Materials, **Texas Agr. Exp. Sta. Bul.** 890, College Station, Texas, March 1958.

to increase the resistance to wear, the use of wax is not generally justified from the standpoint of longer life of the flooring. Wax not only takes much time and effort to put on a floor but removing it and rewaxing it at fairly frequent intervals also requires expense and effort. There is no convincing evidence that a waxed floor requires fewer moppings than an unwaxed floor. From the standpoint of a safe floor and from the effort necessary to maintain a waxed floor, waxing a floor has little to recommend it.

Wall Coverings

Kitchen wall finishes should be smooth, hard, and impervious to moisture and common stains.

Painted plaster is probably the most common finish. Several coats of high-grade semigloss paint should be applied to both walls and ceiling, or the ceiling may be treated with cold-water calcimine and recalcimined when necessary. The ceiling is hard to wash by ordinary methods.

The color used will depend to some extent upon the exposure of the room. A northeast room is more cheerful if warm tints are used—buffs, creams, and warm grays; in rooms of sunny exposure, pale greens, grays, or blues are satisfactory. Two or three colors may be used or several tones of one color. Patchy areas of color should be avoided. Neutral colors are more pleasing as a background than bright unusual colors. Gay colors are, however, satisfactory for accents.

Adhesive-back linoleum is made in a special light weight for walls. It is resilient, does not chip or tear readily, is stainproof and waterproof, and is not affected by sunlight. Ceramic tile is attractive but difficult to apply in the old type of wall construction, and it is too expensive for many homes. Plastic tiles are available and are satisfactory. A number of imitation tiles may be placed directly over a plaster or dry-wall construction; these include asbestos and composition board (painted or enameled), and a light-weight steel or aluminum tile finished in synthetic enamel.

Tiles of copper, stainless steel, and three finishes in aluminum have built-in adhesive areas inside each corner and are easily applied to walls.

There are easy-to-clean vinyl wall coverings with self-adhesive back that look like quality wood, some that resemble building brick and stone, and some that look like leather. There is a cloth-backed vinyl, prepasted and strippable, that is made by printing a pattern on cloth and then coating it with vinyl. It peels off in one strip without steaming or scraping and can be taken from room to room or from home to home. These, too, are washable. Washable, decorating vinyls sold by the yard in many different patterns are widely available. There are also heavy plastic, textured wall coverings which have the look and feel of real fabric—burlap, travertine, grasscloth, and wood grain.

Counter-Top Materials

A variety of materials may be used for counter finish: wood, ceramic tile, aluminum, stainless steel, linoleum, plastic compositions such as Micarta, Formica, Textolite, and other plastic laminates, hardboard, and glass ceramics.

A material that is nonabsorbent, will not spot or stain from the usual household acids, alkalies, and greases, will stand up under the abrasives commonly used, will not scratch, will not be affected by dry and moist heat, will eliminate noise, and will give reasonably long service, are criteria to be considered in selection.

Wood is frequently installed in one section of the counter, often near the range, to provide a block where food may be cut or chopped. It is especially desirable for this purpose, since it prevents the rapid dulling of knives. Birch, beech, and hard maple are the best woods for these counter insets; of these, maple is the most resistant to grease. The wood must be sealed to decrease stain and moisture absorption. Temperatures above 380°F tend to scorch wood.

Ceramic tile is hard and heat- and moisture-resistant, with permanent colors; but it tends to be noisy and may damage dishes handled carelessly. The mortar in the joints tends to deteriorate, leaving dirt-catching crevices.

Aluminum is often anodized, the process that develops the layer of aluminum oxide on the surface. It may be obtained in a variety of colors and in a frosted or plain finish. It is hard and corrosion- and heat-resistant.

Stainless steel withstands heat, liquids, and common scratches but is marked by cutting. A stainless steel counter should always be equipped with a cutting board. Some varieties are stained by acids and some show water marks. Stainless steel is probably the most expensive material to select, but will last indefinitely. It should be remembered, however, that stainless steel and aluminum will conduct electricity, so extra care must be observed when operating electric appliances on them.

Probably linoleum and plastics, in both sheet and laminated form, are the materials most widely used for counter tops.

Linoleum may be purchased in a variety of colors and patterns, it is resilient and durable, quiet in use, and easily cleaned. Linoleum will give good service if a moisture-proof adhesive is used for bonding, if heat protective pads are used under hot utensils, and if the surface is not used as a cutting board.

Plastics

Several plastics are used in laminated counter tops, and sheet vinyl is used by itself. To form the laminated product, a stack of several sheets of special paper treated with synthetic resins is bonded together under heat and pressure. Vinyl or melamine resin may form the surface layer. Melamine used on the high-pressure laminates is tough and one of the most heat-resistant of the plastics. These laminates are rigid and come in a variety of colors and designs and in satin and glossy sheen. The counter edge may be finished by a metal band or a self-material strip; the self edge is widely used at present.

Laminated plastic can be marred by the abrasive action of cleaners and pans, some types can be damaged by certain temperatures of dry and moist heat, and some may dent or even crack under impact. In general, however, laminated plastic has a smooth surface, which is non-porous, resistant to grease, alkalies, and other household chemicals. It will withstand temperatures up to 300°F and is easy to clean. The 1/16 inch, standard quality, counter-top-grade laminate is as resistant to burns as the high-pressure laminate with the sealed-in foil.

Sheet vinyl affords a resilient, smooth, easily cleaned counter top, available in various designs and colors, that is resistant to most stains from household foods and solutions. The Ohio study[5] indicated damage from dry heat as low as 350°F and occasional damage from moist heat. Pads should, therefore, be used, along with boards for cutting. A lighted match or cigarette will cause vinyl to fuse. It is easily cut by knife or scissors and can be installed by the home workman.

Plastic-coated hardboard is manufactured into sheets from wood chips under heat and pressure. It is obtainable in a number of plain colors that are fairly resistant to household stains, are not damaged by heat or lighted cigarettes, but are damaged by abrasion. The material is fairly inexpensive and is easily installed—a good "do-it-yourself" type of material.

[5]E. K. Weaver and V. V. Everhart, Work counter surface finishes for kitchen and utility areas. Ohio Agr. Exp. Sta. Bul. 764, 1955.

Glass Ceramics are heat-proof materials which may be used to cover an area or be recessed within a larger worktop. Its nonstick surface can be used for kneading dough, rolling pastry, and cooling candy.

Liquid Neoprene Rubber

Liquid neoprene rubber finds numerous applications in the home such as repairing leaking roofs and gutters, waterproofing shower stalls, and weatherproofing porches, breezeways, and sun decks. The uses that are most relevant to the present discussion are the mending of torn linoleum and the cementing of loose tiles. It can also be used to protect outside lighting equipment and to repair frayed electric cords.

Experiments and Projects

1. Examine the appliances in the laboratory. Identify the materials used in each.
2. Study the cleaning instructions for items of aluminum, stainless steel, porcelain enamel, copper-clad stainless steel, and for polyimide-coated utensils, Teflon-coated utensils, silicone-coated utensils. Assemble the recommended cleaning materials, and clean an item of each of the different materials. Note the cost of the cleaning supplies.
3. Find examples of each type of plastic that is described in the text.
4. Examine the counter-top in the laboratory. Of what materials is it made? How should it be used and cared for?
5. Examine the specification sheets for ranges, refrigerators, food-waste disposers, dishwashers, air conditioners, food compactors. Where is the insulation placed in each? What type of insulation is in each?
6. Examine the floor covering in the laboratory, in your home kitchen, and in the bathroom. What is the material? Examine it for indentation, cuts, fading or discoloration. How should it be cared for?
7. Examine the exteriors of each of the following appliances: refrigerator, range, clothes dryer, dishwasher. Which have a porcelain enamel exterior? Which are synthetic enamel? Check your answers with the manufacturers' specification sheets. How would you care for each finish?
8. Examine the interiors of each of the following: refrigerator, oven, clothes washer (inner and outer tub), clothes dryer (inner and

outer tub), dishwasher. What materials or finishes are used on each? Check your answers with the manufacturers' specification sheets.

9. Examine plastic containers in the laboratory and at home. Try to find examples of each of the groups discussed in the text.

CHAPTER 3

Kitchen Utensils and Tools

Few awesome ruins remain to recall the world's earliest civilizations, but through the study of pottery fragments and tools, archaeologists have been able to study the origin and progress of cultures of earliest times. Correctly interpreted, they are no less significant than the great pyramids and other enormous and impressive monuments. In examining our present array of utensils and tools, it is interesting to contemplate what they might reveal in research excavations a couple of thousand years hence.

Nonelectric utensils and tools used in preparation of food will be examined in this chapter. For convenience and simplification, equipment will be grouped into (1) utensils used in surface cooking, (2) utensils used in oven cooking, and (3) tools used in food preparation.

The first group includes saucepans, saucepots, kettles, frypans, Dutch ovens, waterless cookers, and pressure saucepans; the second, roasting pans, cake and muffin pans, biscuit- and cooky-baking sheets, baking dishes, and casseroles; the third group, measuring cups and spoons, sifters and strainers, egg beaters, mixing bowls, can openers, cutlery, knife sharpeners, graters, and slicers. Other small pieces of equipment might have been included, but these selected are representative and are most commonly in use.

Every utensil should be judged on construction, efficiency, and care required. It must be well made and of a material fitted to the purpose for which it is to be used. It should be durable, simple in design, and suitable in size and shape. The material and design should contribute to ease of cleaning. It should be designed to accomplish the task for which it was made efficiently, without undue expenditure of effort.

METHODS OF FORMING UTENSILS

Kitchen utensils are manufactured by four basic methods. They may be cast, stamped, drawn, or spun. The tools are usually stamped.

Cast utensils are formed by pouring the molten metal, ordinarily iron or aluminum, into a mold. Metal molds are now commonly used because they are more accurate under high pressure; these molds have to be precision made and are expensive. When the metal has set, the utensil is removed, any surface irregularities ground off, and then polished. A cast product is often somewhat porous and is distinguished from sheet metal utensils by (1) the shank to which the handle is attached being cast in one with the pan, and (2) the rim of saucepan or pot having a straight edge with the bead on the cover. Some utensils that are not cast may have one of these features, but only the cast utensil has both.

More impurities are liable to occur in cast metal; hence there is a tendency for cast utensils to pit more readily than those of sheet metal.

A **stamped** utensil is made in a single operation from a sheet of metal, known as a blank, in a punch and die set, and is simple in form. Tools such as measuring cups and spoons, strainers, and graters, are stamped, as are tinned pie, muffin, and cake pans (Fig. 3.1). Notice that in stamping operation the excess tin is folded over against the sides of the cake pan at the corners. If the metal sheet is fairly thin, the product may be somewhat brittle and, if bent, will tend to crack and break apart.

Utensils of more complex design are **drawn.** This is a series of stamping operations, from 2 to 8, 10 or even more in number, during which the product is stretched into shape. Drawing preserves the position of the molecules in the metal and eliminates any stress or tension that might cause a flaw to develop in the finished product. Occasionally the metal may need to be reheated between steps to reshape the crystalline structure.

In **spinning,** the metal blank is fastened to the rounded end of a mold or chuck, which is then rotated while the metal is shaped over it. The process, once carried out by hand, is now done by machine. Equal pressure must be exerted in all areas to obtain the same gauge. Bowls of steel and aluminum, bun warmers, and similar utensils are manufactured by this method. It is often possible to see the tiny hole in the bottom where the blank was fastened to the mold, and a series of striations on the interior surface is very typical of spun products. This type of finish, however, is sometimes also found on drawn utensils if they have been cleaned with fine steel wool after manufacture.

***Fig. 3.1** Stamped and drawn utensils.* (a) *Stamped: the excess tin has been folded over the end of the pan.* (b) *Drawn: note the number of drawing processes the utensil has required. The next-to-the-last pan shows the shank of the handle welded to the pan.*

Handles and Spouts

Handles and spouts and shanks on cast utensils are an integral part of the casting. On stamped, spun, or drawn utensils, once the main body of the utensil has been formed, a handle and a spout must be fastened on. One of several different methods is employed: bolting or screwing, soldering, brazing, riveting, and welding.

The method of bolting or screwing is used when the attached part may later need to be replaced, if damaged, or when a utensil, commonly used for surface cookery, is to be occasionally put into the oven. Crevices around these devices are hard to keep clean, and the parts themselves are sometimes rough and tend to injure the hand.

Soldering is done at a moderate temperature, using tin or lead as a flux. The join may loosen if subjected to a high temperature or rough handling. In brazing, on the other hand, brass is commonly used as the flux, and the materials are held at red hot heat; this

Fig. 3.2 The use of rivets and welding in fastening handles to utensils: (a) *Outside views: only tiny dimples appear where the spot welding was done.* (b) *Inside views: the rivets project as far on the inside of the pan as on the outside. The welding is scarcely visible on the inside and does not interfere with cleaning.*

method makes a permanent join. Spouts are joined to coffee pots by brazing.

Riveting may be used to fasten the handle to a saucepan. A metal pin with a head at one end is passed through the metal sheets, and the other end is then hammered into a second head. Rivets should be of a size suited to the weight they must lift and should fit the space. Small rivets tend to pull out, and the rivet heads, always visible on the inside of the utensil as well as on the outside, provide places where food may collect. Unless very carefully washed, these places may build up into permanent stains (Fig. 3.2). Fairly large counter-sunk or compression rivets are used to fasten the tang of the best-quality knife blades into the handle. These rivets are flush with the surface and leave a completely smooth finish.

Handles may be attached to pans by the method of spot welding, that is, sending a current of high amperage through the two metal pieces at several distinct spots. The heat fuses the two pieces together, leaving on the surface only very tiny, dimplelike depressions that cause no cleaning problems. These joins are permanent.

Separate spouts are usually fastened to utensils with a completely smooth join. Spouts should be shaped to direct the liquid toward the center of the pouring stream without its spilling over the sides. The edge should be thin and sharp to prevent dripping down the outside of the vessel.

Covers

Most saucepans are supplied with lids of the same material as the pan, or of glass to allow observation of the cooking process. They are flat in shape or domed, occasionally with the center inverted. This last type allows stacking, since the cover knob is below the rest of the lid surface, but does reduce the usable space inside the pan. The dome shape is perhaps the most efficient, permitting the pan to accommodate a larger volume of food. Either pan or lid must be so constructed that the lid will fit closely, to prevent loss of steam when a minimum amount of liquid is used.

Covers for some pans have skirts which fit inside the utensil. With this type of cover, after foods in the pan reach the steaming point, condensation of the liquid from the cover runs back into the pan (Fig. 3.3). On some saucepans the covers are beaded. The bead of the cover fits into a groove between the bead on the utensil and the side of the utensil (Fig. 3.4). The liquid from the condensation of the steam on the cover fills the groove of the utensil and forms a "vapor seal." However, if the heat is not lowered sufficiently after the steam-

Fig. 3.3 Double boiler. The skirt on the cover fits inside the pan. (Revere)

Fig. 3.4 Saucepan and skillet. The bead on the cover fits into a groove between the bead on the pan and the side of the pan. (Revere)

ing point is reached, the cover is lifted by the steam and water is "spit" from the groove onto the range.

SURFACE-COOKING UTENSILS

By definition, accepted both by home economists and by the trade, a saucepan is a cooking utensil with one handle; a saucepot is equipped with two side handles; a kettle has a bail handle.

Capacity of saucepots, saucepans, and other top-of-the-range utensils, except fry pans, is expressed in terms of level full liquid measurements, usually in quarts or fractions of a quart. The "level full" measurement is, of course, different than the usable capacity. For example, a 3-quart saucepan will actually accommodate only about 2 or 2½ quarts of food for cooking. The diameter of the bottom of pans is another aspect of size. This dimension is important in relation to the surface cooking units. For the most efficient use of heat, pans should be the same diameter as the cooking units. Units on gas and electric ranges are 6 and 8 inches in diameter.

Size of fry pans or skillets is stated by the top diameter in inches. The capacity of a skillet, however, is determined by the bottom diameter. Skillets with sloping sides, having a smaller bottom than top diameter, obviously have less cooking surface than skillets with straight sides. However, the sloping sides make it easier to use a spatula for turning food.

MATERIALS USED FOR SURFACE COOKING UTENSILS

The most important single consideration in the selection, use, and care of surface cooking utensils is the material from which the utensil is made (Chapter 2).

Utensils used in surface cookery usually are of aluminum, copper, iron, stainless steel, porcelain enamelware, heat-resistant glass, and glass ceramics.

Heat transfer by conduction and convection are involved in cooking on surface units on top of the range (Chapter 7). Therefore, the selection of materials which conduct heat evenly over the entire bottom of the utensil with no hot spots for food to stick and burn, and one that at the same time is easy to clean and maintain, is of utmost importance.

The relative thermal conductivities of some common materials is shown in Table 3.1. Although silver is an excellent heat conductor, it is seldom used in cooking utensils partly because of its cost.

Table 3.1
Coefficients of Conductivity of Some Materials Used in Household Equipment (in Cal/Cm Cube/°C/Sec)

Silver	0.99	Mica	0.0018
Copper	0.91	Plaster	0.0015
Aluminum	0.49	Brick	0.0015
Iron, wrought	0.15	Water	0.0014
Tin	0.14	Asbestos paper	0.0005
Nickel	0.14	Wood (avg)	0.0004
Iron, cast	0.114	Rubber	0.00045
Steel	0.113	Linoleum	0.00036
Steel, stainless	0.113	Paper	0.0003
Pyroceram	0.0047	Cork board	0.00013
Glass	0.0024	Air	0.00005
Porcelain	0.0022	Cotton batting	0.00004
Concrete	0.0022		

Copper

Copper is an excellent conductor of heat. In copper utensils heat is conducted evenly over the bottom of the skillet or saucepan so that foods cook and brown evenly. They are preferred by many famous chefs and are used in many fine hotels and restaurants around the world. They are being featured in many shops that specialize in gourmet type utensils. However, if foods are allowed to stand in a copper utensil, the copper may react with some food products to form a substance which may cause serious stomach upsets. Copper utensils, therefore, are usually lined with tin to prevent direct contact with the food. Since the tin lining is easily scratched and must be renewed frequently, these utensils are not practical for most households. One manufacturer is now producing a restaurant and gourmet line of cookware of stainless steel with copper metallurgically bonded on the entire outside of the utensil (Fig. 3.5).

Copper is used to increase heat conductivity of utensils made of stainless steel, either by electroplating on the bottom of the outside of the utensil or by using it as a core between two layers of stainless steel.

Copper has a high melting point (about 1080°C). When used on conventional domestic heating elements, there is no danger of melting the copper bottom of a pan or teakettle even when boiled dry and allowed to remain over heat for hours.

Copper discolors when exposed to heat and to the atmosphere.

Fig. 3.5 Solid copper outside. Stainless steel inside. (Revere)

To maintain its warm, attractive appearance, it must be cleaned after each use with a special cleaner or with salt and vinegar.

Aluminum

Aluminum has a coefficient of heat conductivity about half that of copper but more than three times that of iron, steel, or stainless steel. Heat is conducted evenly over the entire bottom of a skillet or sauce pan. There are no hot spots for food to stick and burn. Foods cooked in aluminum utensils brown and cook evenly, The manufacturers of thermostatically controlled surface units on ranges recommend that for best results, flat-bcttomed, medium-weight aluminum utensils be used.

Aluminum is light in weight and does not rust. It has a low melting point (about 659°C), and, if a dry utensil is on a heated unit for a period of time, the bottom will melt. Aluminum utensils will pit if moist food is allowed to stand in the pan. Alkalies found in some foods, hard water, strong alkaline soaps, and dishwasher detergents darken aluminum. Aluminum utensils which have a porcelain enamel or polyimide finish on the outside and a Teflon coating on the inside to protect the metal are not subject to darkening.

Stainless Steel

Because of its hard, smooth, fine-grained surface, stainless steel is easy to clean. It is not affected by most food acids and alkalies, and it is strong and durable.

Stainless steel is a poor conductor of heat, and, when used alone in cooking utensils, it warps, develops hot spots, and produces uneven cooking results except where only water is used in cooking. It will darken if overheated, and this heat stain cannot be removed by methods employed in the home.

Copper is frequently used on the bottom of the outside of stainless steel utensils. The resulting conductivity[1] of the utensil is approximately that of a medium-weight aluminum utensil. Aluminum is also added to the bottom of stainless steel utensils and to the entire outside or inside of stainless steel utensils to improve the conductivity.

Copper and iron have been placed between two layers of stainless steel and utensils formed of these three-ply materials. The copper core stainless steel utensils are good heat conductors and, with stainless steel inside and out, are very easy to clean. The iron core utensils still produce hot spots and uneven cooking.

Cast Iron

Cast iron utensils heat slowly and, because of their mass and poor conductivity, retain heat well. These qualities have endeared cast iron utensils to generations of homemakers. These homemakers' cooking practices are geared to the longer preheating time of cast iron skillets and kettles and to the retention of heat for longer periods of foods cooked in them.

Controlled tests show cast iron utensils do not equal the evenness of heat distribution of heavy gauge aluminumware and copper- or aluminum-bottomed stainless steel utensils.

Cast iron utensils are very heavy and rather brittle. Because iron rusts when exposed to the atmosphere unless the utensil has been given a surface treatment by the manufacturer, iron utensils must be "seasoned" before the first use by heating the utensil with oil to produce an iron oxide to protect the surface.

Porcelain enamel has been added to some lines of cast iron utensils making it more attractive and more expensive.

Sheet Iron and Sheet Steel

Because of their poor conductivity, poor heat distribution, and light gauge, sheet iron and sheet steel utensils warp badly when heated on gas or electric range units. Unless the surface is protected by a noncorrosive material, these utensils will rust.

Glass Ceramic

Glass ceramics are nonporous ceramic materials which are resistant to extreme temperature changes because of a very low coefficient of expansion. This quality which allows direct transfer of a dish from the freezer to a hot oven has had special appeal to many homemakers.

[1]Dependent on the gauges of the metals used.

Utensils have been developed for use on glass-ceramic cook tops. These have carefully machined, flat bottoms to make close contact with the cook top. Glass ceramics are poor conductors of heat and produce uneven cooking results when used on the top of the range except when foods are cooked in water.

Porcelain Enamelware

Porcelain enamel is an applied finish (see Finishes). Porcelain enamel utensils possess some of the characteristics of both the glass and the base metal. When steel is used as the base metal, uneven cooking and scorching of foods results, unless the cooking is being done in water. When aluminum is the base metal, cooking is more even. (Porcelain enamel is usually only applied to the outside of aluminum cooking utensils.) See Figure 3.6.

Fig. 3.6 Heavy gauge wrought aluminum utensils with colored porcelain enamel exteriors. (Wear-Ever)

Enamelware may craze or chip if roughly handled, but the better grades will give long and satisfactory service when used with care. Select utensils that have a smooth, even coating and heavy gauge base metals. Those with fine cracks, thin places in the enamel, and other blemishes are a poor choice.

Research and new processing methods have greatly increased the resistance of enamelware to damage from impact and rapid changes of temperature. Similar technological advances have allowed for a broader choice of base metals and much more attractive styling. Porcelain enamel cookware is now available in a multiplicity of colors and designs on aluminum (both sheet and cast), steel, and cast iron; many feature special nonstick interior surfaces.

Heat-Resisting Glass

One type of heat-resistant glass may be used for surface cooking. Glass is not affected by food acids and alkalies; it will break readily from mechanical impact or from sudden changes in temperature. When glass is used on the surface unit of any electric range, the use of a protective grid is recommended. Glass is a very poor heat conductor and produces uneven cooking results when used on the top of the range except when foods are cooked in water. Glass for surface cooking may be used in the oven but glass for oven use cannot be used for surface cooking.

SELECTION OF SURFACE COOKING UTENSILS

There is such a large variety of surface cooking utensils available to the consumer in the United States in such a wide price range that it is possible for a knowledgeable shopper to purchase exactly the material, strength, and durability she requires, the sizes she desires, and the color and styling which will enhance her kitchen decor.

The choice of a material that will conduct heat evenly and quickly over the bottom is important in choosing a pan for surface cooking. Good heat conductivity is especially vital to the successful operation of the thermostatically controlled surface units, which are calibrated for medium-weight aluminum utensils.

The material and construction of the pan should provide a strong and durable utensil. The kind of material and its gauge or thickness are more reliable measurements than weight in pounds and ounces.

Range-top utensils should be flat-bottomed to make the best contact with the source of heat and straight-sided to minimize radiation heat losses and to use range-top space efficiently.

Covers should fit utensils closely to minimize vapor loss. A domed cover provides extra capacity for cooking bulky foods. If the cover has a vent, the vent should be located away from the cover knob so that the user will not be burned by steam from the vent when removing the cover. Cover knobs should be of a heat resistant material which should remain reasonably cool during cooking operations. They should be attached so they remain tight in use and so that they can be replaced easily, if necessary.

The handle should be of heat-resistant material and of a shape and length comfortable for the hand; handles molded to fit the hand are an advantage. Handles are joined at different angles—some almost at right angles, others at various acute angles. In selecting a pan, the homemaker should test different handles, and choose the one that is found to be the most convenient for lifting. Too long a handle may overbalance the pan or get in the way; too short a handle increases possibility of burns. Be sure that the handle is fastened securely so that it will not wear loose. A detachable handle should be tested for ease of operation and security. If the handle is of wood, a metal shank must connect it to the pan to eliminate any likelihood of the wooden handle being over the flame. Plastic handles should be protected from the heat of the unit where the handle is attached to the pan or provision should be made to attach the plastic handle at a sufficient distance from the side of the saucepan to keep it from the direct heat of the unit. A good quality plastic handle will withstand oven temperatures of up to 350°F. However, it should not be placed under the direct heat of the broiler. Detachable handles are provided by some manufacturers to enable utensils to be used at higher oven temperatures. A ring or a hole in the end of the handle makes it possible to hang the utensil. Plastic parts of the handle should be firmly attached but be removable for replacement, if necessary.

Construction and finish should make the utensils easy to clean. There should be no crevices or seams to harbor food or bacteria. A curve between sides and bottom of a pan simplifies cleaning. If a bead is used on the top of the pan or the cover, it should be pressed close so there is no crevice where dirt may collect. Teflon and silicone finishes prevent food from sticking to the insides of utensils and make cleaning easier. The hard, bright polished surfaces of stainless steel, aluminum, and copper are easily cleaned with appropriate cleaners. Chrome-plated, porcelain-coated, Teflon-coated, and polyimide-coated utensils do not require polishing and can be washed in a dishwasher.

USE

Saucepans, Saucepots, and Kettles

Foods which require boiling are usually started on high heat and then the heat is turned **lower.** Very little heat is needed to keep food cooking after the boiling point is reached.

The size of the pan should be proportional to the amount of food to be cooked. Usually a large amount cooks more rapidly if it forms a thin layer in a large kettle than when it is a deep layer in a small one.

Cooking in a small amount of water tends to preserve the vitamin and mineral content of the food. Green vegetables usually should not be cooked in a tightly covered utensil, or for a long time, since the volatile vegetable acids will change the bright green color to an olive green or brownish hue which is most unattractive. If a cover is used and is lifted once or twice during the first few minutes of the cooking process to allow the volatile constituents to escape, change in color may be avoided. Green vegetables cooked in hard water do not usually show this color change, although the amount of water used in the waterless method may not be sufficient to prevent discoloration. When strong-flavored vegetables such as onions, turnips, and cauliflower are old, they are best cooked uncovered in a fairly large amount of water; but young vegetables of this type may be cooked in a small amount of water if the cover is lifted occasionally.

So-called "waterless cooking" is a misnomer, since two or three tablespoons of water are often added, or at least the vegetables are moist from having been washed. Even when no water is supplied from either of these sources, most food products contain a high percentage of moisture, part of which is extracted in the cooking process and is the source of the steam, and only these products should be cooked in this way.

"Waterless cooking" may be done in any container that has a close fitting cover to hold in the steam, but the heat must be kept very low.

It is recommended that the pan be at least two-thirds full of food.

Fry Pans

To use any fry pan successfully, it should be preheated on medium or medium-high heat before shortening or food is placed in it. Shortening placed in a cold pan may be overheated to the smoking point as the pan is preheating before other food is placed in it to fry.

There are two tests to determine when a pan has been properly preheated to frying temperature on a conventional unit.

1. Drop a teaspoon of water in the pan. If it "skitters" across the surface of the skillet in tiny beads, the pan is at correct frying temperature.
2. Place a bit of butter or margarine in the center of the pan. As soon as it begins to brown, the pan is at correct frying temperature.

The heat should be turned down as soon as the food is placed in the pan and kept only high enough to keep food frying briskly without smoking and spattering. High heat is the principal cause of food sticking in any fry pan.

To preheat a skillet on a thermostatically controlled unit, set the control at 350°F.

When frying eggs, the skillet should not be preheated as described above. The fat should be placed in the skillet and the control set on medium heat just until the fat is melted. The heat should be turned to low as soon as the eggs are dropped into the pan.

Dutch Ovens

The iron kettle known as the Dutch oven dates from colonial days; only recently has it been made of other materials. Dutch ovens were brought to America by the Pilgrims. As is well known, the Pilgrims spent some years in Holland before coming to America in the **Mayflower,** which was a tiny vessel with limited baggage space. The Dutch oven could be used for such a variety of cookery that it took the place of several other pots and pans and, therefore, was a favorite utensil of the early settlers.

The Dutch oven comes in different sizes and has a domed cover which adds greatly to its capacity. The cover fits closely to keep in the steam. The oven may be used with or without a rack. The chicken fryer may double for the Dutch oven very satisfactorily.

Like the range oven, the Dutch oven may be used for cooking several foods at one time. Meat may be browned, placed on the rack, and vegetables placed around it; a separate vessel of fruit may be included. A small amount of water is added to the bottom of the oven and, after steam has formed, the heat is turned very low and the cooking continued for a long enough time to give tender products. Occasionally, more water must be added, but usually this is not necessary if the heat is kept low and the cover not lifted, so that steam is not lost.

Pressure Saucepans

The pressure pan is widely used in meal preparation, since by increasing the pressure, the temperature is increased, thus shortening the cooking time. At 15 pounds pressure the boiling point is raised to 250°F, at 10 pounds pressure to 240°F, and at 5 pounds pressure to 228°F.

Pressure pans are usually of the saucepan type with a single handle. Most of them have a domed cover which increases their capacity. They are made of aluminum, cast or stamped, and of stainless steel. Handles and knobs are usually of plastic materials. Most of them carry the Underwriters' Laboratory Seal, and only those so marked should be purchased. They vary in size from 2- to 7-quart capacity, depending on the manufacturer. Some companies make only one size; others, two or more. The pan has a cover that slides into grooves on the rim of the pan which hold it in position. A rubber gasket on the cover or on the pan provides a tight seal.

A vent tube allows steam to escape. The pressure is registered and controlled by a weight gauge placed over the vent. There are various ways of indicating when the pressure is reached. The pressure is maintained by regulating the amount of heat to the unit or burner. This requires the constant attention of the user. Electric pressure saucepans that are equipped with a thermostatic heat control eliminate this tedious procedure.

A safety device in the cover, a plug of fusible alloy or of synthetic rubber, automatically releases excess pressure or reacts if the kettle goes dry. Cooking time is counted from the moment the desired pressure is reached. Overcooking, even for a very brief time, may result in a mushy, unpalatable product. Cooking in the pressure saucepan, therefore, requires somewhat more attention than cooking by the usual methods. Colors, even green, are not destroyed because of the brief steaming period. This method also tends to preserve vitamins and minerals. The absence of air prevents oxidation of sensitive vitamins, and the small amount of water needed, usually less than a third of that required by ordinary cooking methods, prevents the dissolving of soluble nutrients. Directions must be carefully followed, however.

At the end of the cooking process the pressure is brought back to normal before opening the pan. The usual method is to run cold water over the pan, but the cooker may have a special device to release the steam. When cooking meats and dried vegetables it is desirable to allow the pressure to come back to normal slowly. In any case it is essential to have the pressure reduced before removing the cover; otherwise a serious accident may result. Monroe[2] notes that

[2]Merna M. Monroe, Concerning Pressure Saucepans, **Agr. Exp. Sta. Bul.** 445, University of Maine, Orono, Maine, September 1947.

certain foods such as thick soups, cereals, stews, and dried beans, in fact any foods that tend to thicken the cooking water, lose pressure slowly. As a result the thickened mass may hold sufficient pressure, even though none is indicated, to blow some of the food from the pan when the cover is removed. When cooking such foods Monroe recommends that the pressure pan be placed in a pan of cold water to cool the food at the bottom of the cooker adequately.

It is important to keep the air vent clean. Any obstruction in it may cause the pressure to build up without a means for the steam to escape. Some manufacturers' directions state that the cooker should not be used for cooking stews, applesauce, rice, and similar foods that tend to foam and hence boil up into the vent.

The rubber gasket, whether pure or synthetic, should also be kept clean, free from food and grease. When it becomes worn and no longer makes a tight seal, it should be replaced with a new one.

The cooker sometimes becomes discolored from minerals in the water or in the food itself. Such stains may be removed by boiling a weak solution of vinegar or cream of tartar in the pan and by scouring it with fine steel wool.

Pressure Canners

The renewed interest in home canning of vegetables and meat has increased the sale of larger size pressure cookers which are suitable for canning. The operation of the larger cookers is the same, in general, as the saucepan types. All of the precautions for safe use suggested by the manufacturers should be carefully observed.

Some homemakers have recently put back into use old pressure canners which have been stored for many years. Before use, these canners should be cleaned thoroughly, with special attention given to vent tubes, the tube to the pressure gauge and the areas around the body of the cooker and the cover involved in the locking mechanism. Gauges should be tested for accuracy and gaskets renewed.

OVEN UTENSILS

After the oven is heated, the percentage of radiant heat in an oven is between 60 and 70%, and convected heat is from 30 to 40%. The ability of a baking utensil to absorb radiant heat is the important factor in determining its efficiency in utilizing oven heat. The quantity of radiant heat absorbed by a utensil in the oven is dependent on the material, the color and the character of the surface, the oven finish, the oven temperature, and the oven load. A 400°F oven with black bottom and sides transmits 7 to 9% more radiant energy to all but shiny pans than an oven with shiny sides and bottom. The amount of radiant heat increases with increase in oven temperature. The same quantity of food heats faster in a large pan than in a small pan because of the large exposed bottom area, since the bottom of the oven is hotter and radiates more heat.

MATERIALS AND FINISHES FOR OVEN UTENSILS

The same materials are used for oven utensils as for surface pans and, in addition, tinned steel or iron, earthenware, and heatproof china. To these materials have been added aluminum foil and nylon and polyester cooking bags, although they are not, in themselves, utensils. The material and the finish of utensils used in oven cookery affect the characteristics and quality of the food product. By choosing different materials and finishes, different results in baked products can be achieved (see Material and Finishes). In general, glass, glass ceramics and other ceramic materials, enameledware, anodized aluminum, darkened tin-coated ware, cast iron, polyimide coated utensils, and utensils with dark or rough and dull surfaces are good absorbers of radiant energy. Foods baked in these will bake more quickly and form crisp crusts. Utensils with light and highly polished surfaces, and new tin-coated utensils reflect radiant energy. Foods baked in these will cook more slowly and have soft crusts. Frequently it is desirable to combine a rough dull finish (a good absorber of radiant energy) with a shiny surface (a reflector of radiant energy) in different parts of a utensil to produce the desired results.

SELECTION OF UTENSILS USED IN OVEN COOKERY

Roasting Pans

The roasting of meat in an uncovered pan is now usually recommended. Lowe[3] found that cooking losses are less and meat is more palatable when the uncovered roaster is used. Moist heat, however, is generally preferred for roasting fowl, and some homemakers cook pot roasts in the oven rather than on the surface of the range; in these cases a covered roaster is necessary.

Roasters are oval, rectangular, or round, and of enameledware, aluminum, iron, or steel. Because of its efficiency in absorbing radiant energy in the oven, a dark-surfaced or enameledware utensil with a large bottom area is preferable to a small, polished pan. A

[3]Belle Lowe, **Experimental Cookery,** John Wiley, New York, 4th ed., 1955.

roasting pan with sides approximately 1 to 2 inches high, and with a smooth surface, rounded corners, and as few grooves as possible is easiest to clean.

When a covered roasting pan is used, a dark enamel roaster will cook a bird or roast more quickly than a shiny roaster of some other material.

Boneless roasts should be placed on a trivet to keep the meat out of extracted fat and juices and to prevent sticking. Some roasters have perforated trays to hold the meat; a V-shaped rack is recommended for poultry.

Cake, Pie, and Muffin Pans

Cake pans may be oblong, square, or round; deep or shallow. They are of aluminum, oven glass, tinned steel, Russian iron, or enameledware. Rounded corners and edges make for ease of cleaning. When the pan has a separate bottom, the inset should fit tightly to prevent batter from dripping.

In cakes, breads, muffins, or other products which rise during baking, best results are obtained in utensils with dull, rough bottoms (which pick up the radiant energy) and shiny sides (which reflect radiant energy), allowing the batter to rise before the sides set.

The manufacturers of Pyrex glass recommended for their cake pans a temperature 25°F lower than that used for other materials. This will cut down slightly the amount of radiant energy incident on the glass on sides and bottoms, but the crust on sides and bottom is hard and the cakes are frequently rounded in the center.

The shape and size of the pan affect the product, too. Sharp pan corners tend to cause the cake to be browner at the corners than on the rest of the surface. A shallow pan usually gives a coarser cake than a deep one. Too large a pan exposes the cake to more radiant energy from the bottom of the oven and causes excessive browning.

Certain pan sizes have been adopted as standard by the American National Standards Institute and can usually be found for sale in the stores. Many packaged mixes and cookbooks now specify or at least suggest these standard sizes for the recipe. Dimensions, in inches to the nearest ¼ inch, should be marked on the pan bottom or given on a removable label.

Angel and sponge cakes are baked most satisfactorily in tube pans. The tube should be taller than the height of the side, so that the pan may be inverted on it for cooling the cake, unless legs are provided on the pan for this purpose.

Pie pans are of aluminum, tin, enameled steel, and glass ceramic, with a flat, or juice-catching rim, or a scalloped edge. Aluminum pans with a dull gray anodized finish, darkened tin, enameledware, and glass absorb heat readily and produce crisp lower crusts.

Muffin pans are made of the same materials and finishes as cake pans with similar baking results. Look for muffin pans with dull rough bottoms on the cups and shiny sides. Muffin pans at best are hard to clean. Pans with few joints and creases should be selected.

Baking Sheets

Aluminum, sheet iron, and tinned steel are the materials commonly used for baking sheets. A smooth, somewhat shiny baking sheet will produce cookies lightly browned on top and bottom. Rough surfaces, darkened tin, or dark finished sheets produce cookies overly dark on the bottom and around the edges.

Cookies, rolls, biscuits, and cream puffs rise and brown more evenly when cooked on baking sheets rather than in pans. Sides on pans baffle the heat, and the bottom of the cooky or biscuit becomes too brown before the top is the desired color. This is especially true if the sides are deep. Layer-cake pans may be used for rolls, if necessary, because the sides are low.

Baking Dishes

Any of the materials used for other types of pans may also be used for baking dishes, but covered casseroles and many uncovered baking dishes are frequently of heatproof glass, glass ceramic, earthenware, or china. Since heat-resistant glass absorbs radiant energy extremely well, it is especially desirable for oven use when crisp crusts are desired.

Casseroles come in graduated sizes, the smaller individual ones being known as ramekins. Heatproof china may be plain or have a conventional design or a splash of bright-hued flowers, as the purchaser chooses. Common bean pots make excellent casseroles, suited to a variety of uses. Shallow broad baking dishes give a large surface for browning.

Foods baked in dishes of these wares may be served directly on the table, thus eliminating the cost of an extra container and saving extra dishwashing and storage space. Often the cover is flat on top so that it may be used separately for a pie plate or an "au gratin" dish. A dish with handles, or projecting ears or rim, is more easily removed from a hot oven, but the handles take up a little more space. Casseroles may be set into stands of nickel or chromium to increase ease of handling and to protect the table from the heat.

In selecting a casserole, be sure that the surface is nonabsorbent and free from tiny cracks and flaws which impair its smoothness and make cleaning difficult. Glass will crack if subjected to sudden variations in temperature and should be allowed to cool before cold water is poured into it. Brushing the surface with an unsalted fat before use and soaking before washing minimizes the task of removing remnants of food.

Casseroles should not be used for surface cookery.

Aluminum Foil

Because ovens, roaster, and broiler pans are easier to clean when aluminum foil is used in roasting and broiling, aluminum foil is widely used by homemakers. There seems to be considerable difference of opinion by foil manufacturers as to whether the roast or bird should be tightly covered with aluminum foil and roasting should be done at a 425°F temperature, or if just a tent should be put over the roast and a lower temperature used. The higher temperature and shorter time certainly appeals to busy homemakers and many have been satisfied with the results.

When aluminum foil is used in roasting meat or poultry, the dull side of the foil should be exposed to the oven in order to pick up the maximum amount of radiant energy.

Cooking Bags

The promise of cleaner ovens and cleaner roasting pans, quicker cooking, juicier roasts and poultry, and browning without unwrapping has resulted in a large volume of sales of cooking bags made of nylon and polyester. Most manufacturers have included a caution on the package that the bags are not to be used at temperatures above 425°F and are not to be used in broilers. Some accidents have been reported, however, and most manufacturers now recommend that, not only the bag be punctured with a fork before placing in the oven, but also that meat or poultry be coated with flour before the cooking starts.

KITCHEN TOOLS

Measuring Cups

Measuring cups are of tin, plastic materials, stainless steel, aluminum, or heatproof glass.

Measuring cups of aluminum should be sufficiently heavy to hold their shape without denting or bending. The handle should be welded to the cup or carefully riveted, so that it will not come loose, and should be large enough to allow the use of a holder when hot liquids are measured. A lip on the side helps in pouring. A distinct groove between the bottom and sides, and deep-cut graduations with sharp edges, are to be avoided because of the difficulty of cleaning. Tin cups usually have deep indentations and also have a tendency to rust.

Glass measuring cups are usually smooth on the inside with the graduations marked on the outside surface only. Cups with red capacity marks are especially easy to read. The transparency of glass is an aid in determining whether the substance measured is even with the graduation.

Plastic materials for measuring cups are light in weight and easy to handle, because plastics are poor conductors of heat. They may be purchased in a number of attractive colors to fit into the kitchen color scheme. Certain plastics tend, however, to be somewhat brittle and may crack or shatter under a blow. They may warp if washed at the high temperature frequently occurring in the electric dishwasher.

There are three kinds of measuring cups: those that measure a cup when full, and those that measure a cup about a quarter of an inch below the rim; both of these types marked to indicate ¼, ½, and ¾ of a cup on one side and ⅓ and ⅔ of a cup on the other. (A few types are marked in ounces.) The first type is best for measuring dry ingredients, the second for liquids. The third kind of measuring cup, the single-capacity cup, comes in a set of four: the full, ½, ¼, and ⅓ cups. These single-capacity types are especially accurate for dry ingredients (Fig. 3.7).

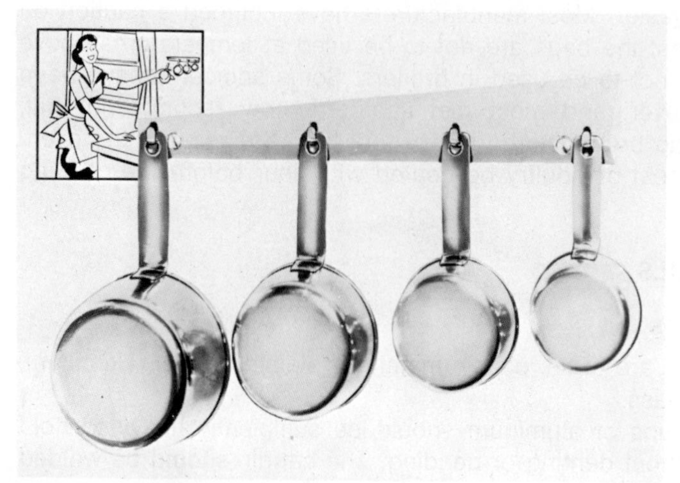

***Fig. 3.7** Individual measuring cups of stainless steel. (Foley)*

A standard[4] developed under the auspices of the American National Standards Institute (ANSI) and sponsored by the American Home Economics Association defines both liquid and dry measures.

> Household liquid measures have a pouring lip and a capacity of one quart, one pint, and one-half pint. Their capacities and subdivisions are defined in terms of quarts or pints, fluid ounces, and cups. The capacity of liquid measures shall be defined by a graduation mark encircling the measure near the top but below the pouring lip. The lip shall be designed as to permit the measure to be filled with liquid to the proper graduation mark while the measure is standing upon a level surface. The subdivisions shall be based upon the relation ½ pint equals 1 cup. Subdivisions of 1 cup shall include only ¾ cup, ⅔ cup, ½ cup, ⅓ cup, ¼ cup. The capacities shall be given as 1 quart equals 32 ounces equals 4 cups; 1 pint equals 16 fluid ounces equals 2 cups; 8 fluid ounces equals 1 cup.

> Household dry measures have a 1-cup capacity (equal to eight fluid ounces level full) or have sets of four measures including 1 cup and the following fractions of 1 cup: ½, ⅓, and ¼ level full. Their total capacities are defined in terms of cups and tablespoons. Dry measures shall be of the following capacities only: 1 cup, ½ cup, ⅓ cup, and ¼ cup. The capacity of dry measures shall be determined by the amount of material contained when leveled with a straight edge of a knife or spatula, and shall be given in cups and tablespoons based on the relation 1 cup equals 16 level tablespoons. The capacities shall be as follows: 1 cup equals 16 level tablespoons; ½ cup equals 8 level tablespoons; ⅓ cup equals 5 level tablespoons plus one level teaspoon; ¼ cup equals 4 level tablespoons.

Measuring Spoons

Measuring spoons may be purchased in a cluster of four—a tablespoon, teaspoon, ½ teaspoon, and ¼ teaspoon. They may also be obtained individually, often with a rack for mounting (Fig. 3.8). The standard tablespoon has a capacity of 1/16 of the standard cup; the teaspoon is ⅓ tablespoon. Measuring spoons are made of aluminum stainless steel, and plastics. Accurate spoons are usually stamped "U. S. Standard" (abbreviated to "U. S. Std."), and only those so stamped should be used.

Can Openers

A good can opener should remove the cover from a round, square, or oval can with the minmum of effort and should leave a smooth,

[4]ANSI, Z61.1—1963, Dimensions, Tolerances, and Terminology for Home Cooking and Baking Utensils.

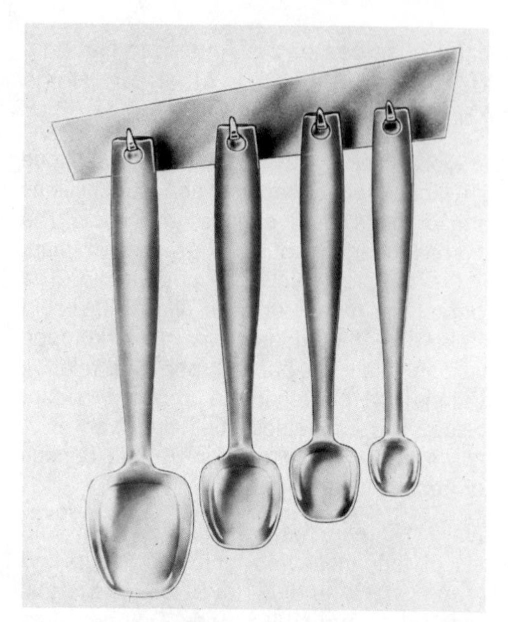

Fig. 3.8 Measuring spoons. The rack may be fastened to the wall or cabinet door. The long handles make reaching into tall cans easy. (Foley)

even edge. Openers that fasten to the wall or table edge usually do this. The can to be opened is held mechanically between two wheels, one with a knife-blade rim, and the cutting is done by turning a crank, which takes only a moderate amount of energy, uses only one hand, and cuts the cover cleanly from the top of the can. Usually only the flat metal support is permanently fastened to the wall; the rest of the can opener may be removed and placed in a drawer. Some can openers are pivoted and will swing back and lie flat against the wall when not in use. The cutting blade of the can opener should be thoroughly cleaned after each use.

Sifters

Depending upon one's choice of a sifter, flour and other ingredients may be sifted once, twice, or even more times in a single operation. The wire screen may be coarse or fine. A fine screen sifts more thoroughly.

The sifter is held in one hand and the sifting mechanism operated by the other, or one hand may both hold and sift. Such a method is very efficient, for the other hand may stir the mixture at the same time.

Sifters are commonly of tinned steel. They should be sturdy in construction with substantial, securely fastened handles. Sifters should be washed in warm soapy water, rinsed in hot water, shaken, wiped as dry as possible, and left in a warm place until thoroughly dry, to prevent rusting.

Strainers

Wire strainers may be used for sifting, as well as for the numerous other operations for which they are intended. The wire screen must be carefully fastened to the solid metal edge, and two or three pieces of heavy metal ribbon beneath the bowl will serve as extra support and tend to increase the life of the strainer.

A type of strainer called a colander, made of tinned steel, aluminum, or porcelain enamelware, has perforated sides and bottom. The holes are drilled and are larger than the individual meshes of a wire screen.

Metal sieves or fruit and vegetable presses, of tinned steel or aluminum, also have drilled holes. The holes are smaller than in the colander and much closer together. The tinned steel sieve is somewhat more efficient than the aluminum because tin can be used in a thinner sheet and the edges of the holes are very sharp. The sieves are often conical in shape and are used with a conical wooden mallet which is revolved against the inner surface to force the product through in a finely divided state. Such an appliance is used for making purées or preparing fruit for whips.

One press on the market is of steel. The perforated seamless bowl has a revolving spiral pressure blade with beveled edges that presses the food through the holes. It may also be used for straining and ricing. A metal finger, attached below the bowl, scrapes off the food.

Beaters

The two types of beaters most commonly found on the market are the whisk and rotary.

The whisk is effective in incorporating large amounts of air and gives maximum volume though a somewhat coarse texture. It may be made of many fine wires, each wire forming a long oval, and all the wires brought together at the top to make a handle; or it may take any one of many spoon shapes. Fine wires give greatest volume.

Rotary beaters are of steel with four circular or elliptical agitators which revolve in a vertical plane.

The agitators are fastened around a support of heavy metal wire, which carries the small pinion wheels. The wire is riveted or welded to the main shaft, to which the large cog wheel and handle are attached. The larger wheel makes one revolution while the beating circles make four or five; in other words, the beater itself will do four or five times as much work as the housewife. Ball bearings increase ease of manipulation. The handle should fit the hand, preferably projecting slightly beyond the hand to prevent cramping the muscles. The rotary beater produces the finest texture.

The beater is always used with some kind of platter or bowl, efficiency of operation often depending upon the shape of the bowl.

Bowls

Broad shallow bowls or platters are used with whisk beaters, narrow deep bowls with the rotary. All bowls should be of a material that will not scratch or chip, of the correct size for the amount of food to be beaten, and appropriate in shape for ease of manipulation. A bowl used with the rotary beater should be heavy enough to stay in place without being held.

Cutlery

Knives, forks, and spatulas of good quality are a worthwhile investment.

A study of the time taken and motions made in preparing three meals a day for ten days disclosed the interesting fact that the homemaker performs some task with a knife on an average of 129 times a day. It may well be said that "the use of the household and kitchen knife in its varied and improved forms marks the dividing line between savagery and civilization; and the more skilled a nation becomes in the preparation of foods, the more attention it pays to the design and workmanship of its cutlery."[5]

The type of steel used in a knife blade determines the cutting quality of the blade and how often it must be sharpened. Many butchers, professional cooks, and restaurant workers demand high quality, high carbon steel knives. A high amount of carbon makes steel hard and, in a knife blade, makes it capable of holding an edge. Since high carbon steels are stained by food, the trend of present manufacturing is to make knives of high carbon stainless steel, which contains chromium and hardening alloys, or from vanadium steel, chrome plated. Tungsten carbide is sometimes fused to one side of a stainless steel blade. Because the tungsten carbide edge is very hard, knives stay sharp a long time. Some inexpensive knives are

[5]**The Most Important Tool in Your Kitchen,** Harrison Cutlery Co., p. 3.

made from sheet steel or sheet iron. They are easily stained and need sharpening often.

Knife blades are made by forging, beveling, or stamping. A forged blade is hammered into shape by hand or machine which develops fine grain in the steel. After forging, the blade is hardened, tempered, and ground—frequently under water to prevent drawing the temper. Some knives, called Frozen Heat knives, of high carbon stainless steel are processed at a very low temperature, 100 degrees below zero, a method that produces a tough, strong blade, very resistant to dulling; they are guaranteed to hold their edge without resharpening for at least three years.

Knives are made also by two other methods. Beveled knives are formed from a steel bar which is thick in the center and tapers toward each edge. Two blades are cut at one time, back to back. Stamped knives are cut from metal sheets and ground to an edge. With new advances in metallurgy, manufacturers claim that good knife blades can be produced by all three processes. The types of grinds which produce the cutting edge are shown in Figure 3.9. The roll grind is used on knives for heavy duty jobs, the flat grind is used on knives for slicing, and the hollow ground knife with its thinner cutting edge for paring or carving.

A knife should have proper "spring and good balance. A blade with spring is ground tapering from handle to point. When such a blade is bent, about one-third of it next to the handle remains rigid, the other two-thirds to the point is flexible enough to form a slight curve. If the blade is of high-grade steel it will usually spring back into position; the iron knife will remain slightly out of line.

Good knives often are recognized by the manner in which the blade is fastened into the handle. In a cheaply constructed knife, the shank of the blade, tang as it is called, is narrowed to a point, which is pushed into the handle and held there by a small nail or brad or

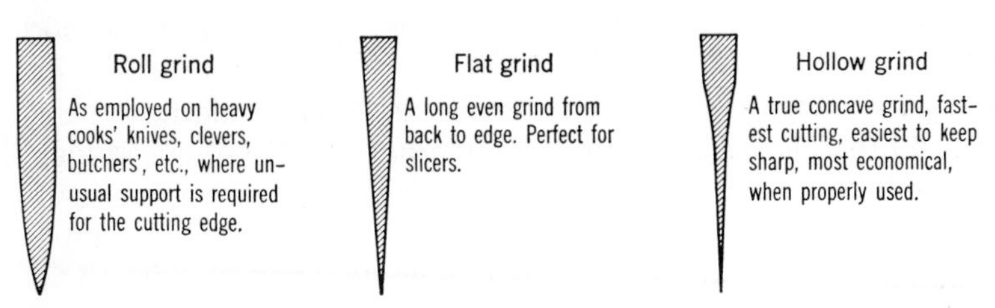

***Fig. 3.9** Cross section of knife edges, showing types suited to different processes. (Russell-Harrington Cutlery)*

sometimes merely by a metal collar. The nail or collar may work loose if the wood becomes softened in water, and the blade may pull out. A better method is to have the tang extend the entire length or at least half the length of the handle and be fastened in by two or three good-sized rivets.

Handles are usually made of wood or a plastic material. When plastic is used the shank is molded into it by heat and forms with it a solid piece. These handles are very strong and durable, there are no cracks to collect dirt, and the material is practically moisture-proof, but they will break with rough usage. Some cutlery manufacturers use rosewood for handles. It is a very hard, close-grained wood that is resistant to moisture and will not stain or warp. Handles should be easy to grasp and feel comfortable in the hand.

Knives are classified according to the shape and length of the blade, and their names usually indicate the use for which they were made. The paring knife has a short blade, 2½ to 3 inches in length, which gives leverage without undue strain on the finger muscles.

The utility knife, 4 to 6 inches long, may have a straight or curved cutting edge. It has many uses: cutting up large vegetables, trimming meat, slicing cold chicken, and cleaning fish—in fact almost any kind of food preparation for which the paring knife is too short.

The carving knife has a fairly long sturdy blade, not too pliable, so that it will sever joints and cut hot, yielding meats with ease. The point has a long, rather thin curve, to assist in cutting around the bones. The slicing knife has a long flexible blade, tapering slightly or rounded at the point, and will give the thinnest slices of cold meat, bread or cake.

To this list may be added the butcher knife, with its thick, heavy, curved blade for strength; the grapefruit knife, curved and serrated, to cut the sections of the fruit from the rind and the enclosing membranes; and the French cook knife, with its straight edge and firm but tapering point. The handle of the French knife is placed in line with the back of the blade, that the hand may not interfere when the knife is used for quantity chopping of nuts and small fruits and vegetables.

Another most useful and efficient knife is the serrated or scalloped-edged bread knife. The serrated edge has groups of fine ridges, running first in one direction and then in another, which are self-sharpening and always cut satisfactorily. Cutting bread dulls other knives rapidly. At present at least one manufacturer makes all its kitchen knives with serrated edges.

A blade should never be heated to aid in slicing, since this treatment destroys the temper of the steel. A knife should also not be used to cut metal, bone, paper, or string.

Transparent plastic knives are stainproof and are recommended for cutting thin slices of vegetables and fruits. They may have serrated edges like the bread knife.

Since good steel cannot be recognized at a glance, the purchaser should buy cutlery made by a reliable manufacturer. It is not so much the number of knives that the housewife has but rather the careful selection of a knife to meet a definite need that will make her kitchen well equipped and her work easier.

All knives need to be sharpened. The knife sharpener may be two finely corrugated steel wheels or two wheels of emery or sandstone turned by a handle. Sandstone wheels are preferred to emery, because they do not wear away the steel as rapidly. If a knife becomes heated in the sharpening process, the temper may be drawn and the knife will lose its ability to keep an edge. Wetting the emery wheel or grindstone prevents this misfortune. Hollow-ground blades should always be sharpened on a fine- or medium-grade oil stone, although a knife sharpener of carbide—the hardest material made by man—is also recommended. A certain technique is necessary to sharpen a knife satisfactorily and may be acquired from watching an expert. Sometimes a gentle smoothing of the roughened edge is all that is required to restore the blade to its original keen quality. A knife will remain sharp longer if food is cut on a wooden board. Do not leave knives in hot water.

Knives should be stored with care. Putting all kinds together in a drawer tends to nick the edges and break the points. Placing them in a wooden rack within the drawer or on the wall, or even hanging them on the wall, will prevent this damage. A new rack of permanently magnetized iron grasps cutlery or other metal utensils placed against it and is a real saver of time and motion.

Forks are of various sizes and have two, three, or four prongs, depending upon the use to which they are put. A long-handled fork is helpful when the kettle is deep or the food product large and heavy. Smaller forks are needed to hold meats when they are cut, and potatoes or other vegetables for skinning. The tines should be firm and sharp. Handles should have the same characteristics as handles on knives.

Parers

Although a conventional paring knife can be used for paring vegetables, many people find they can remove thinner layers of skin with a combination parer and corer, or with a parer with a floating blade. With a floating blade, it is recommended that with carrots and large potatoes, more speed is developed if the parer is used in the direction away from the operator and in small vegetables and fruits if used in the direction toward the operator. With both the combination apple corer and peeler and the parer with the floating blade, it is easy to prepare decorative curls of carrots, cheese and chocolate.

The **spatula,** first cousin to the artist's palette knife, with its flexible rounded blade of stainless steel riveted into a comfortably shaped handle, has many uses. With it, cups of dry ingredients are leveled off, egg whites are folded into mixtures, bowls are scraped clean of batter, and cakes are iced. If the blade is rigid close to the handle, but flexible for the rest of its length, it is more easily manipulated than when flexible for the entire length. The broad, more rigid spatula may take the place of a pancake turner, remove cookies from the baking sheet, and perform other useful tasks. Plastic spatulas are also available.

Spoons

Spoons for mixing, stirring, and serving are of aluminum, tinned steel, stainless steel, enamelware, iron, inexpensive plated silver, or wood. All metal spoons scratch the container, but wooden spoons do not mar, are easy to hold, and do not become warm when hot mixtures are stirred; they will warp and stain, however, and will need to be replaced from time to time. One or two wooden spoons should find a place in every kitchen.

Kitchen Shears

Kitchen shears are useful for many jobs—cutting marshmallows and dried fruits, snipping parsley, chives, and celery leaves, shredding chicken, separating bunches of grapes into serving portions, dividing bread dough into loaves and rolls.

Scrapers

Scrapers of rubber and plastic find extensive use in the home for removing final portions of frosting, whipped cream, melted chocolate, sauces, batters, and similar food products from bowls and pans. The scraper may be molded in one piece or have a separate handle of wood. Those with wooden handles are usually less durable since the handle tends to loosen from the scraper. The rubber scraper should be thoroughly washed in warm, soapy water immediately after use. Fats left on pure rubber, even for a short time, cause the rubber gradually to become spongy and sticky. Synthetic rubber is not affected by fats in this way.

Graters

Graters may be of various shapes: flat, cylindrical, square, or semicircular, but the shape is of secondary importance to the type of hole, which is punched or drilled.

Punched holes have rough uneven edges with four sharp points, and tend to give a mushy product of no distinct form and of compact volume, since much of the food tends to stick in the holes. The drilled holes are round or crescent-shaped, and have a smooth, sharp edge that cuts the grated food into a definite shape, each small sliver separate from the piece next to it. The resulting volume is large.

With either type, care must be taken to hold the product in such a way that the fingers do not come in contact with the sharp edges of the cutters. One safely manipulated grater, a rotary type, may be fastened to a shelf or table and looks a little like a hand-operated meat grinder. Surrounding the cylindrical cutting surface is an outer shell, with an opening on the top, through which food is held in contact with the revolving grater by means of a plunger. The cylinder is turned by a crank. If this grater is to work efficiently the cylinder must be close enough to the outer shell to leave no space in which the food may pack.

Graters may be of aluminum or stainless steel, but are usually of tinned steel. They should be sturdy in construction so that they will not bend out of shape. Drilled graters are more easily cleaned than the punched.

Slicers

Slicers are commonly made of steel. A rotary slicer has two or three blades turned by a crank; a flat slicer has knife plates set in a wooden or metal frame. Blades should be adjustable for different thicknesses of food and removable for sharpening. Slicers should be tested to determine how satisfactory a product they give, and whether they show a tendency to scatter the food.

Limited space does not permit a discussion of the many perforated and lipped ladles and turners, cooking tongs, butter curlers, egg and tomato slicers, molds, and the fancy cutters of all kinds and shapes for garnishes, cookies, and tiny cakes. The same suggestions with regard to fundamentals of selection apply to these smallest of appliances, used chiefly to gain artistic effects, as to the larger pieces of equipment.

Thermometers

There are many operations in cooking which require the accuracy only provided by a thermometer.

A **meat thermometer** by measuring the internal temperature of the meat measures the degree of doneness the meat has attained.

There are so many variables in the cooking of meat—size and shape of the cut, proportion of fat to lean, degree of aging—that the only really accurate gauge of doneness is a meat thermometer.

There are two basic types: dial and column. The dial thermometers consist of a metal probe topped crosswise with a glass-enclosed dial face containing a pointer that moves to indicate the temperature inside the meat. Inside the thermometer there is a strip of two different metals bonded together. Each metal expands when heated but by different amounts; that difference causes the dial pointer to rotate. A column thermometer contains liquid that expands and rises in its glass column as the meat becomes hotter. The level of the liquid, as seen against a temperature scale, tells how done the meat has become.

The dial-type meat thermometers have pointed-tipped metal stems which can be pushed into the meat without first inserting a skewer, as is necessary with most column models to minimize the danger of breaking the glass. One column thermometer has its bulb and most of its stem sheathed in metal and the bottom of the sheathing comes to a sharp point. This model can safely be inserted into the meat, just as the dial-type can, without first making a hole with a skewer.

In using either type, the probe should be inserted into the thickest part of the meat and the point should not touch bone, fat, or gristle. The meat thermometer should be inserted so the dial is turned toward the oven door for easier reading.

When using a meat thermometer for spit or rotisserie roasting, it should be inserted into the end of the meat, parallel with the spit, or at such an angle that it does not hit anything as the spit turns. The spit should not be stopped with the thermometer near the heat source.

A **candy thermometer** will register the correct temperature for a sugar mixture for candy to be of the proper consistency. A candy thermometer will also help in making jellies and in deep-fat frying. It attaches to the side of the pan using a clip on the back of the thermometer. The thermometer should be immersed at least two inches, and the bulb should not touch the bottom or sides of the pan.

An **oven thermometer** may be used periodically to check the accuracy of the oven thermostat. Since temperatures vary from the top to the bottom of the oven, a thermometer should be placed close to where you usually bake food, and read through the glass on the oven door to prevent fluctuations of temperature by door openings.

A **refrigerator-freezer thermometer** will measure temperatures of these appliances. Refrigerator temperatures should be 35 to 40°F; freezers around 0°F. A new refrigerator may be checked with a thermometer to determine how to set coldness control for optimum food storage.

Experiments and Projects

1. Preheat a 10-inch aluminum skillet over medium high heat until a drop of water "skitters" across the surface. Bake pancakes 2 inches in diameter, spaced at even intervals over the bottom of the skillet. Use a pancake batter which has 2 tablespoons of fat per cup of flour. Do not grease the skillet. Place the baked pancakes on a white plate in the same position as they were in the skillet. Repeat using skillets of copper-clad stainless steel, copper-core stainless steel, iron-core stainless steel, and glass ceramic.
2. Study the differences that materials and finishes on oven utensils make on layer cakes. Assemble cake pans, 8 inches in diameter of aluminum with dull bottom and shiny sides, aluminum with shiny bottom and shiny sides, aluminum foil, stainless steel, glass, and shiny tin. Line the bottoms of each cake pan with wax paper. Use a standard cake mix and prepare according to package directions. Set oven control to temperature recommended on the package. Each student will place half (weigh) the batter in an aluminum cake pan with dull bottom and shiny sides and the other half of the batter in a cake pan of other material or finish. Bake with oven rack as near the center of the oven as possible and with cakes on rack at equal distances from sides and back of the oven. Time carefully. Remove from oven and let stand on racks for 10 minutes. Turn out on another rack and examine browning pattern before removing the wax paper. Compare browning in utensils of different materials with other students. Write your observations.
3. Study the differences that materials and finishes on pie pans make in double-crust fruit pies. Assemble 8-inch pie pans of glass, anodized aluminum, stainless steel, aluminum foil, and shiny aluminum. Use a pie crust mix and canned apple pie filling, one can for each pie. Each student will select a glass or anodized aluminum pie pan and a pie pan of another material. Place the two pies on a rack in the center of the oven, at equal distances

from the sides and back of the oven. Bake according to package directions. Time exactly. Remove pies and let cool on racks. Turn pies upside down on white plates and examine bottom crusts. Compare results of pies baked in other materials by other members of the class. What materials or finishes will give you the best results in baking double-crust pies? Why is the undercrust of many frozen pies raw after the recommended baking time is completed?

4. To compare cookie sheets select (1) a cookie sheet with a very smooth shiny top and bottom surface, such as new tin or tin alloy, (2) a cookie sheet with a very smooth, shiny bottom and a dark top surface, such as tin alloy with dark teflon lining, (3) a cookie sheet with a dark top and bottom surface, (4) a cookie sheet with top and bottom surfaces less shiny (but not rough) than the very shiny surfaces used in (1), such as slightly dulled aluminum. Use a commercially refrigerated cookie mixture. Slice uniformly. Preheat oven to recommended temperature. Place cookies on each sheet in the same positions. Bake each sheet separately on the same rack in the same position, in the same oven for the same time, at the same temperature. Compare results. Which cookie sheet gave the best results? Why?
5. To compare baking results in a round and square cake pan use a packaged white cake mix, a round 8-inch cake pan and a square 8-inch cake pan of the same material and finish. Since the bottom of the square cake pan exposes a greater area to the source of heat in the oven, in order to have the same depth of material in the square as in the round pan, have someone hold a ruler in each pan while you pour and spread batter in the pans to the same depth. Bake at recommended temperature. compare results.
6. Examine the cutting edge of a dull knife and a sharp knife under a microscope. Describe how each edge appears. What changes in the edge take place when a knife is sharpened? Try sharpening the dull knife with a stone, a "steel" and an electric knife sharpener. Compare results.
7. To compare parers pare potatoes, carrots, and apples with (1) a conventional paring knife, (2) a combination apple corer and parer, (3) a parer with a floating blade. Have someone time each operation using a stopwatch. Try using each parer to make decorative curls of carrots, cheese, and chocolate.
8. Bake equal quantities of potatoes or carrots in aluminum foil using the drugstore wrap with the shiny side out on one and

the dull side out on another package. Bake for the same time at the same temperature. Compare results.

9. Prepare a beef stew in a pressure cooker: Remove the rack from the cooker. Brown 1 pound of beef stew meat, cut into 1-inch pieces in 2 tablespoons hot fat, stirring occasionally. Pour off excess fat. Add ½ cup water to 1½ cups tomato juice, 1 teaspoon salt, ⅛ teaspoon pepper, 2 bay leaves, and vegetables (carrots, potatoes, onions, celery, parsnips, peas, turnips, as desired) cut in 1-inch pieces. Cook at 10 pounds pressure for 15 minutes. Cool the cooker instantly by setting in a pan of cold water or placing in the sink and running cold water over the cover taking care that the water does not run over the safety plug. (If an electric pressure cooker is being used, it should not be set in a pan of cold water and in running cold water on the cover, care should be taken to direct the water away from the plug and electrical connections.) When pressure is reduced, remove cover from the cooker. Add to hot broth a thin paste of 2 tablespoons flour and ¼ cup cold water. Stir well and cook without cover 2 minutes.
10. To try "waterless" cooking select a covered saucepan. Peel and cut carrots in uniform size pieces. Rinse carrots in cold water. Lift from water into the saucepan making sure that the saucepan is more than half full of carrots. Do not add any liquid. Place the cover on the saucepan. Place the pan on a range unit which fits the bottom area of the saucepan and set heat at "medium." When the cover of the pan is hot to the touch, turn the heat to the lowest point and cook until carrots are tender.
11. Make a list of small utensils and tools, not including saucepans or baking pans, which you think are essential for the average household. Check the total cost of these at your local retail outlets.

CHAPTER 4

Basic Facts About Electricity, Gas, and Household Electronics

Many household appliances are operated by electricity or by gas, and many of them at present are controlled by solid state devices. We shall review basic facts about these three essentials in order to understand how appliances work as they do. This discussion will be brief and may be supplemented profitably by reference to any standard text in physics.

Most homes obtain electricity from a central generating station. The source of this power is the generator, which may be operated in one of three ways, by water power (hydroelectric plant), steam turbines, or by an internal combustion engine, where diesel oil or natural gas is available and inexpensive. Water power will usually provide the cheapest electricity. Electricity can be generated, directed, controlled, measured, and put to work. It may be sent along wires for many miles into town and city homes, farmsteads, and manufacturing plants where it enables the homemaker and working man to perform a great number of tasks with a minimum expenditure of energy.

ELECTRIC CURRENT

The electron theory of electricity has been universally accepted. Scientists have discovered that all matter is made up of a seething mass of tiny particles called atoms. Each atom is composed of still smaller particles, protons, neutrons, and electrons. The protons, each carrying a positive charge, and the neutrons without charge, form the nucleus, the center of the atom. The electrons are negatively charged and move rapidly around the nucleus in a series of orbits, similar to the planets in their orbits around the sun. The electrons in the outer orbits or shells are less tightly held to the nucleus; they may be detached from it and made to move in a definite direction.

To speak of generating an electric current is erroneous. What is generated is the force that starts the electrons moving. This force may be generated by several methods: magnetic, chemical, thermal, and frictional.

Magnetic

The magnetic method used in the electric generator is employed most widely for commercial use. This method was discovered by Michael Faraday in the early nineteenth century.

There exists around a magnet an invisible field of force known as a magnetic field. This field is made up of lines of force that emanate from the north pole of the magnet, pass through the surrounding medium, and enter the south pole of the magnet; if the magnet is of the horseshoe type, the lines of force complete their path inside the magnet itself. Faraday found that by placing a coil of copper wire (a good conductor) in this magnetic field and turning it to cut across the lines of force, the electrons in the copper wire all tended to move in the same direction. This resulted in a difference in pressure at the two ends of the coil. He also found that when these ends are connected by a conductor so that a complete or closed path is provided the difference in pressure will cause the electrons to flow through the path. This flowing movement of electrons is known as an electric current. The rate of flow of electricity is expressed in coulombs per second. One coulomb is 6,300,000 million million electrons. A coulomb per second equals an ampere, the unit for the measurement of electric current (symbol I), named after the French physicist, Ampère.

Voltage. The rotating coil of copper wire is known as the armature, and the magnetic field around the magnet as the field. The armature and the field together make the generator. The difference in pressure between the two sides of the coil in the generator that causes the electrons to flow is called potential difference or electromotive force (emf). Sometimes voltage is spoken of as the pressure, force, or push that tends to cause the current to flow. The use of these terms may lead to the mistaken impression that it is a certain pressure at some point in the circuit that is causing the current to flow when, in reality, it is the difference in pressure between the two ends of

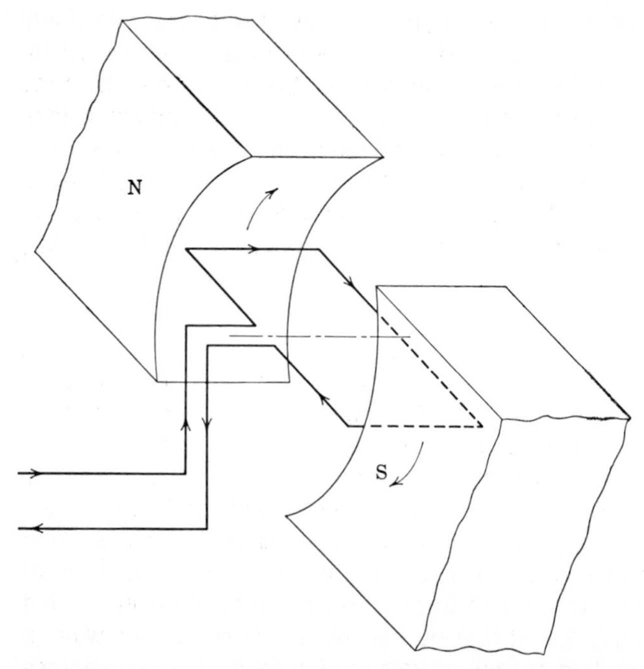

***Fig. 4.1** When a conductor is rotated through a magnetic field, an electric pressure is generated in the conductor and measured in volts. If the conductor forms part of a closed circuit, the pressure will cause a current to flow.*

the copper wire in the armature (Fig. 4.1). The unit for measuring the difference in pressure or the emf is the volt (symbol *E*). The common voltages used in the home are 120 and 240 volts. Although for the sake of simplicity only one coil is shown in the illustration, the armature is made of many coils of wire. As was noted, the armature may be rotated by a turbine which, in turn, is operated by water whenever a dam on a stream raises the water level sufficiently to get enough drop to activate the turbines, as at Niagara, throughout the Tennessee Valley, at Keokuk in Iowa, at Hoover Dam, and in several of the rivers along the West Coast, to mention only a few examples. In communities far from such sources of water power, steam is used. Therefore **a generator is a device, for transforming mechanical energy into electrical energy.**

Chemical

The chemical method of generating an electrical current by wet and dry cells was first employed by Volta, an Italian scientist. He created a difference in electrical pressure by placing two metal strips, one of zinc and the other of copper, in a dilute solution of sulfuric acid, known as an electrolyte. When the two metal strips were connected externally, a current was found to flow from the negative zinc strip to the positive copper strip. This type of cell is known as a voltaic cell. The voltage developed by the chemical method is comparatively small; hence the method is limited in its use.

Two different conductors might have been used for the strips, provided they were not identical; another electrolyte also could have been used if it had been of such a nature that it would chemically attack one of the metals.

Cells are classified into primary and storage cells. The primary cell that is most widely used is the dry cell. Not really dry, it use a zinc cylinder for the wall of the cell and encloses a carbon rod for the positive electrode. The electrolyte of ammonium chloride is combined with manganese dioxide, small amounts of zinc chloride, and often graphite, to reduce the internal resistance of the cell. Together they form a paste which fills the space between the zinc and carbon. Finally the cell is made moistureproof and airtight by sealing it with wax. A new cell has an emf of 1.5 volts. The emf declines—less rapidly if the cell is used intermittently—but gradually just by standing. It cannot be regenerated. Dry cells are used in flashlights.

The storage cell, in contrast, can be recharged by sending an electric current through the cell in the opposite direction to the original current. The common storage cell uses lead and lead compounds for the electrodes and dilute sulfuric acid for the electrolyte. The emf is approximately 2.1 volts. These cells are frequently connected to form a storage battery, which is used in automobiles.

Cells or batteries are also used for the operation of "cordless" equipment. These batteries are usually nickel-cadmium rechargeable types and are built into the piece of equipment permanently. The recharger is in the form of a plug-in electronic device that converts 120 volt AC current from a wall outlet into the low-voltage DC current required by the batteries. The nickel-cadmium battery will withstand severe shock and vibration.

Thermal

When two dissimilar metals are connected at both ends and one junction remains cold while the other is heated, a difference in pressure is created at the two junctions that will cause the electrons to flow. One set of junctions of dissimilar metals is known as a thermocouple and a series of junctions as a thermopile. The voltage developed by this method is too small for commercial use but finds wide application in the measurement of temperatures in ovens and in refrigerators by the potentiometer.

Frictional

In the frictional method of producing an electric current, the difference in potential is obtained by rubbing together two dissimilar materials. The electricity that is generated by this method is frequently referred to as "static" electricity. When a person walks across a rug and touches an object, sometimes the flow of electrons in the form of tiny sparks may be observed.

Circuits

In order for the current to flow it is necessary for a path to be provided from one terminal of the generator through the connecting lines to an outlet in the wall in the house, through the appliance, and back to the other terminal of the generator. The path through which the current travels is known as an electric circuit.

Closed circuit. If the path is complete from one side of the generator to the other, the circuit is spoken of as a closed circuit.

Open circuit. If a break is made in the circuit, either by a broken wire, a loose connection, or a switch, the circuit is known as an open circuit and the current ceases to flow.

Conductors and Insulators

Materials are known as conductors or insulators, according to their ability to conduct an electric current or to oppose its flow. In general, metals, especially aluminum, copper, and their alloys, are good conductors; so is the home water supply with its dissolved minerals. Another good conductor is the human body.

In general copper wiring has been used in house circuits although in recent years an increase in the use of aluminum wire has been observed. With aluminum wire it is more difficult to make satisfactory connections to outlet receptacles and switches than with copper wire. Many electricians do not like to use the aluminum wire, and several states have forbidden its use in the home.

Glass, ceramic materials, plastics, rubber, silk, and cotton are common insulators. Electrons in a conductor are loosely bound to the nucleus and pass easily from atom to atom. On the other hand, the electrons in an insulator are attached rather tightly and resist displacement. Insulators tend to keep the electric current in its proper path.

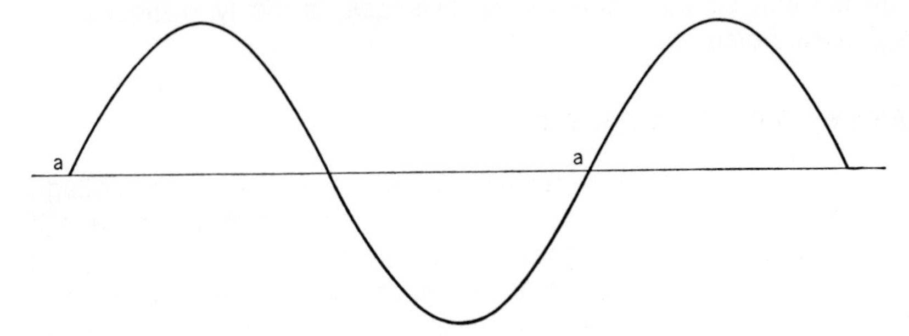

***Fig. 4.2** The cycle of an alternating current.*

Semiconductors have properties between those of conductors and insulators and, therefore, can allow an electric current to flow, or prevent its flow, when desired. Semiconductors will be discussed in detail in the section on household electronics.

Types of Current

There are two general types of electric current: alternating, designated by AC and direct, DC (Fig. 4.2). As the armature revolves in the magnetic field, the voltage and also the current increase from zero to a maximum value—the maximum reached depends on the strength of the magnetic field, the speed of rotation of the armature, and the number of turns in the armature coil—fall again to zero, reverse in direction, then continue to a negative maximum, and once more return to zero. This is alternating current. The distance from *a* to *a* is called a cycle; the number of cycles per second, frequency. Common current frequency in the United States is 60 cycles per second or 60 Hertz (Hz), although a few areas use 50 cycles per second or 50 Hertz (Hz).

Certain manufacturing processes such as refining of metals, charging of storage batteries, and electroplating require direct current. For example, if alternating current were used for electroplating, the plating deposited in the positive part of the cycle would be removed when the direction of the current reversed. Direct current is obtained by placing commutator bars on the shaft of the armature. Brushes impinge on opposite bars and change positions on the bars with the rotation of the armature in such a way that the current always flows in a single direction. Direct current is usually not transmitted long distances, since no economical method of changing its voltage has been discovered. Alternating current, which can be stepped up or down and can be transmitted long distances, is the type most often found in the home.

TRANSMISSION OF CURRENT

In the early use of electricity it was soon discovered that high voltage was necessary if current was to be transmitted long distances and if losses in energy in transmission were to be held to a minimum. It was also learned that different distances of transmission required different voltages and that high transmission voltages were not safe for use in the home. This led to the development of the transformer, which consists of two insulated coils of wire wound on an iron core. When an alternating current is passed through one of these coils, called the primary coil, it causes a changing magnetic field to develop, which induces an alternating voltage in the secondary coil. The two coils have no contact with each other; the iron merely keeps the magnetic lines of force from scattering. The transformer is placed in a container of oil that insulates the coils and absorbs any heat formed during operation.

If the secondary coil has a greater number of turns of wire than the primary, the voltage in the secondary is larger and the transformer is called a **step-up** transformer. If the secondary has fewer turns of wire, a lower voltage develops and the transformer is called a **step-down** transformer. There is a direct relationship between the number of turns of wire in the two coils and the two voltages:

$$\left(\frac{T_s}{T_p}\right) = \left(\frac{E_s}{E_p}\right) \text{ where } T = \text{turns of wire and } E = \text{voltage.}$$

The step up of voltage in the system depends largely on the distance the current has to travel. It may be stepped up ten, a hundred, perhaps even a thousand times. Approximately 1000 volts per mile are used for efficient transmission, to keep the current flowing smoothly through the wires.

In the transformer the power output is equivalent approximately to the power input; i.e., EI output $= EI$ input. If the secondary voltage is ten times the primary voltage, then the secondary current is only 1/10 as large as the primary current and can be transmitted on fairly small wires; in other words, if the transformer increases the voltage by a certain amount, the current is lowered a corresponding amount, and vice versa.

From the step-up transformer, the high-voltage lines (often called hi-lines) carry the current across the countryside to the edge of a city or town, or in rural areas, to the entrance of a farmstead, where, by means of a step-down transformer, the voltage is reduced and the current again increased. This system of wires from the generating plant to the step-down transformer is known as the transmission system. Wires from the step-down transformer to the pole outside the individual home form the distribution system. There is usually more than one step-down transformer, one at the edge of the city where the voltage is partly reduced, and others in various neighborhoods serving groups of homes, where the voltage is finally reduced to the 120/240 volts that can be used safely in the home (Fig. 4.3).

Where polyphase systems are used, the voltages supplied are 120/208 volts. In the two-phase generator, two armature windings are mounted 90° apart, making the emfs also 90° apart in phase. In this case the emf is never at zero point in the cycle. In the three-phase generator there are three sets of windings and emfs, 120° apart. In general the three-phase generator is supposed to have superior operating characteristics. The three coils may form three individual single-phase circuits or may be interconnected in either a Y or delta form (Fig. 4.4). To prevent a short circuit, the three windings must be connected to the outside circuits in such a way that the current will flow in the same direction in each winding. Induction motors use three-phase alternating current very successfully.

In some states all state buildings are required to use 120/208 voltages. This is important to know in buying equipment for school, college, or university food or equipment laboratories. This combination is seldom, if ever, used in one-family homes in the United States, but often is by builders of high-rise apartments because it saves money by requiring small-diameter wires.

Electric ranges sold in volume, designated as builder products, can be furnished in 208 volt models on direct order to the factory. A 240-volt range can be operated on a 208-volt circuit, but it would have slower top-unit heat-up, slower oven preheat, and slow broiling performance. An electric range wired for 208 volts, operating on a 208-volt circuit, would have more satisfactory operating characteristics than the 240-volt range on the 208-volt circuit, but it may be somewhat slower than a 240-volt range operating on a 240-volt circuit.

Electric dryers have been rated at 240 volts for 10 to 12 years. A kit can be obtained to change the dryer element to 208-volt wiring when it is to be used on the 208-volt circuit. A 240-volt dryer will be slow when operated on a 208-volt circuit; the replaced 208-volt wiring will bring it close to its design speed.[1]

[1]Information supplied by Whirlpool Corporation, July 1969.

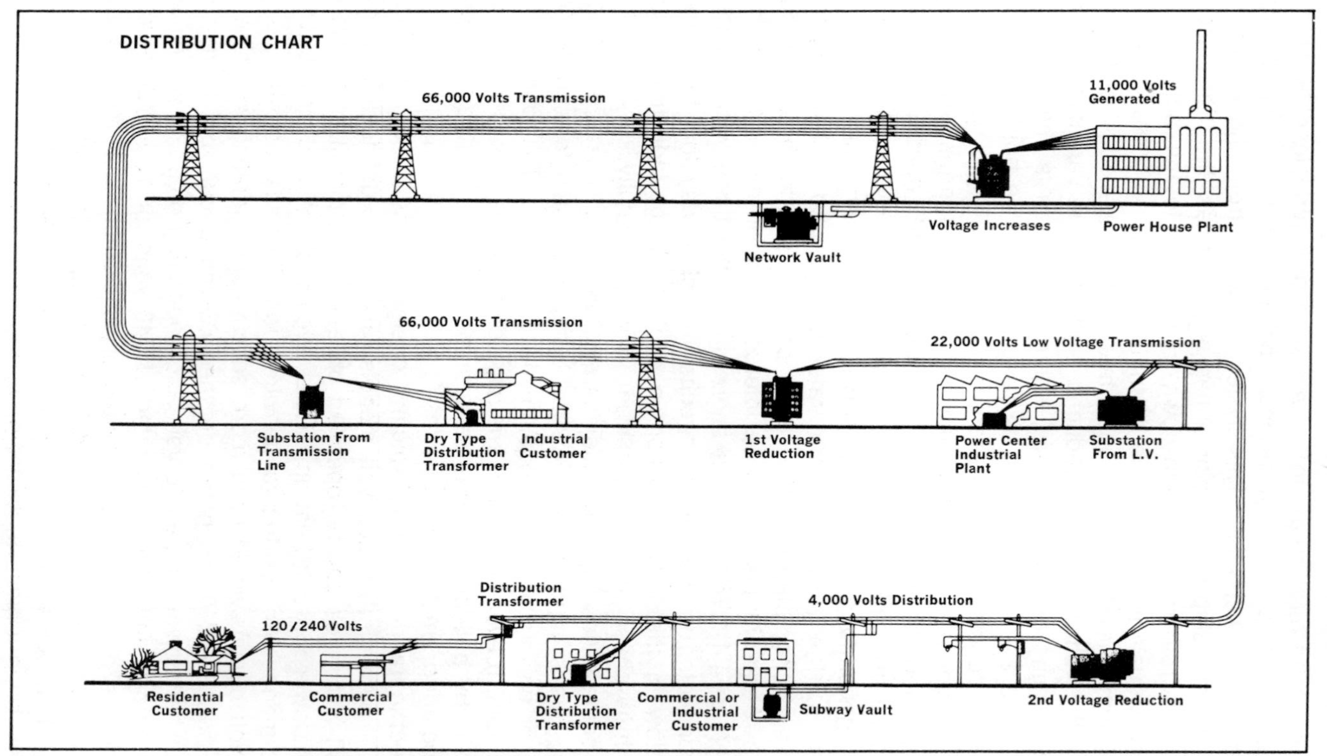

Fig. 4.3 The transmission and distribution of electricity from the generating plant to the home. (Westinghouse)

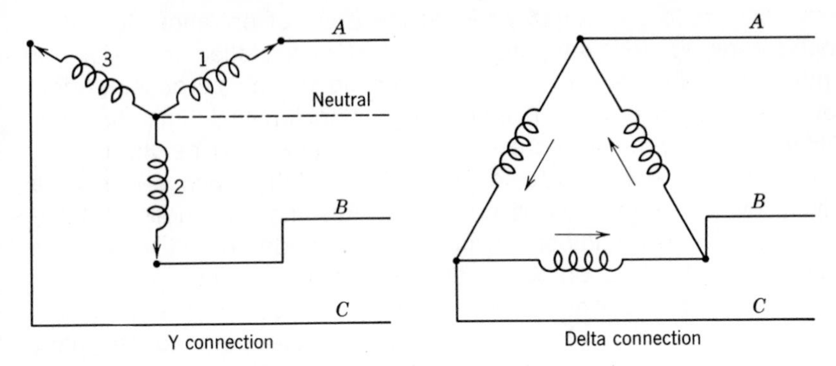

***Fig. 4.4** The three-phase generator has three separate sets of windings connected in a Y or delta configuration. Current in each of the windings must flow in the same direction as in the other two windings.*

Service Wires

At the present time it is customary to install a three-wire, 120/240-volt system in the home. The three insulated service wires are brought, either overhead or underground, from the pole to the house. Overhead wires are attached to insulating knobs at the eaves of the house and then to the house wires (Fig. 4.5). Underground service

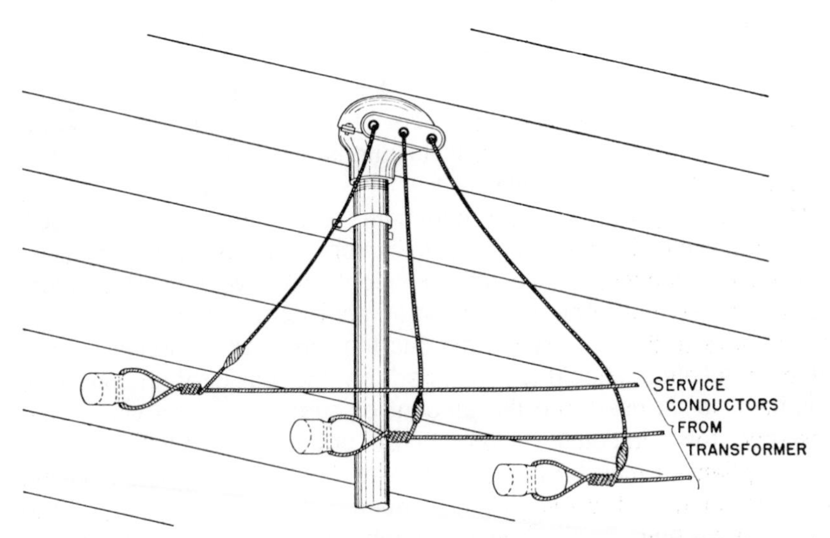

***Fig. 4.5** Overhead service wires to the house.*

wires connect to the house wires at the point of entrance. If underground services are used, the wires must be installed in a duct or conduit, or in the form of a cable approved for the purpose. Underground service is becoming increasingly popular. By eliminating unsightly poles, it greatly improves the appearance of a neighborhood. It may cost somewhat more to install, but if done when a new area is first developed, the cost is not much more than overhead installation. The upkeep is almost negligible, since underground installation is unaffected by tree branches and natural conditions such as ice storms and rigorous winds. Number 2 wires of 100 amperes minimum capacity are recommended, and if the homemaker plans to use many electric appliances, a capacity of 150 to 200 amperes should be provided. According to the National Electrical Code, the three wires must be at least 10 feet above any sidewalk and 18 feet above driveways.

National Electrical Code

The National Electrical Code is a set of minimum requirements drawn up by the Electrical Committee of the National Fire Protection Association, approved by the American National Standards Institute and adopted by fire insurance companies. Its regulations for home wiring must be followed to safeguard persons from electrical and fire hazards from faulty installations. Following these regulations will insure the collection of insurance benefits, should a fire be caused by any electrical defect. The code is revised biennially. Local codes are based on the national code but often contain additional requirements. In rural areas, unless there are state or county requirements, the regulations regarding wiring installations are usually left up to the company supplying the power.

While compliance with the minimum requirements of the code results in installations reasonably free from hazards, these installations are not necessarily the most efficient or convenient. Good service and satisfactory operation of equipment frequently require larger sizes of wire and more branch circuits than the code specifies. Each wiring installation should be planned on the basis of the kind of equipment to be used and the amount of current it needs with some allowance for additional circuits to be added in the future, when other appliances may be purchased. New equipment appears on the market frequently. It is difficult for a homemaker to foresee what she may wish to buy. This planning is the responsibility of the home owner, not the power suppliers.

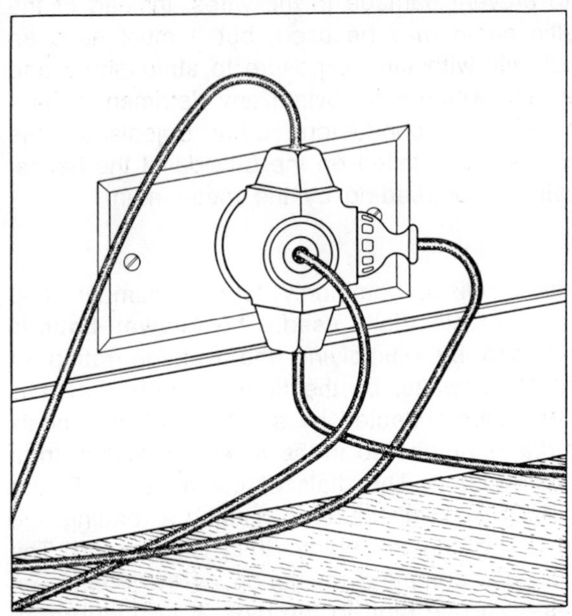

***Fig. 4.6** A sufficient number of convenience outlets should be installed to eliminate the hazard of "octopus" connections.*

WIRING THE HOME

Adequate wiring of the home provides a sufficient number of circuits of the correct size wire to take care of present and future needs, enough switches conveniently located to control the circuits, and enough convenience outlets to which lamps and appliances may be attached to eliminate multiple "octopus" connections (Fig. 4.6).

When planning to purchase or rent a house or an apartment, a person should check these parts of the wiring system to insure future satisfaction. If in doubt, have an electrician check for you, giving him a fairly accurate idea of the portable lighting fixtures and appliances you will be using.

Service Entrance

Entrance wires, connected to the overhead service wires at the eaves of the house, are brought down the side of the house to the point at which they pass into the basement or a hallway—usually within a conduit pipe, a pipe not unlike a water pipe in appearance but with a glazed inside finish to prevent damage to the wires. Instead of the conduit pipe, nonmetallic cable may be used, but it must have an insulating covering which will withstand exposure to atmosphere and other conditions of use, and which will obviate any detrimental leakage of current to any adjacent conductors, other objects, or the ground. The meter is frequently installed on the outside of the house, so that it is readily available for reading by the meter man.

Meter

The meter has a series of dials, commonly four in number, that indicate the kilowatt-hours of electricity used. The **kilowatt-hour** is the unit of energy, obtained by multiplying the wattage rating of the appliance, changed to kilowatts, by the time in hours that it is used. The meter dials are interconnected by a system of cogwheels and consequently each dial is numbered in the reverse direction from the one preceding or following it. The dials of the meter in Figure 4.7 record a reading of 1385 kilowatt-hours. In making a reading, the digit read on each dial is the one the pointer has just passed. The pointer on each dial makes a complete revolution as the pointer on the dial to its left moves from one digit to the next. If in doubt as to whether a digit has been reached or passed by a pointer, look at the next dial to the right to see whether or not the revolution has been completed.

The previous monthly reading is subtracted from the current reading. The hands on the dials are not turned back to zero; this tends to eliminate errors in computing charges.

In some rural areas, kilowatt-hour meters with a cyclometer dial, similar to the automobile odometer, are often used. This type of

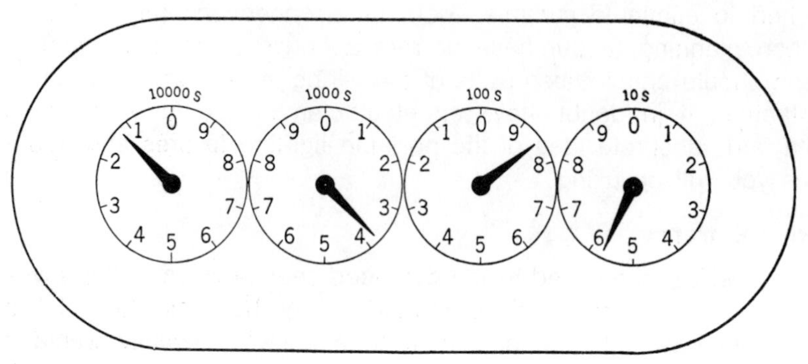

Fig. 4.7 Dials of an electric meter.

meter is more easily read by the householder himself, who mails the reading to the power suppliers.

Main Switch to Fuse Panel

Inside the house, the wires pass through a main switch which is used to disconnect all the wiring inside the house from the entrance wires, should some electrical repairs be needed. Next is the main fuse, or circuit breaker, required only if there are more than six subdivisions or feeders for the house circuits. When the meter is not on the outside of the house, it is located here with the main fuse. Finally, there is the fuse or circuit-breaker panel with one fuse or breaker for each separate 120-volt branch circuit and two fuses or circuit breakers for each 240-volt circuit.

Fuses and Circuit Breakers

Fuses and circuit breakers are protective devices used to prevent the overloading of a circuit, which would cause an excessive or dangerous temperature to develop in the conductors or in the insulation around the conductors. When wires become overheated, the insulation starts to disintegrate, and, if this continues, a short circuit and possibly a bad fire may result.

Fuses are of two types: plug and cartridge. Each type contains a link strip of metal alloy which has a definite current-carrying capacity (Fig. 4.8). When an excess of current passes through the alloy strip, sufficient heat is developed to melt it.

In the plug fuse the alloy strip is enclosed in a porcelain, glass, or plastic cup that screws into a socket. The contact tip of the screw

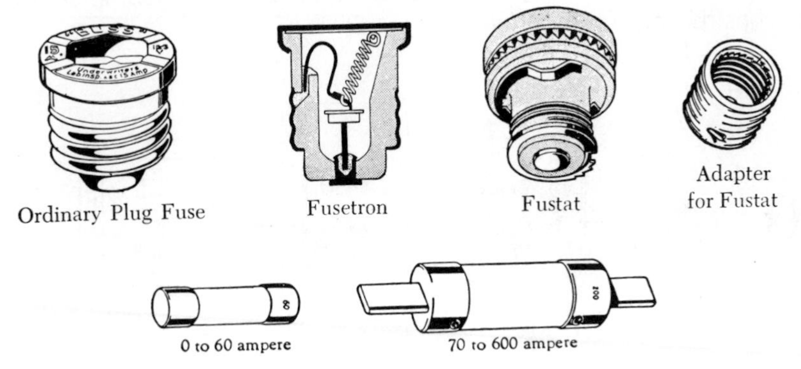

Fig. 4.8 Different types of plug and cartridge fuses. (Bussman)

base is of copper or aluminum and is usually marked with the ampere capacity. A window over the strip makes it possible to see when the fuse has been blown.

According to the Underwriters' Requirements, the plug fuse, in the 1- to 30-ampere range commonly used in the home, must blow at a 50% overload within one minute. An overload would occur if more appliances were connected into a circuit than the circuit would normally carry. Manufacturers have developed a fuse with a greater time lag. Known as a Fusetron, it has a thermal cutout in addition to the fuse link. On a short circuit, the fuse link of the Fusetron blows just as in an ordinary fuse; with an overload, the excessive current causes the cutout to heat, but only if continued will it soften the solder and permit a spring to pull the fuse link from its position in contact with the thermal cutout.

Until 1935 all plug fuses were made for bases of the same size, allowing persons ignorant of the importance of proper fuse protection to use too large a fuse. Inspectors also found people wiping out all protection by bridging *under* a blown fuse with a penny or other material of good conductivity. To guard against this practice, the Fustat was developed. It is made to fit into the standard-base fuse holder by means of an adapter that locks into place. The Fustat prevents the insertion of fuses too large for the circuit, and also makes it impossible to use pennies as fuses. The Fustat can be replaced like any ordinary fuse. It has the thermal cutout so that unnecessary blowouts from the starting of motors are eliminated.

Fuses are rated in amperes; the size used will depend upon the carrying capacity of the wires it protects. Fuses are tested and must meet certain standards of safety to carry the label of the Underwriters' Laboratories. This label is the homemaker's guarantee of safe construction and operation (Fig. 1.6).

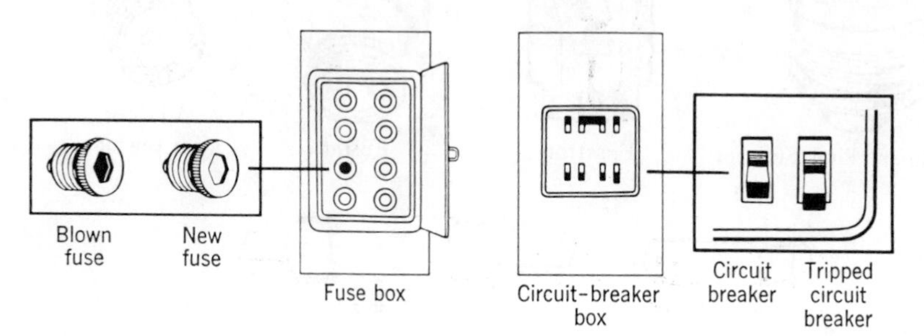

***Fig. 4.9** A fuse blows or a circuit breaker trips to protect the home from overheated wires.*

A circuit breaker is a device constructed with a bimetal strip, an electromagnet, or a combination of the two (Fig. 4.9). Any of these types will open a circuit automatically when the current in the circuit becomes excessive, but the bimetal strip allows for the momentary overload of starting motors. The circuit breaker is reset manually when the cause of the tripping has been removed.

Branch Circuits

Three types of circuits are installed in most homes: general purpose, appliance, and individual. Copper wire is commonly used. All the circuits start at the fuse panel, and the size of fuse to be used with a given size wire is regulated by the National Electrical Code.

Size of Wire

The size of wire used in conductors is dependent on the quantity of current the wire is required to carry. For convenience the sizes of wire are designated by gauge number. The wire gauge most commonly used in the United States is the American Wire Gauge system (A.W.G.). Gauges are marked 0000, 000, 0, 1, 2, 3, etc. As the numbers increase in numerical value, the size of the wire decreases (Table 4.1). The gauges most frequently used in the home are 14, 12, and 10 for general-purpose and appliance circuits and 6 for the range circuits. Larger appliances, such as the water heater, ironing machine, and clothes dryer, should use 12 wire. Gauges 16 and 18 are

Table 4.1
Data for Copper Wire Commonly Used in the Home

Wire Size (AWG)	Diameter (Inches)	Amperes
18	0.040	
16	0.051	
14	0.064	15
12	0.081	20
10	0.102	30
8	0.128	40
6	0.162	55
4	0.204	70
2	0.258	95
1	0.289	110
0	0.235	125

used in flexible cords for portable lamps, and 14 for small heating appliances.

Rate of Work

In order that the discussion on the different types of circuits will be more easily understood, it is necessary to digress slightly at this point to consider certain basic units.

In the use of electricity in the home we are chiefly concerned with the quantity of work it will do for us and the rate at which it will accomplish the work. The rate at which electricity works is known as electric **power** and is measured in **watts.** Since one watt is a very small unit of power, 1000 watts or a kilowatt is the unit most commonly used.

Appliances are rated in watts; motor-driven appliances are occasionally rated in horsepower (hp). One horsepower is equivalent to 746 watts. In obtaining the wattage equivalent of a given horsepower, the power factor must be taken into account, especially in the case of induction-type motors. For this reason most manufacturers at present state the wattage of the motor-driven appliance on the name plate, with perhaps also the horsepower value. A comparison of the wattage with the horsepower converted to watts will show clearly a discrepancy in values, because watts indicate power input, while horsepower shows the output. To obtain power in watts, multiply volts by amperes, $w = EI$.

General-Purpose Circuit

The Code specifies that no wire smaller than No. 14 may be used for the general-purpose circuit and it must be fused with a 15-ampere fuse. This circuit supplies electricity for lights and wall outlets in living room, dining room, bedrooms and halls, and for ceiling, clock, and ventilating fan outlets in kitchen and utility rooms. At 120 volts this will allow a total of 1800 watts for lights and convenience outlets (120 volts × 15 amperes). The Code, however, states that not more than 80 per cent of the total watts allowed should be used at any one time, or approximately 1500 watts. This was sufficient wattage when most household appliances were nonautomatic, rated at 450 to 660 watts. Now, when practically all the small, electrically operated heating appliances are 1000 watts or higher, 1500 watts is most inadequate, and a general-purpose circuit of No. 12 wire is recommended and is frequently required by a municipality. General-purpose circuits of No. 12 wire must be installed in all new houses built in Logan, Utah. Wire of this size is fused with a 20-ampere fuse and provides 24000 watts or an 80% value of about 2000 watts. Circuits longer than 35 feet should always be of No. 12 wire. At least one general-purpose circuit should be provided for each 500 square feet of floor area. An additional circuit to which clocks, radios, and TV may be connected should be installed; one for each 400 square feet of floor area is recommended.

It is usually preferable to have two circuits in each room, with some of the lights and convenience outlets on one and some on the second. This does not mean twice as many circuits as there are rooms in the house, since both circuits may run through several rooms. Such an arrangement distributes the load more evenly, lessens voltage drop, and prevents all lights from going out when a fuse is blown.

Wiring Symbols

Extensive other symbols are available but, with some slight variation, most architects and electrical contractors use the symbols shown in Figure 4.10. The floor plans of Figure 4.11*a* and *b* show typical wiring layouts for a house.

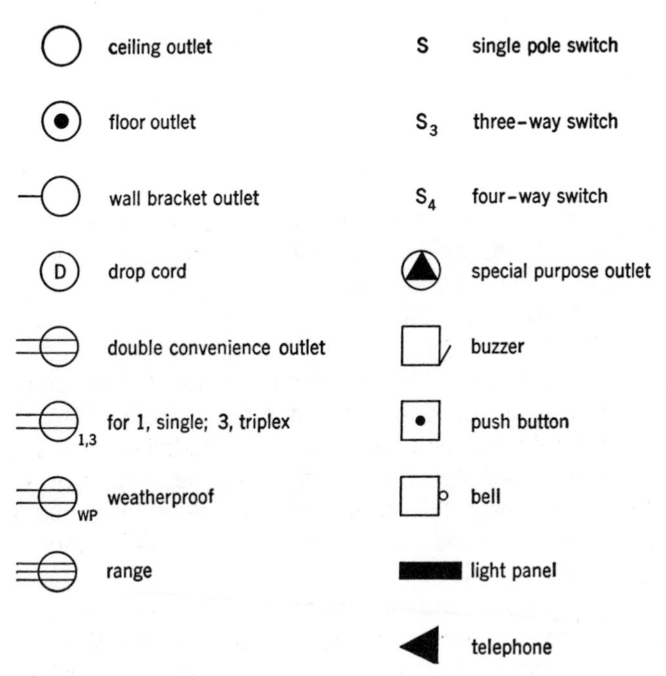

Fig. 4.10 Wiring symbols.

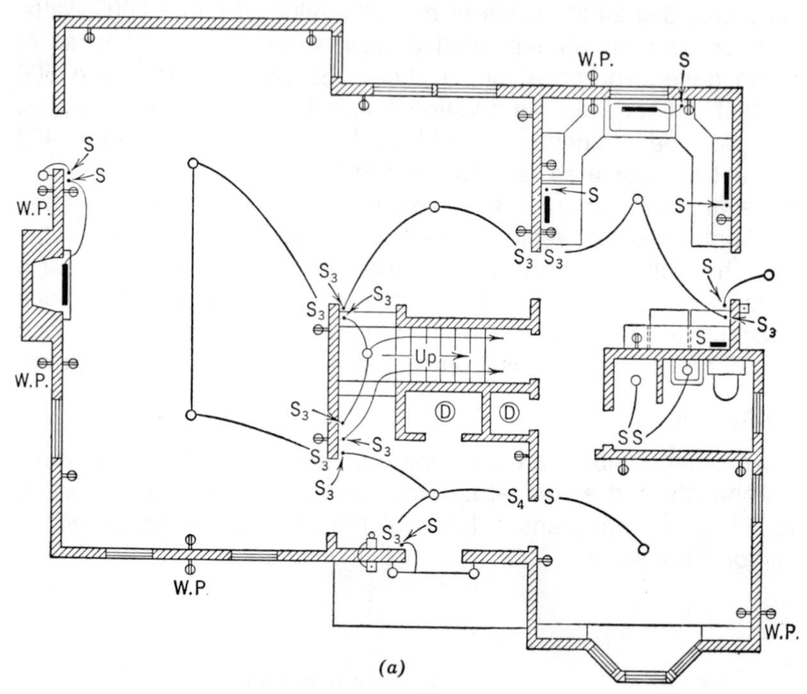

Fig. 4.11a Wiring layout for a house, first-floor plan.

Appliance Circuit

The appliance circuit has, as the minimum, a No. 12 wire conductor protected by a 20-ampere fuse or circuit breaker. A No. 10 conductor fused with a 30-ampere circuit breaker will provide greater flexibility, supplying as it does 3600 watts ($3600w = 120E \times 25I$) and should be installed for runs exceeding 45 feet. These circuits serve appliance outlets in the kitchen, the breakfast room, and often the dining room, and in the laundry, the utility room, and the garage or work room. A sufficient number of appliance circuits should be installed to care adequately for all needs. When lights dim and appliances heat slowly, usually the trouble is not with the equipment but with inadequate circuits of too small wire. Appliances of more than 15 amperes should have their own individual circuits.

Current for general-purpose and appliance circuits may be supplied at 115 volts, but many utility companies try to provide 120 volts. There is always some voltage drop in the branch circuit and, when there is excessive drop, heating appliances take longer than they should to heat, 20% longer, in fact, with a voltage drop of 10%.

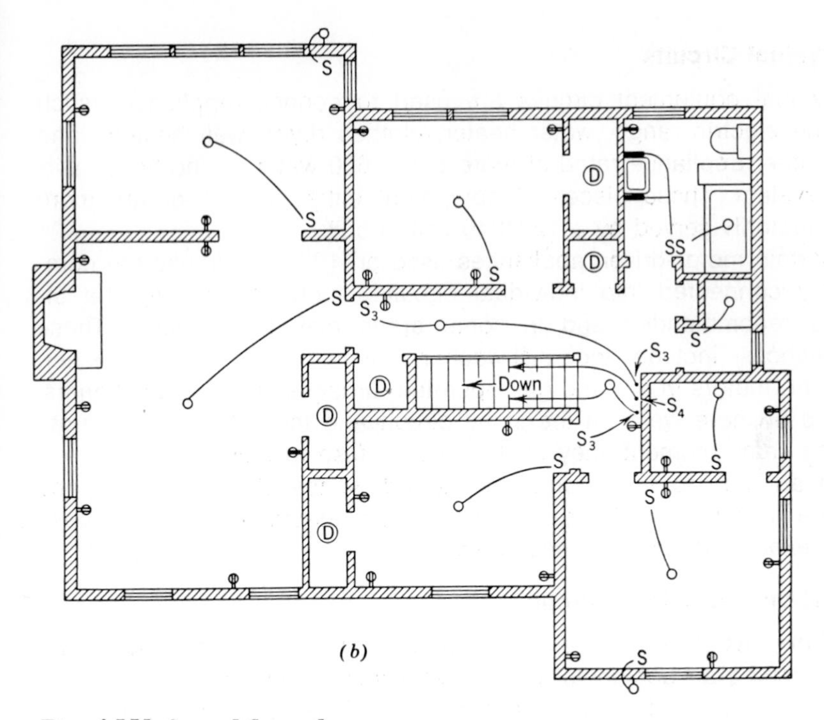

***Fig. 4.11b** Second-floor plan.*

Even more satisfactory than the appliance circuits discussed is a new type installation which uses a three-wire split circuit of No. 12 wire in the kitchen. Between each outside wire and the center wire there are 120 volts, allowing 2400 watts on each side of the circuit or a total of 4800 watts available to the outlets. (Fig. 4.12). The top outlet of each pair is on one side of the circuit, the bottom, on the other side. Some authorities recommend this type of wiring, known as perimeter wiring, for wall outlets in all rooms. A home in a Detroit suburb, wired in this manner, provided 180% increase in electrical capacity at a cost increase of only 21%.

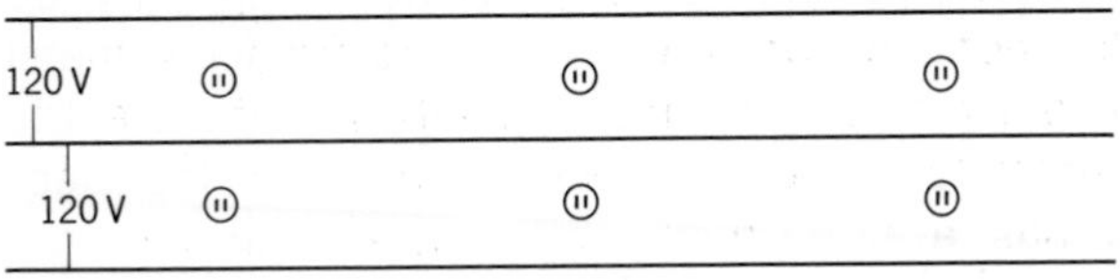

120 V

120 V

***Fig. 4.12** Three-wire split circuit. Equivalent to two appliance circuits but eliminates the cost of one copper wire. Note that there is no connection between the two outside wires.*

Individual Circuits

Individual equipment circuits are used to connect appliances such as the electric range, water heater, clothes dryer, wall heaters, and any other appliance rated at more than 1650 watts, to the house wiring system. These pieces of equipment either require or are more satisfactorily served by a 120/240-volt circuit.

Certain motor-driven appliances used on 120-volt circuits are commonly connected into individual circuits; such connections are always recommended and in some areas may be required. These appliances include the dishwasher, disposer, automatic washer, freezer, motors in oil- and gas-operated furnaces and air conditioners. If two or more were on the same circuit and they started to operate at the same moment, they could cause a fuse to blow.

In addition to the circuits mentioned, space should be provided in the fuse box for future circuits, should electrical needs increase and expansion become necessary.

Short and Grounded Circuits

A **short circuit** is caused by a conductor, usually metallic in nature, establishing an accidental path of very low resistance from one side of the circuit to the other, allowing an abnormally high current to flow, or by the joining of two lead wires on which the insulation has been destroyed. This occurs when the two current-carrying wires of a cord come into contact with each other, so that an intense current passes through the newly formed circuit instead of through the resistance wires of the lighting or heating appliance. Such a condition is usually accompanied by a momentary flash. The excessive current melts the link in a fuse or trips the switch in a circuit breaker, thus opening the circuit. The cord of an electric iron, subjected to frequent bending and twisting that results in abrasion of the cord covering, may occasionally develop a short. Any fabric-covered cord may develop a short circuit if it becomes damp.

The earth is a good conductor of electricity. If a path of conductors is provided from an electric circuit to the earth, the circuit is **grounded** and current flows to the earth. The grounding may be intentional, as when the homemaker connects a wire from the metal frame of a laundry appliance to a water pipe, or attaches the electric mixer on her work counter to a special grounded outlet. If the electrical appliance cord has a three-prong plug, it will fit into the three-hole grounded receptacle in the wall. The grounding prong is longer than the other two blades and, consequently, grounds the equipment before it becomes energized by the flow of current. Sometimes, instead of the three-prong plug, the plug carries a "pigtail" wire which is securely attached to the metal screw holding the face-plate to the outlet. It should be emphasized that the outlet itself must be grounded. Have your electrician check to be sure that this is true. This is especially necessary in moving from one house to another. Most municipality regulations now require the grounding of all wiring in new homes.

These precautions are desirable safety measures. If, on the other hand, a new path of less resistance, such as the human body, is formed between the appliance and the ground, the current will take that path of least resistance. This will result in an electric shock which may even be fatal. Accidents from ungrounded circuits usually occur in a laundry where the person is standing on a wet floor, or in a bathroom where he may come in contact with the metal plumbing. It is not safe to handle an electrical appliance attached to a live circuit when shoes or hands are wet.

Control Centers

Several control centers in the house are an advantage, saving time and steps when a fuse blows and reducing the possibility of voltage drop.

The final step in planning the house wiring system is, therefore, to determine the number of control centers, their sizes, and their locations (Fig. 4.13). If radial wiring is used, a control center is frequently placed near a load center. A main load center should be near the kitchen, since 90% of the current will be used within 10 feet of some location in the kitchen, and since the laundry equipment is usually in an adjoining area. Branch load centers are placed in readily accessible areas on the other floors of the house. From the main center, feeders of No. 12 or No. 10 wire, not over 25 feet in length, run to each branch center. From these branch control centers, comparatively short circuits radiate to the adjacent rooms, reducing heat losses to a minimum with increased efficiency in the operation of appliances.

Types of House Wiring

Conduits may be installed for house circuits, but because of the expense, metallic or nonmetallic sheathed cable is more common. In BX metal-armored cable, the insulated wires are protected by a steel covering, which is spirally wound to make it sufficiently flexible to bend around corners. Nails will not penetrate this covering, and

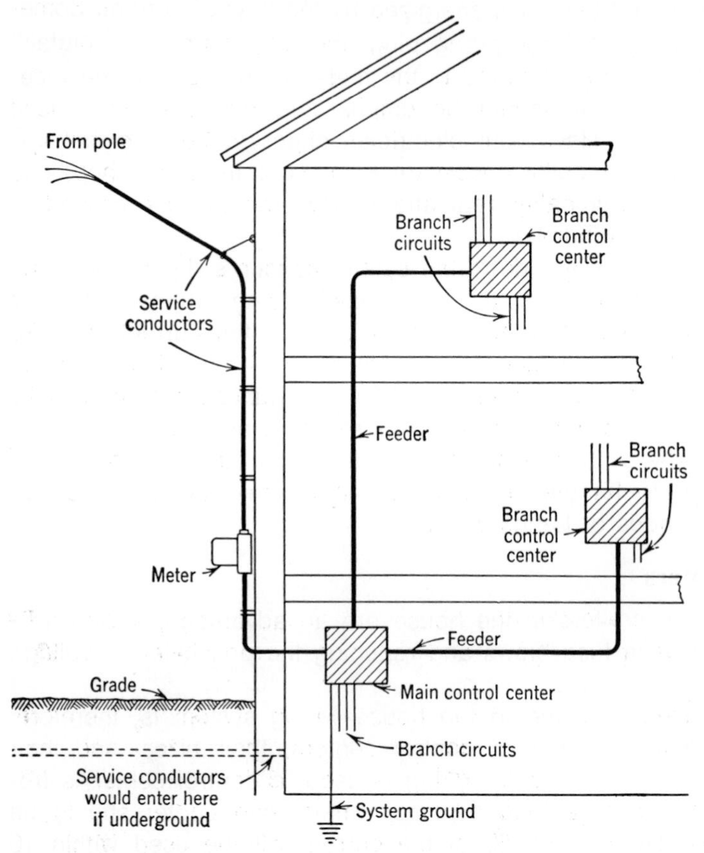

***Fig. 4.13** Schematic drawing of service conductors and radial wiring system. (Westinghouse)*

rodents cannot gnaw through it. It should be grounded. In the non-metallic cable, Romex, the separately insulated conductors are encased in an outer sheath which has been treated to make it water-, fire-, and rodentproof, eliminating the need of additional protection. Sometimes wiring is by the knob-and-tube system in which the conductors are fastened at regular intervals to porcelain insulators except where they pass through the house joists, where porcelain tubes carry the wires and insulate them from contact with the wood. When the conductors are within the partitions, they must be covered with flexible tubing. In general, the knob-and-tube method is not as safe as the other systems, but it is less expensive.

Behind all lighting fixtures and wall outlets are small metal boxes, inside which the conducting wires are connected to the sockets. These boxes prevent any bare wires from coming into contact with flammable materials. The National Electrical Code specifies that the maximum temperature of conductors inside the outlet box shall not exceed 60°C.

Convenience Outlets

There should be enough outlets in any room to permit a lamp or an electric appliance to be used with any piece of furniture placed along the walls of the room, without use of extension cords that cross doorways or traffic lanes. At least two duplex convenience outlets should be provided in each room. Outlets in living room and bedrooms should be located 10 to 16 inches above the floor line and be so placed that no point of any usable wall space is more than 6 feet from an outlet. In the dining room, points may be 10 feet from an outlet; in the kitchen, one duplex outlet should be installed above each work area and an additional outlet provided for the refrigerator. The kitchen requirements may also be met by installing an outlet for every 4 feet of work-surface frontage, exclusive of sink and range. In the laundry, 3-contact, grounding outlets for laundry equipment are placed 40 to 42 inches above the floor. A similar high outlet, designed to take a grounding-type of plug, is installed in the bathroom for the electric razor.

Care should be taken not to place an outlet in the center of a long wall in any room where a large piece of furniture, such as a sofa, piano, buffet, bed, or chest of drawers, may need to be put in front of it.

When the number of convenient outlets is inadequate, but a sufficiently large size of wire is supplied to one accessible outlet, a surface runway or an electrostrip may be installed to increase the number of usable outlets. The surface runway contains built-in outlets at regular intervals; the electrostrip provides outlets at any point on the strip. The plug-in strip is fastened at the back of a work surface or above a baseboard and is connected to the available convenience outlet.

Weatherproof outlets on the outside of the house are used for patio cooking of family meals at any time of day, depending on climate and season; also for occasional special festivities, and for attaching tree decorations at Christmas time. Such outlets are usually controlled by switches on the inside of the house.

Switches

Switches should be sufficient in number and so located that it is possible to light any room upon entering, and to leave the room in darkness, without retracing steps. To do this, any room having more than one entrance should have multiple switch control. Some authorities believe that switches at all entrances are essential only when the doorways are more than 15 feet apart; other authorities say 10 feet apart. But unless two entrances are practically adjacent, some steps must be retraced in darkness when only one switch is provided.

Every room should have at least one light that is controlled by a switch at the entrance. Sometimes it is convenient to have the outlet for a portable lamp so controlled. One type of duplex outlet contains two receptacles separately wired so that one may be connected to a switch.

The switch is placed 4 feet from the floor on the lock side of the door, never behind it. Switch plates may cover several switches grouped together or a switch with a single or double convenience outlet. Such an outlet is especially useful for connecting the electric cleaner. Switch plates and outlet plates are preferably of a non-conducting composition material that may be painted to match the wall. The location of the switch is occasionally marked by a spot of luminous paint, and luminous balls or cylinders are attached to the ends of pull chains. Pilot lights on outlet plates or near the door at the top of the basement stairs indicate when a circuit is closed and help to avoid accidentally leaving appliances or lights turned on.

Because of their easy manipulation by a slight touch of hand or arm, tumbler switches have largely replaced the push button. Mercury switches that operate silently are used in some installations. A switch of this type is placed in a vertical position because its operation depends on the rotation of a button containing an orifice that allows the mercury to flow back and forth to make or break the circuit.

APPLICATION OF ELECTRICITY TO THE OPERATION OF APPLIANCES

Resistance

Even good conductors offer a certain resistance to current flow. Resistance increases with the length of the wire; the longer the wire, the greater the resistance. Resistance is decreased when the diameter of the wire is larger, but increases when the wire is smaller in diameter. Resistance also varies with different materials and with their temperature. Aluminum or its alloys is commonly used for hi-lines, copper for house wiring, and nichrome ribbon or wire for the heating elements of electric ranges and small appliances. Nichrome is a combination of 60% nickel, 15% chromium, and 25% iron. Of these materials, at any common temperature, copper has the lowest resistance, aluminum next, and nichrome ribbon the highest. For any one type of use, however, the kind of material is the same, and its resistance can be disregarded. On the other hand, increase in temperature causes a rapid increase in resistance, especially the resistance of a pure metal, and must be considered. This increase in resistance with rise in temperature is much larger than the corresponding increase in linear expansion. Resistance, volts, and amperes have a definite relationship, which is expressed in the equation $R = E/I$. Resistance is measured in **ohms.**

Heating Appliances

Heat is produced by the resistance of the electric element to the passage of the current. In lighting the home, the resistance of the tungsten wire in the lamp bulb produces sufficient heat to cause **light.**

In designing an appliance, the manufacturer builds into it a wire of the correct length and cross area to give the necessary resistance which will allow the desired current to flow at the rated voltage, thus ensuring enough power to do the job. Most electric appliances are made to operate satisfactorily on a voltage slightly above or below the rated voltage. The resistance is accordingly a part of construction of the appliance and will not change to any appreciable extent during the life of the equipment.

Example

What is the resistance in an 1150-watt electric toaster? Assume that the toaster was constructed to be used on a 120-volt circuit.

$$w = EI$$

$$1150 = 120 \times I$$

then

$$I = \frac{1150}{120} = 9.6 \text{ amp (approximately)}$$

$$R = \frac{E}{I} = \frac{120}{9.6} = 12.5 \text{ ohms}$$

If the toaster is always used on a circuit of 120 volts, the necessary heat is produced to toast the bread in a given length of time. When the toaster is used on a circuit of less than 120 volts, the toasting time will be lengthened; if the voltage is more than 120 volts, the toasting time will be shortened.

Drop in Voltage

A cord used to connect an appliance to an outlet also offers a certain amount of resistance to the flow of the current. If the cord is further lengthened by an extension cord, the resistance may be sufficient to cause a drop in voltage, so that the appliance will heat slowly. Equipment should be connected as near to the place of use as possible. Connecting several appliances into a single circuit can be another cause of a drop in voltage, and, consequently, of slow heating. Homemakers who live near the end of a rural line, or who live near the end of a city distribution system, often have trouble with slow-heating equipment, caused by the drop in voltage which occurs when most of the other consumers on the line are using the current at the same time.

Motor-Driven Appliances

A number of household appliances are motor-driven. Motors are of two types: universal and induction. In each type there is a stationary and a rotating member in which magnetic fields are set up. Interaction between these two magnetic fields results in the motion of the rotor.

It has been noted that power (in watts) is equal to EI, i.e.,

$$w = EI \tag{1}$$

From the equation expressing the relationship between resistance, voltage, and current,

$$E = RI$$

Substitute for E in equation 1,

$$w = I^2R \tag{2}$$

Energy is expressed in power times length of time an appliance is used:

$$\text{Energy} = I^2Rt \tag{3}$$

Careful experiments conducted to relate electrical energy to heat energy show that the heat (in calories) is equal to 0.24 times I^2Rt:

$$H = 0.24I^2Rt \tag{4}$$

The heating of transmission lines is considered a heat loss, since the heat is not used in operating the appliance. Such losses are usually referred to as I^2R losses, or the I^2R drop.

If an appliance is connected to a 120-volt outlet by a 6-foot cord (the usual length of appliance cord), the I^2R drop is disregarded. If, however, the cord length is increased because of the position of the outlet by an extension cord 12-feet long, the resistance in the two leads (the in and out wires in the extension cord which are connected into the circuit in series) increases the I^2R drop of each lead twice and of the two leads four times. This amount of resistance will cause undesirable heat loss in the circuit with consequent drop in voltage and a lessening of the amount of current reaching the appliance. The appliance, therefore, will heat more slowly and, perhaps, may never reach full power.

In the universal motor the armature and field coils are connected in series, and the current from the stationary part (stator) is fed into the rotating part (rotor) through brushes impinging on commutator bars. The universal motor can operate on either AC or DC; hence its name. This motor is used in many small electric appliances; e.g., in food mixers, grinders, hair dryers, shavers, vacuum cleaners, electric fans, and electrically operated sewing machines. Its chief advantage is its ability to adapt to the size of the load; high speed when the load is light, slower speed with a heavy load.

In contrast, in induction motors there is no contact between rotor and stator. The stator is connected to AC current, which is induced from it into the rotor. The magnetic field set up around the rotor reacts with the magnetic field of the stator to produce a rotating motion that is changed into mechanical energy. The AC current used must be of the same frequency as that for which the induction motor was designed. According to Lenz' law, the induced current has such a direction that the magnetic field set up by this current tends to oppose the motion that produced it. In other words, counter or back emf is developed and opposes any change in current flow. Therefore it tends to maintain the flow at a fairly constant rate. When the speed of the rotor decreases, fewer lines of force in the magnetic field will be cut and the current will increase. As the current increases, the magnetic field is also increased, more lines are cut, more back emf develops, and the current decreases again. The back emf acts, therefore, as a current control, increasing or decreasing as the load on the motor requires, and holding the current within safe limits.

Induction motors are found in washers, dryers, ironers, dishwashers, oil and gas-operated furnaces, and air conditioners.

In AC circuits used in home wiring, $w = EI$ because voltage and current are in phase. They rise to a maximum together and fall to zero at the same time. In the induction motor the current and voltage are out of phase, the current being either ahead of or behind the voltage. In this case w does not equal EI. The wattage w divided by EI gives a decimal fraction, usually varying from 0.2 to 0.7, called the **power factor,** its value depending on the type of motor-driven appliances. EI times the power factor gives the true wattage of the appliance.

Motors are occasionally rated in horsepower. One horsepower equals 746 watts. If wattage is calculated from the equation $w = EI$, and compared with the value hp = 746 watts, the two values will be found not to agree because watts indicates power input, and horsepower, the power output.

A motor should not be overloaded; it can be injured by excessive heat within the coils. Overheating may also sometimes result from too great a drop in voltage. The motor may try to start or struggle to reach operating speed without success. Excessive heat may consequently develop within the motor windings and the motor may burn out. At present, most motors have lifetime lubrication; if not, manufacturers' instructions on kind and amount of oil and frequency of application should be followed carefully. Motors stored in cold areas should be allowed to reach normal temperature before use. A clean, dry storage place is requisite for all motor-driven appliances.

Solenoid

Solenoid valves are used as controls in some washers, refrigerators, and doorbells. A solenoid valve may be used in an automatic washer to regulate the off and on action of the agitator and spin cycles. A solenoid is a current-carrying coil wound uniformly in a long helix, often around a soft iron bar which becomes magnetized when a current flows in the coil. The solenoid then acts as an electromagnet with one end equivalent to a north pole and the other end to a south pole. The intermittent action of the solenoid depends on whether the iron bar is inside or outside of the coil.

When an electric current is allowed to flow through a solenoid, the iron core is drawn into the coil and a connection is made to a timer switch that shifts the gears and stops the washing action. If the flow of current stops, the core is released, an opposite shift occurs, and the action starts again. Similarly a solenoid may be used to turn water on and off in the automatic filling of a washer tub, and in starting or stopping a spin cycle.

The primary value of the solenoid is its control of intermittent action. The iron core in the solenoid becomes a temporary magnet; the coil and iron core together often are called an electromagnet, as previously noted. The effect continues only as long as a current flows through the coil. Such electromagnets are used in plants where ranges, refrigerators, and washes are manufactured, to transfer scrap iron or steel from stockpiles in yards to inner furnaces, or in moving large structural parts from one area to another.

Name Plate

Each appliance made by a reputable manufacturer will carry a name plate, either stamped into the metal of the equipment or inscribed on a separately affixed plate. The name plate specifies the conditions under which the equipment is constructed to operate. These specifications include: type of current (AC or DC), frequency of current, voltage and either wattage or amperage, and sometimes horsepower for motor-driven appliances. The manufacturer's name is also given, together with his address, and the model number of the appliance. This information will be needed in case of repairs, especially if no local dealer is available, since the homemaker will then have to write to the manufacturer. Occasionally, special directions for satisfactory operation are found, as, for example: "Do not immerse in water" or "Connect to wall outlet only."

Cost of Operating Appliances

The monthly electric bill is obtained by subtracting last month's meter reading from the meter reading of the current month and multiplying the result by the local rate per kilowatthour. Any homemaker may, if she so desires, determine quite easily the cost of using any appliance for a given length of time. The wattage of the appliance must be changed to kilowatts, and the time must be expressed in hours. The following examples will make this clear.

1. A 1000-watt toaster is used 20 minutes a day during June. What will be the cost of operation at $0.04 per kilowatthour?

    1000 watts = 1 kw
    20 min per day for 30 days = 10 hours
    1 kw × 10 hours = 10 kwhr
    10 × $0.04 = $0.40, cost of operation

2. An electric frypan, rated at 120 volts and 10 amperes, is used on Monday for 30 minutes, on Tuesday for 20 minutes, on Thursday for 1 hour, and on Saturday for 40 minutes. What is the cost for the week at $0.035

per kilowatthour if the thermostat control causes the current to flow only ⅔ of the time?

> $w = EI$ = 120 volts × 10 amp = 1200 watts
> 1200 watts = 1.2 kw
> 30 min + 20 min + 1 hour + 40 min = 2.5 hours[2]
> 1.2 kw 2.5 hours = 3 kwhr
> 3 × \$0.035 = \$0.105, or 10½ cents
> ⅔ × 10½ = 7 cents, cost of operation

3. A ¼-horsepower washing machine is used 1½ hours a week. What will be the cost per month at \$0.05 per kilowatthour?

> ¼ hp (1 hp = 746 watts) = 187 watts = 0.187 kw
> 1½ hours per week for 4 weeks = 6 hours
> 6 × 0.187 = 1.122 kwhr
> 1.122 × \$0.05 = \$0.056, or \$0.06, or \$0.12, if we remember that the fractional horsepower motor is only about 50% efficient.

It will be noted that motor-driven appliances are quite a bit less expensive to operate than heat appliances.

Safety Precautions to Prevent Accidents

The size of fuse should always be used with the size of wire for which it is specified. Using too large a fuse in a circuit, or placing a penny or other conducting material in the socket beneath the fuse, allows an overload of current to flow through the wires and may lead to a destructive fire. Such an overload of current will cause wires to expand because of the excessive heat generated. The heat will tend to bring about a deterioration in the insulation covering the wires so that they finally become exposed and may set fire to any nearby flammable material. A fire in Des Moines, Iowa, took the lives of four people living in an apartment above a restaurant. Investigation to determine the cause disclosed that pennies had been placed under three blown fuses in the restaurant fuse box.

A person changing a fuse should stand on dry wood or on other insulating material. He first pulls the main switch, then uses one hand to replace the blown fuse, keeping the other hand close to the side to prevent it from coming into contact with any conductor, unless it is needed to hold a flashlight.

Consumers' Research findings seem to indicate that fuses are more

[2]Problems are more easily solved if all common fractions are changed to decimal fractions.

rapid in action than circuit breakers and in general are preferable.

House wiring should be checked periodically for damage to insulation from rodents or from the friction that results from uneven settling of the house. Cords should also be checked for frayed and worn places and for loose connection to plugs. They should never be run through doorways, under carpets, across registers, or near water pipes, nor allowed to lie on damp floors, unless cords are rubber covered. A baby boy was electrocuted when he grasped an electric light cord as he was crawling over a cold-air register in his home. His hands were probably damp; small children's hands frequently are.

Such an accident is the result of too great a charge of electricity passing through the body. One side of the residential wiring system is grounded at the power plant. Then any person who comes into contact with current leakage from a defective appliance in the home, and is at the same time in contact with the ground, will allow a parallel circuit to pass through his body, with perhaps fatal results. Water pipes are connected to the ground, and it is therefore safer to keep electrical appliances away from the sink and out of the bathroom; electrical equipment should never be handled when any part of the body or clothing is damp. Since floors of the garage and patio are in contact with the ground, great care should be exercised in using any electrical appliances in these areas; such appliances should always be grounded.

Accidents reported in newspapers emphasize the need for caution. Note the following:

1. On two separate occasions, a child was killed by contact between the metal part of an electric mixer and the sink—a sink is always grounded through its water pipes. One child was leaning directly against the sink, the other against the metal strip along the front edge of the work counter which usually is joined to the sink. The homemaker herself might have received only a shock, but children's hands are more moist than those of an older person and their bodies are lower in resistance.
2. A high school girl took her radio into the bathroom to listen to music while she bathed. She evidently touched it—perhaps to change the station—while in the water and was electrocuted.

No accident would happen were there no current leakage from a defective part, but once it happens, it is too late.

Certain parts of our house are naturally grounded because they are in contact with the earth; for example, the cold water pipes to sink, bathtub, and lavatory, the radiators, gas pipes, heating and air conditioning ducts, and the concrete or terrazzo floors of patios and porches. Anyone in contact with one of these grounded items and, at the same time, a defective, ungrounded appliance may get a severe shock that could prove fatal.

Two means of preventing electric shock are insulation and grounding of appliances. To guard against defective insulation, always select appliances that carry the Underwriters' Laboratories Seal. Then keep the equipment and cords in good repair. Replace cords that are frayed or show exposed wires. Follow manufacturers' directions. Never immerse electrical appliances in water unless they are so designed. Keep motor equipment clean and oiled to prevent the formation of dirt that may cause a current path to external metal parts.

Use a three-prong grounded plug attached to a grounded outlet. However, never ground an appliance with an open-coil heating element, such as a toaster, certain unit electric ovens, and radiant heaters. If you touch the wires and frame at the same time, your body completes a circuit and current can flow through you.

Any appliance that has caused even a mild shock should never be used again until it has been repaired and the cause of shock removed.

When there are young children in the family, the house should be equipped with wall outlets of the type that must be rotated through 90° to make contact. This prevents a child from sticking a piece of metal into the outlet and perhaps receiving a serious burn.

GAS

Gas was first utilized almost exclusively for lighting, but since about 1915 a gradual change has occurred until at present there are seven major uses for gas in the home. It is interesting to note that gaslights have returned to the American scene and have reached such a degree of popularity that at present more than 2,350,000 are in use, about 90% of that number residentially. Approximately 350,000 gas lights are sold annually. These lamps, used primarily for decorative purposes, lend an air of charm to lawns, patios, driveways, swimming pools, cookout areas, and main entrances (Fig. 4.14).

Gas appliances serving United States homemakers now number in the millions. The one that is the most frequently used in the gas range. Gas is also extensively used for house heating and room-space heating, water heating, clothes drying, incineration, refrigeration, and air conditioning. For cooking on the patio, gas, either natural gas or LP, is now frequently used for heating broilers, rotisseries, and grills. In kitchen and outdoor fireplaces gas makes barbe-

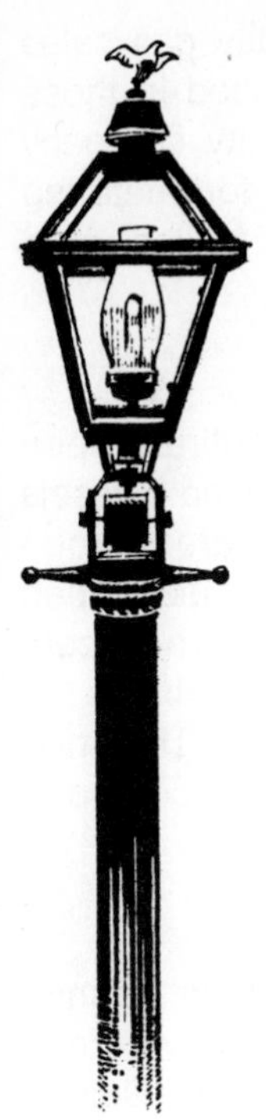

Fig. 4.14 The gaslight. (The Story of Gas, published by American Gas Association, Inc.)

cuing a festive occasion for family and guests. The various appliances and their uses will be discussed in future chapters of this book.

Baltimore, in 1816, was the first American city to have a gas company. Its gas was used principally for street lighting, but was also used in a few public buildings and in the private homes of some of the wealthier citizens. In the early days, gas was sold more or less by the hour, and inspectors were sent out by the company to see that customers used only the amount of gas to which they were entitled. When householders kept their gas lights burning too late at night, they were warned by a rapping on the sidewalk outside the home.

Types of Gas

Four kinds of gas are available in the United States: natural, manufactured, mixed, and liquefied petroleum, commonly called LP, or bottled, gas. Manufactured gas was formerly the most widely used, but the growing availability of natural gas has changed the situation, and today natural gas accounts for 94% of the total utility gas sales to residential customers. Liquefied petroleum gas is supplied in those sections of the country that are beyond the reach of utility company mains, but many of the installations may be arranged for metering the gas within the home. There are approximately 9 million consumers of liquefied petroleum gas.

Manufactured Gas

During the early years, the gas industry relied almost entirely upon coal gas prepared from bituminous coal heated in closed vessels known as retorts. The chief constituents of coal gas are carbon monoxide and hydrocarbons. In 1873, Thaddeus Lowe developed practical methods for utilizing the action of steam on incandescent carbon for the production of carbureted water gas, a mixture of carbon monoxide and hydrogen, both combustible gases. The mixture was further enriched with gasses from the chemical decomposition of a spray of oil.

Natural Gas

The Chinese are said to have discovered natural gas in their country many years ago and piped it through hollow bamboo to where it was used. In certain areas around the Caspian Sea and in Delphi, Greece, natural gas issuing from crevices in the ground was used to mystify the ignorant. These places became centers for worship and the establishment of oracles presided over by priestesses.

Natural gas was first produced in the United States in 1821 from a 27-foot well drilled near Fredonia, New York. At present, it is supplied to every major area of the country.

Natural gas, which is largely methane, is found underground, filling the interstices of porous rocks, known as "sands." The deepest drilled sands today are at least 5 miles below the surface. Above the sands is a thick stratum of impermeable rock that has prevented the gas from working its way to the surface of the earth and escaping. Natural gas is the product, scientists believe, of a chemical action on the marine organisms that were buried in the salt water sands along the seashore millions of years ago. Large volumes of this gas were forced into cavities between the grains of sand and were held there at various pressures known as "rock pressure." This pressure causes the gas to flow mechanically from well to consumer, but it decreases as gas is withdrawn through a well, since the remaining gas expands and fills the crevices.

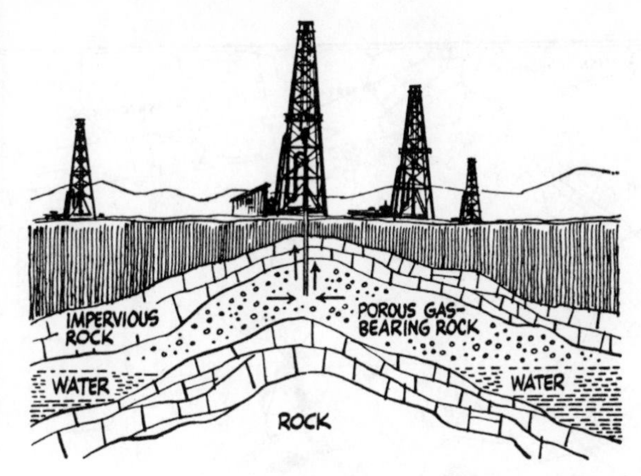

***Fig. 4.15** A cross section of a natural gas field. (*The Story of Gas, *published by A.G.A.)*

When the well has been drilled down into the cap rock above the gas sand, a casing is set in the hole to prevent water from seeping in. Through the center of the casing, drilling continues, and other casings may be set before the desired depth is reached. Finally a pipe is dropped inside the casing and fastened in such a way that the gas must come up through this "tubing" and not escape around the sides (Fig. 4.15).

Gas expands as it comes out of a well and may need to be recompressed to have sufficient pressure to overcome friction of the pipe. Any moisture, natural gasoline, or oil vapors, and, in some cases, sulfur, which may be mixed with the gas, are also removed and the gas is cooled. If gas has to be transmitted a long distance, additional compressing stations are installed along the line about 100 miles apart. These gas lines are patrolled, frequently from an airplane. Since natural gas is odorless, an odorant is added to make any leaks more easily detectable.

Many long-distance pipelines operate at pressures between 300 and 800 pounds per square inch. The highest pipeline pressure reported (1959) was 1600 pounds per square inch, more than 100 times the sea-level pressure of air. To maintain safely such a high pressure requires special high-strength steel for the construction of the pipeline. Lengths of these steel pipes are placed end to end, welded together, and then coated with plastic wrappings of tarred Fiberglas. The pipelines extend more than half a million miles across the United

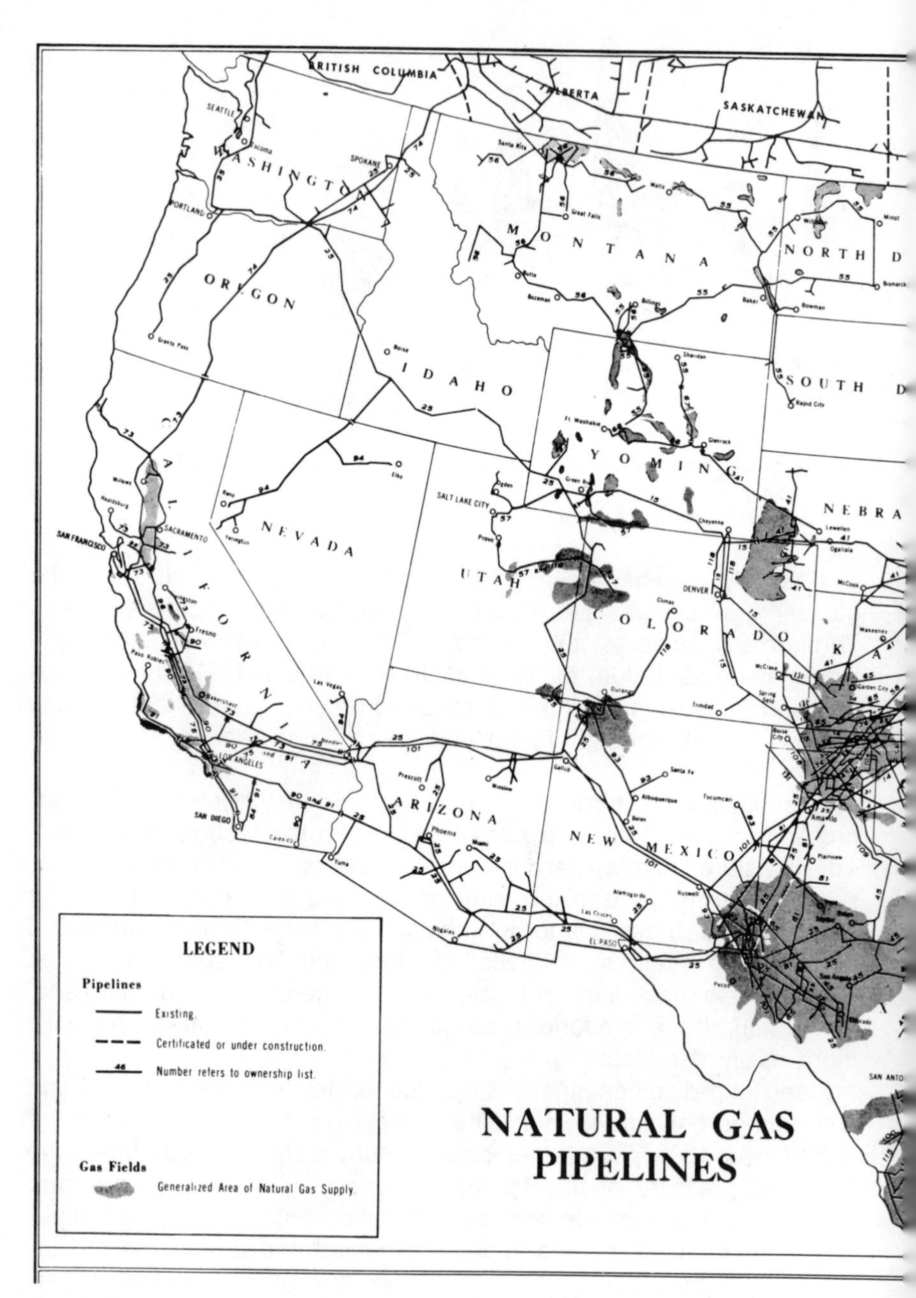

*Fig. 4.16 Major natural gas fields and pipelines. (*The Story of Gas, *published by A.G.A.)*

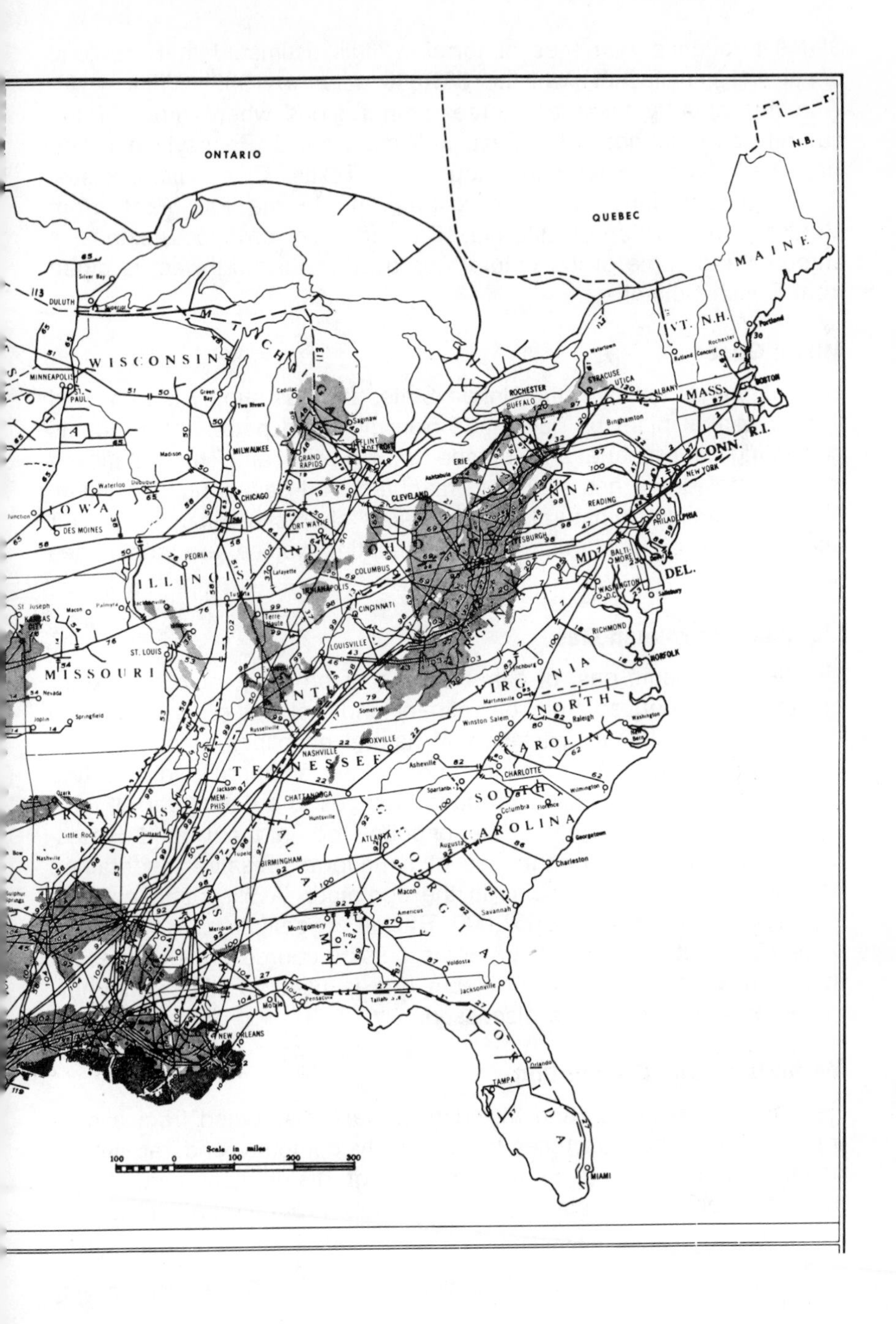
ONTARIO
QUEBEC
N.B.
MAINE
VT.
N.H.
MASS.
R.I.
CONN.
WISCONSIN
IOWA
MISSOURI
ILLINOIS
IND.
OHIO
MD.
DEL.
VIRGINIA
NORTH
CAROLINA
SOUTH
CAROLINA
TENNESSEE
ARKANSAS
GEORGIA
DULUTH
MINNEAPOLIS
ST. PAUL
MILWAUKEE
CHICAGO
DES MOINES
PEORIA
ST. LOUIS
GRAND RAPIDS
DETROIT
FLINT
Saginaw
CLEVELAND
COLUMBUS
CINCINNATI
LOUISVILLE
NASHVILLE
KNOXVILLE
CHATTANOOGA
MEMPHIS
BIRMINGHAM
ATLANTA
CHARLOTTE
Augusta
Macon
Savannah
Charleston
Columbia
Montgomery
Mobile
Pensacola
NEW ORLEANS
Jacksonville
Orlando
TAMPA
MIAMI
RICHMOND
NORFOLK
WASHINGTON D.C.
BALTIMORE
PHILA
READING
NEW YORK
BOSTON
Portland
ALBANY
SYRACUSE
UTICA
ROCHESTER
BUFFALO
ERIE
Binghamton
Raleigh
Little Rock
Springfield
Joplin
Scale in miles
100
0
100
200
300

States through all varieties of terrain. "It is estimated that the total length of gas pipelines will be 831,000 miles in 1969"[3] (Fig. 4.16).

In this country there are three main regions where most of the natural gas is found: in the East; in West Virginia, Pennsylvania, and New York; in the South and Southwest, Texas, Oklahoma, Kansas, Colorado, Louisiana, and Mississippi; and in the Far West, from Montana to southern California. Canada also produces extensive amounts and some of the natural gas used in the northwest is piped from British Columbia.

Mixed Gases

In a number of towns in the United States the gas supply is furnished from a central plant by gas made from liquefied petroleum products consisting of propane and butane, mixed with air. The butane-air mixture provides a gas similar to manufactured gas in heating value, although similar to natural gas in flame characteristics. Butane-air and propane-air mixtures of higher heating values are sometimes used.

Liquefied Petroleum Gas

Liquified petroleum gas, of either butane or propane, or of a mixture of the two, is derived from oil refinery and wet natural gas sources or from the fractional distillation of natural gasolines. It can be liquefied under moderate pressure and is then shipped to distributing centers in special tank cars. Propane changes to a gas under all temperature conditions likely to be encountered in the United States, and consequently a uniform gas at a uniform pressure is obtained as long as any liquid remains in the container. A 95-pound cylinder supplies approximately 2,057,000 Btu. (For definition of Btu, see page 126.) Butane may require some outside source of heat to assist in vaporization. Liquefied petroleum gas is used for the same purposes as the other gases discussed.

Methods of Gas Distribution

Natural, manufactured, and mixed gases are distributed from a central system to the consumer by the utility company and require no home-storage facilities. In the process of distribution, natural and manufactured gases employ storage holders. From the holder, the gas

[3]**The Story of Gas,** p. 15. Educational Services, American Gas Association, Inc., N. Y. 1964.

passes through a system of underground mains in which the pressure is automatically controlled at all times. Gas flows from the mains into small lateral pipes that run into the individual homes.

Large, cylindrical gas holders for manufactured gas were once a common sight near a gas plant; they may still be seen occasionally.

Natural gas also requires some type of storage to increase the capacity of the system and to equalize the peak demand during the colder months of the year with the much lower summer requirements. This need has been emphasized with the increased use of gas for house heating. For some time, depleted natural gas reservoirs have been employed for summer storage of surplus supplies, but they are not ideal, because they are usually located too far from places of large gas consumption. Recently, man-made caverns in the vicinity of maximum use have been found to meet the need. The best depths are between 700 and 1400 feet. A certain volume of gas, known as "cushion gas," must be stored in the underground reservoir at all times to provide needed pressure for adequate gas flow, but a maximum day's output may approach 5 billion cubic feet. There are now 200 such storage pools in 19 states in the vicinity of many of the larger cities of the country. Sometimes steel tanks, similar to pipelines, each one often a mile long, with several laid parallel to each other, are buried underground to serve as storage tanks. Gas can also be stored in the pipeline itself by increasing the pressure. Ready availability of gas is considered to be in the interest of national defense.

Liquefied petroleum gas is delivered to the town and village consumer in cylinders containing 47 to 95 pounds of the liquid. The cylinders are stored outside the home either in some type of outer jacket or are buried underground. The cylinder is connected through a pressure-regulating valve to the house gas-pipe system. Some distributors fill the cylinders at the customer's premises from a tank truck; others replace used cylinders with fresh ones from the central supply. Liquefied petroleum gas is usually brought to the rural homestead by a tank truck, which empties into a large, permanently installed tank, frequently of 1000 gallons capacity. This is the amount needed to supply the numerous farm operations as well as those of the home.

A throw-away 14-ounce propane cylinder for camp or patio stoves and grills is available in many hardware and department stores. Cylinders of three pounds and up for similar use may be obtained from LP dealers. Twenty-five percent of all gas appliance sales per year are for homes in LP gas areas.

Mixed gases are distributed from a central mixing plant through an all-welded steel distribution system.

Physics of Gas

Gas is composed of molecules continually in motion. Since a gas tends to expand and diffuse, the molecules completely fill the containing vessel, and their motion is limited only by the size of the enclosed space. They exert equal pressure in all directions. The tendency of a gas to diffuse is the fundamental essential for gas flow. Just as water flows from a higher to a lower level and heat flows from a hotter to a cooler body, so gas flows from a place of higher pressure to one of lower pressure.

Heating Value

The heating value of a gas is the amount of heat produced when a unit quantity is burned. It is measured in Btu (British thermal unit) per cubic foot of gas. The standard cubic foot of gas, as defined by the American Gas Association, is the quantity contained in one cubic foot of volume at a barometric pressure of 30 inches of mercury, and at a temperature of 60°F. A Btu is a unit of heat energy, the amount of heat required to raise the temperature of one pound of water by one degree Fahrenheit. The amount of heat is equal to that produced by burning an ordinary wooden match.

The heating values of the various gases discussed in this chapter are approximately as follows:

Type of Gas	Btu per Cubic Foot
Natural gas	1000 to 1150
Manufactured gas	500 to 600
Mixed gas	700 to 900
Propane	2500
Butane	3200

Specific Gravity

The specific gravity of all the fuel gases except the liquefied petroleum group is lighter than air. Air weighs 0.0765 pound per cubic foot at 60°F; the fuel gases weigh 0.029 pound per cubic foot for manufactured gas up to about 0.054 pound per cubic foot for natural gas. Liquefied petroleum gases are anywhere from 1½ to 2 times as heavy as air and, therefore, require more careful handling.

Since liquefied petroleum gas is heavy, it always settles to ground level, instead of rising and dissipating. If the gas escapes within the home, from a defective appliance connection, it will settle near the floor, causing a special hazard for children or pets, as well as a fire or explosion hazard. Since the gas will remain below the window level, doors must be opened to remove it from the house.

Gas Meter

Gas piped into the home is measured by a meter, which may be in a case of steel, aluminum, or cast iron.

The basic principle on which the meter works is "displacement." Gas entering one chamber of the meter displaces, or pushes out, an equal volume of gas from another chamber (Figs. 4.17 and 4.18). The meter records the total volume of gas passing through it by counting the number of displacements. When a burner is turned on, the pressure drops slightly in the gas outlet pipe from the meter to the appliance, but remains high in the pipe going into the meter from the street main, causing unequal pressure on opposite sides of the movable diaphragm. This unequal pressure moves the diaphragm and produces a force causing the gas to flow, and the meter starts to record. As long as gas is being used in an appliance, there is unbalanced pressure in the meter and the meter continues to operate.

The flow of gas in and out of the separate chambers of the meter is controlled by sliding valves. Because one compartment is emptied as another is being filled, the flow of gas to range or water heater is smooth and uninterrupted. Each chamber is filled with the same volume of gas every time.

Owing to the interlocking of the gears, every alternate dial of the meter records in the reverse direction. Always read the last number that the pointer has passed. Set down the figures from left to right and add two ciphers to obtain the meter reading in cubic feet.

It is not customary to turn the dials back to zero each month. Instead, the meter reading of the previous month is subtracted from the reading of the current month. This procedure counteracts any error that may occur in a reading.

Future Use of Gas in New Developments

The gas industry predicts improved use of gas in the future. Gas ovens will use a forced draft of air, reducing cooking time to approximately one-half that now required. More extensive infared radiant-type equipment, using gas at low pressure, will be manufactured. This equipment will be more efficient, cost less, and also reduce the time of heat transfer. A new type of gas burner under development

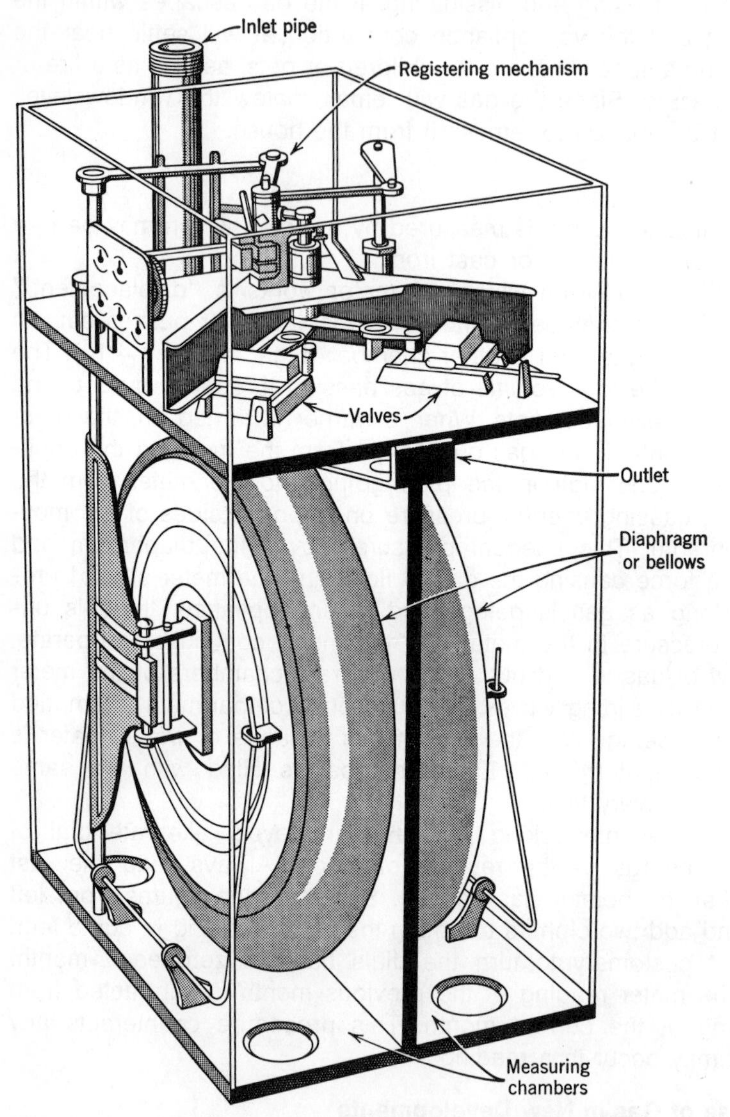

Fig. 4.17 Inside the gas meter. (American Gas Association)

is the "thermocatalytic" burner. It is flameless; the energy is produced by chemical action between air and fuel gas in a narrow tube that can be shaped into any form.

Finally, gas research workers are investigating the possibility of

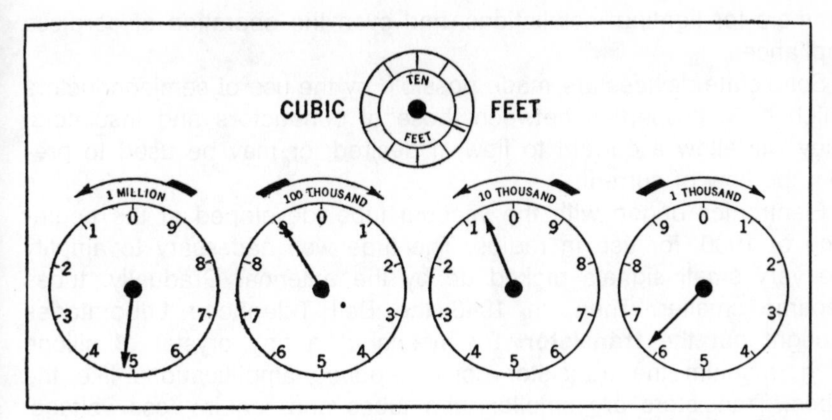

***Fig. 4.18** Dials on a gas meter, indicating a reading of 490,600 cubic feet of gas.*

generating electricity directly from gas right in the home by means of a fuel cell. The fuel cell has two plates, called electrodes, that are immersed in a chemical solution. Fuel gas is led into one electrode, and air, a source of oxygen, into the other. The energy of the reaction is changed into electricity. The cells are almost 80% efficient. A bank of these cells placed in the home will supply all of the electricity that is needed by the household.

HOUSEHOLD ELECTRONICS

Space-age electronics has become a part of the everyday life of the homemaker. Because of their ruggedness, reliability, and very small size, solid state elements are used in appliance control devices that are of much greater versatility and sophistication than would be possible without them. They perform tasks that in the past have been largely done by switches, and work with much greater speed.

Some of their applications include control of speed cycles of agitator and spin in washers, even using light-sensitive devices to keep the washing process going until the clothes are completely clean; sensing moisture content of clothes in dryers; defrost cycles in refrigerators, and the action of the automatic ice-cube maker; the control of heating and air conditioning systems by determining the temperature and humidity at several locations in each room, and holding them at a constant desired level for comfort by adjusting the burners and blowers when necessary. Solid state may also control the speed of sewing machines, the motor action of beaters and blenders, the dimmers for lighting installations, and even the operation of cordless appliances.

Solid state devices are made possible by the use of semiconductors which have properties between those of conductors and insulators. They will allow a current to flow, if desired, or may be used to prevent the flow of current.

Electronics began with the vacuum tube, developed at the beginning of 1900, for use in radios. The tube was necessary to amplify the very small signals picked up by the antenna. Gradually, tubes became smaller. Then, in 1948 the Bell Telephone Laboratories brought out the **transistor.** By means of a tiny crystal of silicon or germanium the transistor could produce amplifications like the tubes. Transistors are smaller than tubes and require less voltage, usually 6 to 12 volts, compared to hundreds of volts for tubes, and so can be operated from batteries (Fig. 4.19).

These crystals of germanium and silicon are the solid state components. The materials occur in nature in the form of oxides, which

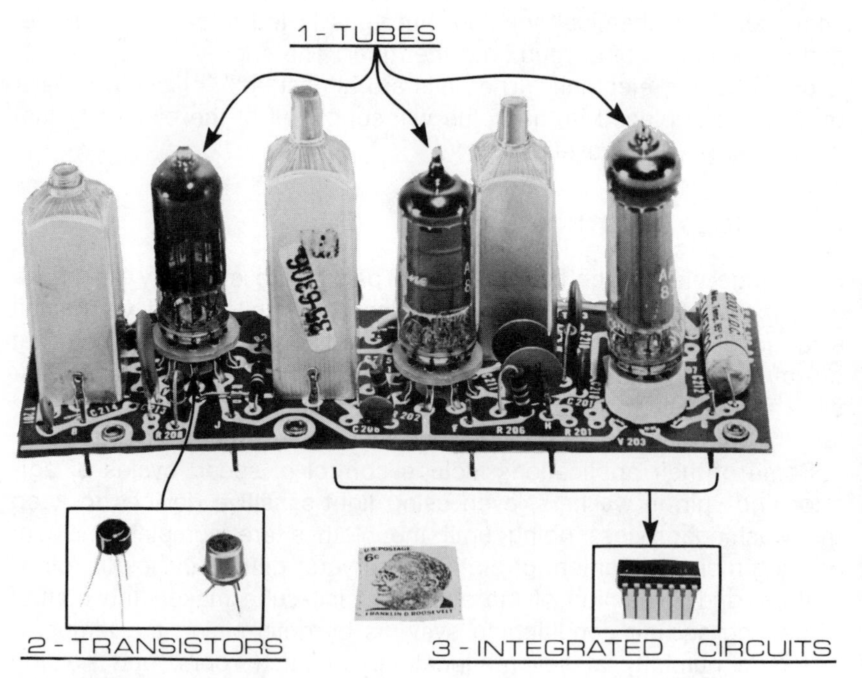

Fig. 4.19 Common electronic components. An appreciation of their size may be gained by comparing them with the size of a U.S. 10-cent postage stamp. (Warwick Electronics, Inc.)

must be passed through several reducing and purifying processes. The necessary electrical properties are then developed in the crystals by adding to them definite quantities of impurities, boron or phosphorus to silicon, arsenic to germanium. The resulting electrical device, about the size of the head of a pin, is encased in a tiny metal can for protection. The metal covering, known as a **heat sink,** is used for dissipating any undesirable heat from the semiconducting material.

Because of their very small size and solid state nature, transistors are physically very rugged and can withstand much more abuse without damage than could a vacuum tube. Their use is somewhat limited by the temperature at which they will operate. The maximum temperature for germanium is 100°C, but silicon controls can go as high as 200°C. In addition, transistors are more susceptible to damage resulting from application of incorrect or excessive voltages than are vacuum tubes. For example, a transistor control device might be damaged by lightning striking nearby power lines. Solid state components are extremely reliable, but in case of failure to function, they must be replaced; repairs are not possible.

Another semiconducting device is the **thermistor.** It is thermally sensitive; as temperature increases, the resistance of the thermistor decreases, allowing more current to flow. It is used as a meat probe in the range oven and in forced-air furnaces to sense the temperature and so control the speed of the blower.

The next contribution to miniaturization was the "printed circuit." The solid state components of the circuit are assembled on a thin sheet of nonconducting material, and the conducting paths between the solid state elements (and other necessary circuit components such as capacitors and resistors) consist of thin bands of copper affixed to the nonconducting backing, instead of wires. The "printed circuit" derives its name from the fact that the pattern of copper bands on the insulating backing is usually manufactured by a photoprinting process.

Another addition is the integrated circuit. It is a very tiny printed circuit, a process of microphotoengraving, in which the components as well as the connections are imprints. The circuit requires no soldered connections and does not deteriorate with age. It is frequently known as a "chip" because of its miniature size. Silicon chips are usually used. By the microphotoengraving method, several hundred identical printed circuits may be built up layer by layer on a single chip until the needed device is obtained.

Transistors and the printed and integrated circuits are used widely in television sets, both black and white and colored, and in radios, with the radio frequently combined with a record player. Some record players contain a solid integrated circuit amplifier right at the sound source. Certain color TV sets also feature solid integrated circuits that improve color clarity and sharpness. Solid copper circuits may replace hand wiring in over 200 possible trouble spots. The etched copper circuits are permanently bonded to a solid base and are assurance of precision manufacture, dependable performance, and long life. Transistors, too, have generally replaced tubes in key circuits for greater efficiency and reliability. The heart of all this successful performance is, as was noted, a solid state device little larger than the head of a pin. In this tiny area the device may contain interconnected transistors, resistors, and capacitors. Another interesting use of integrated circuits is in miniature hearing aids.

Meeting the Energy Crisis

In this chapter we have discussed the sources of gas and electric energy used in the home. The present emphasis is on energy conservation. Specific applied information on this subject is found in the appropriate chapters. Many usable suggestions can be obtained from Small Homes Council Bulletin C1.5, "Living with the Energy Crisis, 1973." It can be purchased for 25 cents from Small Home Council-Building Research Council, University of Illinois, 1 East Saint Mary's Road, Champaign, Illinois 61820.

Terminology

ELECTRICITY

Generator. Consists of armature and field, changes mechanical energy of water power or steam into electrical energy.

Electron theory. The atom is composed of negatively charged electrons orbiting around a nucleus; they can be made to move in a definite direction, and this flowing movement is known as an electric current.

Ampere. Unit of current flow, equals one coulomb per second, symbol **I.**

Volt. Unit for measuring difference in pressure between the two sides of circuit, symbol **E.**

Transformer. Two separate coils of wire wound on an iron core. Current is induced from primary coil into secondary coil; number of turns of wire on secondary determines step-up or step-down transformer.

Circuit. Closed path through which current flows.

Conductor. Usually of copper or aluminum, allows flow of current.

Insulator. Prevents flow of current, keeps current in proper path; glass, rubber, cotton.

Alternating current. Cycles in positive, then in negative direction.

Direct current. Cycles only in positive direction.

Transmission system. From generator to step-down transformer.

Distribution system. From step-down transformer to individual homes.

Service wires. Three insulated wires from pole to house.

National Electric Code. Minimum regulations for wiring the home, safeguards against fire and shock hazards.

Meter. Indicates the amount of electricity used in kilowatt-hours, the unit of energy, symbol **kw-hr** or **KWH.**

Fuse. Strip of alloy metal blows with overflow of excessive current causing heat.

Circuit breaker. Device with bimetal strip. One metal expands more than the other when excessive current flows and breaks connection. Is reset, not changed, when cause of tripping is removed.

Watt. Measure of electric power, the rate at which electricity works, symbol **w.**

Circuits

> **General purpose.** Uses No. 14 or 12 wire, supplying 1800 or 2400 watts, for connection of lights and some low-wattage appliances.
>
> **Appliance circuit.** Uses No. 12 or 10 wire, supplying 2400 or 3000 watts, for connection of portable appliances.
>
> **Individual.** Uses No. 6 or 8 wire, for appliances rated 1650 watts and over and often for motor-driven appliances.

Short circuit. Path of low resistance, allowing too much current to flow; causes too much heat and a blown fuse.

Grounded circuit. Path from electric circuit to the earth.

Convenience outlet. Opening in wall where lamps or portable appliances may be plugged in.

Switch. Means of turning on current to lights or appliances.

Resistance. Opposes current flow and therefore causes appliance to heat. May cause drop in voltage. Varies directly with length of wire and inversely with cross section. Unit, the omh, symbol **R.**

Motor-driven appliances. Interaction between magnetic fields of stationary and rotating members causes motion of rotor. Universal motor can operate on **AC** or **DC.**

Induction motor. Alternating current of stator induced into rotor; interaction of two magnetic fields causes rotation of rotor. Motor changes electrical energy into mechanical energy.

Name plate. Specifies conditions under which equipment is constructed to operate; includes type of current, voltage, wattage, manufacturer's name and address, model number. May indicate special directions "Do not immerse in water."

Safety suggestions. Guard against current leakage. Keep equipment away from contact with water. Keep hands and feet dry when using equipment. Ground equipment whenever possible. Keep cords and equipment in good repair. Never immerse appliances unless so designed. Have any appliance that has caused even a mild shock repaired immediately.

GAS

Types

Natural. Obtained from underground.

Manufactured. Made by passing steam over hot carbon; not widely used at present.

LP. Liquid petroleum, propane and butane, made from oil refinery sources and gasoline; shipped in tank cars, distributed in cylinders. Used as other gases. Is heavier than air, leaks can be dangerous.

Heating values. Expressed in Btu per cubic foot.

Natural gas	1000 to 1150 Btu per cubic foot.
Manufactured	500 to 600 Btu per cubic foot.
Mixed	700 to 900 Btu per cubic foot.
Propane	2500 Btu per cubic foot.
Butane	3200 Btu per cubic foot.

Meter. Two or four chambers, works on principle of displacement, dials read in cubic feet.

ELECTRONICS

Solid state elements. Semiconductors, crystals of silicon or germanium, very tiny.

Transistor. Use of solid state components produces amplification as vacuum tube does.

Heat sink. Tiny metal covering of crystals to protect and to dissipate heat.

Thermistor. Thermally sensitive element—as temperature increases, resistance of thermistor decreases, allowing more current to flow.

Printed circuit. Solid state and other circuit components assembled on thin sheet of nonconducting material and joined by bands of copper instead of wires, manufactured by photoprinting process.

Integrated circuit. Called a chip because of miniature size; a tiny printed circuit, components, and connections are imprints, a process of microphotoengraving.

Experiments and Projects

1. Trace electricity from the generating plant to its use in toasting bread for breakfast. Define the terms you use.
2. Draw dials to indicate 1208 kilowatthours.
3. If last month's kilowatthour meter reading was 1043 kilowatthours, what will be your bill this month, if your rate is $0.04 per kilowatt-hour?
4. Determine the cost of using the following appliances for the month of June at $0.05 per kilowatthour:

   Toaster, 1000 watts, 15 minutes per day
   Washer, ½ hp, 2½ hours per week
   Refrigerator, ⅓ hp, operates ¼ of each 24 hours
   Electric iron, 1200 watts, 1¾ hours per week

   Which appliance is most costly to operate? Why? Consider reasons for variations in the cost of operating a refrigerator, a refrigerator-freezer, a self-cleaning electric oven, an automatically controlled electric surface unit or gas burner, etc. Write a brief discussion of your conclusions.
5. Determine the resistance of the electric iron in problem 4.
6. If arrangements can be made, visit local generating station and gas plant. Report on:

   (a) How electricity is generated, how much the voltage is stepped up, local rate per kilowatthour.
   (b) Source of gas supply, cost per cubic foot, variation in Btu content per cubic foot from day to day.

7. Have an outlet board constructed to resemble Figure 4.20. (This usually can be made by the physics or engineering department.)

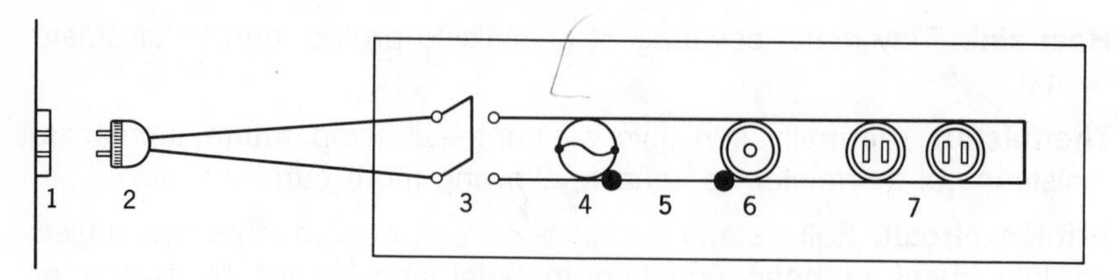

***Fig. 4.20** An outlet board. 1. Wall outlet. 2. Plug to attach board to outlet. 3. Switch. 4. Fuse; 15 amperes at start of experiment, 30 amperes for* (b) *part of experiment. 5. Replaceable wire. 6. 60-watt light bulb. 7. Outlets into which appliances are plugged (suggest toaster, frypan, iron).*

(a) Plug one small electrical appliance after another into multiple outlets until the fuse blows. Explain.

(b) Replace the wire between a fuse and light socket with bare No. 14 wire around which a strip of friction tape is wound. Replace the blown fuse with a 30-ampere fuse, or place a penny beneath blown fuse. Repeat experiment (a). Explain the results and apply them to a situation in your home. **Caution. Stand on rubber mat and open switch when making any changes in the set-up on the board.**

8. Examine the name plates of six appliances that are used in the home. Record the name of the appliance, the name of the manufacturer, the wattage rating or volts and amperes, and any special information given. Examine the appliance and its extension cord for fraying or breaks.
9. Invite a physics professor to give an illustrated talk on electronic components.
10. From the specification sheets sent with household appliances, make a list of those using solid state controls, and the purpose for which the controls are used.

CHAPTER 5
Portable Electric Food Appliances

Many portable electric cooking appliances supplement or substitute for the kitchen range. Most of them provide fast, controlled heat and have relatively high wattages. To use them efficiently, the home wiring system must be adequate to carry the wattages that are indicated on the nameplates of the appliances.

Some portable appliances have permanently attached heat controls, others have detachable controls. Some with detached controls have a water-sealed heat unit, so that they can be immersed in water when they are washed. These heat controls are usually priced separately from the appliance. The same heat control may be used on several different pieces of equipment made by the same manufacturer but are not interchangeable on appliances made by another manufacturer (Fig. 5.1).

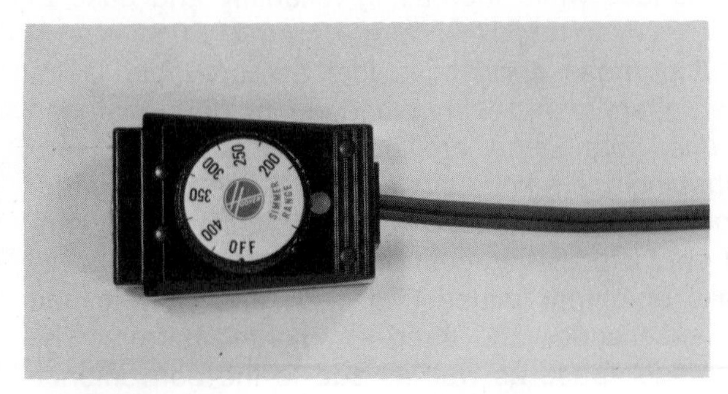

***Fig. 5.1** Detachable heat control. Signal light tells when desired temperature is reached. (Hoover)*

Lack of adequate and competent repair service and the high cost of repairs, when available, is an unpleasant fact of life in most parts of the United States. It is frequently cheaper to buy a new portable appliance than to have an old one repaired. A development in appliance design has been introduced by one manufacturer which should greatly simplify the servicing problem. The manufacturer has designed the portable electric equipment so that it can be taken apart and reassembled by the homemaker. Replacement parts are to be carried by "authorized service centers," a list of which is furnished with the appliance.

Some of today's small appliances are more portable than ever because they can be operated on batteries. Cordless or battery-operated toothbrushes, shavers, slicing knives, and clothes brushes are powered by rechargeable nickel-cadmium batteries. After a 10-to-14-hour charging period, the appliance can be used anywhere until the charge is exhausted. Nickel-cadmium batteries are usually permanently built into a product. The energy in a rechargeable battery is restored by connecting the electric appliance directly with a wall outlet or a recharger. (A recharger is a plug-in electronic device than converts the 115-volt AC power from a wall outlet to a low-voltage DC current required by the batteries.)

It is not feasible at the present time to operate heat-producing appliances such as hair dryers, irons, and toasters on batteries, because of the size and number of batteries which would be required to operate them. They take more power than motor-driven appliances

Solid state controls are being used on some small electrical appliances. In a solid state system the functions of many separate electromechanical operations are combined into electrical circuits which are incorporated into one solid chip of silicone. These chips are very small. Solid state systems make it possible to reduce the size of electronic controls in appliances while increasing reliability and ease of operation.

The number of multipurpose appliances that are available is increasing—ice crusher attachments with blenders, broilers that can also be used as ovens, ovens and broilers with rotisserie attachments, can openers that include a knife sharpener, mixers with blender attachments, coffeemakers that can heat water for "instants," warming trays with a "hot spot" for beverages, electric skillets with keep-warm drawers for rolls or dinner plates (Fig. 5.2), and electric hair dryers with manicure accessories and three-way lighted mirrors. The combinations save storage space as well as add to the convenience of their use.

In using any portable appliance with a detachable cord, the cord

***Fig. 5.2** Electric stainless steel fry pan with broiling lid and warming tray. (Hoover)*

should be connected to the appliance before the plug at the opposite end of the cord is connected to the convenience outlet. The plug at the convenience outlet should be disconnected before the plug at the appliance. This will prevent arcing at the terminals of the appliance.

In disconnecting any appliance, pull the plug from the wall outlet; never disconnect by yanking the cord.

A cord should not be wrapped around an iron or other electric heating appliance while it is hot. Electric appliance cords should not be wound, for storage, in sharp bends.

The numbers of portable electric appliances which aid in the preparation, cooking, or serving of food are increasing each year. The variety of each type shown in dealer displays makes selection difficult, and there is a problem of storage and easy access after they are acquired. Some careful preplanning at home before setting out to buy an electric appliance should result in the purchase of a portable appliance which will add a new dimension to meal or party food preparation rather than another seldom-used piece of equipment to take up space that could be better used for another purpose.

In addition to the general considerations that might be applied to the selection of any piece of household equipment, as outlined in Chapter 1, additional considerations might be applied to the purchase of small electrical equipment for food preparation or serving. Frequently a portable piece of equipment may duplicate the function of a major appliance that is already part of the household inventory. If, for example, the range has a thermostatically controlled surface unit, an electric skillet or saucepan would not be needed. However, if additional cooking capacity is needed or if the electric skillet would provide cooking on the porch or in a recreation area, it might still be a desirable purchase.

If counter space is limited, there may not be room to use and store an additional piece of equipment. An electric mixer or blender should be in position for use at all times. The time and effort necessary to remove it from a storage space and assemble it will remove it from the labor-saving device category and it will be used infrequently. An electric fry pan should be as accessible as an ordinary skillet. Convenient storage, adequate in size and located close to the point of use, is a must for deriving any real use from most small equipment.

The wattage of the appliance must not exceed the circuit capacity of the area in which the appliance is to be used. Usually only one heating appliance can be used at one time on one general-purpose circuit. Most heating appliances use from 1000 to 1500 watts. General-purpose circuits will carry 1800 watts. Underwriters' Laboratory recommends using only 80% of capacity at one time. Most motor-driven household appliances use less power than the heating appliances and more can be used on the same circuit.

Appliances which perform more than one function are likely to be used more frequently and take up less storage than two appliances.

COFFEEMAKERS

Electric coffeemakers are of percolator, vacuum, or drip types. Most cut to a keep-warm temperature after brewing. Some have signal lights, flavor selectors, and reheat settings.

Coffeemakers make the best brew when they are used to capacity. The size selected, therefore, should provide the number of servings needed frequently. Most deliver fewer servings than the rating indicates. To determine the number of servings a coffeemaker delivers, measure the number of ounces of cold water that reach the full mark.

(In a percolator the water should not come beyond a quarter of an inch below the bump on the pump.) Divide the number of ounces by 6. If 6 ounces of cold water are used to start the coffee making, somewhat less than 6-ounce servings will result, since the coffee grounds take up water and some is lost by evaporation during the brewing. The Pan American Coffee Bureau recommends using 2 level tablespoons of coffee to 6 ounces of water for a brew of average strength in all types of coffeemakers.

The Simplified Practice Division of the National Bureau of Standards has set standards for coffee grinds. The one for percolators is the coarsest grind. There is an intermediate grind for drip coffeemakers and a fine grind for vacuum types.

Cold water should be used for starting coffee to brew in percolator and drip coffeemakers. If warm or hot water is used, the extraction will be lessened. Coffee flavor experts recommend that water from the cold water faucet be used in any type of coffeemaker because of the off-flavors that may be imparted to the coffee from accumulations of sediment in the hot water tank.

The handles of all types of coffeemakers should balance the serving pot and be of heat-resistant material. It is often handy to have a knuckle guard to protect a larger hand from becoming burned.

Percolators

The percolator (Fig. 5.3) should have a well-balanced design. The spout should be smoothly welded, brazed, or molded leaving no hard-to-clean places on the inside of the pot.

In most electric percolators the pump of the percolator fits over a well in the bottom of the pot or over a projection from the bottom of the pot containing the heating element. Perking starts when the small amount of water under the pump is heated, forcing the water up the pump stem and over the spreader plate.

Large amounts of sediment and cloudy coffee in a percolator are caused by an undersized basket, an overloaded basket, or a pumping action that is too fast.

Many percolator baskets are not large enough to hold the quantity of coffee required to brew the number of cups stated by the manufacturer. Coffee grounds take up water and require more room when wet. If the basket is not large enough to allow for this, the grounds will go over the top of the basket into the brew.

The size of the holes in a percolator, or a drip coffeemaker basket, are not the cause of grounds and sediment in the coffee. As soon as the coffee grounds are wet, the grounds themselves act as a filter.

If water is pumped too rapidly, most of the water runs off of the

***Fig. 5.3** Stainless steel automatic coffee maker with indicator on handle to show amount of coffee left, with strength selector, and with automatic keep-warm element. (Toastmaster)*

top of the spreader plate rather than going into the basket and dripping through the coffee. This results in a weak brew.

In using the reheat setting, the basket and tube in some percolators must be removed to keep from reperking. In some percolators, an additional resistance is introduced into the circuit after the perking has been completed, which just keeps the brew hot but the temperature too low to resume perking. In some percolators, the serving temperature is maintained by cycling of the heater unit to maintain a temperature lower than the perking temperature.

Coffee flavor is affected by the accumulation of oils in the interior of the coffee maker and the spout. The interior should be carefully cleaned each time it is used with a dishwashing detergent. Most coffee makers are nonimmersible for cleaning.

Vacuum Type

A vacuum coffeemaker usually has a higher wattage than a percolator because all of the water must be hot before it can rise into the upper bowl where the coffee grounds are placed. The heated water is forced into the upper bowl where it mixes with the coffee grounds and brews for proper time.

Fig. 5.4a Electric drip coffeemaker. (West Bend)

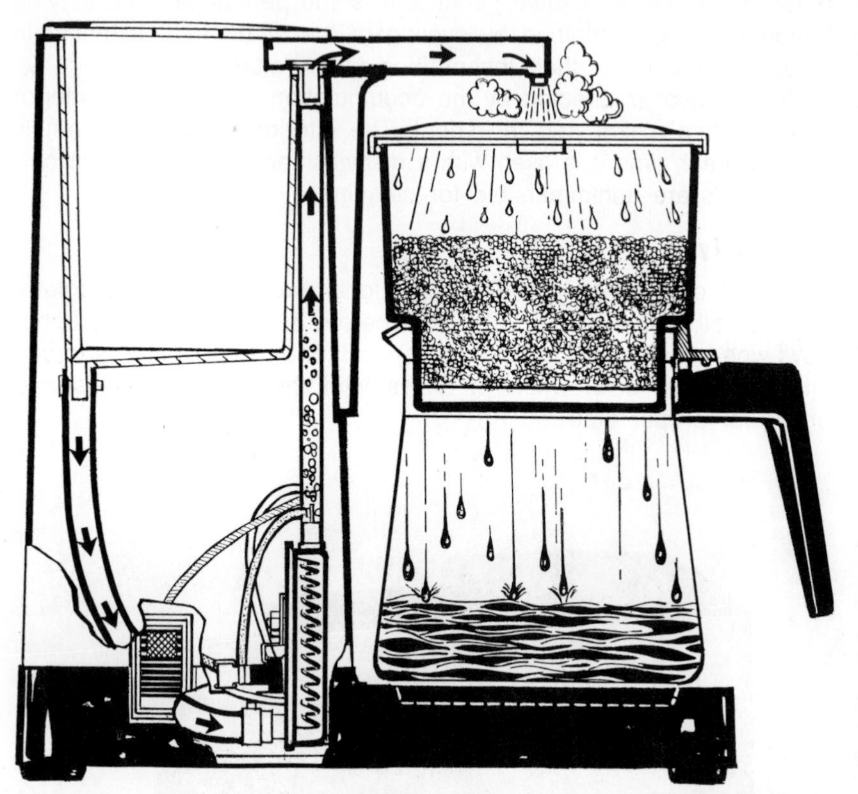

Fig. 5.4b Gravity feeds water from the reservoir to the internal tank heater. Water is heated and forced up the tube and out the dripper spout. The spreader distributes water over the ground coffee through the permanent polyester filter and into the glass carafe. The warming plate keeps coffee serving hot. (West Bend)

There are several types of filters available on the vacuum types of coffeemakers. Filters may be of cloth, stainless steel with very fine openings, or a glass rod. The glass rod is the easiest to remove and wash. The cloth type must be washed after each use to prevent the possibility of imparting an undesirable flavor to the coffee.

Drip Coffeemakers

At the present time there are only a few companies making electric drip coffeemakers. In the one shown in Figure 5.4, the water goes through the coffee grounds only once. It has an eight cup capacity.

***Fig. 5.5** Four-slot toaster with dual controls, two elements for each toast slot, two-pole safety switch, and voltage compensator. (Toastmaster)*

TOASTERS

Well-type or Upright

Most toasters have vertical slots surrounded on two sides by vertically mounted heating elements. One-slice, two-slice, and four-slice toasters are available. The four-slice toasters may be long like a pair of two-slice toasters placed end to end, or may be wide like two-slice toasters side by side. These four-slice toasters sometimes have dual controls (Fig. 5.5). There is also a "slimline" toaster which has a long single slot which toasts two regular slices of bread in tandem or an extra large slice of round-loaf bread.

In automatic toasters there are several types of controls to determine the lightness or darkness of the toast. Some cut off the heat when a certain temperature is reached in the toaster well. Some have radiant control which depends on the heat reflected from the bread itself. This latter type adjusts to the moisture content of the bread. Some use timing only.

Some toasters have levers which must be pushed down to start the toasting and which automatically reject the bread when it is toasted.

Some have levers which must be pushed down to start the toasting and must be raised manually. In some toasters the weight of the slice of bread lowers the bread in the well and starts the toasting. At the end of the toasting period, the bread rises automatically. The bread can only be released before the toasting is completed by changing the position of the browning regulator. A manual release is handy to have in case the bread jams or toast is becoming too brown.

One manufacturer is featuring a toaster that toasts electronically. Toasting is done by a tubular quartz element. A thermistor signals an electronic control to lift the bread when the amount of moisture removed during the toasting process indicates that the color desired has been reached. The opening is adjustable to accommodate sweet rolls, waffles, and thick sandwiches.

Some toasters have a reheat setting which controls an auxiliary heating unit to warm cold toast.

Controls should be located where they are easy to read and to manipulate. Voltage controls, which are found on some toasters, are useful in obtaining better browning if the voltage supplied is consistently different than the voltage requirements shown on the name plate of the toaster.

In well-type toasters, bread should be ejected high enough to remove small pieces of bread. If a toaster is equipped with a single-pole switch, and a fork, used to remove a piece of toast, comes in contact with the heating element, the user will receive a bad shock **if the toaster is plugged into an outlet whether the toaster is on or off.** When the toaster is equipped with a double-pole switch there is no danger of shock **if the toaster is in the "off" position.**

The width of the wells determines the thickness of the bread which can be toasted. Breads with icings or fillings which will melt when heated should not be placed in this type of toaster. A fire can result from toasters gummed up by drippings from filled pastries. Metallic foil-wrapped foods create a shock hazard in this type of toaster when the metal foil touches the electrically live elements.

Crumb trays should be readily accessible for cleaning, and the crumbs should be removed from the tray frequently.

The toaster should have nonscratch feet, and provision should be made on the toaster to deflect heat from the table top.

High wattage gives quicker toasting.

Gate-type

Gate-type toasters have two bottom-hinged doors for holding and turning the bread and one center heating element. Two slices of bread are toasted, first on one side, then on the other. Most are manually controlled and are frequently preferred by people who like to control the degree of dryness of their toast. Some have thermostatic heat controls which shut off the current. The bread must be turned manually by the operator.

Oven Toasters

One reflector-type oven toaster has the heating element at the back of the oven. The top and bottom of the inside of the oven are highly polished to reflect heat to both sides of the bread at one time. The reflecting surfaces must be kept very clean or the toast will not brown evenly. Only a nonabrasive cleaner may be used. An oven-type toaster will take bread of various sizes and shapes as well as sandwiches. The grill size is large enough to hold two slices of ordinary bread. Care must be taken to avoid burns, since the user must reach into the toaster space to retrieve the toast.

Combination Oven and Toaster

One oven-toaster combination has two sealed-rod heating elements, one above and one below the grill, and can toast on both sides at once. The door opens and the shelf moves forward when toasting is completed, so that the hand does not reach into the oven. This appliance can toast outsized food and sandwiches. As an oven, its size limits its use. It is too small for conventional cake and pie pans, and too small for standard TV dinners. It takes twice as long to bake a cake in this oven as it does in a conventional one.

Built-in

These toasters are mounted into a wall between studs. They tilt out to make toast and fit flush with the wall when not in use. They can be lifted out and carried to the table if desired.

FRYPANS OR SKILLETS

Electric skillets are available in large and small sizes in both square and round shapes. Square-shaped skillets hold more food than round ones—a 10-inch square will give one-quarter more area than a 10-inch round. Wattages vary from 800 to 1500 watts. The higher wattages allow the skillet to heat more quickly and recover temperature more quickly when in use. The lower wattage skillets should be considered by purchasers who have limited circuit capacity in the area in which they plan to use the skillet.

Some electric skillets have permanently attached heat controls built into the handle or in the base below the skillet well. These are generally nonimmersible. Other skillets have the heat controls attached to the cord and the skillets may be immersed in water for cleaning (Fig. 5.1). The detachable heat control can be used on other utensils from the same manufacturer. This heat control is usually priced separately from the skillet. One manufacturer makes a heating unit in glass ceramic on which a glass ceramic food container is used as a skillet or a Dutch oven.

Covers for electric skillets are frequently sold separately. Some manufacturers offer a choice of glass or metal covers. Some have a choice of a dome shape or a shallow cover. The dome allows roasting of fowl or large roasts of meat (Fig. 5.2). On some skillets, lids may be anchored in a tilt position. This frees both hands when checking the progress of cooking or the turning of meat and makes it unnecessary to find a place to put the hot cover.

Some manufacturers offer a dome-shaped cover with an incorporated broiler. The broiler cannot be used at the same time as the unit in the skillet (Fig. 5.2).

Some covers are vented to allow steam to escape during cooking and also at the end of the cooking time for crisping. The vent should be located away from the cover knob, so that the hand of the user will not be burned from the steam when removing the cover.

The heat control on a skillet should have a wide temperature range—from 150° to 480°F, to make the skillet usable for holding or keep-warm temperatures to the highest temperatures used in surface cooking. The markings should be legible and easily visible when the control is attached. There should be a signal light to tell when the temperature is reached. (Fig. 5.1).

The feet on the utensil should not mar a table top, and the heat given off should not be high enough to affect a table finish.

An electric skillet may have one or two handles. If it has a single handle, it should be of the right length and shape so that the skillet feels well balanced when it is filled with food. An auxiliary handle opposite the long handle will aid in carrying the appliance when it is hot and full. When an electric skillet is equipped with two side handles it is sometimes designated as a "buffet" server.

Some electric fry pans have a built-in warming tray beneath the cooking surface which makes it possible to keep foods warm while other food is being cooked in the skillet (Fig. 5.2).

Skillets are available with or without Teflon coatings.

PORTABLE OVENS AND BOILERS

Portable ovens and broilers come in a variety of sizes and wattages. Many are equipped with thermostats. Some have a unit for baking and one for broiling. One has a single unit that serves for baking and broiling; the oven itself is turned upside down to broil. If a portable oven size is chosen which will accommodate 8-inch size cake pans, pie pans, and casseroles, they may be adequate for baking needs for one or two persons. They preheat rapidly and may be used where there is an electric outlet which will carry their rated wattage. They use less electricity than a standard range oven and are easier to clean. The sides, top, and back of at least one manufacturer's oven have a catalytic porcelainized finish similar to that found in standard ovens using the continuous type of cleaning. Cleaning takes place continuously while baking (Fig. 5.6).

They are very well suited to heating frozen dinners and casseroles, pies and coffeecakes. Many baking and broiling ovens are priced under $30.

Portable ovens and broilers should be used on surfaces protected by asbestos-lined metal mats.

***Fig. 5.6** Combination portable oven, broiler, and toaster. Top, back, and sides clean continuously while baking or roasting. Glass window is removable for cleaning. (Toastmaster)*

Fig. 5.7a Broiler designed to cut down on spatter and smoke. (Farberware)

ROTISSERIES

On a rotisserie, foods are cooked as they turn on a motor-driven spit. A rotisserie may also be used for broiling. Some, which have a second heating unit in the bottom and a means of closing the spit opening, may also be used for baking. Some of the models have a thermometer on the end of the spit to indicate the interior temperature of the meat. Some have a timer. There should be a switch that turns off the motor when the rotisserie is being used for other cooking. Special attention should be given to the distance from the spit to the bottom, top, and back of the compartment. These distances determine the size of the fowl or roast that can be accommodated. Some units will take a fowl no larger than a small cornish hen or a small rolled roast. If larger items are to be cooked, the spit itself should be strong and have sturdy supports.

Most rotisseries should be used on a surface that is protected with an asbestos-lined mat.

Most rotisseries are difficult to clean, and this factor should be kept in mind when purchasing; one should be chosen that promises to give the least difficulty. Look for easy accessibility to all parts of the compartment, smooth surfaces, and few projections on which grease can collect and burn. One combination broiler-rotisserie has made the cleaning process much easier by locating the drip pan

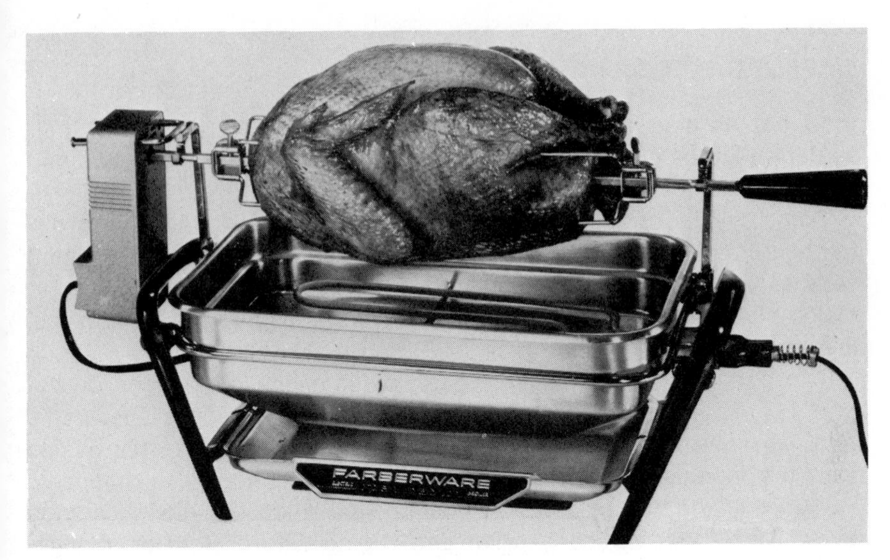

***Fig. 5.7b** Rotisserie attachment to broiler. (Farberware)*

for the juices far enough from the heating element so that juices and grease do not burn on the pan (Fig. 5.7*a* and *b*).

GRIDDLES

An electric griddle should be large enough to eliminate frequent refills but not so large that it is awkward to handle or heats unevenly at the corners and around the outer edge.

Electric griddles are usually made of heavy-gauge aluminum or stainless steel with a heat-conducting core. Some have special finishes on the cooking surface to minimize sticking.

A grease drip container that holds 6 ounces and is large enough at the top to insert a spoon for basting or for removing grease from the container is desirable.

A detachable heat control makes the griddle easier to wash, since the griddle can be immersed in water for cleaning.

Heat-resistant handles should be large enough to keep the hand from being burned when the hot griddle is moved. They should be located so that the griddle is easily transported.

The feet on the griddle should not mar the table top, and the heat given off toward the table should not be high enough to destroy the table finish.

WAFFLE BAKERS

Most homes have had a waffle baker for many years as one of their seldom-used electrical appliances. However, waffle makers have been rediscovered recently.

Some waffle grids have configurations in curves and swirls different from the even projections in rows that are usually associated with waffles. Emphasis is being put on the use of the waffle baker to make party snacks, sandwiches, cornbread, biscuits, drop cookies, French toast, and gingerbread.

The depth and distance from each other of the projections on the grid is related to the crispness of the waffle produced. The waffle grid with deep, far-apart projections usually gives a less crisp waffle than the models with more shallow or closely spaced projections.

With the addition of a Teflon coating, the grids are easily cleaned by soaking in warm soapy water and wiping with a cloth or sponge. For satisfactory results, even with the Teflon coating, seasoning is necessary. Seasoning requires the user to brush unsalted oil or shortening over all grid surfaces and heat. The first waffle is then discarded.

Waffle irons may be square, oblong, or round, and they vary in size. A guide to the size desirable for a family might be to allow the number of sectional waffles made at one time equal the number of servings desired at one time.

The versatility of the waffle iron is increased as the length of the expansion hinge is increased, since this distance determines the thickness of products cooked between the grids.

Most waffle irons have the cord connection in the lower grid section. The heating element wire for the upper grid goes through the expansion hinge between the upper and lower units. These wires should be protected from hinge wear and also from possible fat drippings from the upper grid when upright.

There should be an overflow rim around the lower grid to catch overflow or spill.

Most waffle bakers have an indicator light that tells when the waffle iron is ready for baking; some have adjustments for light, medium, or dark waffles.

The feet on a waffle iron should be mar proof and the amount of heat directed toward the table top should be low enough so that the table top finish is not affected—usually a temperature of less than 175°.

COMBINATION WAFFLE BAKERS AND GRILLS

Waffle bakers and sandwich grills may be purchased separately or as a combination appliance. Most of these combinations have waffle grids which may be turned over to expose flat surfaces for use as a grill. A few have separate sets of grids, one for waffles and one for a grill. Many combinations have Teflon coatings on both the waffle side of the grid and on the grill side. Some have Teflon only on the waffle side. The question arises as to the safety of exposing the Teflon coating to the temperatures of the hot element when both sides of a grid are coated. Underwriters' Laboratories permit a temperature of 752°F on thin coatings of Teflon. What the temperature of the coating is when exposed to the hot elements is difficult to determine. However, if the waffle baker and grill have the UL seal, one would expect that it would have met this specification.

The grill on a waffle baker can be used for sandwiches, bacon, eggs, hamburgers, and other frying jobs. The baker should have a place to take care of excess fat in cooking. Provisions should be made on the upper unit to provide protection for the table top when the grid is lowered to the horizontal position when used as a grill.

DEEP-FAT FRYERS, DUTCH OVENS, SAUCEPANS, AND LOW-WATTAGE COOKERS

Frequently a deep-fat fryer can be used as an all-purpose utensil —fryer, casserole, Dutch oven, corn popper, and food warmer. It should have a thermostat with settings from 150° to 450°F to provide controlled heat for the various uses.

A fryer should have provision for emptying the fat. A spigot is sometimes provided by the manufacturer, but these frequently become clogged and are difficult to clean. A pouring lip at the top of the utensil makes emptying easier. Fryers are usually fairly large in diameter and deep enough to permit immersing of food without danger of the fat overflowing during the cooking. It is helpful to have a marking on the fryer to indicate the full-load limit of fat.

The fry basket should have enough openings of sufficiently large size to permit the fat to drain readily. The handle on the basket should be strong and securely fastened. There should be some provision to anchor the basket on the side of the fryer in the draining position.

Electric Dutch ovens or saucepans offer cooking capacity in addition to the range and have the advantages of mobile utensils. These are probably most suited to cooking operations which require low, moist heat for longer periods of time such as pot roasts and baked beans.

Very low-wattage (70–300) cookers have recently enjoyed considerable popularity. Stews prepared in these may take 10 to 12 hours to cook. Because of the low wattage and consequently the low-heat input, the temperature of the food ingredients rises very slowly. If any pathogenic bacteria should be present in the food when it is placed in the cooker, they could develop rapidly before the cooking temperature is reached, and the ingredients may not be at cooking temperature long enough to destroy the bacteria.

ELECTRIC MIXERS

Portable Types

Portable mixers come in a wide price range and with a wide difference in their capabilities. The lower-priced and lower-wattage models are not powerful enough to mix anything but light batters. The higher-priced and higher-wattage models will handle all but the thickest mixtures. The lower-priced models are seldom built to give full power at each speed. Food mixtures may demand full power but not full speed. The mixer, to qualify as an all-purpose mixer, should have full power at all speeds.

Portable mixers can be hung or kept near at hand without taking up valuable counter space. They can blend, stir, mash, mix, cream, beat, or whip ingredients in a saucepan or a mixing bowl. They must be held in the hand during operation unless they come equipped with a table stand. Most have only three speeds but a few have continuous variable speed controls. More important than the number of speeds is the suitability of the available speeds for the various jobs the mixer may have to perform.

The controls should be located so that the hand that holds the mixer is able to turn the mixer on and off or regulate its speed with the thumb.

The beater should feel balanced when held in the hand with the motor running. The handle should feel comfortable. There should be adequate clearance for the knuckles.

Heel rests should be positioned so that beaters are poised to drip their clinging contents back into the bowl.

The larger the beaters on the mixer, the deeper the mixture that can be handled. The beater should be made of heavy gauge metal that will not rust. A locking mechanism makes it easy to insert and remove beaters.

Portable mixers do not match the performance of a good table model.

Table Models

A stand or table model electric mixer with higher wattage ratings will cope with prolonged mixing of light and heavy batters. Some table models can be removed from their stands for portable use. Some have accessories available such as a blender, chopper, slicer, shredder, or juicer. The better table models have governor-controlled motors to help them maintain their speeds under varying loads.

A table model should handle household mixing jobs with little assistance from the operator other than the addition of the ingredients to the bowl. The bowl and the beaters should be designed to mix ingredients without the necessity of the operator constantly scraping the sides of the bowl. The bowl should be large enough to handle larger volume recipes, such as egg whites for an angel food cake. A mixer is more convenient to use if ingredients can be added to the bowl without stopping the motor. Most table models have two beaters, but one model has only one beater.

Most table models come with one large and one small bowl of heat-resistant glass. Stainless steel bowls for some are available at extra cost. The bowl platform is adjustable to accommodate the different sized bowls. This also allows space for adding ingredients. One table model mixer has only one large glass bowl and, therefore, has no need for platform adjustment.

A timer on a mixer would be a useful addition to facilitate directions common on cake mixes which specify a certain number of minutes of mixing at a certain speed.

Most table models weigh between 12 and 14 pounds. The detached power head, when used as a portable, weighs about 6 pounds as compared to a little over 3 pounds for the heaviest portable type.

One table model weighs about 18 pounds. This model has only one mixing bowl and one whisk type beater which revolves around the interior of a stationary bowl while rotating. The head is not detachable.

The rated wattage shown on the nameplate of the mixer is an indication of the capacity of the mixer to handle heavy mixtures.

BLENDERS

A blender is not a substitute for a mixer. A blender will puree, mix and blend heavy sauces, make mayonnaise and salad dressing, liquify fruits and vegetables in water, aerate fruit juice, remove lumps from sauces and puddings, make crumbs, chop and grind nuts, reconstitute dry milk, and mix drinks. A blender will not whip egg whites, crush ice without a separate ice crusher attachment, chop raw meat, mash potatoes, nor extract juice from fruits and vegetables.

A blender consists of a lower section containing the motor (and, in one case, a heating element) and an upper section which is the food container in the very bottom of which is a four-bladed beater mechanism. In all but one blender, the food container is smaller at the base than at the top so that the blades make contact with the material. The beater blades are made of rigid, acid-resisting material with cuttings edges. The containers are of heavy or lightweight glass, stainless steel, or molded plastic. The capacity of the containers varies. The usable capacity is less than the liquid capacity of the container because of the space needed by the contents when the blender is operated at high speed. The food container should have a firm connection to the base so it will stay in place and not vibrate unduly during the blending operation.

The wattage indicated on the nameplate of the blender is an indication of the load it will carry—the higher the wattage, the more power. The number of motor speeds varies from 2 to 12. Motor speeds vary from 1850 rpm to 17,000 rpm. One blender has seven speeds: 7600 rpm, 9500 rpm, 11,000 rpm, 12,300 rpm, 13,700 rpm, 15,400 rpm, and 17,000 rpm. Some blenders are also equipped with timers.

One blender is provided with a heating unit in the base. The base of this food container unit is wider at the bottom instead of at the top as in other blenders. The cutting blades extend to a larger diameter and revolve at low speeds when compared with other blenders: for blending, 1850 rpm, 2350 rpm, 2750 rpm, 3100 rpm, 3375 rpm; for stirring while cooking, 150 rpm to 240 rpm. This blender was designed to stir and cook as well as to blend. A rotary switch can select "stir" or "blend" action. A rotary thermostat control regulates the temperature from 100° to 375°F. The wattage of the heating unit is low which makes it relatively slow in cooking.

A blender should have a tight-fitting cover to prevent spattering during mixing. Some blenders have a removable central part of the cover to allow ingredients to be added during blending.

Blenders vary in height from 10 to 18 inches. The taller ones are too tall to fit on a work counter under cupboards.

If the cutter assembly is removable, it is easier to remove viscous mixtures from the container after blending.

ELECTRIC SLICING KNIVES

Most electric slicing knives have two ripple-edged blades held flat against each other which are inserted into a large handle containing a motor, with a switch on the underside of the handle. When the power is on, the two blades move rapidly in opposite directions, creating the cutting action. The operator guides the knife with a minimum of pressure. The effectiveness depends largely on how fast the blades move; 2000–3000 strokes per minute give satisfactory results. One knife comes with two sets of blades, one for carving and one for slicing smaller items. One has an arrangement by which the blade may be turned within the handle to permit convenient horizontal slicing of roasts or cakes.

The size and feel of the handle with the blades in motion should be checked by the person who is going to use the knife most often. If the size and weight of the handle and the size of the hand are not compatible, the knife will be hard to maneuver to produce thin, even slices of meat or other foods, such as tomatoes, lemons, oranges, celery, cucumbers, angel food cakes, or stacks of bread for stuffing. These knives are not suitable for cutting through frozen foods or bones.

The blades should fit firmly together so that no food will be caught between them. The shape of the blade may make it more useful in some operations. A narrow and tapered blade may be an advantage in cutting around a bone when carving. Most blades are of stainless steel; a few have tungsten-carbide coated cutting edges, which should remain sharp for a long period of time. Serrated blades cannot be sharpened but must be replaced when dull. Some blades are released from the handle by pressing together spring clips at the base of the blades; a few have a release latch on the handle which ejects the blades.

The location of the switch guard, which protects the switch from being accidentally turned on and serves as a support for the knife when it is set down, should be far enough from the blade to allow the knife to slice all the way to the cutting board.

A safety lock so that the knife cannot be turned on accidentally should be a major consideration in selection and should be on whenever the knife is not in use. During insertion the blades should be grasped from the sides opposite the sharp serrated edges. The blades should be inserted before the cord is connected to the convenience outlet.

An electrically heated knife for cutting frozen foods with a push-button heat selector and adjustable thermostat to prevent overheating is available.

Cordless Electric Knives

Cordless slicing knives are available which are operated by re-chargeable batteries in the handle of the knife. These come with their own recharging units in which the knife must be kept when it is not in use. The charging unit should not be kept near the sink or where it could be accidentally knocked into the water. Anyone at-tempting to retrieve it might get a severe shock. The knife must be charged for at least 10 to 12 hours before peak efficiency of operation is achieved. The cordless knives are more expensive than the cord-connected types.

The cordless knife can be taken anywhere and has no cord to dangle during the cutting or carving operation.

At least one electric slicing knife on the market can be operated either as a cordless or a plug-in type.

In inserting the blades in the cordless types, particular care should be exercised to grasp the blades at the side opposite the serrated edge and to have the safety lock on.

CORN POPPER

A corn popper can be a very versatile utensil as most college students are well aware. Besides its use to pop popcorn, it may be used to heat canned soups, reheat cooked foods, recrisp potato chips and crackers, warm sauces and warm rolls, and cook or steam puddings. All salt residue should be removed after each use to prevent pitting. The inside of the popper should be washed carefully without immers-ing in water. The outside may be cleaned with a damp cloth and polished with a dry cloth.

TABLE COOKERS

Hostesses are finding table cookery an easy way to entertain. It lends itself to simple menus, adds a dramatic flair, and the hostess never disappears into the kitchen. Among the electrical appliances, handsomely styled automatic skillets, griddles, waffle irons, corn poppers, kabob cookers, table ovens, and electric fondue pots are popular for this method of entertaining. Also available are portable gas-fired butane cookers with controlled heat which use any cooking utensil. These are especially welcome in homes and apartments where electrical outlets are few or the current capacity of the circuits is not enough to carry the high wattages of electrical heating appliances.

In addition to the selection tips outlined in the previous discussion of individual portable appliances, consideration should be given to the attractiveness of the utensil both in color and design so that it will harmonize with the decor of the dining area. The cooking stands should be sturdy and sit firmly on a flat surface. The pots should fit securely and snugly. There should be a tray underneath the stand to protect table tops. The handles should be of heat resistant material and large enough to allow an easy grip without danger of knuckles touching the hot surface of the pan. Like all other appliances, they should have smooth seamless surfaces, rounded corners, and no crevices to harbor food particles.

Directions for the use of particular table cookers are contained in instruction booklets which come with the cookers. These should be followed carefully.

OTHER PORTABLE ELECTRIC APPLIANCES

There are many other small electrical portable appliances on the market. Some of these are portable refrigerators, can openers and knife sharpeners, electrically heated dish-drying racks, and many others too numerous to treat individually here.

BUILT-IN FOOD PREPARATION UNITS

Several companies make power units which may be installed in the counter-top or in the wall with one motor for several food preparation attachments. One company provides a separate cabinet unit with the power unit and all food preparation attachments of their manufacture. The cabinet provides space for the use of the individual appliances as well as their storage when they are not in use (Fig. 5.8).

AUTOMATIC CONTROL PANEL

Frequently a homemaker is unable to use her many portable appliances because the circuits to her kitchen or dining area do not

***Fig. 5.8** Cabinet for power unit and storage space for individual food preparation appliances and attachments. (NuTone Division, Scovill)*

have the capacity to handle the wattages of the heating appliances. By adding one 50-ampere circuit—the same as an electric range circuit—an automatic control panel may be installed which will make it possible to use six appliances simultaneously. The control panel can be installed over the counter and under a cabinet. Each of the 15-ampere circuits is equipped with a circuit breaker. Most control panels include a clock for automatic cooking and a minute minder.

USE AND CARE

A few general rules should be followed in using small electrical appliances.

Read the name plate. Know the wattage required. Connect only to a circuit which has the capacity to carry this wattage. Usually only one heating appliance can be connected to a 15-ampere circuit at one time. Know the voltage required. Do not attempt to use an appliance designed to operate on 110–120 volts on a 220–240 circuit. Circuits in many foreign countries for all appliances are over 200 volts.

Read the instruction booklet before using the appliance and then keep it in a handy place for future reference.

Do not use extension cords when using heating appliances.

Do not operate electrical appliances when hands or body are wet. All electrical appliances offer the possibility of shock. Therefore, precautions should be taken in the operation of electrical equipment to eliminate grounding through the user's body. Many appliances may ground through the water pipes when used near the kitchen sink. A piece of equipment with a third prong in the plug and a ground wire in the cord calls for three-prong receptacles which must be especially grounded by an electrician. A toaster or a broiler with an open-coil heating element should not be grounded because of the possibility of the person touching the heating element and the metal shell at the same time with a fork, thus completing the circuit through the body. Coffee cup heaters with immersible heating coils can produce shock if the finger is dipped into the heating water.

Do not immerse electrical appliances in water unless they have sealed units and the manufacturer says they can be immersed.

Experiments and Projects

1. Toast bread from the same loaf in available toasters, using as nearly comparable settings as possible. Compare the evenness of browning and the crispness of the toast. Compare the ease of cleaning the toasters. Compare the time to toast on each setting of each toaster. How do the wattages of the toaster compare?
2. Quick raisin-bread pudding: Toast 3 slices of raisin bread. Spread lightly with butter or margarine. Cut toasted bread into ½-inch cubes; divide among six 6-ounce custard cups. Mix together 3 eggs, ⅓-cup sugar, ½-teaspoon vanilla extract. Stir in one 13- or 14½-ounce can of evaporated milk and 1-cup boiling water. Pour mixture over toast in custard cups. Place custard cups in an electric skillet. Carefully fill the skillet with hot water to within ½ inch of top of custard cups. Set temperature at 210 degrees.

Place the cover on the skillet. Simmer for 15 minutes or until knife inserted in the center comes out clean. Remove cups and cool on wire rack.

3. To compare a waffle baker and a frypan in making French toast. If available, use Teflon coated and uncoated appliances. Cook French toast in waffle maker and in a frypan. Use sliced sandwich bread. Trim top and bottom crust from bread. Mix 2 beaten eggs, ¼-cup milk, ¼-teaspoon salt, 1-tablespoon melted margarine. Dip each slice completely in mix and place one slice in each section of a preheated waffle maker. Repeat, using preheated frypan. Compare desirability of methods, attractiveness of products, and ease of cleaning.
4. To compare a griddle and a frypan in making pancakes: use Teflon coated and uncoated appliances, if available. Make plain or blueberry pancakes from a packaged mix, following the directions on the carton. Bake pancakes on a griddle and in a frypan. Compare the ease of operation, the evenness of browning, the ease of cleaning the appliances.
5. Meal in a skillet: preheat an electric frypan to 350°. Saute (300°) ½-pound pork sausage and ½-pound ground ham until sausage begins to brown but is still light in color, stirring frequently. Sprinkle with ½ teaspoon ground juniper berries. Lower heat to 250°. Add 1½-cups sauerkraut (drained and rinsed) and 1¼ cup sour cream and cook 15 minutes. Sprinkle lightly with salt and pepper to taste. Then sprinkle with paprika and then with ¾-cup garlic croutons.
6. To determine the effect of low voltage on the operation of a heating appliance, connect a variac, ammeter, and wattmeter in the circuit with an electric skillet. Set the variac for the voltage given on the nameplate. Check the time required to preheat the skillet. What was the current reading? Cool the skillet to room temperature. Set the variac 10 volts lower than used in the first part of the experiment. Check the time required to preheat. Compare with the previous time. What effect did lowered voltage have on the current?
7. Bake one 8-inch layer of a white cake mix in a preheated portable oven, and one layer in a conventional range oven. Compare time to bake, current consumption, quality of results.
8. Bake a frozen fried chicken TV dinner in a portable electric oven and another in a conventional oven following the package directions. Compare the results. Compare the energy consumption of each.

9. Low calorie topping: place 2 tablespoons of water and ½ cups of large curd cottage cheese in a blender. Operate the blender until the mixture is the consistency of sour cream. Add 2 tablespoons of lemon juice. Blend.
10. Crumb topping: place 2 slices of lightly buttered toast in a blender. (Break the toast before adding.) Operate the blender until the crumbs are as fine as desired.
11. Vichyssoise: blender chop 2-cups peeled, cubed potatoes with ¾-cup leek pieces. Place in saucepan and cook in 2 cups of chicken broth until potatoes are tender. In the covered blender container, blend 1-cup milk, 1-teaspoon salt, dash of white pepper. Add cooked mixture. Process at "puree" until smooth. Stir in cream. Chill thoroughly. Garnish with chopped chives.
12. Chocolate-peanutbutter shake: in a covered blender container, blend 2-cups chocolate-flavored milk, ½-cup creamy peanut butter and 1 pint of vanilla ice cream until smooth.
13. Compare a portable mixer with a stand-type mixer in making a basic, uncooked butter frosting. Cream ⅓-cup soft butter or margarine on medium speed. Add 3-cups sifted confectioners sugar, dash of salt, ¼-cup milk and 1½-teaspoons vanilla extract. Beat on low speed until blended, then on medium high speed until fluffy. Compare length of time required and effort required on part of the operator of the mixer.
14. Stovepipe bread: mix for ½ minute at low speed in large bowl of an electric mixer, 1½-cups flour and 1-package active dry yeast. Heat until just warm, ½-cup milk, ½-cup water, ½-cup salad oil, ¼-cup sugar, 1-teaspoon salt. Add to dry ingredients and beat on low speed until blended. Add 2 eggs and beat on medium-low speed until blended. Turn speed to low and add 1-cup flour. Beat until smooth. Remove bowl and stir in 1-cup flour with a spoon to make soft dough. Spoon batter into two well-greased 1-pound coffee cans. Cover with plastic lids and let stand in a warm place free from drafts. When dough has risen almost to top of cans, remove lids. Bake in a preheated 375° oven 30 to 35 minutes or until browned. Let cool about 10 minutes in cans before removing to cooling racks.
15. To learn why motors heat when load is increased, connect a wattmeter, voltmeter, ammeter and variac in a circuit with an electric mixer. Record starting current and wattage. Change the speed. Record current and wattage readings at each speed setting. Place 1½-cups flour and ⅓-cup water in the mixing bowl and mix at medium low speed until batter becomes quite stiff,

recording wattage, voltage, and amperage at 30-second intervals. From difference in readings, explain why motors get hot when load is increased.

16. Hungarian stew: put 6 slices bacon in bottom of cooker of deep fat fryer; turn dial to 325° and cook until bacon has rendered most of its fat. Remove bacon and reserve. Pour all but a thin film of fat out of the cooker. Saute 1-cup minced onions, stirring occasionally until limp. Add 1-clove garlic and 1-cup sliced carrots; saute briefly. Return reserved bacon to cooker. Stir in ¼-cup wine vinegar, 2-cups of beef stock and 2½-pounds boneless stew beef (cubed). Season lightly with salt and pepper. Bring mixture to a boil, turn cooker dial down to first M of Simmer. Cover and cook 2 hours or until beef is nearly tender. Turn heat up to 300°. Stir in ¾-cup uncooked rice and one green pepper (sliced), and 1 additional cup of beef stock. Bring to a boil; turn down to R of Simmer; cover and cook 20 to 25 minutes or until rice is cooked. Add more stock, if necessary.
17. Doughnut puffs: Cream 2-tablespoons butter and ½-cup sugar together in the large bowl of an electric mixer at medium high speed. Add 2 eggs and beat well. Sift together 2-cups flour, 2-tablespoons baking powder, ½-teaspoon salt, ¼-teaspoon nutmeg, ½-teaspoon mace. Add flour mixture and ½-cup milk alternately, mixing to make a soft, smooth dough. Chill dough thoroughly. Heat 2-cups salad oil in cooker to 375°. Drop dough by rounded spoonfuls into hot fat, dipping spoon into hot fat each time before putting into batter. Fry until golden brown, turning to brown evenly, about 3 minutes. Lift basket to drain supports. Drain. Put puffs on paper towels. Roll in sugar before serving.

CHAPTER 6
Temperature and Heat

TEMPERATURE MEASUREMENT

Just how reliable are our bodily sensations of heat and cold? We may feel chilly at 80°F in a dry steam-heated apartment and yet feel too warm at 80°F in humid summer weather. In a moist atmosphere we may be comfortable at 65°F and in a dry room feel quite cold at 65°F. If seated in a room in the path of air from a fan, we complain of feeling cold although the temperature of the room is 85°F.

If we touch objects of different materials in a room at 70°F, a metal object will feel cooler to our hand than a wooden object although both are at the same temperature as the room.

Our bodies are usually much warmer than the surrounding air and are cooled by evaporation, radiation, conduction, and convection. Evaporation is greatly reduced in high atmospheric humidity. When we are in direct contact with solid objects, the sense of touch does not register temperature but the rate at which heat is conducted from or to the hand.

We may **see** differences in the higher ranges of temperature by observing color differences. When heated, an electric unit on a range becomes red. An electric coil in a room heater at a high temperature is yellow, whereas the tungsten filament in our incandescent lamps, which are approximately three times as hot as the room heater, appear white. These visual means of differentiating temperature, however, are only possible in the higher temperature ranges.

Therefore, our bodily sensations and our sense of touch and sight are certainly unreliable means of measuring temperature.

Thermometers

Our present methods of temperature measurement are based on the fact that inanimate objects, almost without exception, expand and contract with changes in temperature. The two fixed points that have been selected between which expansion can be measured have been those of melting ice and boiling water. The temperature of melting ice is constant under all conditions. The temperature of boiling water depends on atmospheric pressure. However, the variations for pressures on boiling points can be readily tabulated for reference.

To calibrate a thermometer, the bulb is first placed in a mixture of ice and water and the height of the mercury column is marked on the side of the stem. It is next placed in steam just above boiling water and again marked. These two marks then determine the end points for whatever scale is to be used.

Two temperature scales widely used in the United States have been the Fahrenheit scale and the Celsius (Centigrade) scale. The Fahrenheit scale has been commonly used in engineering work. The Celsius scale is used in all countries for scientific measurement. Legislation has been introduced in the Congress of the United States to change the present system of measurement to the metric system

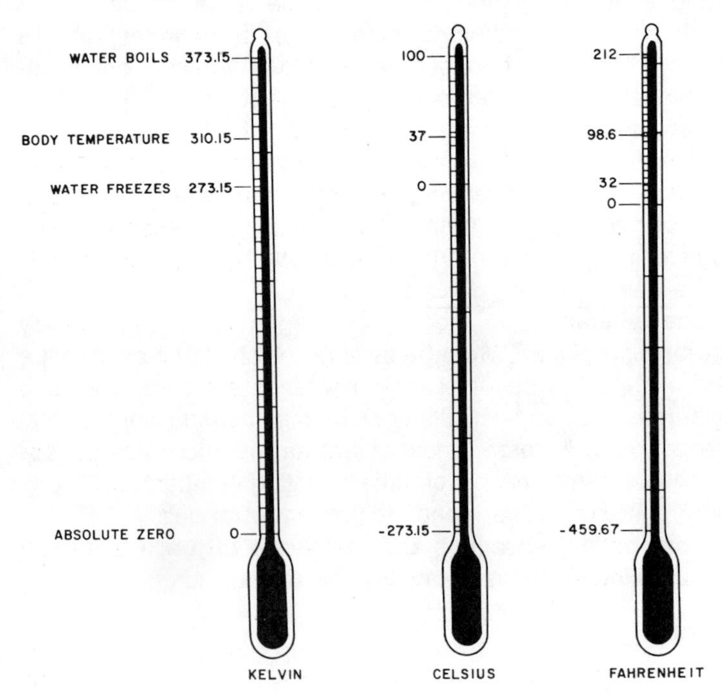

***Fig. 6.1** Relationship of Kelvin, Celsius, and Fahrenheit temperature scales.*

which is now used in all of the larger and more developed countries of the world (Chapter 20).

The Kelvin scale is used to measure temperature in the metric system. The Kelvin scale has its zero point at absolute zero (the point at which all atomic vibration ceases) and a fixed point (273.16 kelvins) at the triple point of water (the temperature at which water exists in all three states: vapor, liquid, and solid).

The Celsius and Fahrenheit scales are derived from the Kelvin scale. The relationship of the three scales is shown in Figure 6.1. Between the temperatures of melting ice and boiling water there are 180 degrees on the Fahrenheit scale, compared with 100 degrees of the Celsius and Kelvin scale. The ratio of these numbers is 9:5. This comparison shows that a temperature rise of 9°F is equivalent to a rise of only 5° Celsius or Kelvin. A temperature of 90 degrees Fahrenheit is 32.2C ($C = \frac{5}{9}(F - 32)$; $C = \frac{5}{9}(90 - 32)$, $C = \frac{5}{9}(58) = 32.2°C$.) Conversely, to change Celsius to Fahrenheit: $F = \frac{9}{5}C + 32$.

Mercury thermometers are not reliable for highly accurate scientific work because of the inconstancy of the rate of expansion of both mercury and glass. Gas thermometers are standard thermometers against which others are checked and calibrated. Gas can be kept at constant volume, and variations in pressures as temperature changes can be measured accurately. Thermometers utilizing the electrical properties of metals are also used for accurate scientific work.

Thermostats

Most present-day household heating appliances are thermostatically controlled. The thermostat is a device for maintaining a fairly constant temperature and so tends to eliminate watching and overheating which may cause a fire hazard.

Types of Thermostats

Two types, the hydraulic and the bimetal strip, have been used for a good many years; the probe tube is of recent application.

Hydraulic. Temperatures of gas and electric ovens and now, also, of certain controlled range surface units and burners, and the temperatures of the freezing coils in refrigerators and freezers are automatically regulated by hydraulic thermostats.

The thermostat has two parts; the dial, graduated in degrees of temperature and located in a place easily accessible for manual setting, and the connecting copper tubing, which ends in a copper bulb filled with a heat-sensitive liquid. This liquid expands when the temperature rises, and contracts again when the temperature falls. As it expands, the liquid presses against a bellows or diaphragm located in the switch, causing the switch to open the circuit at a predetermined dial setting in the case of the range oven and surface unit. Of course, if the equipment controlled is a refrigerator or freezer, the action of the liquid will close the circuit instead of opening it. When the given area cools, the liquid contracts, and recloses or re-opens the circuit, depending on whether it is the range or the freezer that is being controlled.

Certain refrigerator thermostats contain a vapor instead of a liquid. The vapor is the same as the refrigerant used in the evaporator and, therefore, has the same temperature. It is more sensitive to changes in temperature than the liquid.

The thermostat bulb may be located along the side or across the back of the oven or, occasionally, against the oven roof. In the controlled surface unit the hydraulic tube is connected with a sensing element in the center of the unit. The sensing element is spring controlled, so that it is always held in contact with the center of the pan, if the bottom surface of the pan is flat. A pan should be of a material that is a good conductor; aluminum is usually recommended. In the refrigerator and freezer the thermostat bulb is placed close to the freezing unit (evaporator).

The thermostat in some gas ranges does not shut the gas off completely; it causes the valve to close partially and so restrict gas flow to the burner. In others the oven burner cycles completely.

Bimetal Strip. In most small electric appliances, such as hand irons, toasters, waffle bakers, and percolators, the temperature is controlled by a thermostat of the bimetal-strip type; and at least one manufacturer has developed a bimetal control for surface units. The bimetal thermostat, similar in construction and action to the circuit breaker, is made of two metal ribbons, each ribbon being of a different material; the ribbons are fastened together at either end. When the temperature rises above that for which the thermostat dial is set, one piece of metal expands more than the other and breaks the circuit connection. As the appliance cools, the metals contract to their original position, and the circuit is again closed.

Probe Tube. This type of thermostat is attached to the cord of a small electrical appliance rather than incorporated in the appliance itself (see Fig. 5.1). By locating the thermostat separately from the appliance, the manufacturer can seal the electrical unit in the utensil, making it waterproof. It then can be immersed in water for washing. The same thermostat and cord can be used in a family of appliances made by a single manufacturer—a frypan, griddle, saucepan, or percolator. If more than one appliance is used simultaneously, more than one such control will be needed.

The probe tube is filled with liquid silicone; the silicone expands during the heating process and activates a piston that exerts pressure against the electric contacts and opens them.

Thermistors. This type of temperature regulating device uses semiconductors in which resistance decreases as temperature rises. They are being used in some controlled surface units and other types of heating equipment.

HEAT

Difference between Temperature and Heat

Temperature is the hotness or coldness of a body measured in degrees Fahrenheit or centigrade (Celsius). **Temperature** is a measure of the degree but not the quantity of heat (Fig. 6.2).

Heat is a quantity—the amount of molecular energy. All matter is made up of ultramicroscopic bodies called molecules. These are at all times in a state of motion. The nature of the motion and the speed

Fig. 6.2 Temperature and Btu.

of motion depend on the temperature of the substance and the state it is in—a solid, a liquid, or a gas. Molecules move most rapidly in gases, less rapidly in liquids, and least rapidly in solids.

Difference in heat and temperature may be illustrated by observing the difference in the amount of time required for a cup of water at 212°F and a gallon of water at 212°F to reach room temperature. Both the water in the cup and the water in the gallon container are at the same temperature. The water in the gallon container, however, will take much longer to reach room temperature than the water in the cup. In other words, there was a greater quantity of heat in the gallon of water than in the cup of water.

In the conventional system of measurement, heat is measured in British thermal units or in calories. A British thermal unit (Btu) is the quantity of heat required to raise one pound of water one degree Fahrenheit. A calorie[1] is the amount of heat required to raise one gram on water one degree Celsius. The two measurements, Btu and Calorie, differ only in the temperature and mass units used. Since one pound equals 454 grams, and one degree Fahrenheit equals 1.8°C, one Btu equals 252 calories (454 divided by 1.8). In the metric system of measurement, heat is measured in joules. A joule is equivalent to 0.24 calories (1 calorie = 4.18 joules).

Quantity of Heat

One of the most significant things which is witnessed daily in every kitchen is the difference in cooking time of different substances. The difference in cooking time of different foods has taxed the management capabilities of the homemaker since cooking began.

Materials for cooking utensils reach cooking temperatures in different lengths of time. A cast iron skillet on a preheated unit may take 1 minute 10 seconds to reach cooking temperature, whereas a cast aluminum skillet of approximately the same size and thickness will reach the same temperature in 30 seconds. Foods, too, absorb heat at different rates. Three cups of oil may be heated to 212°F in 2 minutes 7 seconds, whereas the same quantity of water in the same pan over the same heat requires 4 minutes 12 seconds. Medium-sized potatoes cut in quarters with ½ cup water added may take 40 minutes to become "cooked" whereas apples cut into pieces of approximately the same size, using the same amount of water, and in an identical pan over the same heating element will take ⅓ to ½ the time. It can also be observed that those substances which require a

[1]A kilocalorie, used in food energy determination, is 1000 calories, written with a capital C, Calorie.

long time to heat seem to retain their heat longer than those which may be heated more rapidly.

The time taken by substances to heat up or cool down within a given range of temperature depends on the mass, certain intrinsic properties of the substances themselves, the difference in temperature between them, and the characteristics of their surfaces.

This different capacity of various materials for absorbing heat, disregarding surface differences, is called **specific heat.** In measuring specific heat, water is used as a reference point. Water is given a value of 1.00 and other substances are given values relative to it. A study of Table 6.1 shows why it requires more heat to raise the temperature of a given quantity of water a certain number of degrees than to raise the temperature of the same mass of almost any other substance. At the same temperature, then, pound for pound, water contains more heat than any other common substance. The quantity of heat depends on the mass, the specific heat and the temperature

Table 6.1
Specific Heats[a] (cal per g°C or BTU per lb-°F)

Water	1.00
Ice	0.50
Steam	0.48
Olive oil	0.47
Sugar	0.27
Porcelain	0.26
Air	0.24
Aluminum	0.22
Crockery	0.20
Brick	0.20
Glass	0.19
Asbestos	0.19
Iron	0.12
Chromium	0.11
Copper	0.09
Silver	0.06
Tin	0.05
Mercury	0.03

[a]The specific heat of a material is the amount of heat required to change the temperature of a unit mass of material 1°.

change. The quantity of heat required to raise 1000 grams of water from the freezing point to the boiling point is:

$$\begin{aligned}\text{Quantity of heat} &= \text{mass} \times \text{specific heat} \times \text{temperature change}\\ &= 1000 \text{ grams} \times 1.00 \times 100^\circ\\ &= 100{,}000 \text{ calories, } 100 \text{ kilocalories (Calories)}\\ &= 418{,}000 \text{ Joules, } 418 \text{ kilojoules (kJ)}\end{aligned}$$

There is another way in which substances may absorb heat and, as strange as it may seem, they experience no change in temperature at all. After the boiling point of water is reached, the temperature of the water will not go higher no matter how much heat is applied.[2] The water boils more vigorously but does not get hotter. Therefore, potatoes in boiling water will not cook more quickly by turning up the heat. When cooking any food in water after the boiling point is reached, heat can be reduced to the lowest point to maintain a gentle bubble-boiling, without increasing the cooking time. With high heat and vigorous boiling, the water is changed to steam (from a liquid to a gas), and if the water in the cooking utensil is completely evaporated the potatoes burn. The quantity of heat required to change the water from a liquid to a gas is called **latent heat of vaporization.** To vaporize 1000 grams of water requires 540 Calories. When steam condenses back into water, it gives up the same amount of heat that was put into it when it was converted from water into steam. This condensation in steam radiators is the reason the heat is maintained in them and why a small radiator can heat a large room.

Not only does water absorb heat continually while it is boiling and not get any hotter, but the same is true of ice as it melts. If a large pan filled with ice cubes and water is placed on a surface heating unit and is stirred constantly, the temperature will remain at the freezing point, 0°C, no matter how high the heat, as long as any ice remains unmelted. Most people apply this knowledge even without understanding it. They have observed that a small piece of ice will cool a cup of coffee much more than a considerably larger quantity of ice-cold water. To melt one kilogram of ice requires 79.6 Calories, or 332.7 kilojoules. The quantity of heat necessary to change a unit of solid to a unit of liquid with no change in temperature is called the latent heat of **fusion.**

The reverse of fusion is **solidification.** The amount of heat liberated by matter on solidification is equal to the amount of heat taken in on

[2]Under steam pressure greater than atmospheric, the boiling temperature can be raised.

fusion. Pure water, on freezing, gives up 79.6 Calories per kilogram (332.7 kilojoules).

Evaporation

When water is left in an open saucepan, it slowly evaporates—changes from a liquid to a gaseous state. In air the molecules are far apart. In water the molecules are closer together and exert cohesive forces forming a restraining layer at the surface. High-speed molecules escape from a liquid surface carrying away with them kinetic energy. The average kinetic energy of the remaining liquid molecules is lowered, which means a lowering of temperature of the liquid. The more rapid the evaporation, the faster the cooling.

HEAT TRANSFER

There are three general ways by which heat is transferred. **Conduction** is a process by which heat is transferred through a substance, or from one substance to another that is in direct contact with it, by molecular activity. **Convection** is a more rapid process of heat transfer involving the motion of the heated matter itself from one place to another. **Radiation** of heat takes place at the speed of light (186,300 miles per second) in electromagnetic waves. No material medium is required for its transmission. Heat transfer by conduction and convection are involved in cookery on top of the range. Radiation and convection are of primary concern in oven cookery.

Conduction

Every time a pan of water is heated on a range, heat is being transferred by conduction. Heat from a gas flame or an electric element in direct contact with the pan sets the molecules of the bottom of the pan into rapid vibration. They strike other metal molecules in the material setting them in motion, thus transferring motion through to the other side of the metal. The metal molecules on the inside of the pan set the layer of water molecules close to them in motion, and they in turn set others moving. Molecular motion, called heat, has been given to the water in the pan. Heat has been transferred by conduction.

If we disregard the rules of etiquette we might cite another useful example of conduction. A cup of coffee, too hot to drink, may be cooled by allowing a silver spoon to remain in the coffee. The bowl of the spoon gets hot from the bombardment of the molecules of the hot coffee. The agitation of the molecules in the bowl spreads to the molecules in the handle of the spoon. These set in motion the molecules of the cooler air surrounding the handle. Heat is conducted from the hot coffee into the handle and into the air, thus cooling the coffee.

Some materials are better conductors of heat than others. Some metals like silver, copper, and aluminum are good conductors of heat. Other substances such as glass, wood, or paper are such poor conductors that we use them as heat barriers or insulators (Table 3.1). The materials between conductors and insulators are called semiconductors. Since in top-of-the-range cookery only the bottom of the utensil is exposed to the source of heat, the conductivity of the material of the cooking utensil is of paramount importance if even browning of meats is to be achieved in a skillet, or scorching and burning in spots is to be avoided in mixtures which thicken in saucepans or skillets.

The relative thermal conductivities of some common materials are shown in Table 3.1. Although silver is an excellent heat conductor, it is seldom used in cooking utensils partly because of its cost.

Copper, aluminum, and combinations of copper and stainless steel and aluminum and stainless steel are the best conductors of heat now available in surface cooking utensils. Iron and stainless steel are poor conductors of heat. Cast-iron utensils heat slowly and, because of their mass and poor conductivity, retain heat well. Stainless steel, when used alone in cooking utensils, warps, develops hot spots, and produces uneven cooking results except where it is used primarily to heat water, such as a coffee maker.

The conductivity of the material of which the pan is made is not important in the oven because there is a film of air around each utensil which offers so much resistance to the passage of heat that it nullifies any effects that the high thermal conductivity of the material might have. This is especially true where metals are relatively thin, as they are in baking utensils. Since the area of direct contact of the utensil with the oven rack is very small, there is almost no conduction from the rack to the pan.

Convection

Convection currents are readily observed when a pan of water is heated. The layer of water at the bottom of the pan is heated first by conduction of heat through the bottom of the pan. Because of the rise in temperature, the water expands. Since it is then lighter than the cold water above it, it rises to the top of the pan allowing cold

water to come to the bottom from the sides. Convection currents keep the water in motion while it heats.

Convection currents are also present in an oven of a range. Air is heated by an electric element or a gas burner in the bottom of the oven. It expands and rises to the top of the oven allowing the cooler, heavier air to come to the bottom of the oven to be heated. In an oven, 30 to 40% of the heat is convected heat.

Radiation

When the sun comes over the horizon in the early morning, heat can be felt by a person who is exposed to the rays of the sun as soon as the sun is visible, but the air surrounding the person can still be cold. This heat, called **radiation,** has been transmitted millions of miles in 8 minutes 20 seconds and is absorbed by the human body. No material medium is required for its transmission.

All bodies, whether they are hot or cold, radiate heat. If a cold object is brought into a warm room, the walls of the room radiate heat to the cold object and the cold object radiates heat to the walls of the room. Since the walls are at a higher temperature, they give more heat per second to the cold object than the cold object gives up. The temperature of the cold object, therefore, rises until it comes to the same temperature as the room. It then radiates and absorbs at the same rate. Objects that are good absorbers of radiation are also good emitters.

When houses are heated by systems consisting of heated wall, ceiling or floor panels, radiant heat from the panels keeps the occupants warm on the coldest days even though the windows are open.

The quantity of radiant heat absorbed by a material is dependent on the material itself, the character and color of the surface, the temperature of the radiating surface, the temperature of the surrounding surfaces, and the area exposed to the radiation.

Different materials at the same temperature absorb different fractions of radiant energy incident on them. Glass will absorb about 95% of the radiant heat incident on it. A blackened surface is an excellent absorber of radiant heat as well as an excellent emitter. If the same body is chromium plated, it becomes a poor absorber and a poor emitter. Materials with rough or dull surfaces absorb radiant heat, whereas bright surfaces reflect it. By varying the character and color of the surfaces of baking utensils, the absorption of radiant heat from the oven can be regulated.

When an oven is heated, the bottom and sides of the oven become hot and radiate heat directly to utensils and foods in the oven. In an oven 60 to 70% of the heat is radiant heat. Therefore, the ability of a baking utensil to absorb radiant heat is the important factor in determining its efficiency in utilizing oven heat.

The material and the finish of utensils used in oven cookery affect the characteristics and quality of the food product. By choosing different materials and finishes, different results in baked products can be achieved. In general, glass, glass ceramic and other ceramic materials, enamelware, anodized aluminum, darkened tin-coated ware, cast iron, polyimide coated utensils, and utensils with dark or rough and dull surfaces are good absorbers of radiant energy. Foods baked in these will bake more quickly and form crisp crusts. Utensils with light and highly polished surfaces and new tin-coated utensils reflect radiant energy. Foods baked in these will cook more slowly and have soft crusts. Frequently it is desirable to combine a rough, dull finish (a good absorber of radiant energy) with a shiny surface (a reflector of radiant energy) in different parts of a utensil to produce the desired results. For example: a cake pan with a dull, rough bottom and shiny sides will produce a cake of maximum volume, fine texture, and tender crust. Radiant energy is picked up rapidly on the dull, rough bottom of the pan causing the material in the pan to become heated and cooked. The shiny sides meanwhile reflect the radiant energy allowing the batter to rise and reach maximum volume before the sides harden.

Another everyday application of the effects of radiant energy is the selection of white or light colored clothes for wear in the sun. Black or dark clothes will absorb the sun's radiant heat. White or light clothes reflect it. A person will, therefore, be more comfortable in white clothes on hot days spent in the sun.

Terminology

Temperature. An indication of the magnitude of the heat energy of agitation of the molecules of a substance. Temperatures in household equipment are measured by thermometers, thermostats, and thermocouples in degrees Fahrenheit or Celsius.

Heat. The amount of molecular kinetic energy measured in British Thermal Units or calories.

Quantity of heat = mass × specific heat × temperature change.

Specific Heat. The different capacity of various materials for soaking up, containing, holding or absorbing heat. In measuring, reference is always made to the corresponding property of water.

Latent Heat. The amount of heat required to change the state of a substance, that is, liquid to a gas with no change in temperature, 100°C water to 100°C steam.

Heat of Vaporization. The amount of heat required to change a liquid to a gas. To evaporate one pound of water at the boiling point requires 970 British Thermal Units. To evaporate one gram of water at the boiling point requires 540 calories.

Heat of Fusion. The heat absorbed when a solid melts. The amount of heat required to melt one pound of ice requires 144 British Thermal Units. One gram of ice requires 79.63 calories to melt it.

Heat of Condensation, The amount of heat liberated when gas condenses to a liquid is the same as that absorbed in the production of that gas from the liquid.

Heat of Crystallization. The amount of heat liberated when liquid freezes equals the heat of fusion required to make crystals into liquid again.

Heat transfer

Conduction. A process by which heat is transferred through a substance, or from one substance to another that is in direct contact with it.

Convection. A more rapid method of heat transfer involving the motion of the heated matter itself from one place to another.

Radiation. The transfer of heat from one object to another without necessarily warming the intervening matter. Speed of transfer is equal to the velocity of light. Heat can be felt almost instantly.

Evaporation. The continual escape of gaseous molecules from the surface of a liquid. In changing from a liquid to a gas, heat is absorbed, giving a cooling effect.

Heat flow. Takes place in one direction—from a higher temperature region to a lower temperature region. Heat cannot spontaneously flow from a low temperature region to a high temperature region.

Calorie. A unit of heat. The calorie written with a small c is the amount of heat necessary to raise the temperature of one gram of water one degree Celsius. The Calorie indicated by a capital C is the amount of heat necessary to raise the temperature of one kilogram of water through one degree Celsius.

British Thermal Unit. A unit of heat, the amount of heat required to raise the temperature of one pound of water one degree Fahrenheit—1 Btu = 252 calories.

Experiments and Projects

1. Touch the following objects in a room:
   (a) A wooden surface.
   (b) A metal surface.
   (c) A glass dish.
   (d) A woolen sweater.
   List them in ascending order of feel of coolness.
   Why do these objects feel different in temperature to you?
2. What type of heat is emitted from a 25-watt electric light bulb? (Low wattage light bulbs are evacuated.)
3. Place a pan of water on a range unit to heat. Describe the activity in the pan from the time the first small bubbles form on the bottom of the pan until the water boils. What types of heat transfer take place?
4. Determine positions of thermostat bulbs in ranges and refrigerators:
   (a) In the laboratory.
   (b) In student homes.
5. Preheat an electric range unit for 5 minutes. Select a medium-weight aluminum saucepan that has the same diameter as the heating unit. Place 2 cups of vegetable oil at 70°F in the saucepan. Place a candy thermometer in the oil. With a stopwatch, measure the time required for the temperature of the oil to rise (1) from 70 to 200°F, (2) from 200 to 300°F. Clean saucepan thoroughly. Place 2 cups of water at 70°F in the same saucepan. Place the candy thermometer (from which the oil has been removed) in the water. With a stopwatch, measure the time required for the temperature of the water to reach 200°F. Explain the difference in time.
6. Heat 4 cups of water from room temperature to the boiling point. Pour 1 cup of water into a small bowl and 3 cups of water into a somewhat larger bowl. Put a thermometer in each. Measure the time required for the temperature in each to return to room temperature. Explain.

7. Place 1 cup of peeled, uniformly sliced potatoes and ½ cup water in a covered saucepan. Place on a preheated electric range unit of the same diameter as the pan. With the switch in "High" position, using a stopwatch, measure the time required to "cook" potatoes. Repeat, but after a full boil is reached, turn heat to lowest heat to maintain boiling. Cook potatoes until tender. Compare results, as to time required, quality of food and condition of the range top.

CHAPTER 7
The Household Range

An unknown author has suggested that man may do without a number of things quite conveniently, but not without a cook. Some type of cooking appliance is a household necessity, and the appliance commonly chosen is the range. The importance of selecting a range that is well constructed and that meets the needs of the individual home is recognized but, with ranges manufactured by a number of companies and each company manufacturing several models, often 8 to 10 or more, the problem of selection is at best a perplexing one.

Since the extension of rural electric high-lines and the availability of liquid petroleum gas in tanks to individual homes, even to those homes located at fairly long distances from railroads, electric and gas ranges have been steadily replacing other fuel ranges. They certainly reduce the time required for maintenance and cleaning needed by the range that uses liquid or solid fuel. In this chapter the discussion will be limited to gas, electric, and electronic ranges. The first section will treat certain features and characteristics common to all ranges; the following sections will deal with the particular properties of the individual types.

Types

Over the years the design of ranges has changed rapidly. The early range resembled the wood and coal stove with the low oven and the overhead warming shelf. Then came the high side-oven, which eliminated stooping. This design was known as the high-oven or conventional-type range. With the modernization of the kitchen, the high oven was gradually lowered and the table-top range became the popular design, so designated because the entire top of the range

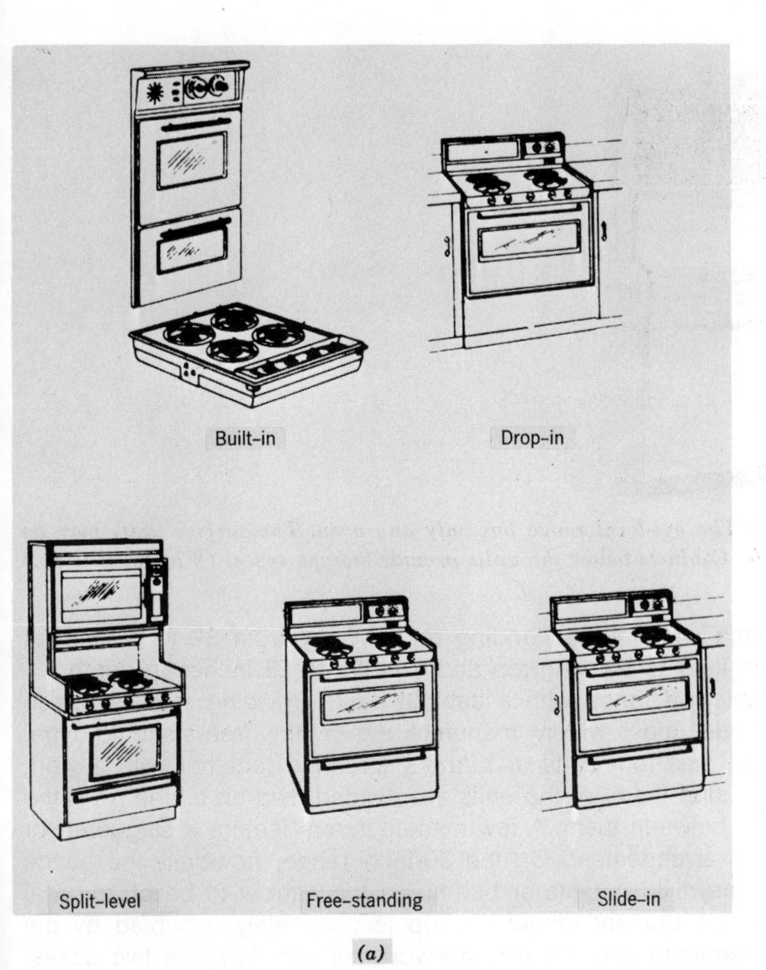

Fig. 7.1a Five different types of ranges. (Gas Appliance Manufacturers Association)

was on a level with the kitchen work tables and provided continuous working surfaces.

At the present time, six different types of ranges are manufactured: free-standing, eye-level, split-level, slide-in, drop-in, and built-in (Figs. 7.1*a* and 7.1*b*). Ranges are available in three different sizes: the standard range, 36 to 42 inches in width, the 30-inch range, and the apartment model, 19 to 24 inches wide. Only a few manufacturers make ranges over 36 inches wide, and most of the production is in

***Fig. 7.1b** The eye-level range has only one oven. The surface units may be retractable. Cabinets below the units provide storage space. (Whirlpool Corp.)*

the 30-inch width. Free-standing ranges vary from 32 to 36 inches high from floor to cooking top and from 25 to 28 inches in depth.

The table-top range with a cabinet base providing storage space is the model most widely manufactured in the free-standing type. This model has four units or burners arranged together with a work surface beside them, or the units are divided, two on a side, with the work area between them. A few manufacturers feature a staggered or inverted V arrangement. On the 30-inch range, however, the space between the divided units or burners is too narrow to be of any real use. In the apartment model the top is completely occupied by the heating elements, and there is no work surface. In these two cases, adjacent counter space of heat-resistant material is an advantage.

The slide-in and drop-in ranges are similar to free-standing ones in appearance with four surface heating elements and an oven below them. The slide-in range has a broiler or, in electric ranges, a storage drawer beneath the oven, while the drop-in model rests on a base of the same material as that used in the kitchen cabinets. Both types are installed in the kitchen counter area between base cabinets.

At the present time the eye-level and split-level ranges are increasingly popular. The first has a single eye-level oven with surface elements installed below it. Frequently the surface elements are retractable and may be pushed in under a counter-top when not in use. The split-level model, also called the double-oven range, has two ovens, one at eye-level height, a second below the surface units or burners. The clearance space between the rear burners or units of the cooking top and the bottom of the eye-level oven varies from 8 to 14 inches. Sufficient clearance should be provided to allow the comfortable use of large saucepans on the rear elements. Separate built-in surface burners and units, and the built-in oven or oven-broiler combination, are widely used, often installed in each new house of a development tract. A variety of arrangements is possible. The cooking top burners and units may be installed in one place or in separate areas in any desired pattern. Some units and burners are hinged so that they can be folded back against a rear panel when not in use, leaving free space on the counter for other work. Electric units and gas burners of this type are constructed to turn off automatically when raised, but the gas shut-off coupling allows the pilots to burn.

The oven-broiler combination can be built into a wall or a cabinet at a height suited to the homemaker. Studies conducted at the Beltsville Research Center on the human energy costs of using built-in ovens at different heights show that, for a minimum expenditure of energy, a wall oven without a separate broiler should be installed so that the bottom of the oven well is 32 inches from the floor. An oven with the broiler beneath it should have the bottom of the well 34 inches from the floor. It is a convenience to have counter space beside the oven.

The built-in surface elements and ovens have the same types of controls, both conventional and special, as in the free-standing range. The oven vent frequently opens into a flue connected to the outside of the house, or into a cabinet supplied with a fan that draws the volatile products through a screening device. The screen may be of three materials, a fibrous material to absorb smoke, charcoal to absorb odors, and spun aluminum or spun glass to absorb grease. If Fiberglas is used, it usually will absorb both grease and smoke. This screen may be removed at intervals and washed with hot soapy water to get rid of collected grease and other deposits.

Construction

Electric and gas ranges have many construction features in common. The frames are usually made of steel with the panels welded together. Most of the outside sections of a range and the oven lining are finished in porcelain enamel or, occasionally, are of stainless steel. Synthetic enamel may be used on the back panel, which often fits flush against the wall, and also on the inside of storage drawers.

The top of the range, and sometimes the entire range, may have a coat of titanium enamel. It is more stain-resistant than ordinary porcelain enamel, and sometimes other stain-withstanding ingredients are also added. The enamel finish is obtainable in a number of different colors. If one desires to have range, refrigerator, and sink all of one special color, care must be taken to see that they match. When the equipment is purchased from different companies, this is often hard to achieve.

The top of the free-standing range is backed by a raised section known as the backguard, a name that clearly indicates its function of shielding the wall from food splatters. On the backguard in the gas range are the range lamp, the opening for the oven vent, and other special parts such as an automatic clock and timer. On the electric range, the backguard does not include an oven vent, but will include the other above-mentioned features, and often, in addition, the switches for the units, although, on most apartment-sized electric ranges, the switches are on the front panel. Many ranges now have recessed bases, similar to the kitchen cabinet base, giving a uniform appearance to the installation and allowing toe space for the homemaker as she stands in front of the range. Leveling screws are placed behind the front panel of the base.

Thermostatically Controlled Surface Units and Burners

At present most ranges have one or more surface burners or units that are thermostatically controlled. A unit or burner of this type has a central sensitive disk backed by a spring that holds the disk in contact with the bottom of the utensil. The control, usually of the hydraulic type, is connected to the switch or valve handle and operates to maintain the temperature for which the dial is set. Such a controlled burner or unit is equivalent in use to a small electric appliance, and tends to eliminate scorching of the food. Since the action of the thermostat is regulated by the temperature of the center of the pan in contact with the disk, it is well to have food cover the center of the pan, and this is especially recommended in the use of frypans for cooking meats. The utensil used on these thermostatically controlled units should be of a good conducting material and have a flat bottom. Manufacturers recommend using a medium-weight aluminum pan. If other materials are used, different heat adjustments will be necessary.

When a thermostatically controlled top burner cycles off and on at low temperature settings, a small tower pilot (in the form of a small tube) remains on to reignite the flame after it has been cut off by the thermostat.

A new owner may find it necessary to experiment with the controlled burner or unit, but once she is well acquainted with its operation, it can be a favorite feature.

Ovens

The size of the range oven is commonly 20 to 21 inches deep and 15 to 18 inches high. In the 30-inch range the oven width varies from 21 to 24 inches, and when the range has two ovens, one of them is usually smaller, not more than 10 or 12 inches high, and often narrow, perhaps only about 12 inches wide.

The oven liner is welded at the seams to prevent leakage of moisture into the insulation, which is usually of Fiberglas. Openings in the liner for the thermostat, oven vent, and electric unit terminals should be fitted with vaportight bushings to prevent leakage at these points. For the same reason the flange of the liner at the door should extend beyond the frame opening. A gasket of silicon-impregnated cord is sometimes placed around the door to increase the tightness of the closing. Self-cleaning ovens use a Fiberglas cord for a seal. The liner itself may be of stainless steel, or of steel finished in chrome-plating, or of a gray coat of titanium enamel. Each kind has advantages. The metal and light enamel finishes will reflect the heat to the food. The dark cobalt coat is an excellent emitter of radiant heat, has approximately the same coefficient of expansion as the steel base, and consequently is not affected by the many variations in oven temperatures that occur when the oven is heated and cooled time after time. Ovens are occasionally supplied with removable back and side panels finished in Teflon, which is easily washed at the sink, or of aluminum foil that may be replaced when necessary. Oven linings of electric ranges are now frequently grounded.

The shelf supports should be an integral part of the side walls but are sometimes removable. The shelves should slide in and out and have positive shelf stops that will hold each shelf in a rigid horizontal position when it is pulled out and so prevent tipping. A rail across the back of a shelf keeps food from sliding off as the shelf is pulled out. Usually, two shelves are provided.

The shelf itself is made of rustproof metal rods or ribbons, which should be spaced close enough for adequate support of small baking utensils. Some shelves are the offset, reversible type, that provide two heights in themselves. The common oven shelf has the side rods that contact the wall supports on a level with the rest of the shelf rods. In the offset shelf there is a 1½- to 2-inch difference in height between the side rods and the central rods on which the utensil is placed. When the shelf is installed with the center rods higher than the side rods, the surface of the shelf is raised. When the shelf is turned upside-down, so that the center rods are lower than the side rods, the shelf surface is lowered.

The oven door should have the same insulation as the walls. Both oven and broiler doors may have a double-panel glass window for observation, or almost the entire door may be of glass, clear, tinted, patterned, or black glass, some of it designed so that the interior of the oven is visible only when the oven light is on. Self-cleaning ovens are sometimes supplied with a shield to cover the glass to protect it during the cleaning process.

The oven usually has an electric lamp that is lighted automatically whenever the door is opened. Some models with the window in the door have a switch on the front panel or backguard by means of which the oven light may be turned on at any time, so that the homemaker may inspect the progress of the cooking operation. Oven lamps are subjected to a special high-temperature exhaust to improve their operating characteristics and increase the life of the lamp at the high external temperatures involved. The lead wires are welded to the base and are separated from each other by an insulation of asbestos to prevent any loose oxide from causing a short circuit. A special cement that will withstand temperatures up to 550°F is used to attach the base.

Temperature in both gas and electric ovens is quite universally controlled by a thermostat of the hydraulic type.

Thermostatic action is different in gas and electric ovens. For example, if the thermostat of an electric oven has been set for 325°F and the switch is turned on, the units will operate until a temperature of about 400°F is reached. This temperature, known as the preheat overshoot, allows the door to be opened and cool food to be placed in the oven without too much loss of heat. At the end of this preheat period, the electricity cycles off automatically and the temperature drops, somewhat lower than the 325°F thermostat setting. The current then turns on again and continues to cycle on and off as indicated in Figure 7.2*a,* usually 10° above and below the 325° setting.

Figure 7.2*b* shows that in heating the gas oven there is also a preheat overshoot. The gas does not shut off completely (unless controlled by an electric ignition system), but the gas valve partially

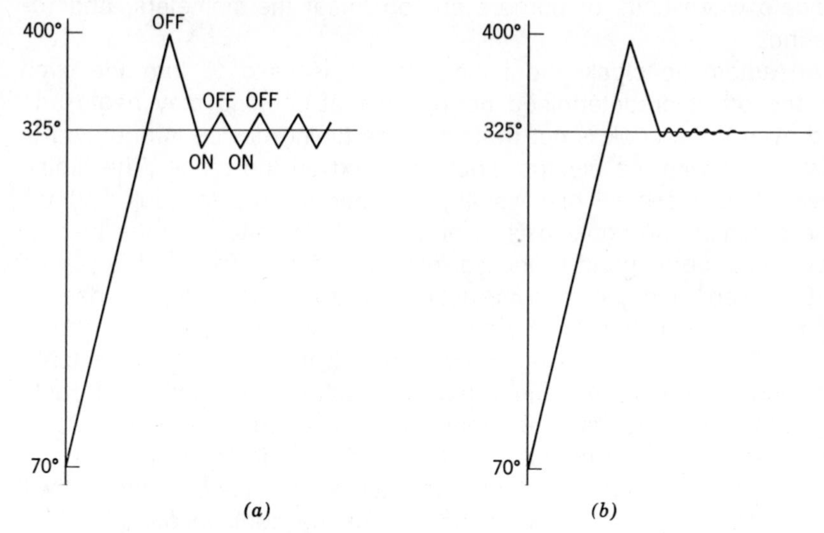

***Fig. 7.2** Heating cycles of electric and gas ovens.*

closes, reducing the flame size to maintain a fairly constant preset temperature. Ranges that feature the low 140°F to 250°F temperatures may use the cycling method, but the curves have smaller amplitudes than in the electric range. Temperature in the food does not cycle but continues to rise until the food is cooked.

When the oven has a Teflon panel lining, the temperature should be lowered to 400°F or below, making a longer broiling time necessary. The panels should be cleaned after each use.

American National Standards Institute requirements state that the broiling area shall be centrally located with respect to the burner, that the distribution of heat shall be uniform over the entire broiling area, checked by toasting bread for 10 minutes or less to an even brownness, and that the average temperature of the broiling compartment shall reach 530°F above that of the room within 16 minutes.

Special Features

Many gas and electric ranges have lamps on their backguards that illuminate the top surfaces of the ranges. Low-end-of-the-line models may not include the lights and the other special features found on the more deluxe types of ranges but will have all the necessary fundamental parts. These special features include automatic clock and timing devices, interval timers, convenience outlets, lights that indicate when units or burners are on, meat thermometers, and rotisseries.

An automatic clock and timing device is used to turn the oven on and off at predetermined hours, so that cooking may begin and end even if the cook is not at home. The timing device is operated in connection with the electric clock installed on the range. The timing device has a special dial, usually indicating hours to cook and the time at which the cooking is to end. The time setting on the electric clock must correspond to the correct time of day. The food is placed in the oven, and the delayed-action controls turned to the desired times. The oven burner or unit is turned on, and the thermostat is set at the desired cooking temperature. The heat will not actually turn on until the predetermined hour is reached. This method of cooking should not be used for foods that deteriorate rapidly at room temperature, and, of course, it will only be satisfactory for foods that can be cooked at one constant temperature. The gas or electricity will be turned off automatically at the end of the cooking period, which can be an advantage if the homemaker's return should be delayed. An automatic clock and timing device may prove convenient for the woman who works outside the home or who is active in club or civic organizations or if she is busy with family or guests. On many ranges, the timing device knob must be turned back to manual (M) when the baking period is over; if this operation is forgotten, the oven will not heat at a later time when the timing device is not used.

Control found on some ranges can be set for immediate start; the control, unless the user selects another setting, turns the oven temperature automatically to 325°F and at the proper time reduces the oven temperature to a holding temperature of 170° to 140°F. To accomplish this change a gear train is driven by a 60 cycle A.C. motor to rotate the temperature control shaft. The entire cooking time set on the control will include not only the time on the 325° setting, but also the time for gradually reducing to the lower setting. Table 7.1 indicates, for recipes furnished with the range, the length of time the product will be held at each temperature, depending on the cooking time specified. If other recipes are used, additional cooking time equal to one half of that given in the recipe must be used to compensate for the reduction in temperature.

Another control added to certain ranges allows slow, low-temperature cooking to tenderize tough cuts of meat when they are roasted. One deluxe model with this control has a delayed start mechanism that will turn on at any indicated time, day or night. The roasting process is commonly 14 hours long. This slow cooking not

Table 7.1

Setting (Hours)	Actual Time[a] at High Temperature (Minutes)	Reducing Time Plus Time at Low Temperature (Minutes)
1	30	30
2	60	60
4–6	135–255	105
7–9	300–420	120

[a]*Frigidaire Product Information,* Volume 68, No. 4 (p. R-68).

only tenderizes the meat fibers, but also reduces shrinkage and spatter.

The interval timer, also called a minute minder, is either electrically or manually controlled and can be set to indicate the time a cooking operation will finish. Some types provide both hour and minute intervals, others, only minutes; some ring or buzz continuously over a fairly long period of time, while others sound only once. This feature enables the cook to attend to other matters without giving thought to the time at which food must be removed from a surface unit or taken from the oven. It may also be used as a reminder for nonfood activities (e.g., when to make a telephone call).

Electric outlets on the range allow any small appliance to be attached, freeing wall outlets for other uses. When two outlets are provided, one of them is frequently time controlled. It may be used to start the coffeemaker ahead of time in the morning so coffee is ready when the homemaker reaches the kitchen, a special boon for one who relies on that first cup of coffee to begin the day.

Modern ranges in top-of-the-line models often have **lights** behind the valve handles and switches, indicating when the burner or unit is in operation; sometimes a different color for different heats. First used only for ovens, they now give the same information for the surface elements. They are especially valuable when a gas flame or an electric unit is turned so low that it might not be noticed, were it not for the signal light.

Another feature on some ranges is a special type of **meat thermometer,** in the form of a metal probe to which a metal cable is attached. The probe is inserted in the meat, and the cable is attached to an outlet in the side of the oven (Fig. 7.3). A roast meter on the probe or on the backguard of the range, over which a pointer moves, indicates different degrees of internal temperature and, there-

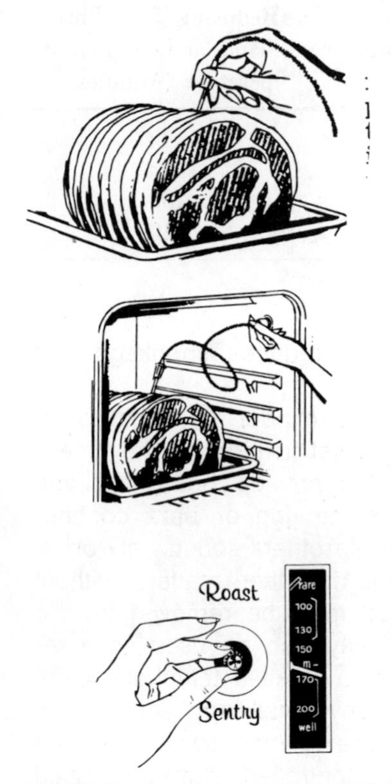

Fig. 7.3 The Roast Sentry will help you roast meat and poultry exactly as you want it done. (Whirlpool)

fore, doneness. The cook may read the meter and remove the meat at the proper time; or, if the range is fully automatic, the roasting process may be stopped automatically, and the cook may be informed by a signal. One manufacturer makes a probe that will start the cooking operation when the homemaker is not at home and, in another type of range, the oven temperature is cut back to the temperature within the roast as soon as the roast has reached the predetermined temperature.

A **rotisserie** increases the versatility of an oven, and is obtainable as an additional feature in both gas and the electric ranges. The spit may be temporarily installed in the conventional oven or it may be a part of the equipment in a special, smaller oven. Not only the single spit but also double and triple spits are now available. A few ranges feature rotisseries on the cooking surfaces of the ranges; these rotisseries have vertical heating elements. The motor that turns the spit is controlled by a switch on the backguard or on the front panel.

Use

As noted in the chapter on utensils, for surface cooking, dull-finished, flat-bottomed, straight-sided pans made of a good conducting material and having tight-fitting lids should be used. The sides of the pans are shiny. The pan should fit the unit or burner or should be slightly larger. A flat pan bottom that entirely covers the heating surface minimizes heat-transfer losses. On an electric range, pans larger than the unit may cause the porcelain enamel around the unit to become overheated; this overheating results in crazing or chipping of the enamel.

If water is the medium for heat transfer to the food, use small amounts. The temperature of live steam is the same as that of boiling water; hence the rate of cooking is the same in both mediums. The lid of the pan should fit tightly to retain the steam.

When satisfactory baking results are not obtained, a number of things may be the cause. The kind of pan used makes a difference. Shiny finished pans are best for cakes; dull finished ones of anodized aluminum or darkened tin for pies. When baking frozen pie, remove it from the pan and place on a cooky sheet to prevent a soggy crust. For any of these baking operations, preheating of the oven is desirable. Pans or cooky sheets must not interfere with the circulation of air in the oven. Free circulation is most easily obtained by placing the pans in alternate positions on the racks and by leaving at least a 1-inch space between the pans or baking sheets and the sides of the oven (Fig. 7.4). Somewhat less space may be left between the pans themselves. A single pan should be placed on the middle or

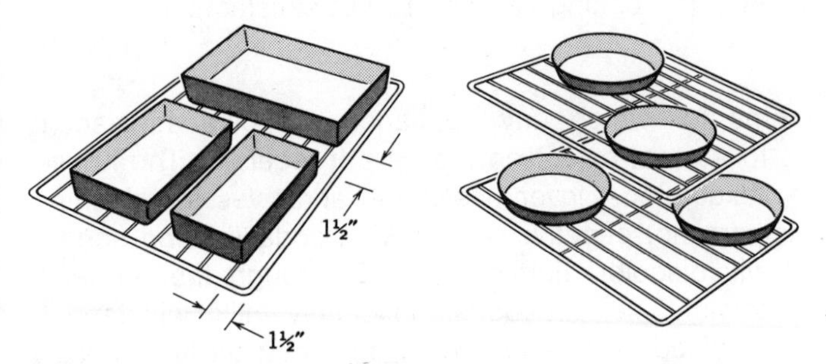

***Fig. 7.4** Proper pan placement in the oven. (Whirlpool)*

top rack and toward the back of the oven to benefit from the circulation of hot air toward the vent.

When a whole meal is prepared, the arrangement of pans is less important than when cooking baked products, since the food for the meal is not cooked by dry heat but by steam.

If containers such as casseroles and deep pans are to be used in the oven for steaming vegetables and fruits or for baking where a thick heavy crust is desired, the material of these utensils should have high absorption power.

For long, slow baking processes, or when a thin crust is desired, utensils should be made of materials which do not have high absorption ability.

Accurate baking temperatures are essential. Too high a temperature causes the cake to be compact in texture with a cracked upper crust; with too low a temperature, the cake has a coarse texture and frequently a soggy layer at the bottom and a sticky, pitted top surface. Increasing the baking period for a few minutes tends to give more uniform browning on both top and bottom crusts.

The range should sit level. Even slight tilting from front to back, or vice versa, upsets heat distribution and causes uneven browning. The homemaker should check the level of the range on an oven rack rather than on the top surface. It is not necessary to buy a spirit level. An adequate means can be obtained by filling a small flat bottle almost full of water and closing it tightly. When it is laid on its side a bubble can be seen on the surface of the water. It should be in a middle position if the shelf is level. Check the level from front to back and also from side to side.

A warped oven rack may have the same effect. Incorrect primary air adjustment in the gas burner disturbs the distribution of heat, and any leakage of air around misfitting doors affects the path of flue-gas travel and seriously alters heat distribution. All these points should be checked when baking results are not satisfactory.

Care

The length of time for which any appliance will give satisfactory service depends to a great extent on the care it receives. The household range is no exception. Never use harsh abrasives for cleaning. If soap and water are not effective, use baking soda (baking soda is harmless, but it will discolor aluminum). Even though the porcelain enamel is acid-resistant, certain foods, noticeably milk, will tend to etch the enamel if spilled on it, and should be wiped off immediately. Once the range has cooled it should be carefully cleaned. The entire surface may be wiped off with a cloth dipped in warm soapy water, then with one wrung from clear water, and dried.

Aeration pans around the burners, reflector pans beneath units, and drip trays may be wiped or removed and washed. Tubular units may be wiped and charred food particles gently removed with fine steel wool. At least one manufacturer features units that may be washed; otherwise be careful not to get water into the terminal connections of the units. Gas burners occasionally may be removed from the range, and the mixer tubes cleaned with a small stiff brush. All parts should be washed in soapy water, rinsed thoroughly in hot water, and dried by being placed upside down in a warm oven. The pilot-light valve may be carefully cleaned with a piece of wire; it can be damaged if roughly handled.

The top of the sensitive disk in the thermostatically controlled surface unit and burner should be kept clean at all times.

Other aids to cleaning (Fig. 7.5) are removable knobs on the control switches of the electric range or the valve handles on the gas range. The cooking top may lift up to enable the homemaker to reach the underlying areas or, in some cases, the entire top is removable. Certain ranges have spill-saver tops; the top surface is recessed and surrounded by an elevated rim to prevent any spill-overs from reaching the floor. The noninsulated bottom of the gas oven may be taken out and washed.

All oven shelves should be removed and washed. Spilled food should be removed as soon as the oven has cooled, using baking soda or a mild abrasive if necessary. Always wipe the entire interior surface of the oven after each use, if possible. The oven walls may look clean, and there may be no visible deposit, but there will always be a brown discoloration on the cloth when the walls are wiped. If the oven is not cleaned until a stain becomes visible, deposits will have burned onto the lining, and will be extremely difficult to remove unless the oven is self-cleaning.

Pyrolytic Method

The self-cleaning oven, available in all types of both gas and electric ranges, eliminates manual oven cleaning and is a boon to the homemaker, who, either from lack of time or lack of interest, does not clean after each use. Before starting the operation, known as the pyrolytic method, a lock arm on the panel above the oven opening is moved to the right. This energizes a solenoid valve which locks the oven door securely and turns on a signal light. Cleaning takes place at a temperature above 750°F, usually between 850°F and

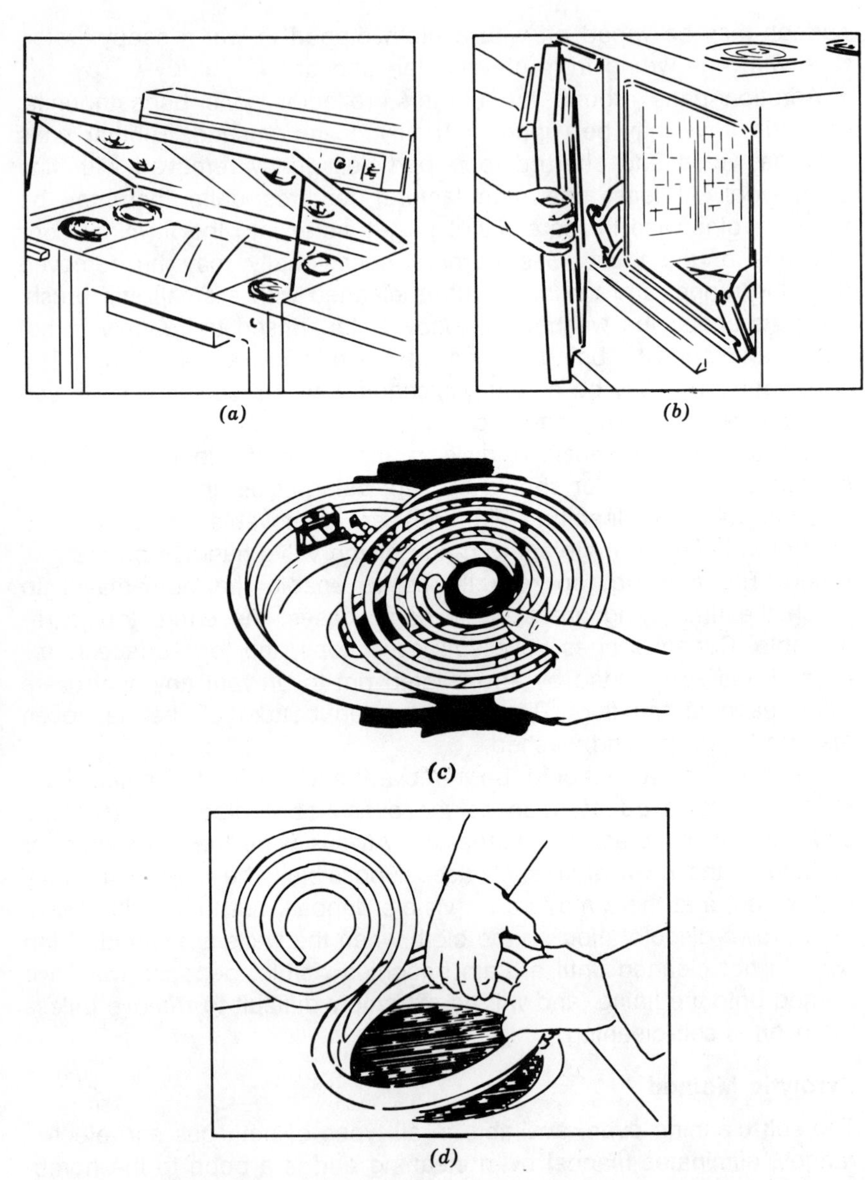

Fig. 7.5 Special features designed to minimize cleaning: (a) *Lift-up top. (Philco)* (b) *Removable oven door. Note nonspill range top. (Westinghouse)* (c) *Removable surface unit. (Whirlpool)* (d) *Monotube unit with hinged bracket. (Frigidaire)* (e) *Pull-out oven lining. (Live Better Electrically)* (f) *Removable lower drawer to clean floor beneath range. (Whirlpool)*

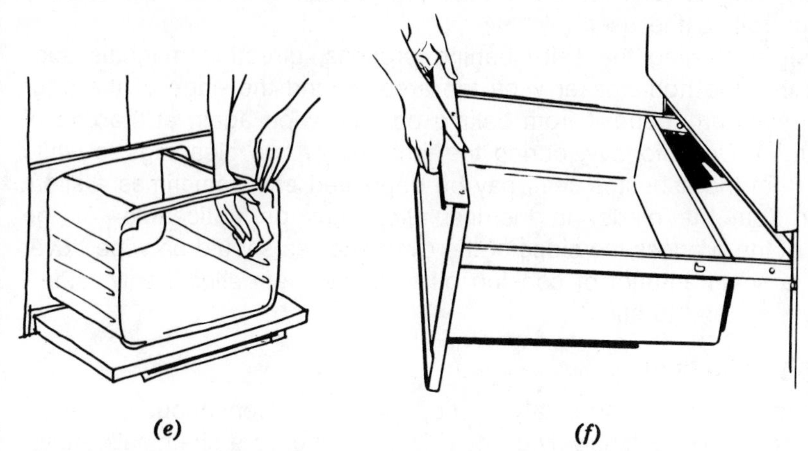

(e) (f)

1000°F, reducing deposits on oven walls to a small amount of white ash, which is easily removed with a damp cloth. The operation requires approximately two hours, costs from 8 to 10 cents, and causes no more heat in the kitchen than would be caused by using the oven for ordinary baking. To maintain the low exterior temperature, two to three times more insulation is used around the oven cavity than in conventional ovens, but it is compressed into the same space. In some range models a blower fan, turned on by the lock lever, runs during the cleaning cycle. The oven cannot be reopened until the temperature returns to normal, indicated by the light going out. The cooling period also may be two hours in length, making the total time four hours.

One manufacturer eliminates the special cleaning switch by replacing the hydraulic thermostat with a new style thermostat charged with helium gas, which permits use of a single control for baking, broiling, and cleaning.

This method of cleaning previously tended to discolor the stainless and the chromium-plated oven shelves; they acquired a purplish tinge. Now racks have been given a special finish that almost entirely eliminates any discoloration. The window in the door also need not be covered. Reflector pans of porcelain enamel, and of stainless and of chromium plated steel, may be placed in the oven for cleaning. Aluminum drip trays, however, should never be cleaned by this method. Another exception is the broiler pan because the homemaker may leave too much grease in it. This would cause more smoke than the system could eliminate without overtaxing the ability of the oven. Consequently the Underwriters' Laboratories will not certify the inclusion of the broiler pan because of the impossibility of controlling the user.

Before starting the self-cleaning process, direction manuals suggest that the homemaker wipe the area around the edge of the door to prevent any deposit from baking on, and wipe again at the end of cleaning. Occasionally, during the first one or two cleanings, a white film from the asbestos seal may be deposited and sometimes a slight odor from the binder in the insulation may be noticeable. At the end of the process the sides of the oven shelves should be wiped over with a small amount of cooking oil to make them slide freely; otherwise they tend to stick.

Catalytic Method

A second method, the catalytic one, called the "continuous cleaning" method, is now extensively used. This method uses a chemical catalyst which is mixed into a coating bonded to the oven walls or panels. The catalyst coating is rough in texture and of a dull, dark color. The catalyst promotes oxidation of any deposits on the walls, changing them to carbon compounds and then to carbon dioxide, a gas that dissipates into the air. The catalyst itself undergoes no chemical change. This process supposedly takes place whenever the oven is heated to 350°F or above, in other words, during the usual, everyday use of the oven. If the oven is very badly soiled, it may need to be heated to 500°F for two hours or more.

However, since the coating is somewhat porous, the food spots tend to spread on the coated surfaces and blend with the color of the coating as the oven heats. Frequently this soil is not completely oxidized and, consequently, soil will build up over a period of time and may even cover the catalyst sufficiently to render it ineffective in action. The promoters of the catalytic process claim that the oven "stays acceptably clean," but that is not "sparkling" clean, as claimed for the pyrolytic method. Moreover, in the catalytic oven the shelves are not coated, the oven door and window are usually not, and often even the oven floor is not. If the broiling compartment is separate from the baking oven, it also may have no catalytic coating, although some broilers are coated. All these parts must then be cleaned by hand. Care must be taken not to scratch the catalytic coating; it is easily damaged, reducing its effectiveness.

Heavy spillovers on the oven floor are a problem where homemakers may get into trouble, since scraping and scouring destroys this finish. Keeping the porcelain enamel oven bottom uncoated may be desirable with other oven areas coated with the catalytic finish. The oven bottom can usually be easily removed and taken to the sink for cleaning. The first continuous cleaning frit was fired on the oven surfaces at 600–700°F. The newer frit is fired at 1200–1300°F, which should make it more resistant to abrasion.

The American Gas Association has set up certain criteria for the design of catalytic ovens; they must conform to all other current standards for ovens and the coating must not be injurious to health during normal usage.

The oven doors of some ranges may be lifted off for cleaning; other doors may be dropped down 180° flat against the front of the range to increase the ease of reaching all parts of the oven. Probably the best type is the oven with linings that slide out or with inner panels of Teflon that may be removed and washed at the sink. If the electric units are taken out when the oven is cleaned, care must be used to push them completely back into their outlets afterwards; otherwise they will not heat when the switch is turned to "on" the next time. At present, however, tubular units in many ovens are hinged, so that the bottom unit may be lifted up and the top unit pulled down for ease in cleaning beneath them.

If aluminum foil is used in the gas oven to catch spillage, place the foil on an open shelf, not directly on the bottom, and cut it only slightly larger in size than the utensil, so that it will not interfere with convection currents. In an electric oven with the lower tubular unit, foil may be placed on the oven bottom below the unit, but care must be taken that the foil does not come in contact with the coil or with the outlet into which the coil is plugged. In general it is considered better practice to place the foil on the shelf beneath the pan. Some manufacturers supply oven foil liners to facilitate cleaning.

After broiling, wipe off spatters of fat with a piece of soft paper before washing. Burned-on fat may be rubbed with steel wool of a fine grade, or a commercial cleaner may be used. However, care should be taken to select a cleaner guaranteed not to damage the porcelain enamel. If the broiler compartment is below the gas oven, the broiler pan should be removed when the oven is heated.

Frequently the bottom storage drawer or base panel may be removed so that the floor beneath the range may be cleaned easily. In many ways the present-day manufacturer seeks to increase the ease of keeping a range in its original attractive condition.

All electrical connections between gas appliances and wiring in the home must conform to the regulations of the National Electrical Code. The code states that all gas appliances using electrical controls shall have these controls connected into a permanently alive electric circuit, not one controlled by a light switch. Gas pipes shall not be used for an electrical ground. No devices employing or depending upon an electric current shall be used to control or ignite a gas supply, if they are of such a character that failure of the current shall result in the escape of gas. Some difficulties have occurred in the use of electric starters. We believe that a manufacturer should guarantee that his starter is of high operating efficiency.

The National Bureau of Standards has established safe temperature limits for the various materials used on range surfaces to help reduce burn accidents: 152°F for metal, 160°F for porcelain enamel, 172°F for glass, and 182°F for plastic. Because many manufacturers may need to make extensive design changes, these requirements may not become effective before mid-1975.

Operation costs depend on local rates, but they may be greatly reduced by the efficient use of the appliance.

Suggestions for Economical Use of Electricity and Gas

1. Keep all parts of the range clean.
2. Use a small burner or unit instead of a large one, whenever possible.
3. Put the utensil on before turning on the heat, so that the heat goes into the utensil instead of into you. Turn off the heat before removing the utensil.
4. Boil only the amount of water that is needed; you speed up the job, save fuel, and prevent heat in the kitchen.
5. When water begins to boil, turn unit or burner to low or simmer position. Slowly boiling water is as hot as rapidly boiling water.
6. Use covered utensils, if feasible.
7. Use the thermostatic surface heat controlled unit for all frying, if such a control is provided. With proper heating fat will not smoke.
8. Do not preheat the oven too long before use.
9. Obtain free circulation of air in the oven by placing pans in alternate positions on racks.
10. When roasting or baking, use the oven to capacity. Cook food for another day.
11. Use accurate baking temperatures.
12. When the oven is well insulated, turn off the heat a few minutes before the end of the baking period, and finish the baking with retained heat.
13. Avoid raising pot covers and opening the oven door during cooking operations.

14. Do not preheat the broiler. The grid is easier to clean if the food is placed on a cold rack.
15. Use electric ignition systems for gas burners, and so eliminate constant-burning pilots.
16. Follow trusted recipes. Use properly specified utensils.

Two general types of tubular units are in use. The first unit developed uses a single coil of resistance wire in a tube of a relatively small diameter. Each unit consists of two of these tubes bent into special forms and supported on a stainless-steel frame, or spider, to which the tubes are anchored so that they are held in shape during the expansion and contraction which occurs in use. In these two tube units, the spirals can be formed so as to provide different heat patterns with different positions of the switch.

In some units, one of the tubes forms the center section of the unit, while a second tube surrounds it. In ranges currently being manufactured, this construction is used on many models for those 8-inch units that have thermostats or indefinite heat controls. With a selector switch, this construction provides for units of two diameters in one. When only the inner coil is heating, a 6-inch unit is available for use with small pans; when both coils are heating, an 8-inch unit is provided for larger pans (Fig. 7.6). One unit has three coils of wire separately controlled that allow the use of either a 4-inch, 6-inch, or 8-inch saucepan as needed.

The second type of unit uses a larger diameter tube into which two or three coils of resistance wire are inserted. All of the electric terminals are brought out at one end, and the other end of the tube is welded tight. The compacted and flattened tube is wound into a single spiral. This type of unit provides an even distribution of heat over the entire surface regardless of the heat being used. This single tube unit, known as Monotube or Radiantube, requires no anchorage to the support as the coil is free to move while expanding and contracting under heat; hence it is not subject to warping in use and maintains a constant contact with utensils. This unit has a hinged bracket that allows the tube to be swung up and to one side for easy cleaning or removal of the drip pan (see Fig. 7.5*d*). The wide, flattened top surface of the unit provides good contact with the cooking utensils. Electric surface units should have an efficiency of not less than 60%.

All tubular-type units have low thermal mass and hence heat rapidly. Since this type of unit heats largely by conduction, best results are obtained by using a pan made of a material having high thermal conductivity and constructed with a flat bottom so as to give maximum contact with the unit.

When using a glass saucepan over an electric unit, the homemaker should place a metal grid beneath the pan; otherwise the direct contact with the hot unit may cause some slight softening of the glass, afterwards resulting in an irregular bottom surface on the pan.

ELECTRIC RANGES

Electricity is a clean and a safe source of fuel for a range. Heat is produced by the resistance of a wire to the passage of an electric current; hence there is no flame. The current is under control, easily regulated, not subject to atmospheric conditions or drafts, and will not be extinguished by liquids boiling over. Heat is largely confined to a small area so that the homemaker has comparatively cool working conditions. Since electric heat is not a product of combustion, gases cannot be formed as by-products.

In selecting an electric range, the purchaser must first decide which type to buy. In addition to personal preference, initial cost, installation cost, and required floor space are factors that should be considered in making a selection. Built-in ranges are usually more expensive to install than free-standing ranges because they require two heavy-duty circuits and the purchase or building of cabinets for the separated oven and surface cooking top. Current models of free-standing ranges require 19 to 42 inches of floor space. Built-in ranges require 48 to 64 inches. Free-standing ranges are available with one or two ovens; a built-in range may be purchased with a single or double oven, or with a single oven having a separate broiler (see Fig. 7.1*a*).

Range Units

In the electric range the source of heat is called a unit and is controlled by a switch. As mentioned earlier, most free-standing ranges come with four surface units. A few models come equipped with surface griddles in addition to the four units. Drop-in surface cooking tops are available with two, three, or four units. A few have four units and a griddle. Most cooking tops provide units of two sizes. The smaller size, usually designated as the 6-inch unit, is rated at 1250 to 1600 watts. The larger size, referred to as the 8-inch unit, has a wattage rating of 2000 to 2700. Most four-unit ranges have one 8-inch

(a)

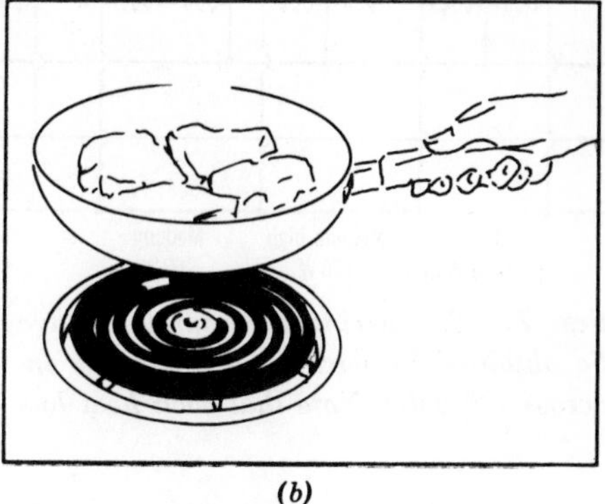

(b)

Fig. 7.6 A 6- and an 8-inch unit in one. (a) ***For small pans it cooks on the inner coil only.*** (b) ***For large pans the entire unit is heated. (General Electric)***

and three 6-inch units or two 8-inch and two 6-inch units. (See Fig. 7.6.)

Surface units on present-day ranges are of the tubular type. A unit of this type consists basically of a tube made of a nickel-chromium-iron alloy, such as Inconel, Incoloy, or one of the high chromium-nickel-stainless steels. Into this tube, one or more coils of nickel-chromium wire are inserted, after which the tube is filled with fused magnesium oxide powder. Nickel-chromium wire is used for the heating element, because it maintains a uniform resistance with a minimum variation and will withstand continuous, high temperatures. After the outer tube has been filled with the oxide powder it is reduced in diameter to compact thoroughly the magnesium oxide, which acts as an excellent electrical insulator at red heat, and also has the property of allowing a free flow of heat from the imbedded resistance wire to the enclosing tube. The heat is available for transfer to the cooking utensils, largely by conduction. To improve the flow of heat by conduction, the top surface of the tube is flattened, after it has been formed to any one of a variety of spiral shapes.

In currently manufactured ranges, the surface units have four, five, seven, or an infinite number of heats. Electric ranges are now rated 240 to 250 volts; it was necessary to design for higher voltage areas, particularly the mid-South and TVA areas (Tennessee Valley Authority). According to manufacturers, low voltage is less of a problem than high voltage.

Most units have two 120-volt coils, equal to one 240-volt circuit.

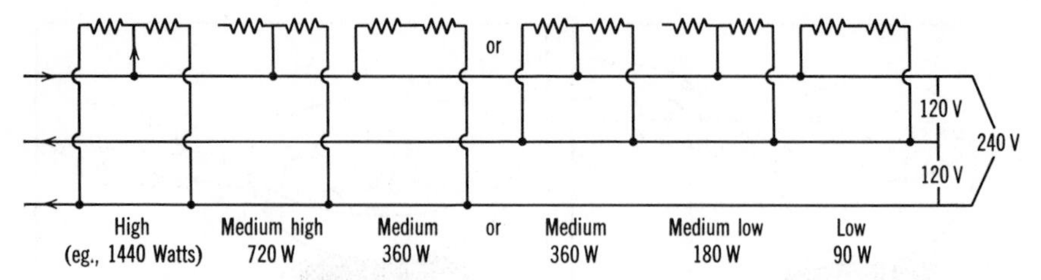

***Fig. 7.7** The wiring diagram for a five-heat electric unit. Medium heat may be obtained by wiring the two coils in series across 240 volts or in parallel across 120 volts. Note that each heat has ½ the wattage of the preceding heat.*

For high heat, the two coils are connected in parallel across the 240-volt circuit. If the two coils are equal in resistance, the next two lower heats are obtained by connecting one coil across 240 volts, and by connecting the two coils in series across the 240-volt circuit. If the two coils are unequal in resistance, the next two lower heats are usually obtained by connecting each coil separately across the 240-volt circuit. The two or three lowest heats are obtained by different connections of the coils across the 120-volt circuit (Fig. 7.7).

High speed units use two 120-volt coils. For example, when a 1250-watt unit is turned on, the coils are connected in parallel across 240 volts, generating approximately 5000 watts for about 30 seconds to give an extra-fast start. The coils are then reconnected in series, and the wattage drops back to its normal rating of 1250 watts.

Since electric heat is flameless, the electric heating units are placed within the oven itself, which is of tight construction. There are two units, an upper and a lower one, tubular enclosed or open wiring. The lower unit is usually a single coil, and the top unit may be a single coil of several loops but it frequently contains two coils; both are used along with the bottom unit for preheating the oven. The entire top unit is used in broiling, and one of its coils may operate with the bottom unit in baking, so that the heat will be more evenly distributed in the oven. One manufacturer uses ¼ the wattage of the broil unit during the bake cycle, and both this section and the bottom unit cycle off together during the off cycle. Every oven, designed by manufacturers' engineers, is balanced for heat distribution in the oven, according to size of oven, selection of materials, mass, type of construction, design and placement of units, and does not change during the life of the range.

No standard method of wiring oven units has been adopted. In most ovens, the top unit consists of two 240-volt coils, and the bottom unit of one 240-volt coil. Ovens vary in the way the coils are connected to produce the different heats for preheating, baking, and broiling.

Oven units vary in wattage; the broiler units from 2500 to 3800 watts, baking units, from 2000 to 3950 watts.

The oven vent opens through the center of the reflector pan beneath one of the back surface units, and it directs volatile cooking products away from kitchen walls.

Broiling means cooking with radiant heat. Therefore the electric unit is especially adapted to this type of cookery. As was noted, the top unit in the electric oven is used for broiling. In a few ranges a heat-resistant glass plate is placed below this unit to prevent fat from spattering onto the coil; in others, a removable glass plate is placed above the broiler coils. This position allows direct heat from the broiler and, at the same time, helps to keep the top surface of the oven liner clean. One of the oven shelves is used to support the broiler pan, and the speed of broiling is regulated by the distance the pan is located from the unit, frequently 3 to 4 inches below the coils. The power of the broiler unit is sufficiently high for the food to be placed under it as the unit is turned on, and cooking starts immediately. During the broiling process in the electric oven, the oven door is usually left ajar, allowing any moisture to escape and giving the crisp product desired. Some electric ovens have controls that permit selection of Hi-Lo Broil. At the lower setting the top unit cycles to regulate broiling temperature.

The broiler pan is especially constructed to keep liquid products from smoking. In both the electric and gas range, the pan itself and its grid may be of chromium-plated metal, of stainless steel, or occasionally of aluminum; some are of steel with a coating of porcelain enamel. The pan is often 1½ to 2½ inches deep and sometimes has grooves on the bottom that direct the flow of liquids to a front corner depression, away from contact with the heat. The grid has narrow slits through which the juices and melted fats drip; these juices are then protected from the heat by the wide solid surfaces of the material between the openings. Such a broiler pan is called a smokeless-type.

Unit Controls

Controls of several types are used to produce the different heats of the range units. One is the rotary switch. The positions of the switch are designated on the dial; sometimes by numerals, other times by words, depending on the manufacturer. Other manufacturers use push-button controls for both surface and oven units. A separate push-button is used for each heat with the markings directly below or above the button. Many rotary and push-button controls have an indicator light that glows when the current is on. For greater flexibility in heat control, most manufacturers include at least one surface unit that is operated by a thermostat. This type of unit has already been discussed. The control dial is usually marked in degrees; the manufacturer provides a cooking chart that shows where to set the dial for different kinds of cooking.

The controls for the surface units and for the oven are mounted on what is commonly called the control panel. On free-standing ranges, this control panel is usually on the backguard, although some manufacturers suggest that the purchaser install it on the front of a range hood. On most apartment-type ranges, the controls are on the front of the range, directly below the cooking surface. On drop-in cooking surfaces, the control panel is frequently mounted flat on the surface with the units.

Oven Thermostats

Electric-range ovens are equipped with automatic oven-temperature regulators. The thermostat universally used is of the hydraulic type. The bulb of liquid is placed in the oven, and, as previously described, the expansion or contraction of the liquid controls a switch that, in turn, opens or closes the circuit (Fig. 7.8).

In some models the oven thermostats and switches are separate. The switch usually has four positions: "off," "preheat," "bake," and "broil." The thermostat dial is marked with the degrees of temperature. To preheat the oven, turn the switch to "preheat" and set the thermostat for the desired temperature. When the oven is preheated the switch is turned to "bake." This turns off the top unit, while the bottom unit or the bottom unit and a portion of the top unit maintain the baking temperature.

On many current ranges, the switch and thermostat are combined. The design of the control dial and the methods of marking vary with different manufacturers. In one design the control is marked "off" and "broil" in addition to the degrees of temperature. To preheat the oven the control is turned to "broil," which turns on the top unit. It is then reset at the required baking temperature. This turns on the bottom unit. When the oven is preheated both units automatically go off. Baking temperature is then maintained by the bottom unit and a part of the top unit. There may be single or dual lights to indicate when the oven is preheated. The thermostat dial is located on the control panel.

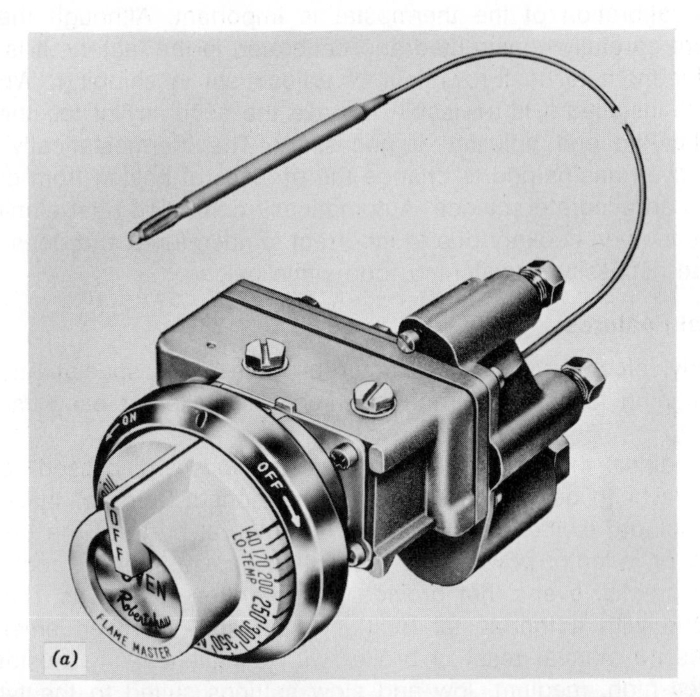

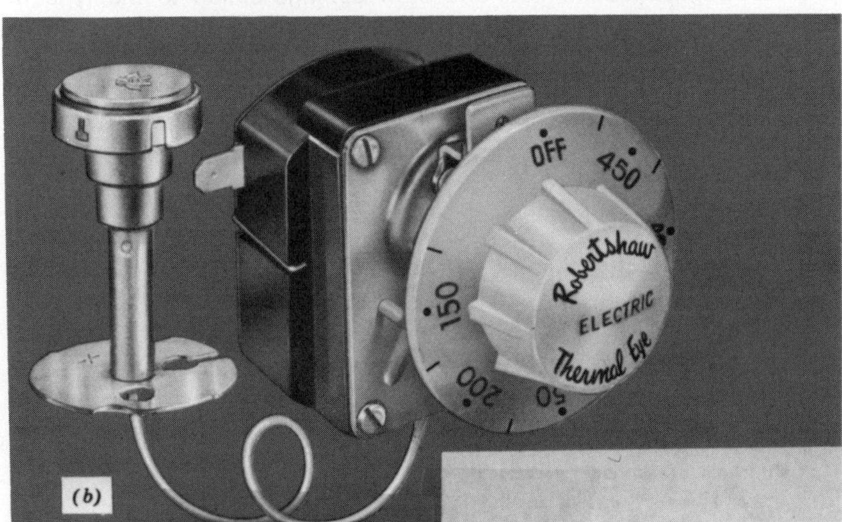

Fig. 7.8 (a) *Hydraulic type thermostat. The bulb is installed in the oven. (Robertshaw)* (b) *An hydraulic sensing element combined with a bimetal anticipator, used to control any single coil surface unit up to 2700 watts. This Thermal Eye sensing element is spring biased against the utensil bottom, resulting in fast heating, minimum overshoot, very rapid recovery. (Robertshaw)*

The calibration of the thermostat is important. Although thermostats are carefully constructed and calibrated in the factory, it is possible for them to be thrown out of adjustment in shipping. When a range is installed it is advisable to have the accuracy of the thermostat checked and adjusted if necessary. The thermostatically controlled oven has helped to change the process of baking from guesswork to an accurate science. Automatically controlled heat eliminates failures in oven cookery due to incorrect temperatures and does away with the necessity of watching food while baking.

Special Features

Recently, electric-range manufacturers have given special attention to designing features for ranges which will make for easy use and care.

In addition to the special features previously discussed, others that have been developed should be mentioned. Some of these features include: element rings built in as integral parts of the cooking surface or reflector bowls, thus preventing grease and dirt accumulation in cracks, ovens that provide temperatures as low as 140°F, a rotisserie with a thermostat built into the spit, an oven timer that requires no manual reset, a broiler with a multiheat thermostat that provides high, medium, low and slow settings suited to the type of food being broiled, and at the same time eliminates raising and lowering the grid, a broil unit that broils both sides of the food at the same time (Fig. 7.9), a high-walled broiler pan used over a shallower pan containing water to prevent spatters on oven walls (Fig. 7.10), infrared heat lamps installed below the upper oven or beneath the wide top of a high backguard to keep food and dishes warm. In the use

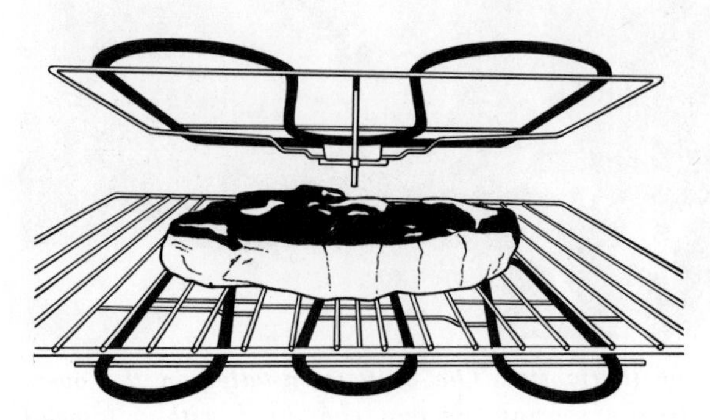

Fig. 7.9 "No Turn Speed Broil"® unit broils both sides of the meat at the same time. (Westinghouse)

***Fig. 7.10** Radiant-wall spatterfree broiler grill. Deep sides of pan keep grease spatters inside, off oven walls. (Frigidaire)*

of these heat lamps, the food, especially meats, may tend to become dry and, therefore, should be covered with cellophane.

A range is also featured with a built-in water heater. Extra attention must be given to safety conditions in such a model. One range contains an especially designed Braille control panel that has raised dots to enable a blind person to adjust surface units and oven controls to various temperature settings. A Braille instruction book is furnished with the range.

In the electric range, the cooking utensil usually sits in direct contact with the surface of the unit; in the gas range, because of the need for air to supply the necessary oxygen for correct combustion of the gas, a grate is placed over the burner, on which the utensil is set. Surrounding the gas burners are aeration pans or bowls and beneath the electric units are reflector pans, each employed in the proper functioning of the respective heating elements, but also used to catch any accidental spillage. In addition it is desirable to have removable trays beneath the bowls or pans to hold any excessive spill-over. The tray should have a raised edge, tight corners, and an easily cleaned, rust-resisting finish. Ordinarily, built-in range tops do not have these trays.

Counters that Cook

In addition to the conventional built-in surface units, "counters that cook" are on the market. They are obtainable in two sizes, two or four units. One type, manufactured by Corning, has the range top of glass-ceramic, a very durable, easily cleaned material. Sunburst designs, indicating heat areas on the surface, turn yellow when hot (Fig. 7.11). Beneath these sunbursts are heating elements and thermostatic sensors, a separate thermostat for each unit, that ensure even heat (Fig. 7.12). Control dials with on-off indicator lights allow infinite settings from 150° to 475°F. The areas of the top that are not in use remain cool. Corning recommends using special utensils, "Cook-mates," that have exceptionally well-ground, flat-bottomed surfaces

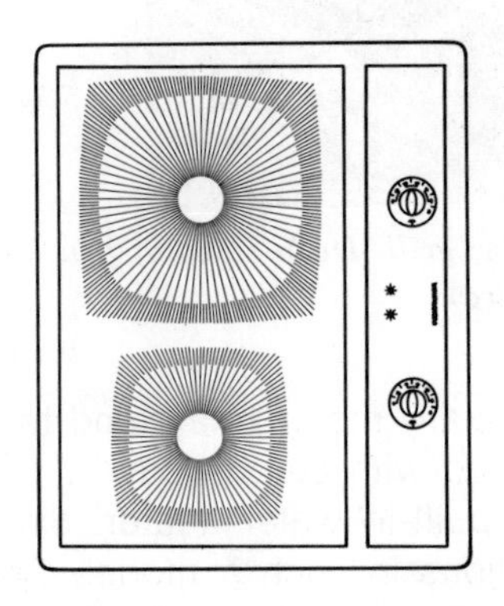

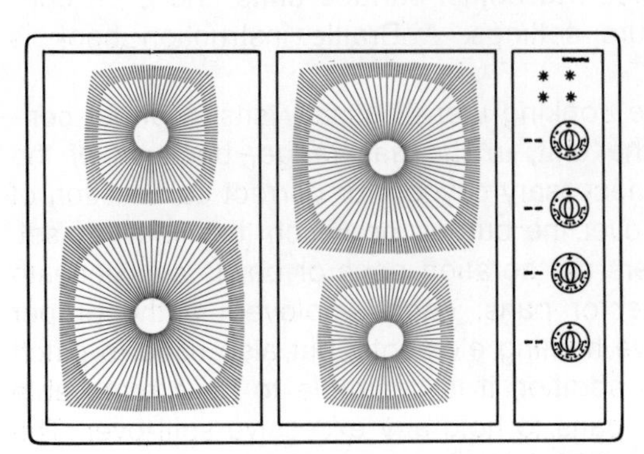

Fig. 7.11 Two models of the Counter That Cooks. (Corning)

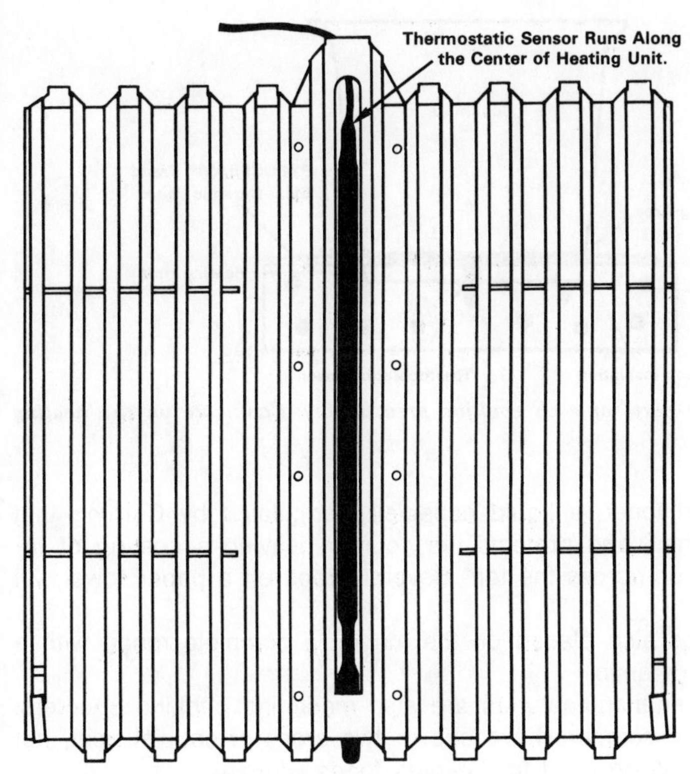

***Fig. 7.12** The 6-inch heating unit, showing the flat wire grid and the thermostatic sensor on a mica card. The element wire is an iron-chromium-aluminum alloy ribbon of precise resistance. (Corning)*

to make perfect contact with the cooking top. One advantage of these utensils is their use as serving dishes, and even storage dishes in refrigerator and freezer, eliminating the need for other containers. According to Corning, speed of heating is comparable to the latest in electric units on rangetops. Each 8-inch unit provides 2000 watts, each 6-inch unit, 1200 watts.

Insulation below the heating elements prevents heat from going downward into the base cabinet. The first layer is high-temperature aluminum silicate fiber, the second a layer of Fiberglas that completely fills the remaining space in the 3-inch heater box (Fig. 7.13).

During cooking and afterwards, the counter may be wiped off with a damp cloth to remove any spatters, with no danger to the top, since glass ceramic is not affected by sudden changes in temperature. It is recommended that once a day the surface be rubbed over with the

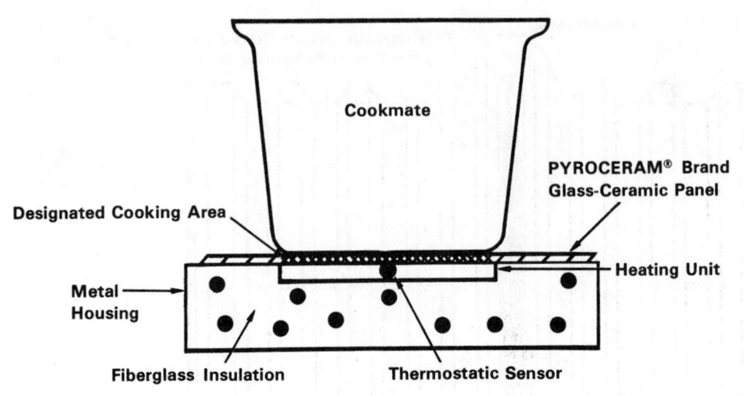

Fig. 7.13 The parts of each cooking area, with a Cookmate on the heating unit. (Corning)

Cleaner Conditioner, a liquid especially formulated by Corning that cleans, polishes, and protects the counter, leaving a coating of invisible silicones across the top. Several drops on a paper towel will do the job.

Corning has also placed on the market a complete range with a top of glass-ceramic.

Certain other manufacturers are also marketing cooking counters. Some of these counters have one or two areas of special heat, but other areas may be used for keeping foods warm.

Many companies are now featuring ceramic-top ranges, some even a whole line of them in different type models. Some ranges are supplied with a signal light that indicates if the ceramic surface is too hot to touch, if the temperature exceeds 120°F.

The Cool-Heat countertop range, is unlike any conventional range; it cooks without the usual heating elements, but by the application of electromagnetic induction within the cooking container itself. The method is said to be cooler, cleaner, safer, more efficient, and better controlled than is possible by conventional methods (Fig. 7.14 and 7.15).

This range, with a black or white ceramic surface, has solid-state electronic circuits within the range, indicated by the four patterns on the range top. An alternating current sent through the coils creates an oscillating magnetic field that induces a second electric current in the bottom of the cooking utensil. The metal of the utensil must offer some resistance to the passage of the current so that it will heat. Glass and ceramic materials offer too great resistance to current flow and will not heat; aluminum and copper are such good conductors that they do not provide sufficient heat for cooking. But

***Fig. 7.14** The Cool/Heat counter-top range cooks food without conventional gas or electric heating elements. Under the smooth, white-ceramic top are induction heating coils. When a pot or pan of magnetic materials such as iron or steel is put over a coil, where indicated, the pan couples with an oscillating magnetic field created by the coil, and the pan heats. The range surface does not have to be heated to heat the vessel. The unit is safer, cleaner, and faster than conventional ranges. It can also be used as a cutting surface even while cooking. (Westinghouse)*

it has been found that utensils made of steel including stainless steel and porcelain enameled steel and those made of cast iron or enameled iron possess sufficient resistance to the flow of current to provide

Fig. 7.15 The surface of the Cool/Heat counter-top range stays so cool that a steak is pan fried through a towel without even singeing the fabric. Infinite and constant heat selection makes it possible to keep liquids and foods at the exact temperature desired. (Westinghouse)

an easy path for a magnetic field and therefore give a desirable range of cooking heats. Units are the drop-in type fitting into openings in a counter top; each unit functions independently. Controls are permanent magnets of the slide-rule type and activate power switches inside the range permitting infinite heat selection from high to off. Cooking starts rapidly and heat decreases quickly at the end of the operation.

As an aid to cooking effectiveness a set of cooking vessels is supplied with each range. They have a stainless-steel exterior, a carbon-steel core to absorb magnetic flux, a thick aluminum layer to spread the heat evenly, and an inner coating of stainless steel or of Teflon.

The range operates on standard 220 volt, 60 ampere, AC power, the same power supplied for conventional electric ranges. An inverter circuit increases the frequency to a value above the range of human hearing (the upper limit of audibility is 20,000 cycles per second).

Safeguards provide against misuse of the range. A pilot light indicates when a unit is on, but if no utensil is over a unit or if it boils dry, a control automatically shuts off the range. If a utensil of copper or aluminum is erroneously used and the current in the induction coil begins to climb, the current is limited automatically to a lower safe value.

The range top never becomes hot; it may be warm from heat given off from the cooking utensil. Any spills or stains cannot burn on—they wipe off. The magnetic controls lift off for ease in cleaning. The entire top can be used as a counter work surface, even as a cutting board, ceramic-glass is so tough; it is not damaged if utensils are accidentally dropped on it (Fig. 7.14).

Another new type of range is called the Touch-N-Cook range. Instead of switches to turn heat on and off, the homemaker touches control pads on a panel of tempered glass that covers the entire backguard of the range and a logic circuit similar to those used in a computer does the rest. The numerical controls on the panel indicate time and temperature settings for the cooking operations (Fig. 7.16*a*).

The panel is divided into three areas; on the right the controls for surface cooking (heat values of 1 through 9 are provided below the surface controls), in the center number pads give the settings for temperatures and times, on the left are the oven controls with a clock and 12-hour timer. (Fig. 7.16*b*). A surface timer can be preset to turn off the surface cooking areas or to hold food warm for serving. A preheat cycle is automatically programmed with use of the oven, an automatic broil cycle controls heater wattage and broiling time for varying degrees of doneness. It is claimed that this touch system provides greater accuracy than the conventional method.

Any error in operating a control which is beyond the limitation of the range is indicated by a flashing light.

Any pan may be used with the range; a pan with a flat bottom is most efficient for use on a surface unit. It gives even heating and therefore best cooking results.

Just two steps are needed to set a control: touch the control pad and then the number or numbers required for the setting. For example, in cooking bacon for breakfast, touch "right front," then "5" or "6" for medium heat. When the bacon is done, touch "right front," then "off."

When the oven needs cleaning the homemaker latches the door and touches "clean" on the panel. The oven heats for a 2-hour cycle, then turns off automatically; after an hour for cooling, it is again ready for use. The ceramic top is easily wiped off at any time.

Fig. 7.16a New electric range with unique "touch" controls will bring computer technology into the kitchen for the first time on a practical basis for everyday use by the homemaker, simplifying home cooking operations. It makes use of a glass touch panel, integrated circuits, and solid state components. (Frigidaire)

The smooth control panel with no projections or indentations is washed with a sudsy cloth, then rinsed and dried. In case of need this panel is easily removed and replaced by a service man.

Installation of an Electric Range

The cost of installing an electric range depends largely on wiring conditions. The National Electrical Code requires that where only one range is to be installed in a private residence, the wires in the branch circuit for the range must be of sufficient size to carry the connected power load of the range. When the range is to be installed in a residence where ample wiring in the entrance feeders is already provided, the installation cost of the range is comparatively low, as only service wires from the meter to the range must be added; but,

***Fig. 7.16b** Close up of panel. (Frigidaire)*

if the range is installed in a house in which the entrance feeders are for the lighting load only, a new service circuit must be installed from the secondary main to the meter, and this additional wiring considerably increases the cost.

The size of wires used in range installation depends upon the type of wiring and the capacity of the range. In wiring a service circuit from the meter, however, it is good practice to use wires of sufficient size to carry a large-sized domestic range, in case of future need. A No. 6 wire, of 120/240 volts and 60 amperes capacity, will usually meet the installation requirements.

An electric range should always be used on a circuit in which the voltage is approximately the same as that for which the range is designed. Modern high-speed heating elements should not be subjected to voltages above the maximum operating voltage rating. In a range designed for 118/236 volts, the maximum operating voltage is 124/248 volts. Increased voltage increases the temperature of the heating element, and experiments have shown that, as the temperature increases, the element's life is shortened. When the line voltage is below the design voltage, the normal speed of heating is decreased, but there is no harmful effect on the life of the unit. The life even may be increased, but the efficiency of the range will be reduced, it will be slow in operation, and the oven may give unsatisfactory baked products.

Electric ranges should be grounded, as a protection against electric shocks. The usual practice is to ground the range frame on the neutral wire of the range circuit. The ground wire may have to carry a large current, and wire smaller than No. 10 copper wire should not be used.

The National Electrical Code requires that each appliance rated at more than 1650 watts shall be provided with some means of "cut-off" from the electric circuit. In single-family dwellings the range may be connected to an outlet by a heavy cord and plug, or it may be permanently installed in its own circuit, which is supplied with a switch to turn the current on and off if necessary. The switch is more economical, but the plug connection to an outlet eliminates the necessity of calling a mechanic every time the range is disconnected or moved. A plug also provides a quick, safe means of disconnecting the range in case of fire and in servicing.

An electric range should carry the Underwriters' Laboratories Seal of Approval (see Fig. 1.6).

It is recommended that the homemaker read and follow the manufacturer's instructions that come with the range. The instructions take into consideration the materials used, the construction and design of the units, and the intended use of the range.

In 1972 the Association of Home Appliance Manufacturers (AHAM) published a "standard to establish a uniform procedure for determining the performance characteristics of household electric ranges under specific test conditions and to establish certain minimum requirements."[1] This standard was approved by the American National Standard Institute and adopted as an American National Standard.

The information obtained was based on tests on all parts of the range, many of the tests requiring special apparatus and special areas of controlled temperature and humidity, conditions not always available in the college equipment laboratory. Representative information includes:

1. Nameplate. A nameplate shall be installed whenever possible in front of the range, visible with the oven door open, or on the range in a

[1]AHAM ER-1, "Household Electric Ranges," p. 7.

convenient location accessible for inspection without moving the appliance after it has been installed. It shall be made of suitable materials.

2. Marking. Each heating unit shall be permanently marked with a plate or stamping showing its rated watts input, rated voltage and identification necessary to facilitate replacement.
3. Overcurrent protective devices. Plug fuses or circuit breakers shall be easily accessible without moving the range from its installed position and so located that easy replacement or resetting by the user is facilitated.
4. Wiring diagram. Electrical diagrams to all the circuits within the appliance shall be attached to the range.
5. Range lamps. All range lamps shall be easily replaceable without the use of special tools. Mounting shall be such that adequate finger room is available to grasp the lamp for insertion or removal.
6. Rated watts input. When a test is to be conducted at rated watts input, the watts input shall be maintained within plus or minus 2%, unless otherwise specified.
7. Electrical measurements. All electrical measurements shall be accurate within plus or minus 1%.
8. Oven Doors. All oven doors shall be equipped with a closing device which will hold the door firmly closed. Drop-opening doors shall be counterbalanced and shall have a point between 30 and 60° where the door will remain open.
9. Cooktop, oven lining. Suitable materials and finishes for the exposed surfaces of these parts are porcelain enameled metal, metal finished with a heat and corrosion-resistant metallic coating, and heat- and corrosion-resistant metal.
10. An oven switch shall be provided unless the oven thermostat embodies an "on-off" switch.
11. Each oven shall be equipped with a thermostat which will provide automatic control of the oven unit and permit cooking operations in the oven to be accomplished by maintained temperature. The thermostat shall have an accessible calibration adjustment.
12. Each surface heating unit shall be individually controlled by a switch. The switch shall provide at least five heat control positions. The various positions shall be legibly and permanently marked on or immediately adjacent to each control.
13. The heating efficiency of surface units (including deep-well cookers) shall be not less than 55%.
14. The oven-air temperature shall reach 330°F (166°C) above room temperature in not more than 10 minutes when operated in accordance with the manufacturer's instructions.

Tests on baking and browning performance are included in the experiments.

An appendix, not a part of the standard, was included to give guidelines for the preparation of use-and-care booklets for electric ranges, to avoid the following shortcomings of many existing booklets.

Subject matter not understandable to user.
Content does not encourage reading.
Omission of answers to often-asked questions.
Language and illustrations too technical.
Too many models covered in one book.
Appearance unattractive.

Some suggestions for improvement were the following:

Cover of booklet should be attractive.
Size small enough for easy handling, large enough for good presentation.
Print clear and sharp, easy to read.
Language easily understood by nontechnically oriented women (and men).
Separate booklet for each range model.
Be brief, but complete, all in one booklet, no additional sheets.

THE ELECTRONIC RANGE

Home-size electronic ranges have been on the market since 1955. All previous methods of cookery applied heat to the outside of the food; in electronic ranges the microwaves penetrate the food, setting the food molecules in motion, an action that generates heat within the food and brings about the cooking results. The process is very rapid, frequently two to ten times faster than by conventional methods, depending on the amount of food.

The electronic oven per se is not designed for quantity cookery. The cooking ability of the electronic oven is comparable to that of a large surface unit, a fact that should be taken into account when planning the use of the oven. For example, a large casserole for eight or ten persons can overtax the electronic oven and should be baked in a conventional range.

The electronic oven is used for cooking most foods usually cooked in gas and electric range ovens, including foods commonly cooked on surface units. It is also used to thaw frozen foods, to heat left-overs, to freshen slightly stale bread products, to warm the baby's bottle, and to melt butter and chocolate without danger of burning.

Microwaves themselves are sources of energy, not heat. They may be reflected, transmitted, or absorbed (Fig. 7.17). Metals **reflect** the waves; consequently the usual kitchen utensils of aluminum, stainless, and iron cannot be used. Glass, ceramics, paper, and several

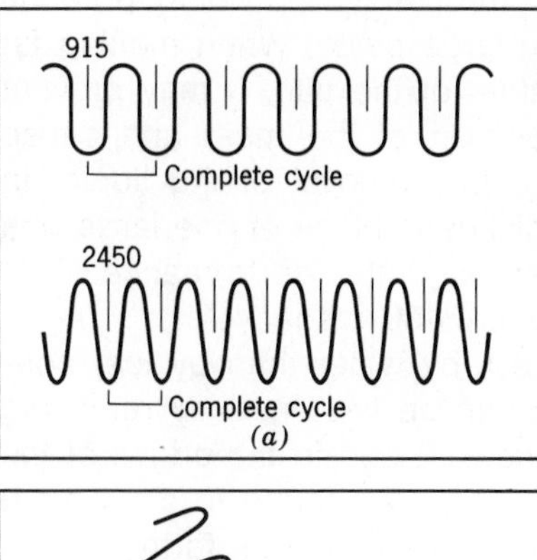

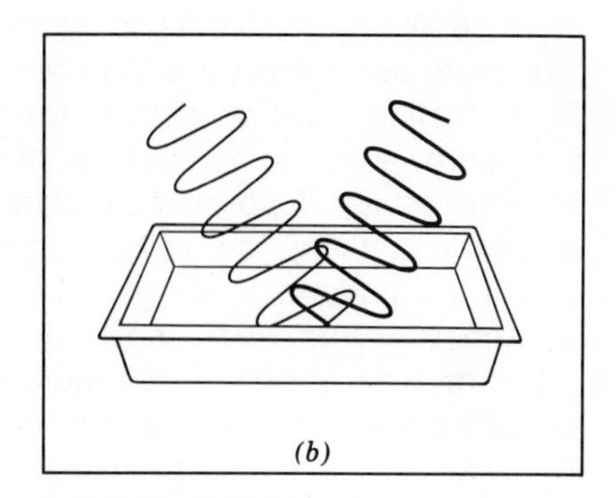

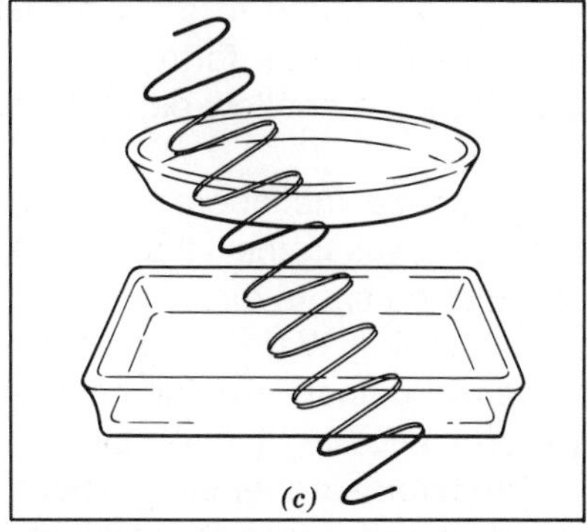

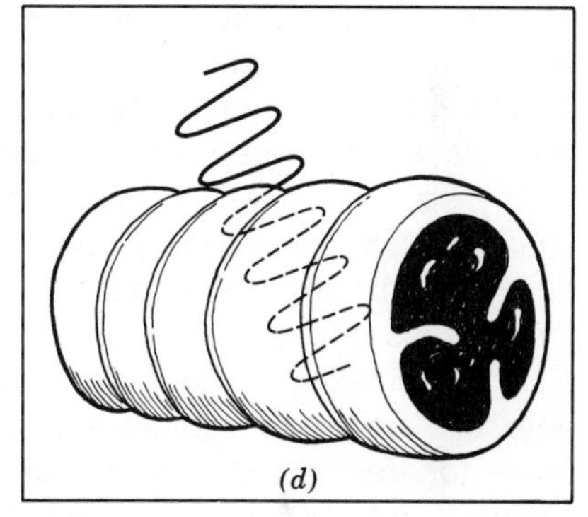

Fig. 7.17 (a) ***Comparative lengths of cycles of 915 MH_Z and 2450 MH_Z.*** (b) ***Metals reflect microwaves.*** (c) ***Glass transmits the waves.*** (d) ***Food absorbs the waves. (General Electric)***

forms of plastics **transmit** microwaves and are the materials commonly used for food containers in electronic cookery. The foods themselves **absorb** the radiations.

Most foods may be cooked by the electronic method, but some foods cook more satisfactorily than others, and certain methods tend to improve the results. For example, lining a glass cake pan with paper gives a larger volume; the moisture given off in the baking process is absorbed. On the other hand, glazed sides of the glass pan tend to leave the surface of the cake wet, so that it does not rise as high. Pies are best when baked on porous paper plates. Many foods cook uniformly; vegetables and fruits are included in this category and give desirable products without the use of additional water. Since the microwaves penetrate only 2½ to 3 inches, roasts preferably should be rolled and be long in comparison to width. Uneven penetration of the waves may be improved by turning the meat over once or twice during the roasting process; frequently, individual portions give more satisfactory results than one large piece. When melting fat from roasting meat collects in the bottom of the pan, it may prevent the microwaves from reaching that portion of the meat and cause uneven results. If possible, provide for the removal of the liquid. In casserole cooking, too, several small dishes in place of one large one give more uniform heating and, therefore, better performance. This is especially true when the frequency is 2450 megacycles.

Microwaves are absorbed more readily by water than by ice; consequently the thawing of frozen foods can be hastened by removing the water as it forms and moving the thawed food to the edges of the container. In the 2450 megacycle oven, food was found to cook more evenly in a moderately thin layer than when in loaf form several inches thick, i.e., a meat loaf should be spread in a shallow pan rather than placed in a deep, narrow one.

The size and shape of the pan does influence the operation. Use shallow pans rather than deep ones, small pans rather than large ones, round rather than square or rectangular pans. As already noted, Pyrex glass and glass-ceramic cookware (Corning Ware®) are useful dish materials. Corning makes glass-ceramic dinnerware for household use under the Centura trademark and another line of dinnerware for restaurant use under the Pyroceram trademark. These products are not recommended for microwave use, since these glass-ceramic materials differ from those used in Corning Ware products and they may break during microwave cooking. Certain plastics, polyesters, Teflon, and silicone plastics also make desirable utensils. In cooking large portions, rest periods are valuable. This is especially true if the food cannot be stirred during the cooking process. It may be removed from the oven to the counter for 10 to 15 minutes or longer, then returned to the oven. The rest period allows the heat to penetrate toward the center of the food and results in more even cooking. Microwave ovens may be used also for barbecuing meats and poultry. They are subsequently smoked in a separate unit. The total time for the process is from 9 to 12 minutes, compared to 36 to 42 minutes by conventional methods. A tasting panel rated all of the microwave barbecued products as good as those barbecued over hardwood coals.

Microwave heating is a radiant process, a part of the electromagnetic spectrum. The penetration of the waves into food depends on their frequency. The term frequency has become familiar to us in its application to the electric current used in the home, which has a frequency of 60 and, occasionally in certain areas, 50 cycles per second. In contrast, microwaves in one domestic range on the market have a frequency of 915 megacycles, this is, 915,000,000 cycles, per second, and other electronic ovens 2450 megacycles, or 2,450,000,000 cycles, per second. These microwaves are the same type as the Hertzian electromagnetic waves used in TV and radio, and the United States Bureau of Standards has adopted the term Megahertz in place of megacycle, i.e., 2450 MH_z, instead of 2450 mc. The length of a radio wave is approximately 0.3 of a mile, compared to 4.8 inches for the length of the 2450 MH_z.

Magnetron Tube

The generator that produces the waves is a magnetron tube (Fig. 7.18*b*). The term electronic has been associated with products that use vacuum tubes in their operation. The magnetron is also a tube, a specialized one used to develop microwave energy. The energy is directed into a tunnel, or wave guide, from which antennae inject it into the oven cavity. The magnetron must reach a certain temperature before it can have high voltage applied to it. To reach this temperature requires from 3 to 90 seconds, depending on the model of the range. At the end of this time a light appears on the range panel beside the time control to signal that the oven is ready for use. Microwaves travel in straight lines and should strike the food from all sides. To aid in this process, the oven is lined with stainless steel, chromium-plated steel, or aluminum to reflect the waves back to the food from all directions.

To help distribute the energy uniformly in the oven a rotating shelf or a fanlike stirring device is used. In a late model the rotating shelf is discontinued and replaced by a vaned stirrer in the roof of the oven that distributes the microwaves. The wave penetration is related to the frequency. Heating at the 2450 frequency is largely peripheral; with the lower frequency of 915, the penetration is deeper, sometimes spoken of as "core" penetration.

Allan Scott notes that the 2450 MH_z frequency gives more uniform distribution of heat and is used by the majority of manufacturers of electronic ovens. Scott also states that the power output of any tube depends on the load in the oven. "The load varies with the amount and location of food in the oven, and with the instantaneous position of the slowly revolving fan within the oven which distributes the microwave energy in a constantly changing pattern to provide even cooking of the food. Approximately 90% of the power delivered

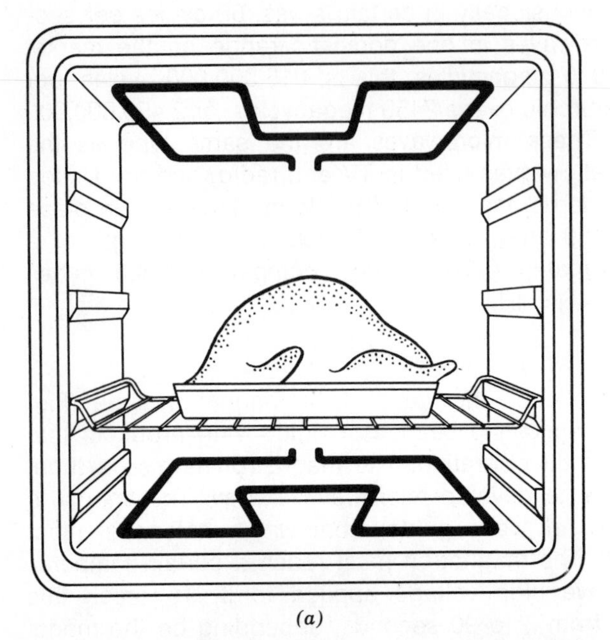

(a)

Fig. 7.18 (**a**) ***Turkey roasted in the conventional electric oven.*** (**b**) ***Similar turkey cooked in the 915-MH_Z oven. The food is placed on a rotating shelf connected through the bottom of the oven to the magnetron. The antenna distributes the microwave energy in all directions. (General Electric)*** (**c**) ***2450 MH_Z energy emitted by the magnetron antenna (A) strikes a slowly revolving fan (B), which reflects the power, bouncing it off walls, ceiling, back, and bottom of the oven. The power enters the food from all sides to cook it evenly throughout. (Amana Refrigeration, Inc.)***

to the oven is absorbed by the food."[2] Heating is seen to be proportional to the size of the load; for example, one average potato will bake in four minutes, but four potatoes require eight minutes.

Types of Ovens

Portable or counter-top units operate on 115-120 volts, the standard household current, and require no special wiring. Larger, free-standing and built-in units, if combined with conventional range units, will be installed on 120-240-volt circuits. The portable units are very popular and are used not only in the kitchen but also in the family

[2]Allan W. Scott, Designers' Guide to Microwave Oven Tubes, **Appliance Manufacturer,** April 1969, p. 71.

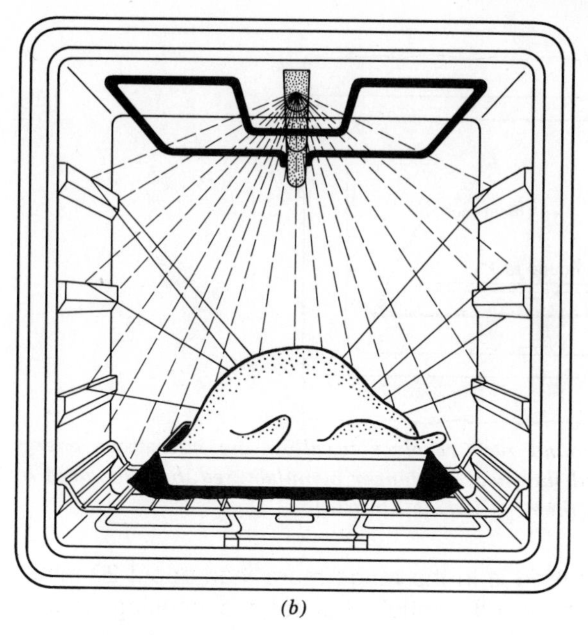

(b)

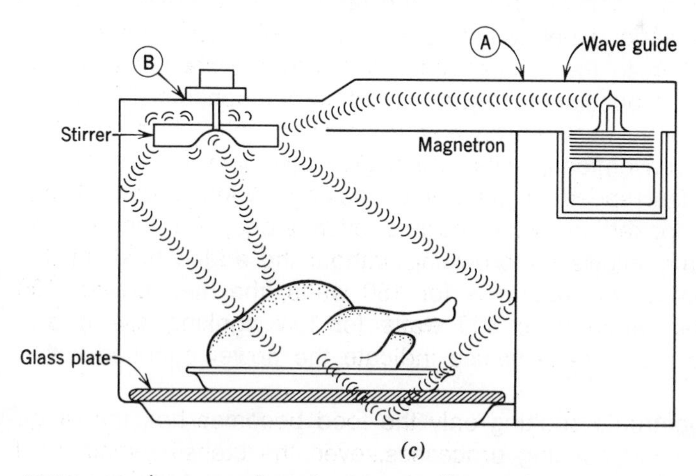

(c)

room or den or placed on a rollaround cart to be taken to the patio or back yard picnic area.

One company includes a broiler unit in the top of the electronic range, designed to prevent the unit coils from interfering with the movement of the microwaves. Other companies add conventional electric ovens to the electronic cooking centers.

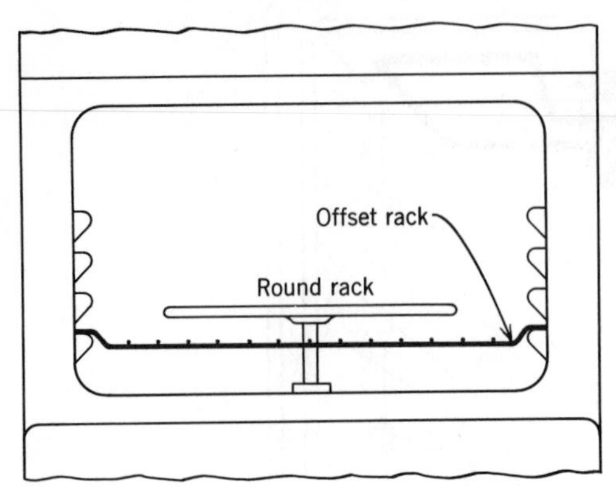

Fig. 7.19 The round rack revolves to allow the microwave energy to be distributed around the food. No longer manufactured, but still used by homemakers who own them. (General Electric)

Usually, if the food is in the range more than about 20 minutes, the surface will be browned enough, but for the cooking of pies, rolls, and quick breads, all short processes, a standard, thermostatically controlled oven is considered essential to give the customary golden-brown that people prefer to see.

One model is a conventional electric range with two ovens, one at eye level, the second below the surface units. The lower oven contains, in addition to the regular units, a removable shelf that makes contact, when in position, with the electronic components (Fig. 7.19). This oven may be used as an electronic oven only, or in conjunction with the conventional units, which give desirable surface browning with the fast speed of electronic cooking. When larger masses of food are cooked in the electronic oven alone, the longer time required usually allows for browning without the added heat. In this range the low power setting is for 150 watts, the high power, 700 watts. Another range uses 500 watts for low cooking speed and 850 watts for high. These values indicate the power supplied within the range.

Since in electronic cooking only the food becomes hot, the range itself and, in short cooking processes, even the utensil remain cool to the touch. The food may therefore be removed from the oven without using a pot holder. But if the food cooks longer than a few minutes, heat will pass to the utensil, and it will also become hot. When the food stands on the kitchen counter, it will continue to cook, a fact that should be remembered, or the food may be overcooked. In this case, too, heat from the food will be transferred to the utensil, which may become hot enough to require the use of a pot holder. Moreover the food may tend to stick to the utensil, a situation than can be avoided if the food is put immediately into a serving dish.

In several electronic ovens two controls are provided; one offers a choice of high or low power, the other is the time control. There are no thermostatic controls. The oven setting turns the operation on and off, and at the same time sets the limits of the cooking period. At the end of the indicated time, the microwaves shut off automatically, and a signal is given. The opening of the door also stops the production of microwave energy instantly, preventing any possible danger. The metal screen in the door of certain models is scientifically designed to keep the microwaves from passing through it.

Tested recipes are furnished with each oven to aid the beginner in successful operation.

Care

The electronic range is easy to clean. Since the lining does not become hot, food particles cannot burn on or stick to it. The interior may be wiped occasionally with a cloth wrung from warm suds, and then dried.

To maintain a close fit the door seal should be wiped frequently with a sudsy cloth, then rinsed and dried; if any change in door-seal operation is observed, have it checked immediately by a competent service organization. Never operate an empty oven. Keep a dish of water in the oven between times of use.

Limitations

Certain foods, notably soufflés and angle cakes, have not been baked very successfully in the electronic oven. Research experiments on electronic cooking methods have been carried on by Van Zante and Nakayama in the household equipment laboratories at Iowa State University, with the use of agar in cylinder form. These experiments showed much greater unevenness in the internal heat patterns of agar heated in the electronic range, than in that heated in the conventional electric range oven. Van Zante assumed that similar results would be obtained in foods cooked in the electronic oven, and classwork with meat patties, although not so carefully controlled, showed similarly uneven cooking. Other foods have shown similar irregularities in evenness of doneness.

Van Zante also noted that load size and the position of the individual portions influenced both the cooking time and the degree of heating in the separate items. A special problem in the laboratory, using potatoes for the medium, confirmed these results. When eight potatoes were baked at the same time, they had to be moved around shifting, the potatoes in certain positions were better done at the end in the oven occasionally to obtain even cooking. In spite of this of the period than potatoes in other positions.

In experimental work in the Equipment Laboratory at Utah State University, Chatelain used the 915 MH$_{Z}$ electronic oven in a comparative study of cooking certain foods by three methods. The same foods were cooked entirely electronically, with a combination of conventional electric heat and electronic waves, and in the usual conventional electric oven. Preliminary tests to determine the evenness of heat in the different ovens used water in plastic drinking cups as the load. The results indicated unevenness of heat in the electronic oven, even though the cups were placed on a rotating shelf. Similar unevenness was observed when the water was heated in the conventional electric oven. Chatelain obtained comparable results in the electronic oven when cooking cakes in glass custard cups, placed around a cake in a larger container; the cakes varied in doneness at the end of the baking period depending on their position in the oven. Apparently the large cake baffled the heat in such a way that it did not reach all the cakes in the custard cups equally well.

It is of interest to learn that the Federal Trade Commission assigns definite wave lengths for the operation of electronic ranges, so that the use of these ovens will not interfere with radio or television reception.

Safety

In the spring of 1973 a consumer testing organization questioned the safety of microwave ovens, and this led to Senate hearings on U.S. radiation-control laws. One difficulty arose from failure to distinguish between ionizing and nonionizing radiant energy. Exposure to ionizing radiation-X rays, gamma rays, and cosmic rays has a damaging effect on cells and chromosomes, and the effects tend to build up through repeated exposure. Nonionizing radiant energy, of longer wave length, includes radio waves, microwaves, infrared light, and visible light and is not harmful to plants or animals, including humans. Microwaves are located between radio waves and visible light.

Over 700,000 microwave units are in use without a single record of harm to persons. These ovens must meet federal safety standards developed by the Bureau of Radiological Health, the Food and Drug Administration, and the U.S. Department of Health, Education and Welfare in cooperation with U.S. industry under the leadership of

AHAM. Ovens are also tested and approved by Underwriters' Laboratories. These strict standards indicate that exposure to microwave oven radiations must not be above 1/10,000 of the level that might cause injury, a very large safety margin. Leakage from the oven can be no more than one milliwatt per square centimeter measured two inches from the oven at the factory and not more than five milliwatts ever afterward. Two leading scientists told a Senate committee that one has about as much chance of being harmed by rays from a microwave oven as of getting a sunburn from moonbeams.

Moreover, two, and in some cases three, special safety interlock switches prevent the opening of the door when the oven is in operation and shut off the energy automatically when the door is opened. A special door seal keeps any chance high level radiation inside the oven.

THE GAS RANGE

Most gas ranges are conventional four-burner models, but six-burner ranges are available for large families. These larger ranges will have two or more thermostatically controlled burners and may feature a controlled built-in griddle at one side or in the center of the range top, frequently finished in Teflon. This griddle may be constructed to be replaceable when necessary with an extra-large grate; the super-large burner beneath can then be used for heating out-sized utensils. The trend, however, is toward a portable griddle rather than the built-in type.

Burners

In the gas range the source of heat is known appropriately as a burner, to which the flow of gas is directed by the turning of a valve handle.

Although burners differ somewhat in form, the principal parts are the same, with mixer head, mixer throat, and burner head cast in one piece to prevent gas leakage (Fig. 7.20). The mixer head carries an air shutter and an opening for the gas orifice. The orifice is the aperture in a needle through which the gas flows and by means of which the flow is controlled. Most of today's ranges are equipped with double coaxial orifices. The orifice in the needle is sized for use with liquefied petroleum gas, and sets of outer orifice hoods are supplied for the other gases. One of these sets of hoods, usually the set of natural gas hoods, is in place. All hoods are clearly marked to indicate the gas for which they are to be used. Orifice hoods are sized

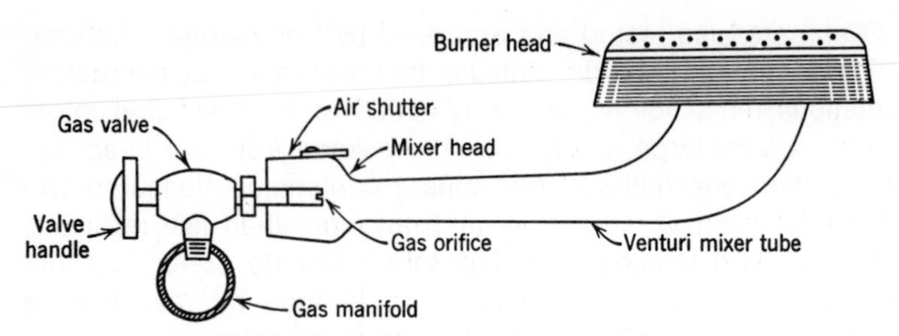

***Fig. 7.20** Gas burner construction; gas and air are mixed as shown in the modified diagram. (American Gas Association)*

to within 15% of their respective burner ratings at 4.0 inches water column pressure.

Ranges complying with optional natural gas requirements are marked for use with natural gas only and have fixed orifices. Similarly, ranges for use with liquefied petroleum gas are equipped with fixed orificies by the manufacturer.

Openings in the burner, known as **ports,** are drilled horizontally on the outer rim and horizontally, or at a 45-degree angle, in the center of the burner. The inner ports, four or five in number, heat the central portion of the utensil bottom and frequently form the simmer section of the burner. Ports may be an integral part of certain die-cast burners. Instead of ports, some burner heads have continuous slots to give rings of heat.

The horizontal pipe through which gas flows from the fuel line to the different orifices is called the manifold. On present-day ranges the manifold is usually concealed behind the front panel of the cooking top, but it may be behind the backguard. Attached to the manifold are burner valve handles which direct the gas through orifice and mixer head into the burner. Forced through the orifice at a velocity of 100 to 160 feet per second, the gas develops sufficient suction to draw air through the partly open shutter. This air, called primary air, mixes with the gas, before it is ignited, in a porcelain-lined venturi tube, the smooth finish of which increases the injection of primary air and gives a clean, sharp flame. The rough interior surface of the common cast-iron burner tends to cause gas turbulence. Primary air amounts to about 60% of the air needed for complete combustion.

The gas-air mixture now flows through the ignition port on the side of the burner head. This port is connected to the pilot light by a flash tube, so named because the tube contains a proportion of air too large for normal combustion and the flame flashes back to the burner and ignites it. The top of the pilot flame should be level with the center of the flash tube. The mixture will not burn without the presence of secondary air that surrounds the burner head and supplies the additional oxygen required.

Several manufacturers are now featuring solid state ignition. The method creates sparks that ignite the gas when the burner is turned on, and then the electric system turns itself off. An electric ignition system is also used more and more frequently to ignite gas pilots, and sometimes burners directly.

On divided-top ranges, the pilot light is commonly placed between two burners, but in this range, and in many others, there may be a pilot light beside each burner. The individual pilot lights are miniatures, so tiny that a minimum of heat is given off. In this arrangement no flash tubes are used. All burners should light within 4 seconds, and the flame should carry immediately across all ports.

Burners are of three sizes; **giant** for rapid heating and for large utensils, **standard** for general cooking in average-size pans, and **simmer** for low heat (Fig. 7.21). The simmer burner is usually a part of a larger burner but with its own mixer tube; when the valve is partly closed, the gas is shut off from the main burner and the simmer alone supplies heat. When no simmer position is provided, the gas may be reduced at all the ports to the amount of heat required.

The most efficient and economical method of operation is to have the gas turned on full. The simmer is, therefore, to be preferred to a larger burner turned down—the simmer is burning at full speed and only the size is reduced. This small flame prevents excessive escape of heat into the kitchen and is an asset in waterless cookery.

According to specifications set up by the American Gas Association the burner ratings in Btu (British thermal units) per hour are:

Giant—not less than 12,000 Btu per hour.
Regular—not less than 9,000 Btu per hour.
Simmer—not less than 1,200 Btu per hour.
Pilot—not more than 300 Btu per hour.
Minipilot—not more than 125 Btu per hour.

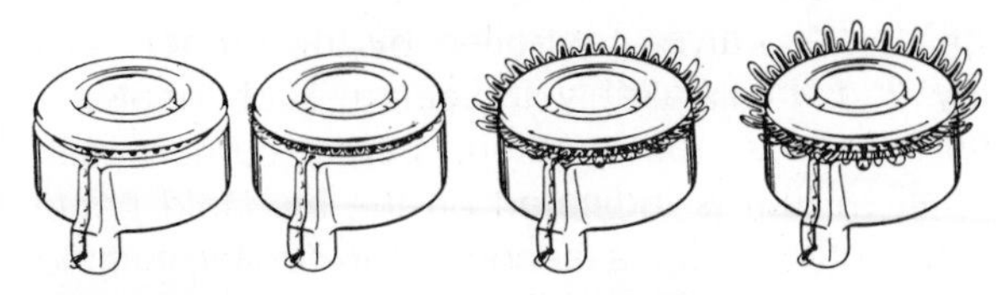

***Fig. 7.21** Gas burners make available the exact heat needed, from the tiny "keep warm" flame to "gentle simmer," "medium high," and "high." (Roper)*

For natural gas this rating is equivalent to 11 cubic feet per hour for the giant burner and an average of 8.3 cubic feet for the standard burner. The burner of slot construction is a high-speed type with a heat capacity of 18,000 Btu per hour. A standard or giant burner, or a section thereof, classified as capable of simmering operation, must be able to maintain a stable flame even when gas is flowing to the ports at only 500 Btu per hour.

Flame from the horizontal ports projects outward for an inch or two; it is then lifted upward against the utensil bottom by the inrush of secondary air around the burner, and by the natural expansion of the gaseous products of combustion. The flames extend an inch or more above the grates and should have a 6-inch spread under the utensil on a standard burner and a 7¼-inch spread on a giant burner. These flame types have a wide range of heat adjustments and may be turned fairly low without being extinguished. The simmer flame should not extend above the grate and should maintain water—in amounts up to 6 quarts—at boiling temperatures when the vessel is covered. In thermostatically controlled burners, the outside border of the flame should be an inch from the edge of the pan.

For minimum operating cost, the flame should be as close to the bottom of the utensil as possible without the formation of carbon monoxide gas. The usual distance between grate top and burner is one half to three fourths of an inch. The short distances have proved especially satisfactory on ranges using a fixed orifice with a governor to maintain constant pressure. Some utilities secure greater efficiency by installing an individual pressure regulator with each range which eliminates fluctuations in pressure and results in more even combustion of the gas.

According to A.G.A. Approval Requirements, the thermal efficiency of top burners shall not be less than 40%.

Valve handles must be so designed that they can be attached in the correct position only. A top valve handle shall be distinguishable from the oven and broiler handles, and be clearly identified with the burner it serves. On any given range, all handles shall have the same direction of rotation. At a distance of 10 feet the handle shall indicate whether the valve is open or closed.

The "off" position of the gas valve, controlled by the burner valve handle, must be clearly and unmistakably indicated. High heat and simmer positions should likewise be marked. Some burners also have a medium heat position that is indicated on the manifold panel, and several manufacturers provide for a second stop for the simmer burner, where the flame is so small that only enough heat is produced to keep the food warm. Several models are supplied with safety push-in valve handles so that the gas cannot be turned on accidentally; this is an especially desirable feature where there are young children.

Burner grates are usually of cast iron, occasionally of steel. They should be sturdy, firmly supported to prevent any shifting in excess of an eighth of an inch, and light in weight to absorb as little heat as possible. Minimum contact with the range top prevents heat transfer away from the cooking vessel and increases efficiency. A good deal of variety in design is apparent in the grates on new ranges, especially in grates over burner bowls. ANSI Approval Requirements for gas ranges state the grate arms shall support a utensil 2¼ inches in diameter, placed centrally over a burner.

The gas oven is heated by a burner beneath the oven floor. The floor plate has openings at the corners or along the sides through which the convection currents rise to circulate throughout the oven. Food in utensils placed on the hot oven shelves is heated partly by the convection currents, and partly by the radiation that occurs once the lining of the oven becomes heated. An oven vent carries off products of combustion and volatile food fats and moisture; it opens on to the front surface of the backguard and so directs the vapors away from the wall and into the room. A range hood or kitchen ventilating fan is an advantage in removing these volatile materials before they are deposited on the walls. Because oxygen must be available for the combustion of the gas, the gas oven is not as tightly constructed as the electric oven.

The oven burner, with a gas input of 21,000 to 24,000 Btu per hour, is similar in construction to the surface burner. It has a double row of ports on the upper or lower side of the tube, so spaced that the flame runs from port to port when the burner is lighted at one point. The flame must not flash back when the temperature regulator is moved up or down and shall not be extinguished by the opening or closing of the door. A self-latching valve handle is mandatory on all gas ovens.

The pilot, usually the low-temperature type, is a safety device. Gas comes to it directly from the main supply, and is not affected by the condition of the thermostat. Sufficient gas passes to keep the pilot burning continuously whenever the burner is lighted. If the burner flame should be extinguished accidentally, the automatic pilot would re-ignite the escaping gas and prevent an explosition. The automatic pilot, now installed on all ranges displaying the Blue Star seal, must meet the ANSI Approval Requirements, which set definite

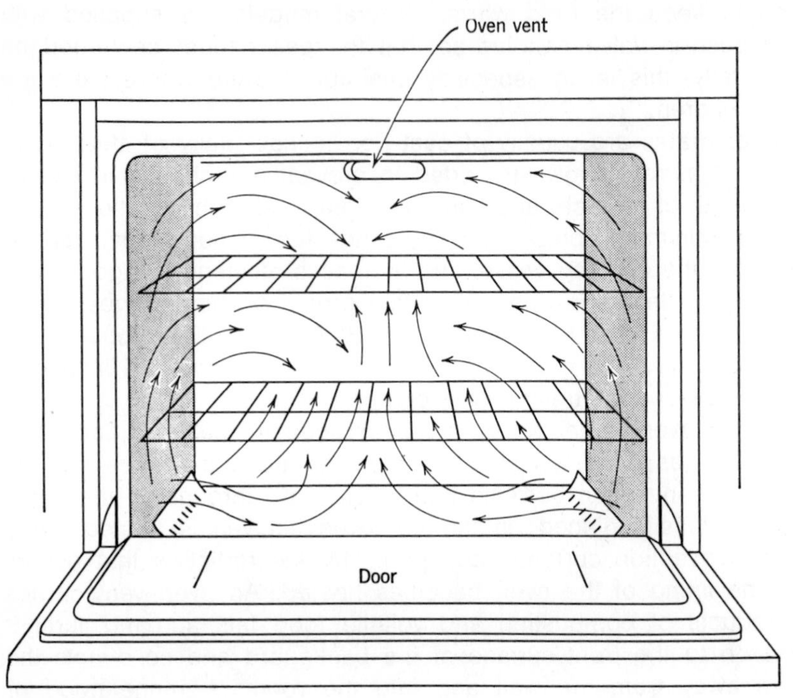

Fig. 7.22 Circulation of heated air in the gas oven. (A.G.A.)

lengths of time within which the pilot shall turn on or close the gas supply. These requirements also include times for the action of the electric pilot igniter in lighting the gas pilot when such an igniter is employed (Fig.7.22). Time allowed is 36 seconds.

Types of automatic oven-ignition systems:

1. **Standing pilot.** A small standing pilot beside the oven burner ignites the gas when the oven control is turned on. If this pilot goes out, it must be reignited with a match, but this rarely happens.
2. **Electric.** When the oven control is turned on, a glow coil (hot electric wire) lights a pilot, which then ignites the oven burner.
3. **Solid State (electronic).** When the oven control is turned on, tiny sparks light a pilot, which ignites the oven burner.

When an automatic clock is used with the automatic oven pilot, it controls the shut-off valve. The mechanism opens and closes the valve at the times indicated and, with the aid of the pilot that heats an expansion element, ignites or turns off the oven burner.

According to the ANSI Approval Requirements, ovens and their controls shall be so designed that the oven temperature can be increased from room temperature to 400°F within 10 minutes. The change in a predetermined oven temperature shall not exceed 25°F as the result of simultaneous operation of the oven and any other section of the range for one hour.

The temperature in the oven is controlled by a hydraulic-type thermostat. As the temperature increases, the liquid in the tube expands, partly closing the valve and decreasing the supply of gas to the burner.

To prevent the flame from being extinguished when the main valve is nearly closed, a by-pass valve always permits enough gas to enter the burner to form a bead flame about ⅛ inch high. The flame must be just large enough to hold the temperature at the dial reading without allowing it to creep gradually higher. It is sometimes difficult to maintain the desired low temperature with a ⅛-inch flame when the oven is well insulated. The flame must not go out if the door is opened suddenly or accidentally slammed shut. Frequently electric ignition is used for all burners, so no pilot is required. Many ovens in present-day gas ranges cycle off and on to maintain accurate selected temperatures down to 140°F.

Modern methods of installing oven burners reduce the amount of secondary air so that less gas is required to maintain a given temperature. The thermostat works most effectively if the gas is always turned on full. In all thermostats, the gas valve is designed to allow almost the full flow of gas to continue until the oven temperature is within a few degrees of the reading on the dial wheel; this full flow of gas means that the homemaker cannot shorten the preheat period by setting the dial temporarily for a temperature higher than the one desired.

Supports for oven shelves should be spaced at each 3½ inches of oven height.

With the increasing emphasis on the dietetic value and attractiveness of broiled foods, the broiler has become a necessary feature of the range.

In the gas range, the broiler compartment is often below the oven so that the same burner is used for broiling and baking, although not at the same time since broiler heat is too intense for most baking operations. In large ranges, the broiler may be in a separate section, on a level with the oven. The broiler, often called a waist-high broiler, may be built in the form of a drawer, like the one below the oven, with supports at several different levels on the sides for the broiler pan, or the broiler may be attached to the door frame so that it swings out. In either case the burner is above the broiler pan, of course, and it is usually assumed that any volatile products given off in the broiling process are consumed as they rise into the flame. Most gas broiler compartments are provided with three support positions into any of which the broiler pan may be placed.

In present-day gas range models the broiler burner frequently is in the oven itself, at the top, and may be supplied on either side with steel-mesh panels that become red hot and so give off radiant heat. When the broiler is located within the oven cavity, there has been some difficulty in keeping the broiler pilot lighted. Robertshaw has developed a new "Unidial" system to overcome the trouble. A mercury vapor safety pilot acts as a gas diverter. The diverter has two outlets, one to the bake burner, the other to the broil burner. Normally the diverter is open toward the bake burner, but when the dial is turned to broil, the sensing bulb of the diverter is heated, and

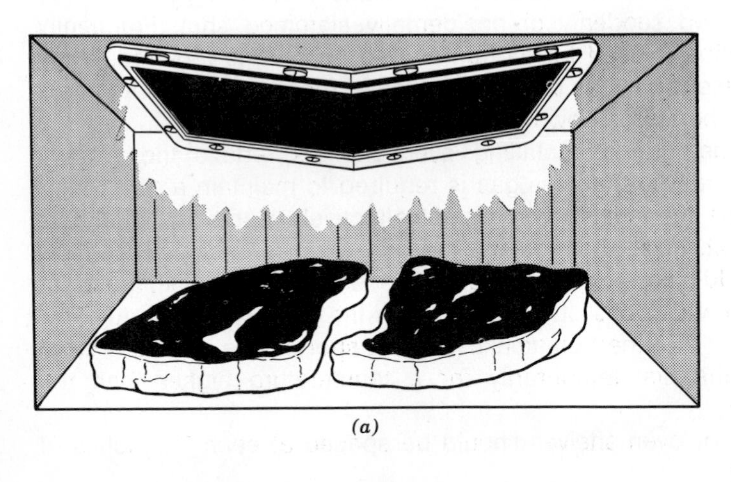

(a)

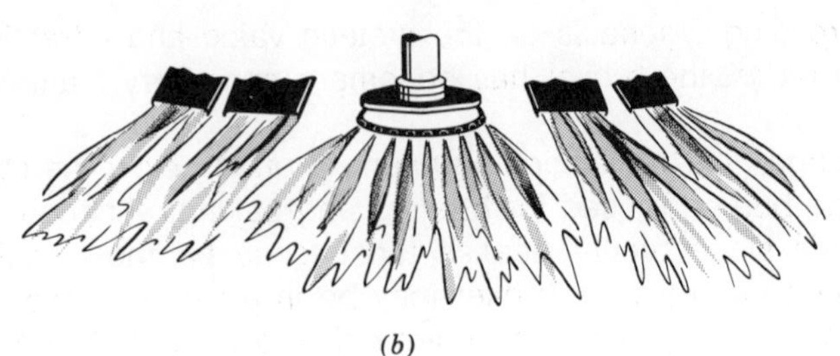

(b)

Fig. 7.23 (a) *An air-gas mixture ignited inside the gull-wing burner heats the ceramic to 1600°F, generating infrared rays which are focused directly on the food. (Hardwick)* (b) *The Blanket-O-Flame burner spreads heat uniformly over the entire broiling surface. (Whirlpool)* gas is shut off from the bake burner and allowed to pass to the broil burner. Gas cannot be supplied to both burners at the same time. Consequently, there is no broil pilot to be extinguished because of the heated oven reducing the supply of needed oxygen.

In some ranges, the broiler pan may be raised or lowered by turning a handle on the panel, an arrangement that prevents accidental burning of the hand when the homemaker wishes to turn or change the position of the food.

One manufacturer advertises an eye-level "Microray" broiler. This broiler focuses infrared rays directly on the food. The broiler burner is of ceramic material which is heated by the gas-air mixture to 1600°F. The infrared rays thus generated travel in straight lines at the speed of light, 186,000 miles per second, giving faster, cooler, more economical, and cleaner broiling operation. Other types of infrared broilers are also available.

Another manufacturing company features a "blanket" of flame burner that sends heat to every area of the oven or broiler for uniform cooking and even browning (Fig. 7.23).

Low-temperature broiling is often desirable, and the broiler valve handle on the gas range may have a click stop for a low flame in addition to the full-on position.

Special Features

Several ranges have disposable aluminum aeration pan liners that protect the surface finish on the pans and may be replaced when they become soiled. A griddle or grill may be put over two surface burners; with a cover it may be used to bake, roast, and steam. One manufacturer locates the rotisserie on the top of the range between two vertical radiant burners. When the spit of the rotisserie is removed, it may be used for a vertical broiler.

Top burners in either the standard or the giant size may have flame-height selectors to suit the size and material of the utensil. A low flame setting is used for small pans and for pans made of other materials than aluminum, and also for foods requiring low temperatures for cooking.

Ovens with constantly burning pilots also have safety shut-offs in case the flame is accidentally extinguished. In some models, there are "oven-ready" lights to indicate that the desired temperature has been reached.

The Robertshaw Flame Master oven control gives precise oven temperature control as low as 140°F, instead of the 250°F-low limit of standard controls. At these low temperatures, the main burner is cycled off and on to hold the temperature at the selected heat; this low heat is used to thaw frozen foods or to keep cooked foods warm until serving. Cook-Hold oven programming is proving more popular than the Delay-Cook-Hold method.

Roast meters have various refinements: one type lights an indicator on the control panel when the roast is done, another turns the oven off automatically at the end of the cooking period, and a third, with a single setting, starts the oven, roasts the meat, turns off the heat, and then rings a chime. Built-in rotisseries, in addition to the spit, may have attachments for holding kabobs and for basting the food while it is revolving.

Waist-high broilers are obtainable, with the burners placed in either the horizontal or the vertical position.

One manufacturer makes a gas range with a flat top of thermo-resistant glass that has a zero coefficient of expansion. Four 8-inch circles on the glass are heated by gas burners hidden beneath the glass panel. The burners are controlled by sliding valves; the farther they are pulled out, the more intense the heat, thus providing an almost unlimited number of cooking temperatures. A built-in warming shelf, heated by air vented from the burners, may be used to keep food at serving temperature. The range top fits flush with the counter top, leaving no crevices or corners and, therefore, is easy to clean.

Heater Range

In homes without central heating, a heater range is sometimes used to warm the kitchen as well as provide means of cooking. This range has a gas room-heater as part of its standard equipment. The heater is located at the side of the oven, behind a grilled end panel through which the heat flows into the room. An automatic valve which turns the gas on and off is controlled by a thermostat set for the desired room temperature. In one model, the burner has an input of 35,000 Btu per hour for natural gas and 30,000 for propane (liquefied petroleum). A safety shut-off prevents accidents if the pilot is extinguished. The rest of the range is similar in construction to other gas ranges.

The American Gas Association (AGA) has set up standards for the construction of gas ranges and also developed performance tests for surface burners, ovens and broilers. These standards have been approved by the American National Standards Institute. Some of the construction standards have been included in the previous discussion. Others are:

1. Construction of all parts in accordance with concepts of safety.
2. Burners shall be easily removable, not bolted to their supports.
3. Top burners shall be centered within ⅛ inch with respect to aeration bowl openings.

4. Means shall be provided for automatic ignition of all burners.
5. Automatic pilot igniters (electric hot wires or electrodes) shall function only to light pilot burner.
6. Drip trays shall be of rust-resistant finish and easily removable.
7. Main oven shall be equipped with a thermostat.
8. Oven bottom or section of bottom shall be removable and replaceable.
9. Supports shall prevent racks from tilting when partially withdrawn.
10. Broilers shall have at least three rack positions.
11. Cord to connect range to source of electricity shall be a three-conductor flexible-grounding type.
12. Domestic gas ranges shall have permanent metal nameplate containing the following information:

    Manufacturer's or dealer's name.
    Trade number of appliance.
    Normal hourly Btu input rating for main burners.
    Minimum clearance between back and sides of range and combustible material.
    Type of gas for which to be used.
    Voltage, frequency, and current input.
    Wiring diagram.
    Symbol of organization making tests for compliance with standards.

Performance

1. A change in oven temperature of not more than 25°F as a result of simultaneous operation of any other section of range with oven for one hour.
2. If oven has programmed cooking system to hold food at serving temperature, it shall maintain temperature specified by manufacturer within ±15°F when in "hold warm" position.
3. Heat distribution in the oven shall be so uniform that plain layer cakes baked at 375°F (±10°F) for 25 to 30 minutes shall be evenly browned—tops and bottoms of cakes not more than 58% nor less than 25% reflectance, determined by reflectometer; overall range of reflectance not to exceed 26%. Range of reflectance of cakes from one rack not to exceed 18%. Reflectances are referred to magnesium carbonate as 100%. Baking of other types of food has verified the conclusion that a range that will bake the cake satisfactorily will bake other foods equally well.
4. The broiler temperature of a separate broiler shall reach 530°F within 12 minutes, 16 minutes if the broiler is heated by oven burner or separate burner located in oven compartment.
5. Heat distribution in the broiler shall be uniform when bread covering the entire broiling area shall be toasted to even brownness in not more than 10 minutes. The broiling area equals 80% of grill area.
6. Maximum surface temperature on exterior of range, door, and valve handles shall not exceed the following figures.

Area	Max. Rise Above Room Temp., °F
Metal door panel	95
Exterior sides, top, back	110
Glazed door panel	110
Frame	125
Door handle	40
Thermostat knob	40
Valve handle	40

7. Temperatures on surfaces immediately above top pilot burners shall not exceed 70°F above room temperature of 70°F.

Recipe for the cake batter is given, also type of mixer to be used, pans to be used, and baking method specified.

Standards are subject to frequent revision. These approval requirements must be met if Blue Seal is to be used.

Terminology

GENERAL

Range. An appliance that performs surface cooking and baking and roasting and broiling operations. The name stove is no longer used professionally.

Free-standing range. A floor-supported assembly consisting of a cooktop with one or more ovens designed for use without supporting cabinets. May have four to six units or burners with or without a griddle.

Split-level. Two ovens, one at eye-level, a second below the surface elements.

Eye-level. One oven above surface burners or units; surface units may be retractable.

Slide-in. A variation of free-standing, rests on floor, top edges squared to fit between base cabinets. Cooktop may be recessed, with or without a backguard.

Drop-in. Surface elements and oven and broiler rest on base of same material as cabinets.

Built-in. Cooktop separate from oven and broiler permits flexible arrangement, installation at convenient height. Surface elements installed in counter, oven and broiler built into cabinet or wall.

Backguard. Raised section at back of cooktop to protect wall. May hold range light, automatic clock, and timer.

Thermostat. A device activated by temperature changes which automatically regulates the temperature with which it is connected.

Thermostatically controlled surface burner or unit. Has central sensitive disk that contacts center bottom of utensil, maintains temperature for which dial is set, eliminates scorching of food.

Oven. Usually lined with porcelain enamel, has two racks with stops to prevent tipping when rack is pulled out.

Vent. Opening from range oven to room, allows volatile cooking products to escape from oven.

Automatic clock and timing device. Turns oven on and off at predetermined times. Programmed cooking—may be Delay, Cook, Hold or Cook, Stop or Cook, Hold.

CLEANING OVEN AUTOMATICALLY

Pyrolytic method. Self cleaning in locked oven at temperature between 850 and 1000°F; reduces deposit on oven walls to small amount of ash. Operation requires approximately four hours, costs 8 to 10¢.

Catalytic method. Continuous cleaning, catalytic material mixed into porcelain enamel coating of oven, oxidizes food soils as they occur.

Preheat overshoot. In preheating oven temperature tends to reach a point 25 to 50°F higher that that indicated on thermostat. Allows door to be opened and cool food to be placed inside without too much loss of heat.

Smokeless broiler pan. Narrow openings in grid allow fat to fall through into pan where it is protected from heat.

Rotisserie. Spit controlled by motor turns food above a heated coil or a gas flame.

ELECTRIC RANGE

Surface unit. A heating unit mounted in cooktop on which utensils may be placed for cooking food. May be two or three units in one, acting as 4, 6, or 8-inch unit.

Control. Usually named a switch, may be rotary or push button for individual heats.

Vent. Opens through center of reflector pan beneath one of back units, directs volatile cooking products away from kitchen wall.

Reflector pan. Placed beneath unit, sends heat flowing downward from unit back upward.

Counters that cook. Made of glass ceramic with designs indicating heat areas. Used with flat-bottomed utensils that make close contact with unit.

Cool-Heat range. Cooks by induction, current from unit induced in bottom of steel or cast iron utensil cooks food. Set of cooking utensils provided with range.

Touch-N-Cook range. Control pads on back panel, supplied with solid-state components, indicate time and temperature settings.

ELECTRONIC OVEN

Magnetron tube. Changes electric energy to microwave energy.

Microwave. Very short waves of a frequency of 2450 MH_z or 915 MH_z Length of radio wave is approximately ⅓ of mile; length of 2450 MH_z is 4.8 inches.

Frequency. Number of cycles per second.

Cycle. Length of cycle measured from starting point back to starting point.

Wave guide. Channel through which microwaves travel from magnetron tube to antenna.

Antenna. Emits microwave energy to oven in all directions; may revolve.

Stirrer. Sometimes designated as a fan; energy emitted by the antenna strikes a slowly revolving fan that reflects the power, bouncing it off walls, ceiling, back, and bottom of oven.

Safety of electronic oven. Over 700,000 microwave units in use. No report of any injury. Ovens must meet very strict federal safety standards.

GAS RANGE

Manifold. Horizontal pipe through which gas flows from fuel line to different burners.

Burner valve handle. Attached to manifold, directs gas into burner.

Orifice. A predetermined-size hole, an adjustable orifice that can be used on any type gas has a hollow needle with an adjustable hood. A fixed orifice is designated for one type of gas only.

Shutter. Opening on mixer head through which primary air flows to mix with gas before it is ignited. Primary air is 60% of air needed for complete combustion.

Venturi tube. Porcelain-lined tube in which gas and primary air mix; the smooth finish of tube increases injection of primary air and gives a clean, sharp flame.

Ignition port. Opening on side of burner through which gas flows to pilot.

Pilot. Gas flame at end of flash tube which ignites the gas-air mixture. It flashes back and lights the burner.

Burner. Flat head at end of Venturi tube in which gas is ignited to heat surface or oven or broiler utensils. Oven and broiler also have burners.

Ports. Opening on side of burner through which gas flows and where it is ignited. Instead of ports, some burner heads have continuous slots to give rings of heat.

Pressure regulator. Maintains gas pressure to the appliance at required level, frequently at 4 inches of water column pressure.

Standards. The American Gas Association (AGA) has set standards for construction of gas ranges and developed performance tests for surface burners, ovens, and broilers. The nineteenth edition, published in 1972, was approved by the American National Standards Institute, Inc.

Experiments and Projects

1. Make a table to record the following information on electric and gas ranges:
   (a) Nameplate information. How useful is it?
   (b) Arrangement and sizes of surface units and burners.
   (c) Is controlled burner provided? Type of control, where located?
   (d) Position and type of switches and valve handles, number of heats?
   (e) Size of oven or ovens, number of shelf supports, number of units or burners, how controlled, position of vents?

(f) Type of broiler, construction of broiler pan?
(g) Special features. Evaluate importance to you as a homemaker.

3. Controlled surface elements:
   (a) Demonstrate automatic controlled gas surface burner: Add 2 cups of water to a 1-quart glass saucepan. Place over an automatic burner set at the boiling temperature for your area. Bring the water to a vigorous boil. Note the height of the flame. Add four ice cubes and again note the height of the flame. Explain results.
   (b) Use a thermostatically controlled electric unit:
      1. To pop corn: To a 2-quart covered saucepan add 2 tablespoons of vegetable oil, and ½ cup of popcorn. Cover the pan and place it over the unit with the temperature set at 375°F. Time: 5 to 8 minutes. Repeat the process in an electric frypan.

      It is stated that a controlled surface unit or burner is a satisfactory substitute for portable electric appliances. Do the results of this experiment justify this statement?

      2. To cook pancakes: Use a commercial mix and follow the directions on the package. Cook on a conventional griddle over a controlled range unit.
4. Compare desirability of cooky sheet material: Use cooky sheets of aluminum, shiny tin, and darkened tin. Preheat the oven to 375°F. Use icebox sugar cooky rolls. Cut the cookies in ⅛ inch slices. Bake 4 cookies on a darkened tinned sheet at 375°F for 8 to 10 minutes until golden brown on top and bottom. Using the same length of time, bake 8 cookies on each sheet. Compare the results. Make a rule for the use of cooky sheets of different materials, and finishes.
5. Compare the evenness of heat in ovens of gas and electric ranges. Prepare the cookies as in Experiment 4. Use all cooky sheets of the same material, aluminum or tin. Place the oven shelf in the approximate center of each oven. Bake the cookies the same length of time in each oven. Compare the evenness of heat.
6. Use of broiler:
   (a) Spread of heat over broiler pan: Cover the grid of the broiler pan with slices of white sandwich bread. Place the broiler pan in the center of the broiler compartment, approximately 4 inches from the unit or ports, as close as the shelf supports will allow. Turn on the broiler burner or unit, and toast the bread until the center slices are golden brown. Compare broilers.
   (b) Cook S'mores on broiler: Place in centers of soda crackers

½ inch squares of semisweet chocolate. Cover with ½ large marshmallow (cut with kitchen shears). Place in broiler comparment far enough below source of heat that chocolate will be melted before marshmallow becomes too brown.

Note. Heat for broiling is usually controlled by the distance of broiler grid from the heat source.

When measuring and testing instruments are available:

1. Check the accuracy of the gas oven thermostat set at 350°F. Suspend an iron-constantan thermocouple as near the center of oven as possible and connect to potentiometer. Record temperature from a cold start, noting preheat time, temperature overshoot, and cycling period. If necessary, adjust thermostat. Repeat for two additional ranges. For comparison plot data for the three ranges on one sheet of graph paper.
2. Repeat (1) in an electric oven.
3. Baking-browning performance of gas and electric ovens:

   (a) Biscuits:

   Place oven rack in position nearest to center of oven. Use Pillsbury Hungry Jack Refrigerated Biscuits or equivalent. Use aluminum baking sheet of a size to permit clearance between sheet and oven walls of 1½ inches, one biscuit per every 10 square inches. Time required to open oven door, put in baking sheet, and close door shall not exceed 15 seconds. Bake biscuits according to label instructions, remove from oven, remove from sheet, cool 5 minutes, measure reflectance at center top and bottom of each biscuit. Tops and bottoms of all biscuits of three successive loads shall not have a reading lighter than 39% nor darker than 20%; difference between top and bottom on any individual biscuit cannot exceed 15%. Break each biscuit in half to check for doneness.

   (b) Cakes:

   Place one rack in top position in oven, other rack in lowest position. Use four 22 gauge round aluminum cake pans, 8 inches in diameter, 1½ inches deep. Grease and flour inside bottom of each pan. The cake recipe is given in AHAM ER-1 Standard, page 16 and in American National Standard Z21.1, 1974, or commercially prepared cake mixes may be used if results correlate with those of the standard cake. Use 0.8 pound (383 grams) of cake batter in each pan. Stagger pans in oven (see Fig. 7.4). The time required to open door, place cake pans in oven and close oven door shall not exceed 15 seconds. Bake cakes 25 to 30 minutes, remove from oven, allow to stand in pan 10 minutes, turn out on cooling rack, cool to room temperature, and measure reflectance. Tops and bottoms of the four cakes shall not have any reading lighter than 58% or darker than 25%. Difference between maximum and minimum readings shall not exceed 26%; range of reflectance for cakes from any one rack shall not exceed 18%.

4. Broiler heat distribution:

   Oven door shall be at broil stop position or closed if so specified. Thermostat shall be set at maximum broil setting; preheat for 6 minutes. White sandwich bread, not more than two days old, unrefrigerated, non-frozen, freshly removed from package in one-half inch slices with crust removed shall be spread over grill area and placed not less than 2 inches and not more than 5 inches from the heating element. The bread shall be toasted not more than 90 seconds. Reflectance readings of 80% of the bread shall be between 10 and 40%.

5. Efficiency of range unit:

   Instruments required:

   Recording wattmeter, recording voltmeter, and current transformer to protect instruments, rheostat to hold voltage constant.

   Use a straight-sided, flat bottom pan that fits the unit. Use 1000 grams of water. A thermometer is inserted through a cork in a hole in the center of the cover of pan with the bottom of the bulb placed 5 centimeters from the bottom of pan. Voltage is held constant, and watts and volts are recorded. Water is heated to 90°C; the temperature recorded every minute. Plot a time-temperature graph.

   Calculation of efficiency:

   1. Watts × seconds = wattseconds.
   2. Wattseconds/4.2 = calories of energy supplied.
   3. Weight of water × weight of utensil × specific heat of utensil × change in temperature = energy utilized in calories.
   4. Energy utilized/energy supplied × 100 = percent efficiency.

CHAPTER 8

Refrigerators and Home Freezers

The most important aspect of refrigeration is the removal of heat, heat that is usually undesirable because of its deleterious effects.

Refrigeration dates back to very early times, if it is defined as simply the cooling of food below the temperature of the surrounding atmosphere for the purpose of preserving the food. When primitive man had killed, he probably ate the fresh meat to repletion, then discarded the remains of his feast and, when hungry, killed again. Such a manner of living was possible only when the number of people in a given area was comparatively small and when the supply of game was abundant. With an increase in the inhabitants of a land, the natural food supply became less, and man was forced to find some method of storing any temporary surplus. The first refrigerators were doubtless caves or cold springs, and these are still sometimes used today.

Later, ice was harvested from rivers and ponds in the winter, stored until summer, and used to keep food cool.

The first home delivery of ice in America was in 1802. This necessitated the building of a form of ice chest, a heavy wooden box large enough to hold both ice and food together in a single compartment, the forerunner of today's refrigerator.

Attempts to bring about refrigeration by mechanical means were made as early as the middle of the eighteenth century. It was not until after 1890, however, that artificial ice production and mechanical refrigeration made rapid progress. In that year, an unusually mild winter caused a severe ice famine the following summer.

Need for Refrigeration

Our early ancestors recognized the need of preserving food from spoilage, although the cause of the spoilage was unknown. It is now known to be due to the presence of tiny plant microorganisms—molds, yeasts, and bacteria. Molds and yeasts may be seen when they have increased sufficiently in number, but bacteria, which cause the most harm, are always microscopic in size. As the organisms grow and multiply they cause certain chemical changes in the food, which alter the flavor and may be detrimental to health. Experiments have proved that, in addition to the presence of suitable food, these plants require a certain amount of moisture and warmth for growth. Ordinary atmospheric conditions in most sections of the world supply the necessary moisture and warmth for at least a part of each year. Tests show that bacteria multiply rapidly at high temperatures; therefore all modern, efficient, automatic refrigerators are constructed to maintain temperatures of not more than 40°F, but at present somewhat lower temperatures are usually indicated—between 32 and 35° for milk, 29 to 31° for fresh meat, and 20 to 25°F for cured and dried meats because the curing process lowers the freezing point. These lower temperatures are obtained by circulating currents of extra cold air around the compartment in which meat and vegetables are kept, and in some models by use of a center drawer. The additional cost of maintaining such low temperatures is probably negligible.

Alternate freezing and thawing of foods is to be avoided. Vegetables, fruits, and meats are made up of many tiny cells, containing watery fluid surrounded by fibrous walls of cellulose. When these foods freeze under ordinary conditions the liquid in the cells expands and tends to rupture the surrounding walls. When the food thaws, the texture is no longer firm but mushy and undesirable. Foods should be kept cold, therefore, but should remain at a temperature above freezing. Records of the U. S. Weather Bureau show that in the United States there is an average of only 19 days when atmospheric temperatures are neither too high nor too low for the safe keeping of foods. The need for some method of artificial refrigeration is at once apparent.

Construction of Refrigerator

Refrigerator cabinets are made of sheet steel, welded to form an inner and outer shell with insulation between. Experiments have proved that 80 to 90% of the heat that gets into the refrigerator comes through the walls. Good insulation is, therefore, a true economy. An efficient insulating material must be heat- and moisture-resistant, non-destructible, and odorless; it must maintain its position between the walls and not settle or sag, leaving air pockets to conduct heat.

Fiberglas, made into slabs or matted into blankets, is now widely used for refrigerator insulation. As a further precaution against any possibility of moisture being absorbed, the Fiberglas is treated with a plastic binder, or sealed into the walls with an odorless asphalt mixture. High-density Fiberglas, used in some models, does a satisfactory insulating job in using little space; it increases the storage capacity of the refrigerator without using additional floor space. Several new types of insulation have been developed for refrigerators.

In some models Freon 12, the common refrigerant used in the refrigerating system, is combined with Fiberglas, in others the Freon is mixed with certain chemicals in the cavity between the outer and inner shells. The chemicals react to generate heat, changing the Freon liquid to a gas that forms close-celled Urethane foam. This foam expands to fill every bit of space between the walls, and as it does so, entraps the Freon in millions of tiny cells. The foam becomes rigid and bonds to all the surfaces, blocking any possible path for heat leakage. It is moisture-resistant, and 1¼ inches of it is equivalent to 3 inches of conventional insulation. Polystyrene foam may also be used as a core between sheets of polystyrene and Fiberglas cloth. It forms a rigid panel that may be cut to any desired shape and size, is light in weight, noncombustible, and moisture-resistant.

These insulations provide a strong, rigid structure that will make possible the use of lighter-weight materials in the frame, while at the same time preventing even slight distortion in alignment of the walls during manufacture and shipping. This is a definite asset, since a slight change in shape may affect the sealing of the door gasket.

The door should be as well insulated as the walls.

Refrigerators have flush-door construction. The opening is lined, both on the door and on the cabinet, with a breaker strip of polystyrene plastic or other material to prevent the heat leakage that would occur if the metal lining came into direct contact with the outer frame. The entire door lining is sometimes of plastic. The material should resist cracking, and grease and oils. One manufacturer uses Permalon for the door liner, a rather expensive but long-lasting material, very resistant to ill effects from oils and greases.

A gasket around the outer edge of the door completes the seal, and should be fungus resistant. Specification sheets of information indicate that vinyl plastic is widely used for the gasket; it has an advantage over rubber gaskets in that it is not damaged by grease. A

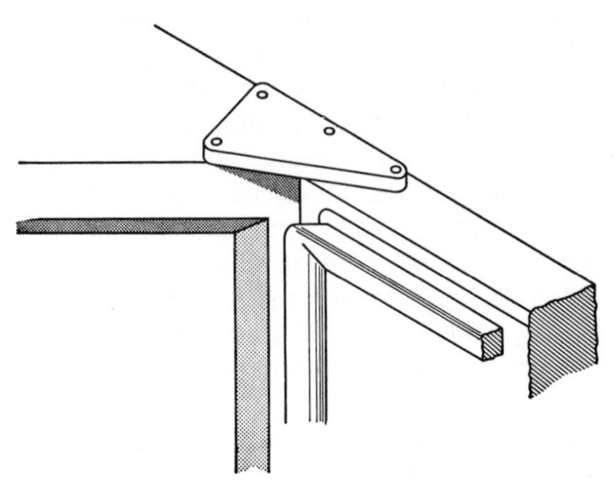

Fig. 8.1 Flush-hinge door allows refrigerator to be placed between cabinets on either side.

continuous magnet is sometimes imbedded in this vinyl seal to hold the door tight on all four sides. In some models, however, the magnet does not extend to the hinge side. The door is easily opened by the touch of a finger, even from the inside, thus preventing the tragic accidents of children being shut into discarded refrigerators and smothering. The law now requires that the door open from inside or out by a pushing or pulling force of not more than 15 pounds. The U. S. Public Health Service has issued a bulletin on entrapment hazards and how to correct them.[1]

Whatever type of latch and hinge combination is used, it should hold the door securely. It is a convenience to have a latch that does not require hand manipulation—one that will hold when the door is pushed shut with arm or knee, in case the hands are filled with supplies taken from the cabinet. Some doors can be opened with foot pedals.

The seal of the door may be checked by shutting a dollar bill or a piece of paper of similar texture between the door and the frame at different places. If the door is properly sealed, you should not be able to pull the paper out. If the paper can be removed, it means that warm air can get into the food chamber; the latch should be adjusted until it maintains the seal. Flush or offset hinges are now frequently installed (Fig. 8.1). They permit the door to swing open at

[1]U.S. Public Health Service, **Preventing Child Entrapment in Household Refrigerators,** Publ. No. 1258, 1964. For sale by Supt. of Documents, U.S.G.P.O., Washington, D.C. 20402.

a 90-degree angle, which is within the cabinet limits. They are especially desirable when the refrigerator is placed adjoining a counter, wall, and kitchen cabinets.

Styling tends to change from year to year, and current models have trim, straight lines and square corners. These new styling trends are due, no doubt, to the current preferences of homemakers for built-in refrigerators and free-standing models that can be placed between cabinets and counters, leaving no space on either side.

Refrigerators are of several different types (Fig. 8.2). The conventional refrigerator has the evaporator within the food compartment, across the top of the cavity, usually with an air space on either side separating it from the inner walls of the compartment. Many refrigerators, however, are combination types with the food compartment separate from the freezer section, each maintaining its own individual temperature.

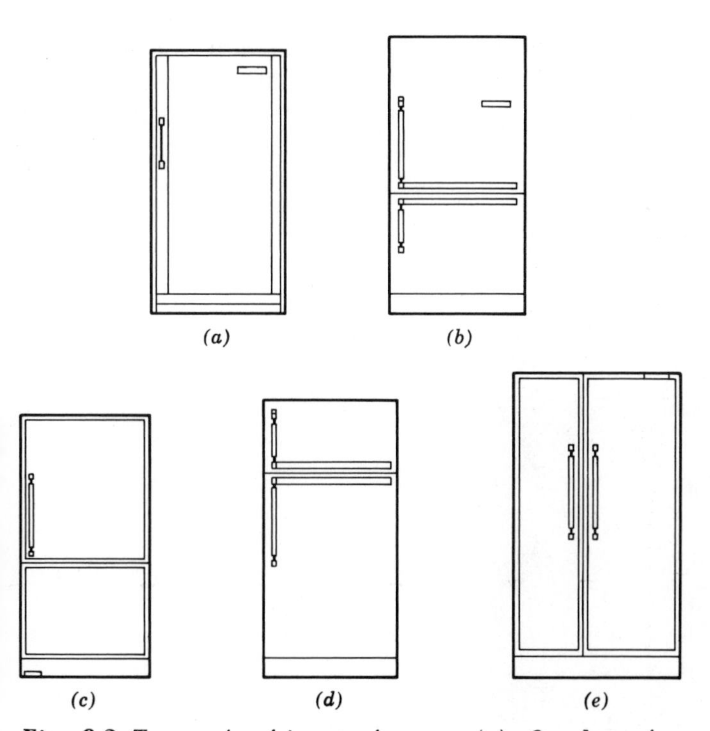

***Fig. 8.2** Types of refrigerator-freezers.* (a) *One-door, freezer inside with separate door.* (b) *Freezer below food compartment.* (c) *Freezer below in form of roll-out drawer.* (d) *Freezer above fresh food compartment.* (e) *Side-by-side refrigerator freezer. (Frigidaire)*

Seventy homemakers out of each 100 buy the 2-door combination refrigerator-freezer. The average size purchased is 14 cubic feet, and the trend is upward. A larger size makes fewer shopping trips necessary. To shop once a week for a family of two persons, a combination model is recommended, with a fresh food section of 9 cubic feet and a freezer section of 2 cubic feet; add one cubic foot to each section for each additional member of the family.

The freezer area may be either above or below the fresh food compartment. Each section may have a door; instead of the lower freezer section with a door, some models feature a roll-out drawer. Occasionally the refrigerator may have a single door covering both fresh food compartment and freezer, and the freezer has an inside door of its own. When the freezer compartment is at the top, it is easier to load and ice cubes are more readily reached. At the present time (with one exception) equipment companies are not manufacturing refrigerators with the freezer below the food compartment, but past models may be found in homes.

A popular model has a side-by-side vertical arrangement of refrigerator and freezer, with two narrow doors that occupy less space when opened than does a single door. One manufacturer has made

Fig. 8.3 Refrigerator-freezer with dispensers. (General Electric)

Fig. 8.4 Closeup of dispenser in use. A single household water connection supplies both the icemaker and water reservoir. (General Electric)

possible the interchange of hinges and latches, so that the two doors may be opened in several different ways to suit the needs of the homemaker. Several manufacturers provide reversible hinges on two-door top freezer refrigerators. No one uses them on single-door models. Adjacent counter space is recommended; reversible hinges make it possible to adapt the door opening to the location of that space.

One side-by-side model has dispensers added to the freezer side that automatically fill your glass with chilled water or ice without opening the door (Figs. 8.3 and 8.4). Either side of the panel is supplied with a cradle, similar to those used in cafeterias for dispensing water, milk, or fruit juices. Press the glass against the water cradle and the glass fills, against the ice cradle and two cubes at a time drop into the glass. You may use either dispenser separately or both, one after the other.

The newest type refrigerator is a three door side-by-side model with two freezer sections (Fig. 8.5). The upper freezer, supplied with an automatic ice maker, gives easy access to often used items with minimal loss of cold air. The lower freezer is designed for general frozen food storage and has easy to clean tempered-glass shelves that retain cold and a large glide-out basket. It also features a chilled water dispenser without opening the door. In the refrigerator section cold air is circulated to provide proper temperature and humidity.

Tests performed in the equipment laboratory at Iowa State University indicate that the partition between the fresh food and freezer compartments of the side-by-side model may reach a temperature below freezing on the food side, causing any fresh food placed too close

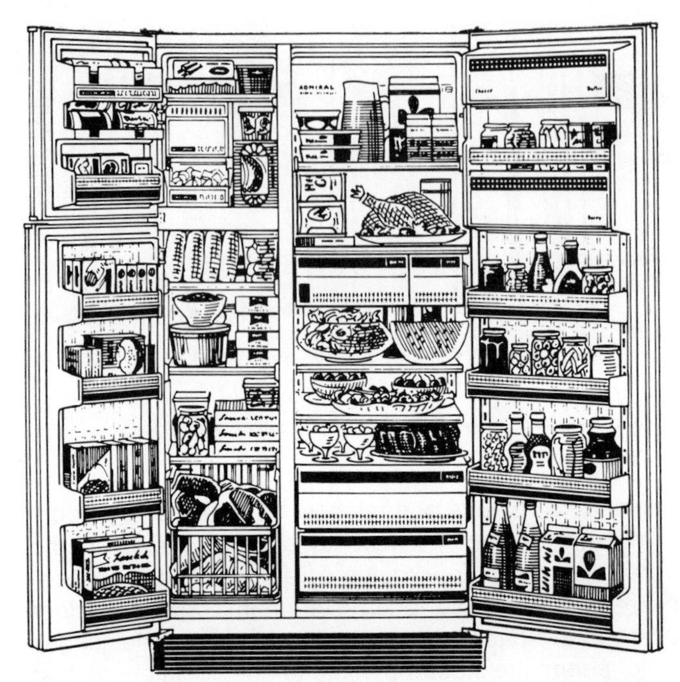

***Fig. 8.5** A three-section, no-defrosting freezer/refrigerator, featuring chilled water dispenser, automatic ice maker, and cold-can carousel. It has a 24-cu ft capacity, 15.31-cu ft refrigerator. The dual freezers total 8.65 cu ft; it holds 302 lb; and it has tempered glass refrigerator shelves. (Admiral)*

to the wall to freeze. The shelves in this model are naturally somewhat narrower than in other refrigerator types, and may cause certain limitations in the storage of food.

The refrigerator may be finished, both inside and out, in porcelain enamel, often in color or with decorative panels. Baked-on enamel, under a variety of trade names, is frequently used on the outside. One manufacturer uses titanium porcelain enamel for the entire finish of a deluxe model.

Because of the present increased use of plastics for refrigerator storage-drawers and door liners, Fern E. Hunt of Ohio State University carried on a research project on three plastics in current use and porcelain enamel to observe their resistance to abrasion, food stains and odors, food acids, microbe survival, soiling and cleaning methods. With the exception of effects of food acids, the tests showed that porcelain enamel was a much more desirable material for refrigerator liners and storage drawers than the three plastics tested. Fern Hunt concludes, "We see a strong argument from the standpoint of consumer satisfaction for continuing to use porcelain enamel."

The lining should have rounded corners and unions, to simplify cleaning, and the bottom should be slightly depressed to catch liquids accidentally spilled. Shelf supports may be integral with the lining but are commonly built as separate knobs and glides. Shelves are made of corrosion-resistant metal rods or bars, near enough together to hold small storage dishes without likelihood of their tipping, but spaced widely enough apart for efficient air circulation. Stainless steel is the material most frequently used, but some shelves are of anodized aluminum. A single glass shelf over the crisper drawers is found in a few refrigerators, but glass shelves may be broken if containers of warm food are placed on them. Some recent models, however, use glass shelves exclusively instead of the metal ones, but the shelf is of tempered glass, guaranteed to be break-resistant.

Models may have one or two sliding shelves. Sliding shelves should have positive stops. A rear guard rail prevents dishes from falling off at the back when the shelf is pulled forward. At least one manufacturer places guard rails on all sides of certain shelves, making a "basket" shelf designed primarily for storage of small containers.

Shelves that are adjustable to more than one level increase the flexibility of food-chamber space and arrangement. Similar advantages are obtained from a divided shelf, part of which may be removed when larger space is needed for storage of oversize articles. Narrow shelves recessed in the door afford storage for many items not easily accommodated on the inner shelves—milk, fruit, bottled drinks, cheese, butter, eggs. The shelves for eggs are molded or perforated, so that each egg may be stored separately. Eggs should be placed with the large end up, since the air cell is in that end. Special covered compartments accommodate cheese and butter. Some manufacturers disapprove of storing butter in a container that maintains a spreading consistency in larger quantities than one-fourth pound and limit the recessed butter holder to this size.

A refrigerator should be placed in as cool a location as possible, out of direct sunlight and away from the range and house-heating equipment.

The refrigerator should be leveled, so that the door may seal correctly. Leveling screws are usually provided at the two lower front corners. If the refrigerator is not level, there will tend to be vibration causing wear on the motor.

In addition to the free-standing refrigerator, built-in, cabinet-based, and wall-hung models are available; the choice will often depend on the space where the refrigerator must be located.

Heat and Cold Exchange in Refrigerators

Cold is the absence of heat. The theory of refrigeration is based upon certain fundamental laws of heat, a form of energy generated by the vibration of molecules. Understanding the principles of refrigeration requires a knowledge of methods of heat transfer and of the terms latent heat and specific heat.

Heat always passes from a warmer to a cooler body (Fig. 8.6). This process happens in one of three ways: by radiation, by conduction, or by convection. Since the room air in which the refrigerator is located may be at least 30 to 40° warmer than the air inside the refrigerator food chamber, heat is conducted through the cabinet walls to the cooler area. The larger the heat gradient between outside and inside, the greater the tendency for heat to pass from one section to the other.

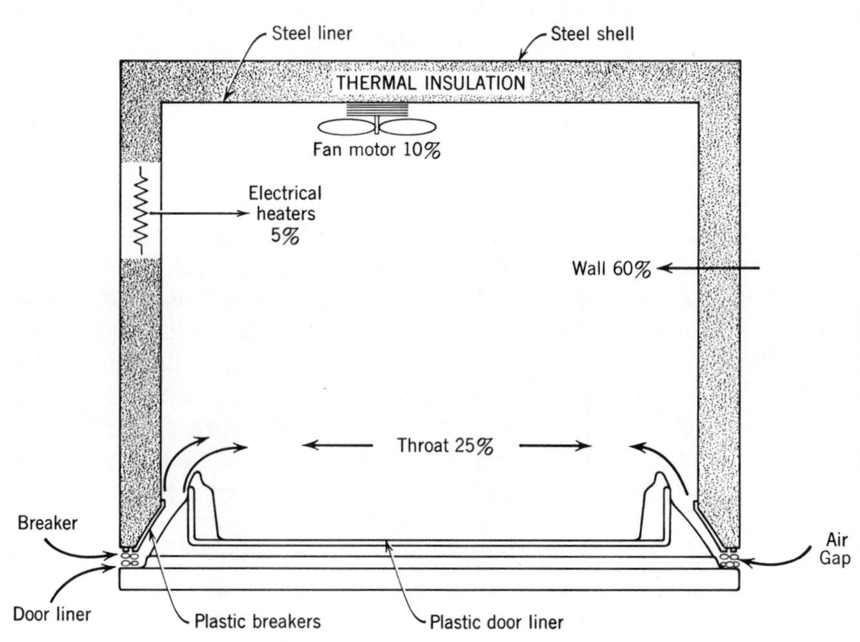

Fig. 8.6 Typical heat load. (Courtesy of Nicholas J. D'Aleandro, Chief Engineer, Kitchen Appliances, Fedders Corporation)

When food at room temperature is placed within the refrigerator, the warmth it contains passes into the surrounding air and is carried by convection currents to the cooling unit, where the heat causes the refrigerant to change from a liquid to a gas. The heat absorbed to bring about this change of state from liquid to gas is known as latent heat of vaporization and cannot be measured by a thermometer. This heat is used in separating the molecules of the liquid which are packed more closely together than are gas molecules.

When the gas is again liquefied, it gives off an amount of heat equal to that which was absorbed, but this process is carried on outside the refrigerator cabinet.

The specific heat of any substance is its capacity to absorb or retain heat, compared to the capacity of an equal weight of water. Water is taken as the standard because of its large heat capacity. In constructing the refrigerator cabinet, it is necessary to know the specific heat of the materials used, and also the specific heat of the foods to be stored in the refrigerator, in order to determine the size of cooling unit to build to absorb the heat from the various substances and maintain the desired temperature. Table 6.1 gives specific heats of certain materials used in refrigerator construction. The specific heats of representative foods are given in Table 8.1.

Table 8.1
Specific Heat of Some Foods Commonly Stored in the Refrigerator

(Fresh) Beef	0.75	Chicken	0.80	Fish	0.80
Berries	0.91	Celery	0.96	Milk	0.90[b]
Butter	0.60	Cream	0.68	Watermelon	0.92
Cheese	0.64	Eggs	0.76[a]		

[a]One dozen eggs weigh 1.34 pounds.
[b]One quart of milk weights 2.1 pounds.

It is to be noted that the specific heat of those foods containing a higher percentage of water approaches more nearly the specific heat of water, 1.0. Since the specific heat of milk is 0.90, it would require 0.9 Btu to raise the temperature of 1 pound of milk 1°F.

Compression System

Automatic refrigerators operate primarily upon the two fundamental laws previously referred to: (1) when a liquid changes to a vapor or gas, heat is absorbed, and (2) when the vapor is again liquefied it gives off the heat it has taken up. An electric motor or heat from gas or oil supplies the energy that starts the cycle (Fig. 8.7).

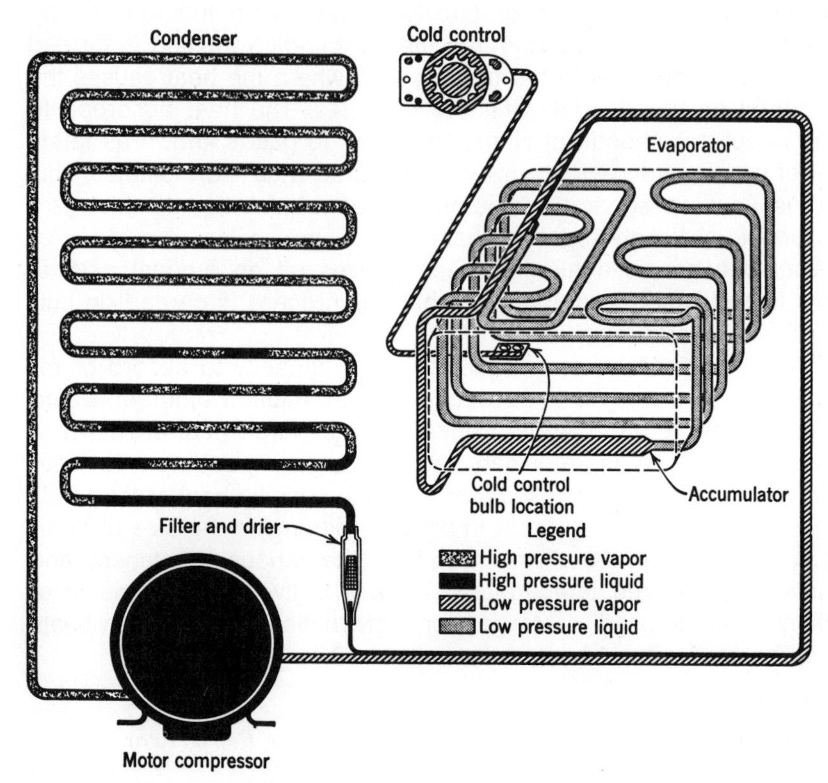

***Fig. 8.7** Diagram of the compression system. Note that the capillary tube is fastened to the suction tube. (Norge)*

The operation in electric refrigerators goes on a continuous and controlled course. The system is closed so that the refrigerant, Freon 12, may be used repeatedly. The essential parts of the compression system are: (1) the compressor, which gives the system its name; (2) the condenser; and (3) the cooling unit, also called the freezer unit or evaporator. In addition, there are: the motor, which drives the piston in the compressor cylinder, the thermostat, which controls the off and on periods of the motor according to the temperature of the food chamber, and the tubes, which connect the various parts, both capillary tubes and larger sizes. The size of the motor usually varies between ⅕ and ⅛ horsepower, but the frost-free models which employ fans to circulate the chilled air usually have larger motors, perhaps as large as ⅔ horsepower. The additional cost of operating this type is usually about 50 cents per month.

The evaporator is located within the food chamber, at the top or against the back wall. It is made of stainless steel or of anodized aluminum. The conventional type, a U-shaped receptacle, may be located at one side of the chamber or centered at the top. Now, however, the evaporator is commonly the freezer-chest type, extending across the entire top of the refrigerator, above the food compartment. In combination refrigerator-freezer models, the food compartment is usually cooled by a refrigerated plate on the back wall. Although the conventional and freezer-chest evaporators provide space to store frozen foods, they are not freezers in the commonly accepted meaning of the term. Their temperatures fluctuate, and foods stored in them will not remain market fresh for more than a week. The desired temperature is maintained by a hydraulic thermostat, with the bulb either attached to the cooling unit or placed close to it. A number of new models are featuring a different type of thermostat.

When the compressor is the reciprocating type, it draws the gaseous refrigerant from the evaporator with each downward stroke of the piston, then compresses it—a process that increases its temperature as it discharges the refrigerant into the condenser coils. If the condenser is to function efficiently in changing the refrigerant back into a liquid, the temperature of the refrigerant must be raised high enough—perhaps as high as 120°F—so that heat will flow into the cool air surrounding the coils. A capillary tube connects the condenser to the evaporator. This capillary has sufficient resistance to aid the compressor in building up the necessary pressure in the condenser. For adequate cooling, the condenser coils may be brazed to a large finned plate or metal screen attached to the rear outside wall of the refrigerator cabinet (Fig. 8.8). If the condenser is below the food chamber, the coils are supplied with fins to dissipate the heat, or a fan forces the warm air into the room through a front grill at the

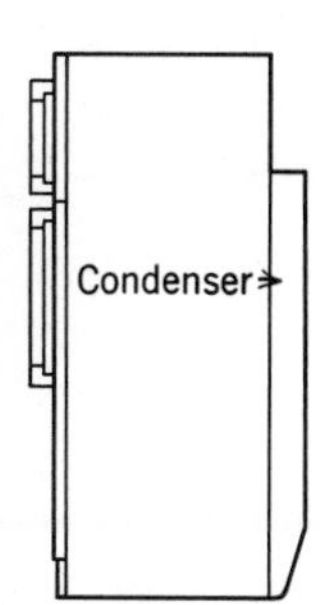

***Fig. 8.8** Condenser at rear of refrigerator cabinet.*

bottom of the front panel. The bottom location is more common in models that are installed flat against the wall or are built in.

The condensed refrigerant passes back to the evaporator through the capillary tube. This tube has either an expansion or a restrictor valve (depending on the manufacturer) to control the flow of the refrigerant. In one system, however, the flow in the capillary is directed correctly by forming the tube into a half loop. The capillary is frequently soldered to the suction line that brings the gaseous refrigerant from the evaporator back to the compressor. This construction acts as a heat exchanger, as well as a support for the small capillary tube.

A great deal of study of capillary tubes has been made to determine the correct length and bore to give the best performance over the range of operating conditions usually found in household refrigerators. The method of testing capillary tubes has been approved by a committee of the ASHRAE (American Society of Heating, Refrigerating, and Air-Conditioning Engineers).

Rotary Compressor

A number of electric refrigerators use a rotary compressor instead of the reciprocating type. Figure 8.9 shows the five positions of the Rollator cycle. The refrigerant enters the chamber at the bottom, is compressed by the rotating movement of the crankshaft, and is discharged through the vertical pipe. Note that the movable blade is always in contact with the sleeve and separates the chamber into two

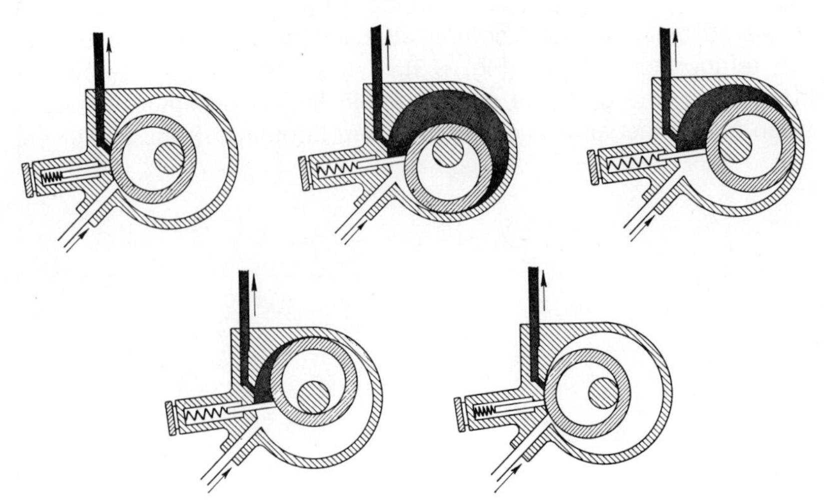

Fig. 8.9 Rollator cycle. (Borg-Warner)

sections. Intake and compression take place simultaneously and not in separate movements as in the reciprocating compressor. In other respects the system is similar. Freon 114 is the refrigerant used.

High and Low Side

The refrigeration system between the valve where the vapor enters the compression cylinder and the valve through which it sprays into the evaporator is known as the "high side" because of the high pressure maintained; the remainder of the system is known as the "low side." To maintain this difference in pressure so that there will be correct liquid flow from the high to the low side, a system of valves is essential.

The motor and compressor of the compression system are usually hermetically sealed. Both are within the same housing, and the whole system so interconnected that there is almost no possibility of leakage. Since the motor is permanently lubricated, additional oiling is eliminated.

Temperature Controls

The motor and compressor are activated by the temperature of the evaporator. A thermostat placed in contact with the evaporator sleeve closes the circuit whenever the temperature of the evaporator becomes warmer than the setting of the control indicates, and opens the circuit again when the air and food in the food compartment are sufficiently chilled. Originally the thermostat bulb contained a liquid which contracted and expanded with decrease and increase in temperature. At present, refrigerator thermostats frequently use a vapor similar to that used in the evaporator. Therefore it will have approximately the same temperature as the refrigerant and is more sensitive than the liquid. The bulb has been replaced by small diameter tubing, called a feeler tube, with one end closed and the other end attached to a bellows or sometimes to a flexible diaphragm. The closed end is clamped to the evaporator sleeve. A switch brings about the action. The contact points of the switch are made to close and to open by snap action to prevent arcing. The thermostat also has an adjustment screw for high altitudes (Fig. 8.10).

Most no-frost refrigerator-freezer combinations provide separate controls for the refrigerator and freezer compartments. No other types of refrigerators have separate controls. There is only one refrigerator on the market that has any kind of a humidity control, and it is not automatic. There is no humidity sensing device available which is suitable automatically to control humidity in a refrigerator.

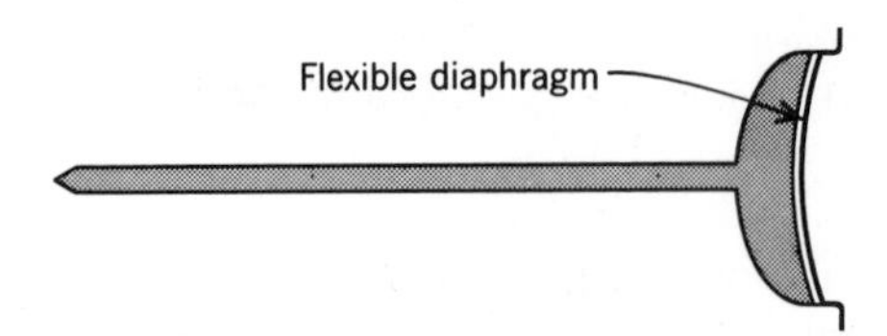

Fig. 8.10 Flexible diaphragm. The control-bulb shape has been eliminated by most control manufacturers and replaced by a plain closed-end tubing. It is then called a feeler tube. The bulb shape is not necessary since the coldest point of the tubing determines the gas pressure in the tubing and bellows. The bellows is sometimes replaced by a flexible diaphragm. (Whirlpool)

The difference between the cut-in and cut-out temperatures of the thermostat is known as the temperature differential. With the exception of the combination refrigerator-freezer models with an evaporator plate in the food compartment that defrosts during each cycle, the temperature differential is usually 10 or 12°. In the cycle-defrost models the differential may vary from 32 to 50°. The cut-in temperature is always above 33° to allow for complete defrosting of the plate during the off cycle. When the control dial is turned colder or warmer, the cut-out and cut-in usually will move down or up together, thus maintaining the same differential.

Absorption System

In the absorption system, gas generally is used to produce the necessary heat to activate the refrigeration cycle, and an additional part is needed—the absorber—hence the name of the system. Ammonia is the refrigerant commonly used. The action of an absorption system depends upon the ease with which ammonia gas dissolves in cold water, and on the fact that the combination is at best an unstable one, easily broken down by heat. The solution is slightly stronger than ordinary household ammonia. The process is continuous, not intermittent as in the compression system. There are no valves or moving parts.

Figure 8.11 shows a simplified somewhat stylized diagram of the refrigerating mechanism. It is made up of a series of steel chambers connected by steel tubes, all welded together. The principal parts are the generator, the condenser, the evaporator, and the absorber. In reality the absorber is not the vessel indicated but consists of several, perhaps as many as ten circular turns of unfinned tubing. And fins are placed around the "ammonia-return" tube between the separator and condenser to prevent any water vapor from rising with the ammonia vapor into the condenser.

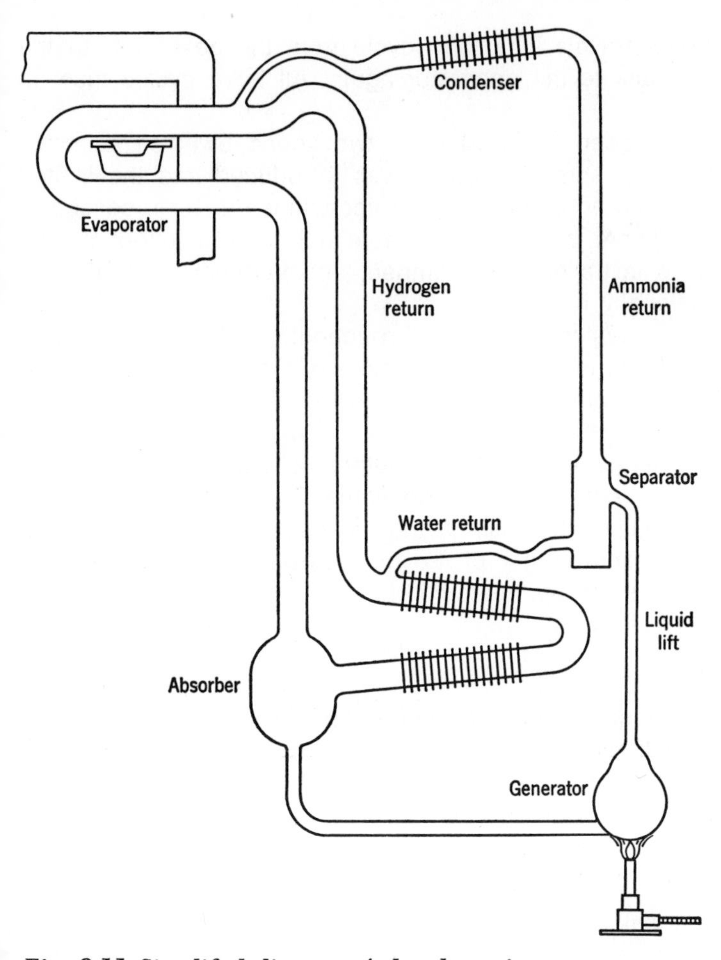

***Fig. 8.11** Simplified diagram of the absorption system.*

Because liquid ammonia does not readily change to a gas under the pressure conditions in the absorption system, hydrogen, the lightest gas, is introduced into the evaporator to help in bringing about the change. The pressure of the ammonia in the evaporator is then less than its total pressure by the amount of pressure exerted by the hydrogen. Change of ammonia to a gas at this reduced pressure, with the absorption of the heat of vaporization, is rapid.

When the refrigerator is in operation, three separate cycles take place at the same time within the unit. The ammonia travels in one cycle from the generator, to the condenser, to the evaporator, to the absorber, and back to the generator. The hydrogen travels in another cycle from the absorber, to the evaporator and back to the absorber again. The water travels in a third cycle from the generator, to the absorber, and back to the generator again. All three cycles meet in the absorber.

Originally developed in Sweden by two young Swedish scientific investigators, the absorption system was introduced into the United States in early 1900 and attained wide acceptance. Later it ceased to be manufactured here, but now is available in 3- to 4-cubic foot models sometimes used in trailers and campers and frequently operated by LP gas, occasionally by kerosene.

The absorption system is sealed throughout, so there is little possibility of leakage. It is noiseless, because of the absence of moving parts, and, for the same reason, wear on the mechanism is reduced to a minimum.

The gas burner is fitted with an automatic cutoff, a disk with a finger attachment extending into the flame. This finger cools if the flame is accidentally extinguished and closes the gas valve immediately, eliminating any danger of fire or explosion. When lighting the burner, the disk must be heated before it will open the inlet and permit the gas to flow.

Refrigerants

As noted in the discussion of refrigerator systems, Freon 12 (dichlorodifluoromethane) is the refrigerant used in many household units. The question arises as to why this refrigerant was selected; in other words, what characteristics are desirable in a refrigerant and how nearly do the ones in use meet the requirements?

A refrigerant should be:

1. Nontoxic under all conditions.
2. Nonflammable.
3. Nonexplosive by itself or in any mixture with air.
4. Of a characteristic but nonirritating odor.
5. Easily detected in leaks by a simple test.
6. Stable, to avoid disintegration during repeated compression, condensation, and evaporation.

The refrigerant should have:

1. A noncorrosive action on metals.
2. A fairly high latent heat.
3. Comparatively low condensing pressure.
4. Evaporating pressure close to atmospheric pressure.
5. Little or no effect on lubricating oils.
6. A boiling point considerably lower than the temperature of the space to be refrigerated.

Freon 12, used almost exclusively in the present-day household electrical refrigerator, was developed in the United States as a result of extensive chemical research for a safe, practical refrigerant and was introduced in 1931. Freon 114 (dichlorotetrafluoroethane), Freon 21 (dichloromonofluoromethane), and Freon 11 (trichloromonofluoromethane) are other members of the Freon group used in certain systems. Another member, Freon 22 (monochlorodifluoromethane), is used in some home freezers, because it is primarily a low-temperature refrigerant.

Freon 12, either as a vapor or as a liquid, is nonflammable, noncombustible, and nonexplosive when mixed with air. It is nontoxic in concentrations as high as 20 per cent by volume for an exposure of 2 hours, and the vapor in any concentration does not irritate the mucous membranes.

It is a stable refrigerant, withstanding repeated changes of state indefinitely, and has no corrosive action on any metal commonly used in refrigeration units. Magnesium alloys must be avoided, however. Mineral oils are used for lubrication. They mix with Freon 12, and consequently there is no separation of oil in the evaporator, but they must be dehydrated before use. In general, chlorine compounds are good oil carriers and assist in lubrication.

Freon products are colorless and odorless in concentrations below 20 per cent by volume. Freon 12 has a boiling point of −21.7°F at atmospheric pressure, a condensing pressure of 93.2 pound gauge at 86°F, a vaporization pressure of 11.8 at 5°F, and a latent heat of vaporization of 69.5 Btu per pound at 5°F.

At this rather low heat of vaporization, a fairly large volume of the refrigerant must be circulated, with the advantage of using less sensitive valves, which have a more positive regulating mechanism, and less chance that the system will become clogged.

Freon 12 dissolves natural rubber used in gaskets but has no action on synthetic rubber. The vapor is not absorbed by foods stored in the cabinet and has no effect on their odor, color, taste, or structure.

Ammonia, a colorless gas of characteristic odor, is flammable and explosive under certain conditions but not commonly under those found in the home; it has no corrosive action on iron or steel but does corrode copper, especially in the presence of moisture. Leaks may be detected by a sulfur candle.

Ammonia boils at −28°F at atmospheric pressure, has a latent heat of 565.0 Btu per pound at 5°F, a vaporization pressure of 19.57 pound gauge at 5°F, and a condensation pressure of 154.5 pound gauge at 86°F. The fairly high condensation pressure necessitates cylinders and pipes of heavy construction. Ammonia is very soluble in water, about 900 volumes of ammonia being soluble in 1 volume of water at ordinary temperatures. It will burn under certain conditions, and it is detrimental to food.

According to one investigator, certain properties of a refrigerant may be determined from the chemical formula. If the formula has no hydrogen component, the refrigerant will not burn. When a refrigerant does contain hydrogen, it is usually combustible if mixed with air. A refrigerant containing hydrogen along with chlorine and fluorine, either or both, is harmful to man if breathed. If chlorine or fluorine is present, and the refrigerant burns, the products of combustion are harmful. Refrigerants containing chlorine and fluorine are decomposed in a flame to form harmful products.

Temperature of Evaporator and Food Cabinet

Just as it is necessary for a radiator to be hotter than the room in order to heat the room, so it is essential for effective cooling that the evaporator be colder than the atmosphere in the cabinet. The thermostat is usually set at the factory where the refrigerator is manufactured for the conventional evaporator to reach a maximum temperature of 24°F, and a minimum of 0 to 4°F. Commonly the temperature in the evaporator compartment of a single-door refrigerator is between 15 and 20°F. At this setting, the motor operates from 15 to 30% of the time, or from 3 to 7 hours out of 24. Lower temperatures may be obtained by setting a manually controlled dial. In some units, additional coils soldered to the bottom of a shelf in the evaporator cause a concentration of refrigerant at that section and automatically permit rapid freezing without excessive cooling of the food compartment. In such an arrangement, the shelf is permanently attached to the evaporator sleeve. This construction may interfere with the flexible storage of frozen foods.

Almost all manufacturers point out in their Use and Care books that frozen foods should be stored only a week or two in this type refrigerator.

The coldest location in the food chamber of the conventional refrigerator is immediately under the evaporator, but new methods of circulating the air in the food chamber by a fan tend to keep all areas at approximately the same temperature. One model uses an automatic damper control to maintain a constant temperature in the food compartment. Should the damper be manually controlled, seasonal adjustments may be necessary, but usually very little is gained by changing the controls. It is important to know the maximum and minimum temperatures that a refrigerator will hold under varying conditions; the average temperature is of comparatively little importance.

Special Features

An interior light in a refrigerator is not new, but more and more these lights are recessed in the rear or side wall of the food compartment where they do not interfere with the storage space. At least one refrigerator uses split-level lighting.

Crispers for fruits and vegetables are made deeper to accommodate heads of cauliflower and cabbage more easily. Meat keepers are also deeper. Some models have a chiller drawer for meat beneath the horizontal freezer unit; this drawer maintains a temperature 7 to 15° colder than the storage shelves. In another model, a flow of chilled air from the freezing coils is directed to the meat storage container.

At least one or two manufacturers feature a special storage area in the refrigerator. Within the walls of this compartment very cold air circulates, but does not pass into the interior of the section. Consequently the air is still, and almost 100% humidity is maintained. Foods are kept moist and fresh without being wrapped or covered for days longer than by other methods of storage. Another model has a glide-out center drawer, providing two sections, one for 21 pounds of meat, the other for a half-bushel of vegetables. The temperature is low enough to keep unfrozen meat, even hamburger, fresh for a week, and to keep vegetables at optimum temperature and humidity conditions. This arrangement allows the homemaker to store a week's supply of food, which eliminates more frequent shopping trips (Fig. 8-12).

An ultraviolet lamp is used sometimes to deodorize and prevent mold growth. One unit ceases to operate whenever the refrigerator door is opened and, in several "no-frost" models, the opening of the food-compartment or freezer doors automatically stops the fans used to circulate the cold air and so prevents loss of cold air blown into the room. Especially desirable is the continued construction of refrigerators with increased cubic foot capacity which, nevertheless, occupy the same floor space as did the smaller models.

A recent refrigerator is constructed so that it allows for the installation of an icemaker mechanism within the evaporator at any time, thus converting the model from conventional to automatic ice making. Manufacturers who market these models do not offer the same model with a factory-installed icemaker, so an accurate comparison of cost is not possible. This refrigerator can freeze and store

***Fig. 8.12** Some side-by-side refrigerator-freezers have Hi-Humidity Compartments. The compartment is surrounded by cold air from the freezer section, shielding the vegetables from the damaging effects of dryness and premature overripening. (Amana Refrigeration, Inc.)*

as many as 120 cubes of ice in 24 hours, an equivalent of 5 pounds. There are two kinds of ice cube controls on the market. One of them, used in the model under discussion, regulates the weight of the cubes deposited in the ice storage bucket, allowing the homemaker to increase or decrease the amount of ice which the icemaker will produce before it shuts off. The other kind varies the size of the cubes by regulating the amount of water that flows into the icemaker.

Shelves and drawers are frequently supplied with nylon rollers, greatly increasing the ease with which they may be pulled out, bringing back-shelf foods within comfortable reach. Shelves may also be adjustable to different heights.

In the past, cleaning behind and beneath a loaded refrigerator has been a problem for the homemaker, but now many refrigerators roll out on wheels permanently installed beneath the cabinet. Between cleaning times, safety stops lock the wheels into position to prevent their moving. In some models front wheels are adjustable for cabinet leveling.

Use of the Food Chamber

Milk, and sauces and desserts containing this product, and other protein foods such as meats, poultry, and meat broths are most liable to spoilage and should always be kept at 40°F or below if they are to be held for more than a few hours. Eggs and small fruits—strawberries, cherries, raspberries—are given the next coldest place. Berries should be picked over, and the bruised fruit removed before storage. Salad vegetables occupy space near the fruits. In the warmest location, strong-flavored vegetables and fruits find their places when it is necessary to store them at refrigerator temperatures.

Do not clutter up the refrigerator by storing foods that do not require refrigeration or will later be discarded. Jellies, pickles, commercial salad dressing, peanut butter, mustard, vinegar, etc., come under the first heading. Carrot, celery, and radish tops, outside lettuce leaves, store wrapping-paper, and cardboard cartons are examples of the second group.

Although cold air can hold less moisture than warm air, foods stored uncovered at any temperature will tend to dry out unless the relative humidity is approximately 100%. Manufacturers have tried to solve this problem by providing covered containers to preserve the crispness of succulent vegetables.

Containers are of various kinds: crispers of porcelain enamel, with or without openings for ventilation; opaque glass dishes in a variety of shapes and sizes; bags of polyethylene or plastic, and covers of the same material for use over bottles and bowls; seal sacks with zipper or slide fastenings. Aluminum foil is also widely used because of its ability to take any shape. They all help to decrease the evaporation of moisture from the food and also aid in preventing spread of odors. Less moisture is lost if the container is comparable in size to the food to be stored. Moisture-vapor proof bags are, accordingly, especially satisfactory for storing small amounts of succulent vegetables and have the added advantage of requiring the least space in the refrigerator.

The covered crisper pans should be of noncorrosive material. They are usually of porcelain enamel on steel, with covers of the same material, or of glass, aluminum, or plastic. They maintain a humidity of approximately 90% when the temperature is 40°F.

Extensive tests in the household equipment laboratories have shown that fresh, succulent vegetables, unless protected with a skin, should be stored covered in all types of refrigerators. Vegetables containing a high percentage of water, such as lettuce and celery, keep best when tightly covered; others—beans, carrots, etc.—may be placed in a ventilated container. Peas and lima beans should be left in their pods, fresh corn in its husk. Brussels sprouts and asparagus should not be washed until just before they are cooked. Although most investigators recommend that raspberries, strawberries, and similar fruits be stored uncovered, one investigator affirms that berries placed in a covered glass jar will keep fresh for weeks, if held at a temperature of approximately 38°F. Peaches, plums, and pears are protected by their skins against dehydration. Strong-flavored fruits and vegetables tend to contaminate the air of the refrigerator and should, accordingly, be kept in covered containers or placed in polyethylene bags. Raw meat should be covered lightly with waxed paper. Other protein foods and cream and butter are always protected with a covering. Leftovers remain fresh longer if covered.

The location of crisper drawers is flexible. In some models, they are placed at the bottom of the food chamber; in others, they are placed one above the other in racks below the evaporator. The crispers normally located across the bottom of the compartment may also be stacked, leaving extra space for bulky food. Occasionally the crisper is placed on the door and will drop down for ease in removing food. Many of these humidity drawers are of glass, or have glass covers, increasing the speed with which any given food may be selected. Sometimes the lower surfaces of the drawers are ridged to keep excess moisture from the food.

Moving the crisper around to various locations in the refrigerator has little effect on temperature distribution in the refrigerator, because all refrigerators must be capable of maintaining proper temperatures regardless of where the user places the various containers of food. The vegetable crisper has no more effect than large bowls or other food storage items.

Some manufacturers recommend putting hot food into the refrigerator to cool it rapidly, claiming that modern refrigerators have sufficient cold capacity to take care of the situation.

Use of the Evaporator

Ice trays are commonly made of aluminum and now frequently of plastic. The single divisions are of various shapes: cubes, perhaps tiny cubes, rectangles, and thin slices. Plastic trays are flexible and permit the easy removal of cubes, an advantage no longer emphasized since metal trays have been fitted with shelf releases and ejectors for freeing cubes from the sections one at a time or by the trayful. The metal tray may be permanently wax-treated to increase ease

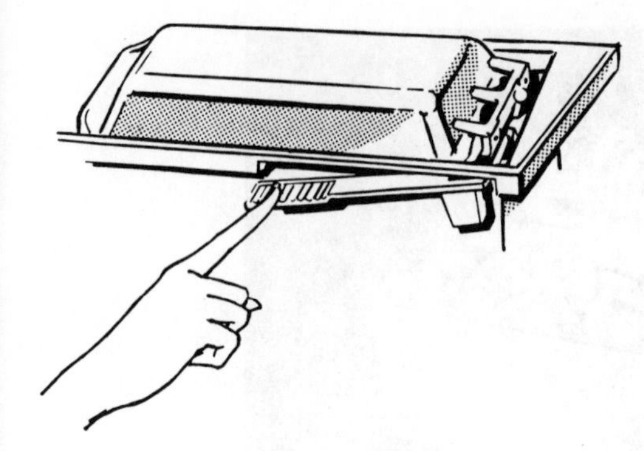

***Fig. 8.13** Automatic ice cube ejector. Place ice cube tray on ice bucket, press lever, and cubes drop into bucket. They remain dry and are ready for instant use. (Admiral)*

of cube removal. The waxed surface is affected by hot water, and moderately warm water is recommended. In case of accidental removal of wax, a solution is available on the market for retreating the trays. Trays should be filled to within ¼ inch from the top edge to allow space for expansion of the ice and, in some trays, to permit operation of the cube release.

A number of trays are now emptied by a special ice cube ejector; pushing a lever causes all the cubes to fall at one time into a container, in which they may be stored until needed (Fig. 8.13). Many top-of-the-line refrigerators make cubes automatically without the use of trays and, by the action of a sensitive signal arm, keep the portable ice-keeper supplied. Automatic icemakers are very popular. A connection to the water supply is essential (Fig. 8.14).

Instead of the common utilization of a cam, lever, or rake, one manufacturer uses a conveyor belt system. Water entering at one end is frozen by a stream of zero cold air, and comes out at the other end as ice cubes.

Ice trays are frequently used for making frozen desserts. Frozen desserts are classified into ice, sherbet, ice cream, mousse, and parfait, depending upon the method of preparation. When they are frozen in the trays of the mechanical refrigerator, they often tend to be coarser in texture than when frozen in the freezer fitted with a dasher. Certain precautions must, therefore, be taken to ensure small crystals. It is recommended that the mix be stirred or preferably removed to a chilled bowl and beaten when it has frozen to a firm consistency but still is not hard. The adding of small amounts of

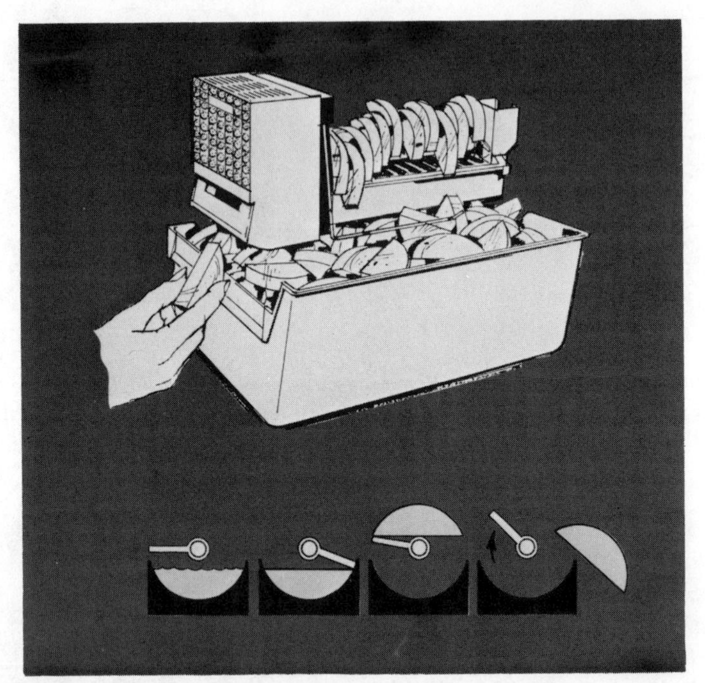

Fig. 8.14 Automatic ice cube maker. (Whirlpool)

interfering agents also assists in preventing the growth of crystals. These substances include gelatin, cornstarch, marshmallow, tapioca, cream, eggs, Junket—in fact, any material that makes the mixture more viscous and so coats the crystals and prevents their growth. Sugar should be dissolved, since it is itself crystalline, and can form the nucleus around which a crystal may grow. Cream should not be beaten to a stiff consistency—it should hold its shape but still flow from the bowl. Day-old coffee cream is better to use than a heavier cream. In general, the homemaker should make only those frozen desserts that cannot be readily purchased. In most parts of the country many ice creams and sherbets are less expensive to buy than to make, even disregarding the time and effort involved.

Care

Since the preservation of food is the main purpose of the refrigerator, it must be kept spotless at all times, clean and dry. The outside finish may be wiped off as frequently as the range or kitchen table is wiped. Special waxes are available for polishing. Food spilled in the inside should be cleaned up immediately and shelves and walls kept free from moisture.

The refrigerator should be thoroughly cleaned at the time of defrosting. The freezing unit and inside walls should be washed with soda water—2 tablespoons of baking soda to a quart of water—then dried. Ice trays should be emptied and washed before new ice is frozen. The cubes may be preserved by being put into a plastic bag (or left in the ice tray) and then wrapped in several layers of paper. Wash the trays in warm soapy water, rinse, and drain. Some trays are finished in silicone, which may be removed if too-hot water is used. Wash the gasket with warm soapy water, rinse with clear water, and dry. Food and grease should be wiped from the gasket as soon as deposited; grease will cause natural rubber to become soft and sticky. Even oily secretions from the skin will cause it to deteriorate, so fingers should not touch it in opening the door. Gaskets made of synthetic materials are less easily damaged. Vinyl plastic is widely used.

Defrosting

Frost varies in density and structure, depending on the way it is formed and on its age. The first deposits of moisture freeze to a spongy brittle mass, but as more and more condensation occurs the density increases and gradually a coating of solid ice is produced. The accumulation of frost may be diminished by keeping foods of high water content in covered containers.

Defrosting is advisable before the deposit is ¼ inch thick. Ice on the coils acts as an insulator, cuts down the efficiency of the evaporator, and forces the motor to run a higher percentage of the time. Some refrigerators have a small red indicator on the outside of the evaporator. When it is no longer visible, defrosting is recommended.

Units in conventional refrigerators may be defrosted by shutting off the current or by filling the ice-cube trays with warm water and placing them inside the evaporator sleeve. The latter method is more rapid but it should not be followed unless advised by the manufacturer of the model in question. When warm water is used, the switch should be turned to the off position. Neither method will greatly increase the temperature of the cabinet, since ice in melting absorbs heat—probably not more than 8 or 10° when defrosting takes place overnight. Some models have special cycles to which the refrigerators are switched when defrosting is necessary. Defrosting takes place somewhat more slowly in this case, but the temperature inside the food chamber never rises above 50°F.

The drip water should be emptied as soon as defrosting is complete. Otherwise it may evaporate into the cabinet air and cause new frost to form immediately when operation begins again.

Because many homemakers find the defrosting operation bothersome, manufacturers have devised a number of different systems by which the evaporator coils are defrosted automatically. The need for such a method has become greater with the extensive use of the horizontal evaporator of larger size, in which frost is harder to detect and from which it is more difficult to remove.

Several methods have been used. The first one employs a time clock, which interrupts the refrigerating cycle at a predetermined time, usually between 2 and 3 A.M., for a sufficient interval to allow the frost on the evaporator to melt. The process is fairly long, and foods stored within the evaporator may tend to soften. Ice cream especially seems to be affected, and acquires a grainy consistency. In a second method coils of an electric heater are attached to the evaporator. The compressor does not operate during the defrosting period. This is a comparatively short process, but the temperature may rise above 32°F. Instead of being fastened to the evaporator, the heater may be placed in contact with the capillary tube outlet, heating the refrigerant as it enters the evaporator. The frost deposit is melted rapidly and, in this case, the compressor operates during defrosting. In a fourth method, the cycle is reversed and the warm, gaseous refrigerant is circulated through the evaporator, which temporarily becomes a condenser. In all two-door refrigerator-freezers, the fresh food section is defrosted automatically, but the freezer section may require manual defrosting.

In "no-frost" models, both fresh food chamber and freezer are defrosted automatically. The frost deposit is changed by sublimation to a vapor by moving currents of cold air without passing through the liquid stage. In some models, the moisture that has accumulated on the evaporator plate is whisked away quickly before frost can form, but more commonly frost does form on the evaporator. Since, however, the evaporator is out of sight behind the wall of the fresh food compartment, the frost is not visible and does not interfere with food storage. In some models two or three times daily at 12- or 8-hour intervals a defrost timer control located behind the base plate activates a heating element that quickly thaws any thin film of frost that has formed. A 2:00 A.M. and 2:00 P.M. schedule is suggested. This clock control must be reset to the correct time of day when the house current has been interrupted. Another type of clock activates the defrost cycle after a specified period of compressor run, without regard to time of day. In certain models one manufacturer employs a control that defrosts only when defrosting is necessary, which may be several days apart or even at weekly intervals. This control uses infrared energy that radiates heat to all parts of the coil when defrosting is needed. Several other systems use radiant defrosting. As previously noted, in the combination refrigerator-freezer, the food chamber may be cooled by means of a refrigerated plate attached to the rear wall of the compartment. This plate defrosts automatically on each cycle as soon as the temperature has reached 36 or 38°F, and before the unit starts to operate again. This method eliminates a build-up of frost and cuts down temperature fluctuations.

The water left by defrosting may flow into a container within the food chamber, or it may be piped to a tray in the compressor compartment, where heat from the compressor and the condenser will evaporate it.

One of the disadvantages of automatic defrosting is neglect to clean the refrigerator at regular intervals. Surveys have indicated that some refrigerators have been left for months without being cleaned. As a result off-odors develop. The drip tray in the compressor compartment is out of sight, and a special effort must be made to remove it frequently and wash it in warm soapy water.

In any but the "no-defrosting" method, frost may accumulate on top of ice cubes and tend to contaminate them. If for any reason defrosting is not completed before the refrigeration cycle begins again, an ice deposit may build up in spots from which it is difficult to remove. None of the systems maintains a guaranteed constant freezing compartment temperature, and the effect on foods of any extensive fluctuation in temperature is uncertain.

If accessible, refrigerator condensers should also be cleaned if the air is to circulate freely around the coils. Dust and lint depositing between the baffles will shut off the circulation and hence will hinder the cooling action. Check the condition of the coils regularly. A stiff brush or the flat tool of the electric cleaner will do a satisfactory job. Always disconnect an electric refrigerator before starting to clean the condenser. One manufacturer obviates the need for cleaning the condenser by constructing the metal fins sufficiently far apart so that dust does not collect between them.

Cost of Operation

The cost of operation must be considered as well as the initial purchase price.

The 9-cubic foot electric refrigerator uses about 350 kilowatt-hours a year, a 14-cubic foot frost-free model, 1200 kilowatthours per year. Combination refrigerator-freezer types, depending on size, use more electricity than conventional refrigerators. With the addition of other features it is expected that kilowatthour demand may easily double within the next 10 or 12 years. Even now many families tend to buy large refrigerators with a big freezer compartment held at zero temperature, and it holds such a variety of tempting foods—soft drinks, ice cream, snacks, that the door is opened frequently, causing the motor to operate more continuously. But if it gives more pleasure to the family, it is probably worth the extra cost.

Tests carried on year after year in the course in refrigeration at Iowa State University indicate that kilowatthour consumption is less when the refrigerator door is opened less frequently, even if for longer periods of time, than when opened more often for short periods.

It is advisable, therefore, to remove all the foods needed at one time, instead of taking out one or two articles and then repeating the operation in a minute or two. The same procedure should be followed in putting food into the refrigerator. The rise in the temperature of the food chamber was always more rapid during the first 30 seconds the door was open than in the succeeding seconds.

Initial cost depends in part upon the accessories supplied with the refrigerator, such as water pitchers, dishes for leftovers, and egg baskets. They may increase the cost out of proportion to their value. Low-end-of-line models, so called, are just as carefully constructed and have the same kind of units as the deluxe cabinets. A model may have only one crisper, but many foods are now purchased in plastic bags, and these furnish sufficient protection; they may have no tray releases or divided shelves, but all the basic essentials are there. The homemaker must decide for herself whether the special features are worth the extra cost because they meet a particular need.

Guarantee

Refrigerators are usually guaranteed. Most manufacturers guarantee the cabinet against defects in material and workmanship for 1 year and the refrigerating system for a period of 5 years. The purchaser should read the guarantee carefully to know exactly what it includes.

In selecting a refrigerator-freezer, choose a brand name manufactured by a reputable company that has been in business for many years, one that is sold by a reliable dealer and for which qualified service is readily available. Request a complete demonstration. Note the quietness of operation. Quality appliances are sound insulated.

They are tested as produced under a control program, and should have a final complete test before being shipped from the plant.

For the same approximate cost, today's refrigerator is about 96% larger than those manufactured 20 years ago; because of improved, thinner insulation, they do, however, occupy the same space; for example, an 18.8 cubic foot model fits the same space as a 10 cubic foot 1948 model.

Certification

Appliance manufacturers, through their national trade association, the Association of Home Appliance Manufacturers (AHAM), sponsor a program for measuring precisely the cubic feet of usable refrigerator and freezer space and the square feet of shelf area, so assuring that the buyer will obtain a unit of the desired size. Measurements are made according to standards in nation-wide use and are published in Certification Directories distributed to dealers every six months. Every certified unit is identified by a distinctive seal (Fig. 8.15).

FREEZERS

A freezer may be a separate appliance, or it may be combined with a refrigerator. In standard freezers, two different cabinet designs in a variety of sizes are available: a chest type with top-lid or lids, and the upright model with the front-opening door (Figs. 8.16 and 8.17).

Construction of Cabinet

The separate freezer cabinets are made of aluminum, stainless steel, enameled steel, or galvanized steel. Depending on the base material,

***Fig. 8.15** A dark blue and silver AHAM seal certifies the net refrigerator volume and net shelf area for refrigerator-freezers and for upright freezers. A light blue and silver seal certifies net refrigerated volume only for chest freezers. With both seals AHAM declares that the certification is accurately stated and determined in accordance with the requirements of the certification program.*

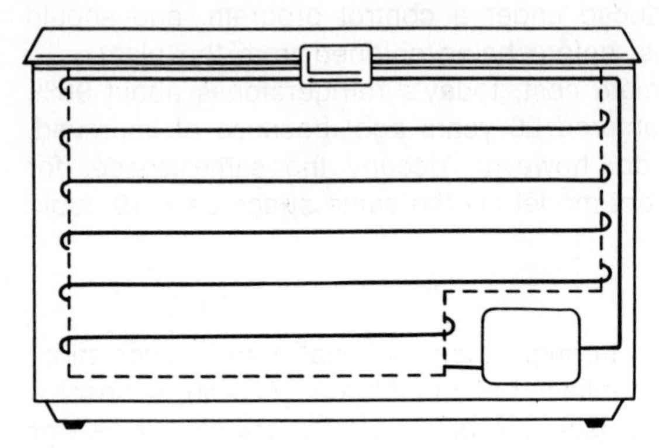

Fig. 8.16 In chest freezers the coils are frequently wrapped around the box.

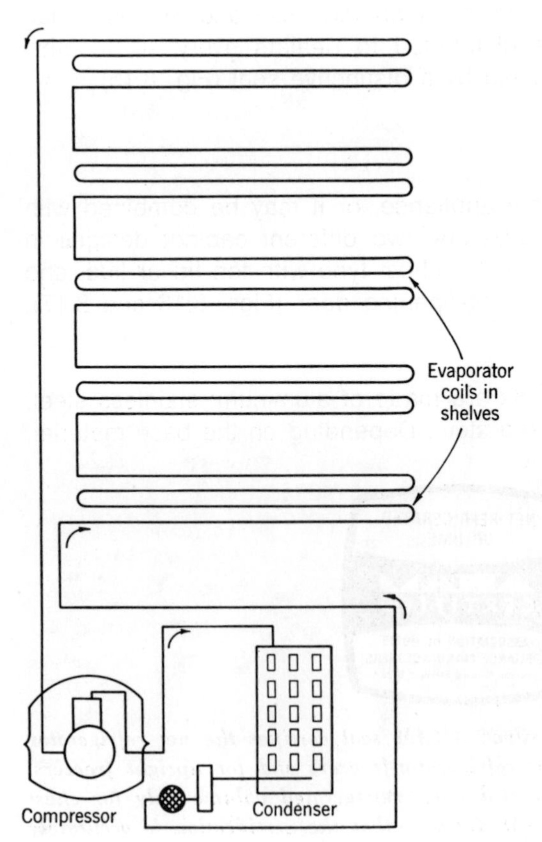

Fig. 8.17 In many upright freezers each shelf is supplied with coils for more rapid freezing.

the cabinets are finished on the outside with synthetic enamel, on the inside with synthetic enamel, porcelain enamel or aluminum. Plastics are used for door linings and breaker strips. The breaker strip on the freezer is important, since it must prevent heat conduction to the inner walls and sweating on the outside of the freezer. The doors of a number of freezers are held closed by magnets embedded in the gasket.

The insulation varies from 4 to 6 inches in thickness; it must provide maximum protection against transfer of heat and at the same time obviate excessive bulk. High-density Fiberglas meets these requirements. Added insulation is necessary when the condenser coils are placed against the exterior walls of the cabinet. Adequate insulation should maintain below-freezing temperatures for a minimum of 36 hours in case of power interruption. The hardware should be chromium plated and positive in action to maintain a very tight seal between the lid or door and storage compartment.

Freezers are usually supplied with leveling screws for adjustment to uneven floors. When loaded, and sometimes when empty, the freezer is a heavy appliance, and if not properly adjusted, will tend to warp and throw the door or lid out of alignment. The lid of a chest freezer should be counterbalanced.

On an upright freezer, one outside door is preferred. The temperature gradient between room and cabinet space is large, much larger than in the household refrigerator, and the tendency for heat to pass from the outside to the inside of the box is accordingly increased. The greater the number of doors, the more the chance of heat leaking into the box. Loss of cold is more probable from vertical models; the cold air flows out the opened door toward the floor.

The present trend is toward sealed-in units. The evaporator coils in the chest freezer may be within the insulated walls and may entirely surround the cabinet, or they may be stamped into plates that become the inner walls of the cabinet or that divide the sections. In specification sheets freezers with from five to seven refrigerated plates are mentioned. The chest-type often has a sparate space for quick freezing, in which the temperature may vary from 0 to −30°F, the lower temperatures being more satisfactory for rapid freezing, but also more expensive to operate. Storage sections should be held at 0°F as the maximum.

In the upright freezer the coils are frequently brazed to horizontal shelves on which the food packages are placed. In some freezers, however, the freezer shelves are of open-grid construction to allow better circulation of the cold air. Combination refrigerator-freezer models have the coils in the surrounding walls, and sometimes, special fans are installed in the side walls to send blankets of zero-cold air against the food packages for fast freezing. In any type, the objective is to bring as much food as possible into direct contact with a cold surface for more rapid freezing (Fig. 8.18). Freon 12 is the most widely used refrigerant, but Freon 22 is used in a few models. Freon 22 is especially adapted for very rapid freezing, where space and time are limited.

Freezers vary in size from 3 to 32 cubic feet. Storage capacity may be estimated approximately from the generally accepted value that 1 cubic foot is equivalent to 35 pounds of frozen food. Shelves on door and lid do not increase storage capacity, but usually add to convenience of use. Chest models hold more food than upright freezers of comparative price.

> The rated storage capacity in pounds of frozen food is an arbitrary figure which is standard for the entire industry. It does not tell any individual user actually how many pounds of frozen food she can store in that compartment. The standard figure used is 35 lbs. per cubic foot, which is based on an arbitrary selection of a variety of frozen foods that vary in weight from a few pounds per cubic foot to 70 to 80 pounds per cubic foot. For example, pastries weigh considerably less than ground meat. Because of this wide variation, it was felt by the industry that some kind of an average figure should be selected and used by all, so that even though it would be impossible for an individual consumer to store exactly the quantity advertised, the consumer is able to compare one product with another, and know that one with a larger advertised capacity will hold more frozen foods than one advertised with a smaller capacity.[2]

The smaller freezers up to 6 cubic feet are usually intended for storage only. They may be used to supplement the drawer at the locker plant, for storage of commercially frozen foods, or for storage of foods prepared at home and frozen in the locker quick-freeze section. A few packages may be frozen in this freezer, if necessity arises, but freezing of any large amount of food will cause the temperature of the rest of the food to rise beyond safe limits and should be avoided.

The freezer section in the combination refrigerator-freezer is usually about 1.5 cubic feet, although some manufacturers advertise a capacity of 3 cubic feet, and, in an 18 cubic foot combination, as much as 8 cubic feet may be available for freezing. In at least one model of the side-by-side refrigerator-freezer the freezer section has

[2]Private communication, General Electric Company, August 1969.

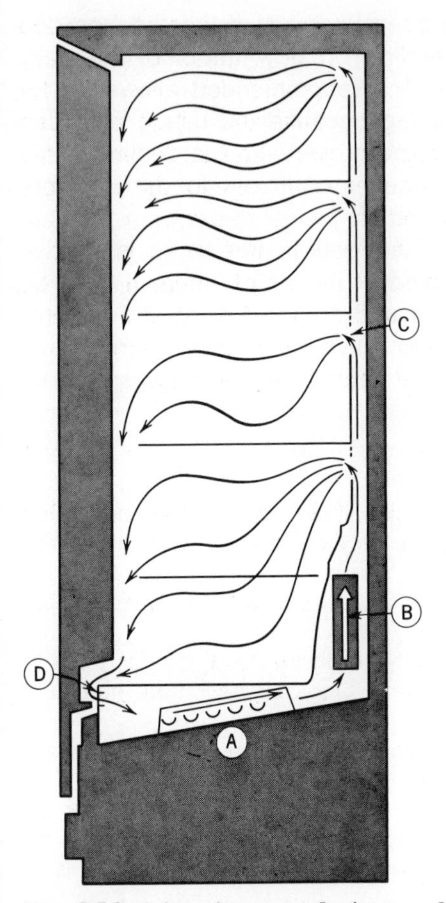

***Fig. 8.18** A frost-free upright freezer. 1. Cooling coil (A), hidden from view, is below and completely apart from the storage compartment. A blower (B), located behind the storage compartment, draws air through the coil where it's chilled and dried. 2. From there, it is blown up through a duct in the back wall of the storage compartment, and into the compartment through shelf-level vents (C). 3. The cold, dry air circulates throughout the cabinet—absorbing all frost-causing moisture and heat. 4. Air then passes out of storage compartment through a grille at the bottom, front (D), and returns to the cooling coil, where the process begins again. Once every 24 hours, the blower stops and electric heaters melt the frost from the hidden cooling coil. The heat of the compressor evaporates melt-water into the air outside the freezer, assuring a totally efficient Frost-Proof freezing system. (Frigidaire)*

a 9 cubic foot capacity. Some authorities, however, suggest the purchase of a separate freezer if more than 4 cubic feet are needed for all-time use. The size of freezer needed is sometimes based on the requirements per person and has been variously estimated from 2.3 to 4 cubic feet per person, depending on how much of the daily menu is supplied from the freezer. The recommended allowance for freezer space per person is increasing because the usage of frozen foods is increasing. Even though home preservation is declining, the added convenience and growing varieties of frozen foods has more than offset the storage space required for home freezing.

Another dual-temperature unit is the walk-in freezer. It is built in sections which are usually assembled at the point where it is to be installed. It affords space for chilling and aging of meat and for bulk storage of fruits and root vegetables, such as apples, carrots, and potatoes, but also has a quick-freeze section and storage space for frozen foods.

Freezing of Foods

1. Food is no better after it is frozen than it was before freezing. Select varieties most suitable for freezing in the local area and of the correct maturity and highest quality.
2. Choose foods that your family prefers. Do not use valuable freezer space for foods that will not be eaten.
3. Prepare food for freezing immediately. Keeping food too long before freezing causes lower quality and flavor. One half hour from garden to freezer is a good rule to follow.
4. Freeze products in meal-sized packages to prevent waste. Date and indicate contents of each package.
5. Freeze foods quickly by placing the packages flat against the refrigerated area, or in path of cold air circulated by fan (Fig. 8.18). Usually the control should be turned to the lowest setting while the food is in process of freezing.
6. Do not refreeze foods that have thawed completely, but if food is still firm and contains ice crystals it may be refrozen in an emergency.
7. Even at 0°F or below, food will tend to lose moisture. Protect food with airtight coverings impervious to moisture. Fluctuations in temperature may draw the moisture from the food and deposit it in the form of frost on the inside of the package. Such fluctuations may occur when too large an amount of unfrozen food is placed in a partially loaded cabinet. The manufacturer should include in his instructions for a given size freezer the maximum number of pounds of food that should be frozen at one time. Keep a freezer thermometer inside freezer to check on temperatures.
8. An alarm or signal light should be provided to notify the homemaker of failure of freezer operation, resulting in warmup.
9. The more rapid the turnover in products stored in the home freezer, the less the cost of operation per pound of food stored. A variation of

28 to 13 cents per pound is noted by the United States Department of Agriculture. There is a limit, however, to the amount of turnover that is possible.

Packaging materials to be used in preparing food for freezing should be moisture-vapor proof, stain, and grease-resistant, odorless, nontoxic, and pliable enough to be easily molded. Among the materials in common use are waxed cartons, with and without liners of Pliofilm or cellophane, glass and plastic jars, aluminum boxes with snap-on covers, polyethylene bags and sheets, aluminum foil, and laminated papers. Kraft papers treated with wax or paraffin may also be used for short storage periods of two weeks or less. A bulletin from Sears suggests, if the freezer is fairly large and used frequently, that you invest in pint- and quart-size straight-sided plastic freezer containers with airtight close-fitting lids. They are space saving and stackable. Use heavily waxed or plastic-coated milk cartons for short-term storage. Desiccation of vegetables and meats and development of rancidity is usually caused by improper packaging, inadequate storage, or both. The packaging material and method of storage has more effect on flavor than moderate fluctuation in temperature.

It has been suggested that the load frozen in the freezer at one time should not exceed one tenth of the total capacity of the freezer, and that one fifteenth is even better. Another authority says approximately three pounds per cubic foot of freezer capacity. In any case, the load should freeze within 24 hours, since slow freezing tends to impair the quality. It is estimated that an energy consumption of 0.1 kilowatthour is required to freeze one pound of food and to lower its temperature to zero. Research experiments carried on in the household equipment laboratories of Iowa State University indicate that freezing of food in the chest may be speeded up if packages farthest from the walls are interchanged with those in contact with the walls once or twice during the freezing process. The food at the center of the top section of the chest is usually several degrees higher in temperature than any in the upright, even though most manufacturers now wrap the evaporator coils around the chest almost the full height of the walls.

Usually packages placed on the door shelves of the upright freezer should be stored there only after they are frozen. If they are placed on the door shelves for freezing, they should be exchanged occasionally with packages on the main shelves to speed the freezing process.

A few upright freezers now feature automatic icemakers.

Location of Freezer

The kitchen is undoubtedly the best location for the freezer to save time and motion, but the heat there will cause it to operate a larger percentage of the time. A cool, dry basement or back porch is suggested, provided the temperature never drops below about 40°F. The systems of some freezers will fail to operate if the temperature falls as low as 35°F.

The upright occupies less floor space, but, since this means that the weight is concentrated in a small area, care should be taken that the floor is braced adequately. There are reports of freezers falling through floors into basements. The upper surface of the chest-type may be used for a work surface. It may be more difficult to remove packages from a chest freezer than from an upright, although this depends to a great extent on the methods of storing used by the homemaker. At present, baskets are a part of the accessories of a chest freezer; there even are some that roll back and forth on metal tracks inside the freezer for easy access to the food below without the need of lifting the basket. Some upright freezers have shelves that pull out, and some have shelves on doors and lids to keep small packages within easy reach.

When the lid of the chest type is open, the cold air, being heavy, remains in the freezer; in the upright it tends to fall out the open door.

Sweating on the exterior of the cabinet tends to occur whenever the relative humidity is sufficiently high. This condensation may be prevented by maintaining the temperature of the outer walls slightly above ambient temperature. To do this a number of manufacturers place the condenser coils against the inner surface of the outside walls and also across the bottom of the freezer. These coils produce just enough heat to prevent condensation of external moisture and at the same time offer a larger condensing area for the gaseous refrigerant inside the coils. If no sweating occurs when the relative humidity is 88% the insulation is considered satisfactory. A slight condensation that disappears during the "on" cycle of the unit is allowable. Some chest freezers have vented bases that keep the bottom of the freezer and the floor beneath dry, preventing damage to the floor and rusting of the freezer parts.

Defrosting

Many new freezer models never need defrosting. Once or twice in 24 hours during an off period of the unit, heat in one form or another melts from the evaporator any ice that has formed. Some models are frostproof.

Other home freezers are usually defrosted only once a year, at a time when the least amount of food is stored. Food remaining should be temporarily moved to the refrigerator evaporator. The frost may be removed from the inside freezer walls with a cold-water spray. Never use warm water since it will tend to build up too high a pressure within the evaporator. Between defrosting periods, excess frost is removed from the inside surfaces by scraping with a blunt instrument such as a wooden or plastic putty knife. Place the freezer in a dry, cool, well-ventilated location, and frost will be reduced to a minimum. It is generally thought that a deposit of frost tends to increase the temperature of the cabinet, the percentage of operating time, and consequently energy consumption. McCracken and Fisher,[3] however, found only a slight increase in time of operation and kilowatthour consumption, but they did find that frost reduced the amount of usable storage space, and interfered with the removing of packages and with the opening and closing of the door. Frost may also interfere with the sealing action of the gasket.

Cost of Operation

The chest type with evaporator coils around the walls refrigerates the food load more satisfactorily with lower energy consumption and shorter operating time than the chest with the coils in the divider plates. Upright freezers with refrigerated shelves freeze more rapidly with less kilowatthour consumption than other types. The compressor should operate continuously while food is being frozen, but no more than 50% of the operating time should be required to maintain the storage temperature. The size of the motor varies from ⅛ to ½ horsepower.

Test Specifications

Standard methods for testing household refrigerators have been set up by the American Society of Heating, Refrigerating, and Air-Conditioning Engineers and approved by the ANSI. Similar standards have been adopted for both refrigerators and home freezers by the Association of Home Appliance Manufacturers (AHAM). As a result of the test procedures, information is obtained on energy con-

[3]E. C. McCracken, and Marilyn G. Fisher, Frost Deposit in the Home Freezer, **Refrigeration Engineering, 61** (166), 1953.

sumption, percentage of operating time, number of cycles per 24 hours, average temperature of cabinet air, and total pounds of ice cubes per freezing. Methods for determining the food-storage volume and shelf area of refrigerators are also outlined, and the presence of the AHAM seal attached to the appliance is assurance that the claims of the manufacturer are reliable (Fig. 8.15).

The Underwriters' Laboratories test for safety, that is, fire, life, and explosion hazards. The purchaser should look for their seal on any refrigerator or freezer under consideration.

Terminology

Refrigerator. A means of cooling food below the temperature of surrounding atmosphere to prevent spoilage from presence of molds, yeasts, and bacteria.

Construction. Two shells of sheet steel, insulation between, fitted with shelves, some in door.

TYPES

Compression. Compressor, condenser, evaporator (freezing unit), connecting tubes, operated by electric motor.

Absorption. Generator, condenser, evaporator, absorber, connecting tubes, operated by heat-gas or oil flame, occasionally by an electric unit.

REFRIGERANTS

Freon 12. Dichlorodifluoromethane, is nonflammable, nonexplosive, nontoxic, and stable.

Ammonia. Flammable, detrimental to food, very soluble in water, easily separated from water by heat.

Latent heat. Heat absorbed to bring about a change of state; heat of vaporization-liquid refrigerant changes to gas in evaporator, absorbing heat from stored food.

Specific heat. Capacity of substance to absorb or retain heat compared to capacity of equal amount of water. Water has specific heat of one; all other materials less than one. It is necessary to know specific heat of materials used in construction of refrigerator and of stored foods to determine size of cooling unit.

Special features. Interior light, crispers for fruits and vegetables, ice trays, some ice cubes made automatically.

Defrosting. Often automatic; if manual, by time deposit is ¼-inch thick.

Freezers. Chest, upright, 3 to 32 cubic feet; 1 cubic foot stores 35 pounds frozen food.

Packaging material. Must be moisture-vapor proof, odorless, and stain and grease resistant, pliable. Examples are waxed cartons, plastic containers, pliofilm, polyethylene bags.

Certification of volume and shelf area by AHAM.

Experiments and Projects

1. Make a table to record the following data on refrigerators and freezers:
   (a) Size of motor.
   (b) Method of opening and closing door. Ease of operating door latches.
   (c) Thickness of insulation (i.e., distance between outside and inside walls).
   (d) Material of breaker strip and gasket.
   (e) Type and arrangement of shelves. Use of shelves on door.
   (f) Type of freezing unit, method of defrosting, position of container to collect defrost water, location of any fans, effect on fans of opening door.
   (g) Position of condenser, construction to permit circulation of air.
   (h) How to distinguish between a refrigerator, a combination refrigerator-freezer, and a freezer.
2. Evaluate locations provided in refrigerator for storage of meat, vegetables, milk, cheese, butter, and baked custard.
3. To compare variations in temperatures in the refrigerator, suspend a thermometer in a flask of water by inserting it through a hole in a cork with which the flask is closed. You may need to check different areas on several different days.
4. Comparison of methods of freezing: Turn temperature control to the coldest setting. Make lemon sherbet. Mix together until sugar is dissolved ⅔-cup lemon juice, grated rind of one lemon, 1½-cups sugar, one quart half-and-half. Divide into two portions. Freeze one part in the refrigerator tray, stirring thoroughly with a wooden spoon once or twice during freezing period. When this portion reaches mushy stage, freeze the second part in freezer equipped

with dasher. Compare the two products on texture, color, flavor, and desirability. Explain the cause of the differences. Turn the control to a normal setting.

5. Make fruit ice in freezer: Cook a quart of cranberries in 2 cups of water. Put the pulp through a sieve. To the pulp add 1¼-cups sugar and stir until completely dissolved. Add the juice of one orange and of one lemon. Freeze. When half frozen, remove ice to a chilled bowl and beat with a chilled beater. Compare the ease of using the freezer and refrigerator for freezing. What would be a desirable use for this product?
6. Freeze ice cubes: Select ice cube trays of different shapes if possible. Compare the time required to freeze water and products made in Experiments 4 and 5. Explain the reason for the differences. Suggest uses for different shapes of cubes.
7. Which type of refrigeration appliances most appropriately meets your needs and preferences? Would your parents select a different model? Associate each model with precise family characteristics and situations.
8. To determine the speed of freezing ice cubes, the cost in KWH, and the effect of opening the door (when measuring and testing instruments are available), use the following apparatuses:

   KWH meter
   Potentiometer
   Multiple switch board (see Fig. 8.19)
   Thermocouples
   Scale
   Stopwatch
   Thermometer

   Connect KWH meter into refrigerator circuit. Connect three thermocouples through the multiple switch to the potentiometer. Turn control of freezing compartment to coldest setting.

   Weigh two ice cube trays. Fill each tray with 454 grams of water at 75°F. Place across middle of each tray a piece of two-inch-wide refrigerator tape and insert a thermocouple in center of one cube area. Suspend the third thermocouple in center of food compartment. Place Tray 1 in the freezing compartment with two sides touching two walls; place Tray 2 in center of freezer not touching any wall. Record temperature of each thermocouple every 10 minutes. Check rate of freezing every 30 minutes, recording time refrigerator door is open. While open, check temperature of food chamber every 15 seconds.

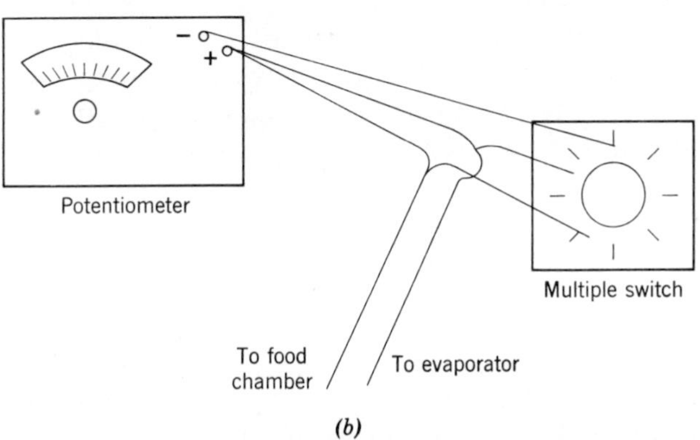

(b)

Fig. 8.19 (a) ***Multiple switch used to connect several thermocouples to the potentiometer. (b) Setup when one thermocouple is connected to evaporator, a second placed in food chamber.***

Record initial water temperature, measure of water, weight of water, KWH consumption, time to freeze ice cubes.

Determine effect of opening door on temperature of food compartment. Make a series of diagrams of each tray, indicating progress of freezing.

9. To determine effect on energy consumption of opening the refrigerator door during a two hour period for
   1. 30 seconds-12 times

2. 60 seconds- 6 times

Set up refrigerator with KWH meter, place thermocouple near center of food chamber and connect to potentiometer, close door and wait 20 minutes before starting tests. Open refrigerator door, keeping door open for 30 seconds at 10 minute intervals. Read temperatures:

(a) Before test begins.
(b) Immediately before each opening, and all during time door is open.
(c) Immediately after each closing of door and at 30 second intervals for 4 minutes after door is closed.
(d) At close of test.

Read meter at close of two-hour test. It may be necessary to count revolutions of KWH meter to obtain an accurate record of consumption.

Repeat experiment at next class period, keeping door open 60 seconds at 20-minute intervals.

Draw graphs indicating temperature changes and showing the comparative KWH consumption for the two times of opening.

10. To determine effect of derfosting on temperature in refrigerator, and time to bring temperature back to 40°F, use the following apparatuses:
    1. Potentiometer
    2. Multiple switch
    3. Thermocouples
    4. Stopwatch

    Place junction of one thermocouple on evaporator (fasten it in place by means of a strip of moist cloth); suspend second thermocouple in center of food compartment. Wait 20 minutes. Switch off refrigerator, noting time. Read temperature every 10 minutes until temperature reaches 50°F. Reconnect refrigerator and record temperatures every 2 minutes until temperature reaches 40°F.

    Draw graphs indicating the relationship between temperature and time. Repeat experiment with other refrigerators and compare results.

CHAPTER 9
Dishwashers

Automatic dishwashers offer the consumer much in flexibility, convenience, and performance in the cleaning of dishes, pots, and pans. Improved designs provide effective wash action, better water distribution, greater convenience in loading, and improved flexibility in use. For the most effective results with mechanical dishwashing, the user needs an understanding of the following:

1. Features of the dishwasher.
2. Knowledge related to soils and stains and how they respond to wetting.
3. Dishes and their design as related to effective care in the dishwasher.
4. Detergents that may be used most effectively in the washing process.
5. Water conditions in the home.
6. How to load the dishes for most effective exposure to water during the wash cycle.

Principle of Operation

The principle of operation of the automatic dishwasher is the directing of hot water and detergent at high speeds over dirty dishes. The mechanical action is provided by a motor-driven impeller or rotating spray arm plus other auxiliary water distributing devices. When the washes and rinses are completed, air warmed by a heating element is directed over the dishes for drying.

The impeller dishwasher, in which a motor-driven impeller distributes water in all directions inside the tub, is one of two general systems in current use. The other is the spray arm dishwasher, in which water is recirculated by a separate pump, emerging from the nozzles of the spray arm, which rotates by recoil and sends water onto the dishes.

In both types of dishwashers, the washing is accomplished by the combination of the mechanical force of the water striking the dish and removing the soil, and the chemical action of the hot, detergent water on grease and other food soils. To accomplish both functions, the dishes must be arranged in the tub so they will drain properly and water will not be trapped.

Detergents

Detergents designed for use in the dishwasher are low sudsing and highly alkaline. They are not suited for hand dishwashing. The ingredients found in dishwashing detergents are designed to perform the following functions:

1. **Surfactant.** Lowers the surface tension of the water so that it will quickly wet the surfaces and the soils. Lowering the surface tension makes the water sheet off dishes and not dry in spots. The surfactant also helps remove and emulsify fatty soils like butter and cooking fat. Nonionic surfactants are used because they have the lowest sudsing characteristics.
2. **Builder.** Combines with the water hardness minerals and holds them in solution so that neither the minerals nor the minerals in combination with food soils will become attached to the dishes and leave insoluble spots or film.
3. **Corrosion inhibitor.** Helps to protect machine parts, prevents removal of patterns from china and the corrosion of specific metals as aluminum.
4. **Chlorine compound.** Helps break down protein soils like egg and milk, aids in removing such stains as coffee or tea, lessens spotting of dishes and glasses.
5. **Perfume.** Masks odors from chemical base of cleaning product as well as odor from stale food.
6. **Processing aids.** Acts to allow active ingredients to be combined into usable form.

The quantity of detergent to be used in a load of dishes must be adequate to soften the water effectively, suppress foam from food soils, provide the necessary cleaning and suspension of soil, and protect the materials being washed. Underuse of the product results in poor cleaning, redeposition of soil, spotting, filming, and possible damage to some items being washed. Less than the recommended amount of detergent will be needed with very soft water or smaller loads of dishes. A wise guide is never to use less than 2 tablespoons because of the protective functions served by the detergent.

To cope with the low phosphate or nonphosphate products all phases of the dishwashing procedure become more critical. Prerinsing of dishes may be needed; loading must be done with more care;

proper water temperature of 140 to 160° is needed as well as the quantity and type of dishwashing detergent.

The ingredients that perform these functions differ in content and amounts. For this reason, one dishwasher detergent may be more satisfactory than another in certain areas. Because water quality varies, different detergents should be tried to determine those that are best for local water conditions. Each ingredient performs a different role. An illustration of the ingredients which may appear in a detergent and their function is as follows:

1. Sodium silicate—effective on fats, holds soil in suspension, inhibits corrosion and tarnish of metals.
2. Sodium phosphate—ties up the hardness minerals, calcium and magnesium.
3. Chlorinated trisodium phosphate—found in many brands of dishwasher detergent. Enables rinse water to sheet off, reduces water spotting.
4. Sodium carbonate—increases washability and improves soil suspension, so there is less redepositing of soil on dishes.
5. Wetting agent—helps penetrate soil more quickly. Low sudser.

Dishwasher manufacturers provide two detergent injections for the two wash phases of the cycle for loads of dishes with heavy soil, such as the family load of dishes and pots and pans.

Soils

Food soils commonly found on dishes fall into two classes—those soluble in water and those insoluble in water. The insoluble soils may be dispersed by water, swelled by water, melted by water, or may remain unaffected by the water-detergent solution. An understanding of the behavior of the soils in a detergent-water solution and the design of the dishwasher will enable the consumer to determine the degree of rinsing and scraping necessary before washing.

Water

For most effective dishwashing, the water should be of the proper quality and quantity. High quality water is free of undesirable minerals and of the proper temperature.

Minerals

Soft water (less than 3 grains hard) or water free of calcium and phosphorus is preferred. Some authorities recommend that if the water supply is more than 10 grains hard, a dishwasher should not be used unless a water softener is installed in the home. Rinse additives may be used to overcome streaking or spotting when water is hard. Many dishwashers have the rinse additive dispenser for this purpose.

Iron is another mineral that may cause problems in automatic dishwashing. Water that contains more than 1/10 p.p.m. of iron will stain dishes.

Temperature

Water of the correct temperature for effective dishwashing ranges from 150 to 160°F. Modern dishwasher designs may include heating elements designed for raising the temperature to the desired level for washing or rinsing. The cycle does not continue until the water has reached a desired temperature. In house designs which locate water heaters some distance from the dishwashers, a small heater close to the dishwasher may be used to supply water of the correct temperature for this purpose.

Amount of Water

Dishwashers use an average of 4.5 gallons of hot water for the rinse and dry cycle to 18 gallons for the pots and pans cycle.

Water Pressure

Water pressure may vary from 15 to 100 pounds per square inch. The preferred pressure should be between 40 and 70 psi. Pressure is important in providing the correct amount of water in the washer for the specific phases of the cycle. Force of water during operation is not determined by the pressure but rather by the washing action.

Water Fill

The dishwasher fill may be controlled by a timing device or by a meter. The time fill is operated by a valve which is set for a definite period of time at a certain pressure to give the correct amount of water. The fill generally lasts for approximately 60 seconds and admits approximately 12 pints of water flowing at a rate of 1½ gallons per minute. When the water pressure is lower than the desired pressure, an inadequate amount of water will be present when the valve closes.

The metered fill permits water to come in until a certain predetermined weight is in the tub. Even when the pressure is low or high, the correct amount is always metered into the dishwasher.

Basic Types

The five basic designs of dishwashers are the built-in, free-standing, portable, convertible, and dishwasher sink. The built-in is an under-

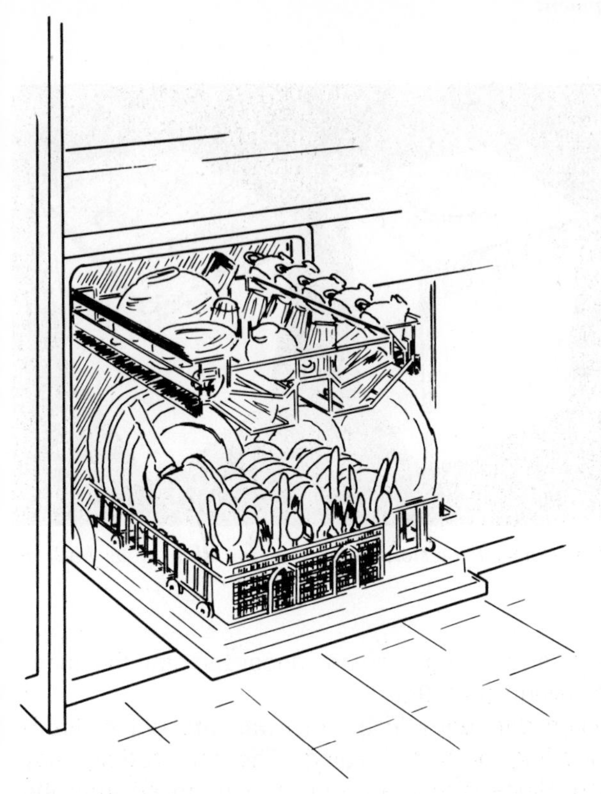

***Fig. 9.1** Undercounter dishwasher. Separate racks increase ease of loading and unloading. (Frigidaire)*

counter design that is permanently connected to a hot water line, drain pipe, and electric circuit. The dishwasher is integrated with the cabinets (Fig. 9.1).

A free-standing dishwasher is similar to the built-in undercounter type. It is designed to go into a vacant space, usually at the end of a row of cabinets.

A portable dishwasher requires no installation. It is ready for use when the hoses are snapped to the kitchen faucet and the electric cord is connected to a wall outlet. This dishwasher drains into the kitchen sink.

The convertible is a front-loading portable dishwasher and is ideal for those who plan to move or remodel later. It can be used as a portable immediately. The casters may be removed later, and the washer may be installed undercounter or as a free-standing model (Fig. 9.2).

A dishwasher sink design has the sink top and bowl, cabinet, faucets, and dishwasher, all in one compact unit. The unit requires

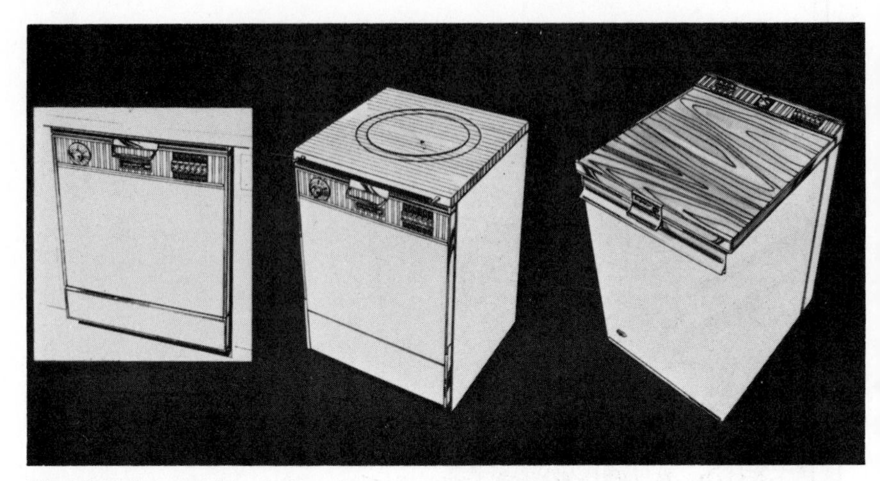

Fig. 9.2 Convertible dishwasher. Provision for use as portable or as built-in unit. (Whirlpool)

the water supply, drain, and wiring that are necessary for the function of all parts of the design (Fig. 9.3).

A second basic design difference in the dishwashers is the direction of loading: top-loading or front-loading. The top-loading may be portable or installed. Since dishes are put in or removed from the top, all or part of the upper rack must be removed during loading. Some current models have racks attached to the lid, which may be lifted back to permit easier loading. This model may have top racks which fold back or to the side and provide greater flexibility in loading items of various sizes in the lower rack.

The front-loading dishwasher may be a portable or installed model. The door opens down and the racks roll out separately. This dishwasher should have special attention given to balancing to prevent tipping when the loaded racks are pulled out.

Cycles

The cycles or combinations of conditions to be found on the dishwasher are designed to accomplish the cleaning and drying needs of specific dishes, pots, and pans in the load. Each cycle has a specified arrangement of wash, rinse, and dry phases. The more effective cleaning results from the dishwashers which have the two separate washes, each with fresh water and detergent. The detergent dispensers may be metered, so the correct amount is released for each wash cycle. The numbers of rinses in the cycle may range from one to

***Fig. 9.3** Built-in undersink dishwasher with soft food disposer. Designed for small kitchens. (General Electric)*

four. These may be quick sprays or longer flushes of water which are agitated or recirculated over the dishes.

In addition to the basic cycles, some of the extras which may be included on the dishwasher are: a prerinse before the regular cycle; a rinse-and-hold cycle which provides a short rinse to soften soil and keep it soft while storing dishes for wash at a later time; a short rinse and dry for those dishes which have not been used recently and have some dust to be removed—oily soil will not respond too favorably to this cycle; a utility cycle for utensils, which may use more vigorous action and usually has a shorter drying time; a short cycle, which may use less water; a heavy cycle for extra soil or grease; a sani-cycle, which provides a temperature of approximately 150°F in the final rinse for more effective control of bacteria; and a plate-warming feature to warm clean dishes before a meal.

A study of the needs to be served by a dishwasher in the home will enable the homemaker to determine which of these features she will use. For the homemaker who chooses the model which may offer only one wash phase in the cycle, the dishes should receive more care in prerinsing. Rinsing may well be done with cold water, as hot water in the wash can then remove the greasy soil. Foods which may become redeposited on dishes should be removed before washing. Even the most effective filtering screens may permit some fine food particles to pass through back onto the dishes.

For those with a partial load who desire to hold the dishes until the dishwasher is full, the rinse-and-hold cycle serves this need. This is especially convenient and economical for the small family or single person. This cycle is also used if one desires to cancel any cycle already started, or to drain water at any time.

Electrical Requirements

The dishwasher may be connected to a separate 115-volt, three-wire, alternating current, 15-ampere circuit. The motor is rated at ⅓ horsepower. The heaters are tubular units which range in wattage from 100 to 1100 watts for the dry phase of the cycle. The wattage applied to the wash and rinse phases of the cycle range from 1000 to 1400 watts. One specific design employs 1400 watts for the wet phases and 600 watts in the dry phase. The heating element may be used for three purposes—to maintain or achieve desired temperature of water for the wash, to raise rinse water temperature to desired level, and to dry dishes.

Wash Action

Wash action in the dishwasher may be provided by water coming from as many as five different positions (Fig. 9.4). Some levels are especially located for the cleaning of certain items such as the silverware. In each design, the water should be able to move freely over and around the items in the load.

The wash action may vary from one cycle to another. Regular

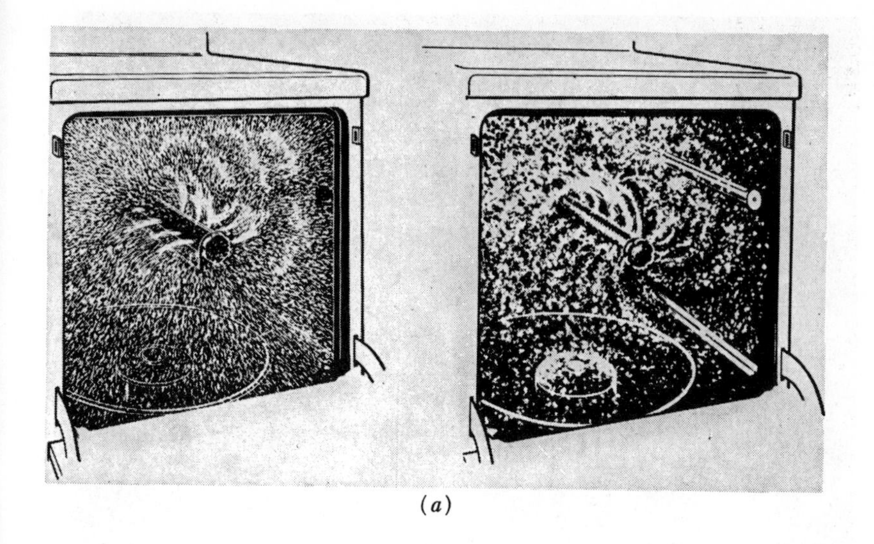

(*a*)

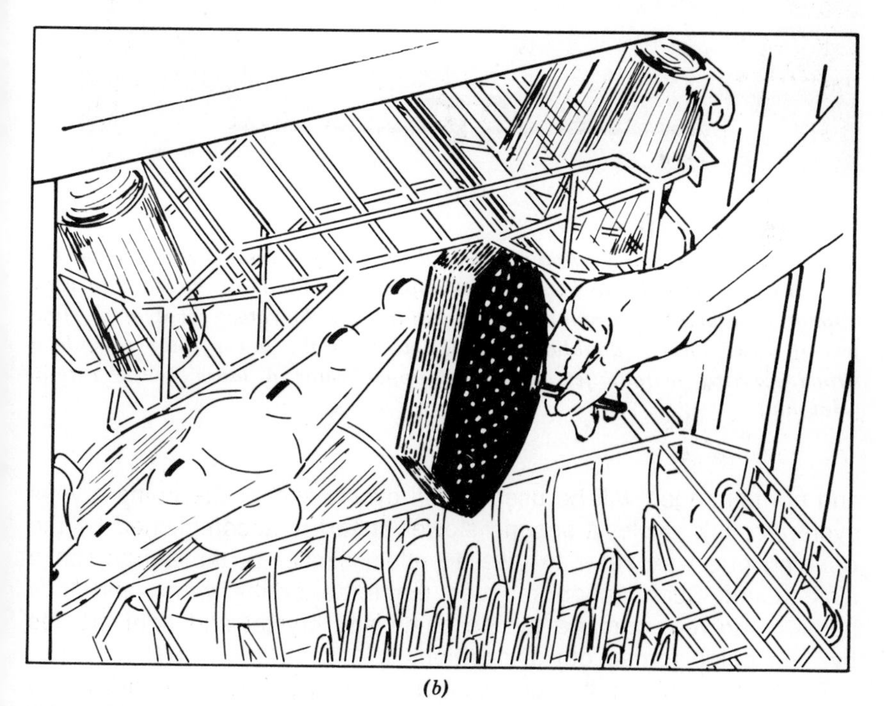

(*b*)

Fig. 9.4 (a) *Pattern of spray is thrown by horizontal mechanism of dishwasher. (Frigidaire)* (b) *Mechanism rotates to throw a vertical spray of water. The strainer may be removed and cleaned under a faucet. (KitchenAid)*

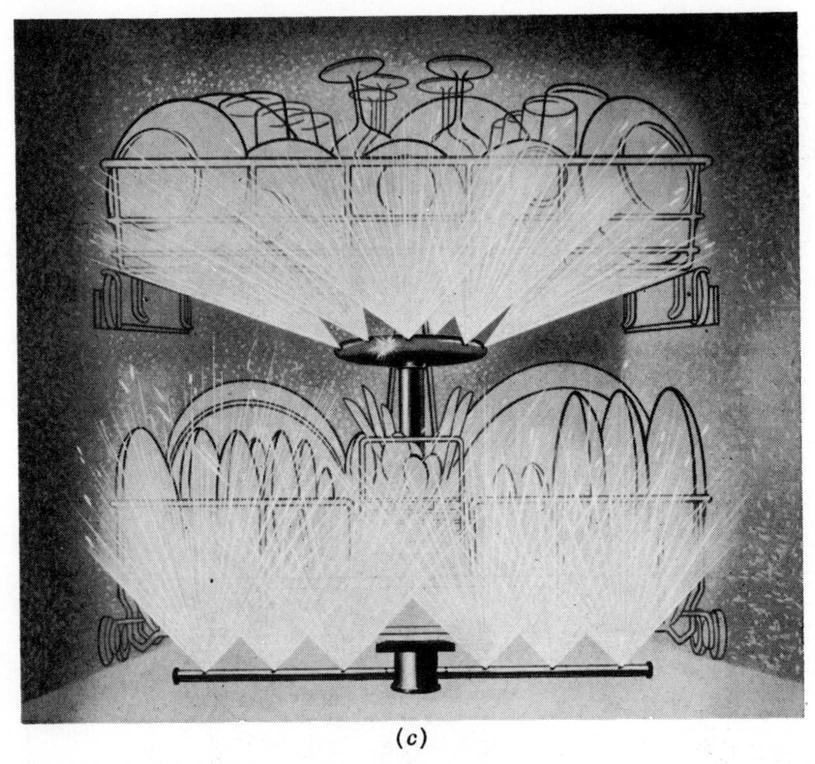

(c)

Fig. 9.4 (Continued)
(c) *Double-deck washing action is obtained by two rotors, one directly below the lower rack, a second one beneath the upper rack. Water is forced into each rotor from a circulating pump beneath the tub. As the rotors revolve, whirling jets of water form a double fountain. Jets from the lower rotor provide a strong washing action; jets from the upper rotor a more gentle action. (Hotpoint)*

and normal speed will be used for all cycles except the china-crystal cycle. For this cycle, a second slower speed is accomplished by the opening of an air valve to give an effect similar to the aerator on a sink faucet. Such action reduces the force of the water and thus helps to prevent damage to fine china and crystal and light articles by their being forced out of position.

Filtering action in the dishwasher may vary from one design to another. One design traps the food in a filter as it is washed from the dishes. This prevents the soil from returning through the spray arms

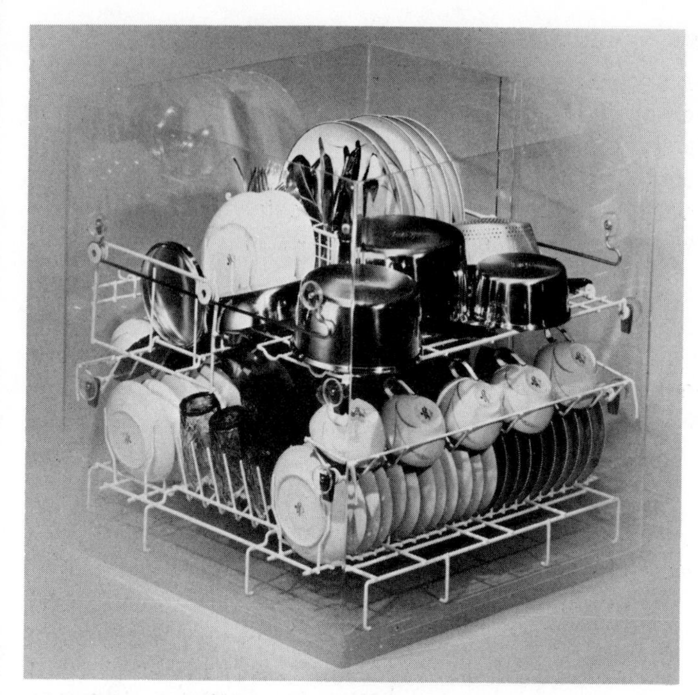

Fig. 9.5 Flexible rack design in dishwasher. Lower rack designed to hold small items; top rack may be rotated out of way when not in use; folding utility rack with prongs upholds larger dishes, silverware, and extra glasses. With prongs down may be used for pots and pans. (Maytag)

to be recirculated. The reverse flow of water sweeps the soil from the filter down the drain during the drain period.

Loading of the Dishwasher

Items to be washed in the dishwasher should be loaded as directed by the manufacturer. The rack design or loading plan recommended by the manufacturer will insure that the soiled side of each item will be fully exposed to the flow of water-detergent solution. This may require a specified location for each item or may be completely at random. Especially important to the removal of soils which need to be dissolved is rapid wetting and high water temperatures. (Fig. 9.5).

Another factor related to effective loading is the design of the dishes. Dishes which have deep bottoms or curved sides should be placed on an angle so water will drain off the bottom or from inside.

Safety Features

The dishwasher should have safety features such as a motor overload protector, water overflow protector, and a safety switch. The motor overload protector prevents the motor from burning out if it becomes overheated. The water overflow protector prevents overflow of water should the fill valve become defective. A safety switch stops the flow of water when the dishwasher door is opened.

Problems

A checklist on dishwasher problems as developed by a manufacturer of water softeners may prove to be effective in the analysis of dishwasher results.

Problem	Cause	Solution
Spots and film on dishes	Low water temperature	Water temperature should be 150° in the dishwater. Adjust water heater control to compensate for heat loss in pipes from water heater to dishwater. (Approximately 160°.)
	Low water pressure causing machine to underfill or rinse improperly	If low pressure is a local condition: schedule use of machine when water is not in use elsewhere in home; if machine is of the time-fill type, check with manufacturer to see if conversion to weight-fill process is possible; water pressure booster pump may be advisable.
		Clean and replace strainers in water fill lines.
		Be sure flexible hose on portable machine is not kinked.
	Improper drainage from clogged drains	Clean drain screen and water outlet line.

Problem	Cause	Solution
	Blocked jets and water passages	Check and clean if blocked.
	Mineral deposits from hard water	Install tank-type water softener for permanent prevention of deposits. For interim prevention use water conditioner with the dishwasher detergent.
	Spots and film on dishes	To remove, put dishes in dishwasher, wash as usual; rinse; omit drying cycle. Remove and hand towel metals. Place 3/4 cup chlorine bleach in cup on lower rack. Place two tablespoons water conditioner in bottom of dishwasher. Run through wash cycle again; DO NOT add detergent. Stop machine; place two cups vinegar in bowl on lower rack. Allow machine to complete rinsing and drying cycles. (Note: too frequent use of this method may harm machine interior.)
	Improper kind of detergent used	Use only those detergents formulated for automatic dishwashers. General use detergents will foam, hamper machines' washing and rinsing actions.
	Improper quantity of detergent used	Follow manufacturer's directions for amount to use. Overuse can cause spots.
	Detergent not dispensing completely in the wash	Be sure detergent cup is operating properly. *Do not overfill detergent cup.* Do not add detergent until time to start machine. If cup retains moisture, wipe dry.

Problem	Cause	Solution
	Inadequate scraping or rinsing	Remove excess food residues.
	Dishes block rinse sprays	Follow manufacturer's instructions for racking. Do not nest dishes (cups inside cups). Alternate large and small plates.
	Dried-on-soil	Scrape dishes right after using. If dishes will not be washed for an hour or more rinse with cool water in addition to scraping before loading machine.
Dishes and flatware still soiled after washing	Low water temperature	Water temperature should be 150° in the dishwasher. Adjust water heater control to compensate for heat loss in pipes from water heater to dishwasher. (Approximately 160°.)
	Low water pressure causing machine to underfill or rinse improperly	If low pressure is a local condition: schedule use of machine when water is not in use elsewhere in home; if machine is of the time-fill type, check with manufacturer to see if conversion to weight-fill process is possible; water pressure booster pump may be advisable.
		Clean and replace strainers in water fill lines.
		Be sure there are no kinks in flexible hose on portable models.
	Detergent not dispensing completely in the wash	Be sure detergent cup is operating properly. *Do not over fill detergent cup.* Do not add detergent until time to start machine. If cup retains moisture, wipe dry.

Problem	Cause	Solution
	Slow drainage from machine	Clean filter screen over drain. Open clogged waste drain from machine to sewer.
	Improper quantity of detergent	Follow manufacturer's directions for detergent quantity.
	Improper kind of detergent used	Use only detergents for automatic dishwashers. Their cleaning power is much greater than that of general-use detergents.
	Inadequate scraping or rinsing	During the meal, soak cooking utensils with cooked-on food in conditioned water. Rinse tableware that has come in contact with egg, mayonnaise.
	Dishes and flatware improperly loaded	Alternate large and small flat dishes. Invert bowls, cups, glasses so that water will drain off. Do not obstruct spray heads.
Bronze tarnish on silverplate	Detergent solution in combination with high temperatures in the machine affects areas of exposed "base" metal where silver is worn away	Soak in vinegar for ten minutes to remove bronze hue. For permanent protection, have silver replated.
Sulfide tarnish on sterling silver and silverplate	From prolonged contact with eggs and egg products	Rinse silver immediately after use. Use silver polish to remove tarnish.
	From hydrogen sulfide in air or dissolved in water (characterized by "rotten egg" odor)	Use silver polish to remove tarnish. Store in protected box or bags.

Problem	Cause	Solution
Discoloration of stainless steel	High heat resulting from contact with flame or heating element causes blue haze *in* the metal	No cure—cannot be polished out.
Dark spots or pits on metal utensils	Concentrated detergent deposits on metal	Do not sprinkle or allow dishwashing compound to fall directly onto metal.
Etching and/or darkening of aluminum	Combination of detergent action and impurities in the water	Hand washing is recommended for highly polished, decorative or anodized (colored) aluminum ware.
		Machine washable aluminum should not be placed directly beneath detergent dispenser, nor should detergent be sprinkled directly onto aluminum. Scour with soap-filled steel wool pads to remove discoloration.
Dishes do not dry	Low water temperature will hinder drying action	Water temperature should be 150° in the dishwasher. Adjust water heater control to compensate for heat loss in pipes from water heater to dishwasher. (Approximately 160°.)
	Rinse water does not drain off	Avoid nesting of articles. Invert cups, bowls so that water can drain off.
		Clean drain strainers and lines.
	Dishwasher heater not functioning properly	Correct mechanical problem.

Problem	Cause	Solution
Loose knife handles	High temperatures in dishwasher	Test wash a single knife for several weeks. Professional recementing may be needed for already damaged knives.
Chipped dishes	Rough handling	Load machine carefully.
Rack marks on dishes	From older, metal racks	Replace with new plastic-coated racks.
Plastic dishes or articles lose shape	High temperatures in dishwasher	Only specially formulated plastic is safe in automatic dishwasher's high temperatures. Wash all others by hand. Consult manufacturer if in doubt.
Walls on insulated cups and tumblers separate	High dishwasher temperatures	Wash by hand.
Brown stains	Tea or coffee	Remove with a little water conditioner on a damp cloth. If water supply does not contain iron or manganese, diluted chlorine bleach will remove stains from china cups. DO NOT SOAK in bleach. DO NOT USE BLEACH ON PLASTICWARE. Do not scour plastic with steel wool or cleansers—surface glaze may be damaged.
	Precipitated iron and manganese in water	A special filter to remove these impurities from the water can be installed on the service water supply.
		To remove iron stains, place one cup of citric acid crystals in a container on lower rack of dishwasher. Run the complete cycle

Problem	Cause	Solution
		without detergent. Water must be at least 140° for maximum effectiveness of citric acid. Caution: citric acid can damage machine or china if used in excess. If citric acid does not work, a stronger acid may be used. Place ½ teaspoon of oxalic acid crystals in detergent cup. Do not add detergent. Run dishwasher through complete cycle, but skip drying operation. Reset machine, add detergent, run through complete cycle. Caution: oxalic acid is a poison. It can also be harmful to dishwasher interiors and china. Handle with care and store out of children's reach.
Fading pattern on china	High water temperatures—detergent action	Wash china by hand unless specifically guaranteed by manufacturer as safe for machine washing.
Warping of wooden ware	High water temperatures	Wash by hand.
"Etching" of glassware	High water temperatures, detergent action may "etch" soft leaded glass or crystal (more expensive types)	Wash expensive glass or crystal by hand. Make sure proper amount of detergent is used in machine. Check to be sure rinse cycle operates properly.
Melting of rubber utensils	High water temperatures	Wash by hand unless manufacturer guarantees results of machine washing.
Lime deposits on dishes	Improper kind/ quality of detergent used	Run dishes through complete machine cycle. Fill machine for second wash cycle, adding no deter-

Problem	Cause	Solution
		gent. Add one pint vinegar and allow wash cycle to be completed. Rinse and wipe off deposits with damp cloth.
		See "Spots and film on dishes" problems.
Rusting of cast iron utensils	High water temperature, detergent action removed seasoning	Wash by hand. To reseason, coat utensil with unsalted fat, place in "slow" oven one to two hours.

Source: The Dishwashing Book, Calgon Center, Pittsburgh, Pennsylvania.

Experiments and Projects[1]

1. Demonstrate the sanitization of infants' feeding equipment in the dishwasher.
2. Emphasize ways that using the dishwasher for children's dishes and utensils is a safeguard against the spread of germs.
3. Ask a homemaker with small children to tell of ways her dishwasher has given her more time to care for them.
4. Discuss how children can help in loading the dishwasher—example—their own drinking glasses.
5. Relate importance of using extremely hot water in obtaining cleaner dishes with fewer bacteria.
6. Teach safety in the use of electrical appliances which are connected with the water supply.
7. Discuss properties of dishwasher detergents and how they aid in removing soil and sanitizing dishes.
8. Study water conditions as to hardness and temperature. Establish standards to give the best results and determine how to correct undesirable conditions.
9. Have each student select a portable or built-in model for the home kitchen, justify his choice, and draw a scale plan, fitting it into the kitchen arrangement.
10. In studying house planning, encourage girls to plan for space, outlets, and water supply for the dishwasher.

[1]**Source.** General Electric Company.

11. Compare the advantages and disadvantages of a built-in dishwasher with portable ones.
12. Conduct experiments to show differences in time and energy spent in hand dishwashing versus machine dishwashing. Compare the end products.
13. Study different manufacturers' models by using specification sheets and trips to different appliance dealers.
14. Teach proper selection, use, and care of the dishwasher by using the owner's manual, equipment texts, and other references such as those listed in this publication. Plan a field trip to an appliance store.
15. Illustrate how the dishwasher can be used to store dishes and utensils, and discuss when this is desirable.
16. Have students survey friend and neighbor dishwasher owners to determine other uses of the dishwasher; for example, washing seldom-used dishes.
17. Name items which are not recommended for automatic dishwashing such as wooden articles, lead crystal, and "granny's hand-painted tea set." Bring out science principles which support these recommendations. Also, list points to consider in selecting dishes and glassware if one is to wash them electrically.
18. Discuss the cost of washing dishes electrically versus washing by hand in terms of equipment, supplies, water, power, etc.
19. Point out how the dishwasher can be used to keep the kitchen clean by loading dishes in the dishwasher as they are used in meal preparation.
20. Discuss the amount of preparation needed with some dishwashers. Study features which make this possible.
21. Discuss the need for long-range planning for household equipment by couples who are getting married.
22. Use the dishwasher to clean dishes and utensils following foods classes and demonstrations.
23. Invite a utility company home economist to demonstrate the correct use and care of the dishwasher in the foods laboratory, or give this demonstration yourself.
24. Study dishwasher design. Soil eight place settings of dishes for use test. Use the following procedures for soiling:
    (a) **Dinner plates.** Divide plates into thirds. Soil one-third with 1T. broccoli flowerettes, one third with 1T. tuna fish combined with mushroom soup, and one third with 1T. frozen squash.
    (b) **Salad plates.** French dressing ½t.
    (c) **Sauce dishes.** Beaten egg yolk ½t.

(d) **Glasses.** Milk.
(e) **Cups.** Coffee with cream and sugar.
Add silver to the load. Let soiled dishes stand for two hours. Wash as recommended by the manufacturer. Examine to determine the number of dishes that have been effectively cleaned.

25. Test water hardness and temperature to determine water quality at the dishwasher.
26. Study dishwasher designs of different manufacturers to gain a better understanding of their features.
27. Use a dishwasher with each cycle that is provided. Evaluate the results.
28. Discuss their appliances with homemakers who own dishwashers.
29. Discuss dishwashers with an appliance service man to determine the common problems he encounters with dishwashers, and how to prevent these problems.
30. Have a service man talk to the class on dishwasher problems.

CHAPTER 10
Waste Disposer, Trash Compactor

A waste disposer designed to handle the food wastes in the home is generally attached to the kitchen drain. Food wastes are fed into the opening of the waste receiving cylinder. Basically six types of action are employed in the waste disposal process—hammering, shredding, cutting, grinding, rubbing, and pulverizing. As waste reaches the bottom of the shredding compartment, the whirling impeller comes in contact with the food. There, centrifugal force, acting on the waste, presses it against the impeller arms, the fibrous waste cutting blades, and the rind positioner. As the action continues, waste is finely divided into particles small enough to pass through into a lower evacuation chamber. From this point, the particles enter the drain and flow to the stack of the drain system (Fig. 10.1).

Waste disposers may be of two types—continuous feed and batch feed. The continuous feed permits the addition of food during the grinding. An antisplash guard prevents food wastes from splashing out. A batch feed design may be filled loosely and actuated by dropping the cover into place, or by a remote switch as is used for the continuous feed. The capacity of the disposer varies from 1 to 3 quarts.

The effectiveness of the design of the disposer is directly related to the rate of flow of cold water. Cold water should be flowing during the entire process of handling food and several minutes afterward to clear the line of the waste. Cold water must be used to harden fat or grease so it may be cut and flushed away with all other types of waste. When warm water is used, grease may be melted and as it comes in contact with the lower waste pipes, it may harden and cause a stoppage. Often, a water actuating switch in the cold water supply

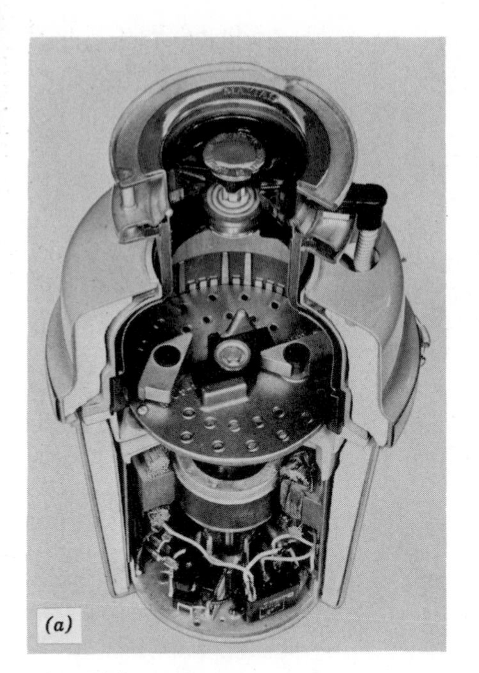

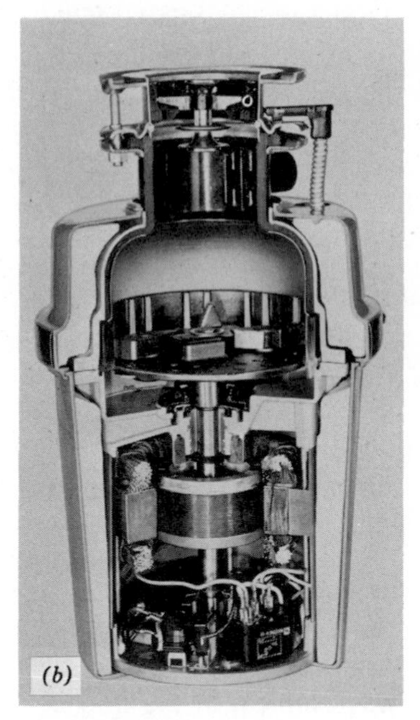

***Fig. 10.1** Food waste disposer includes swivel mounted impeller arms and an exclusive positive-pressure water seal.* (a) *Polyurethane impeller cushions absorbed noise from the impeller arms as they rebound from hitting hard objects. Because the arms are swivel mounted, the unit is virtually jam free.* (b) *The spring-loaded water seal below grinding chamber is designed to protect against water leaks to the motor area—a major cause of disposer failures. (Maytag)*

line prevents operation until the water is flowing in the tap at a desirable rate—approximately 1 quart in 8 seconds. Periodic flushing—weekly or even daily—with a sink filled to 3 or 4 inches of cold water will be valuable in cleaning the drain as rapidly as possible.

Other factors affecting performance are the types of foods combined for disposal, and the loading of the disposer. An understanding of the behavior of food wastes when placed in the disposer in combination with water is important both in the loading and removal of food from the plumbing system. Foods may be classified by their weight and the rate at which the food will grind. Foods that grind slowly move into the drain in small quantities when there is an adequate flow of water. A flow faster than 1 quart in 10 seconds is desirable. Foods representative of those that grind slowly are corn husks, animal fats, chicken skin, and ham rind.

Foods which grind swiftly and produce tiny particles may collect in a clump and be moved along the drain line with a fast flow of water. Foods representative of this classification are roots (carrots, beets, potatoes), celery, cabbage, onions, lettuce, melon rind, and leafy green vegetables. With a long pipe line or one with little slope, the force of the water is reduced as it travels from the sink and may form a blockage in the drain line. Permitting the water to continue to flow for 30 seconds after the grinding stops will help to prevent the stoppage. Regular flushing is recommended for this type of problem.

A third group of foods, classified as heavyweight, will tend to collect in the trap unless the water flow is at least 3 gallons per minute (1 quart in 5 seconds). If the sink is not used regularly, the drain should be flushed daily to avoid deposits of the following kinds of heavyweight wastes: animal bones, sand, nuts, dried legumes, egg shells, and olive pits.

In loading the disposer, a mix of the three types of food wastes—those that grind slowly, those that grind swiftly, and the heavyweight products—will modify the characteristics of the others to some extent. Regardless of the mixture, some of the heavyweight foods will always remain in the trap unless the drain is flushed periodically.

For satisfactory use, the disposer should not be packed with food; the food should simply be dropped in. Such a method is possible with the open top of the continuous feed model. It is especially desirable to feed fibrous material gradually, since the disposer usually requires a longer operating time before this material is flushed away. With bones and fruit pits, the rubber disk may be partly closed with the sink stopper; if the sink has two compartments, the drain opening in the adjacent sink should be closed. Corn cobs, melon rinds, and citrus fruit rinds should be cut into pieces before being put into the waste cylinder. Do not add any large amount of liquid grease at one time; allow it to harden first. Viscous foods, like cooked oatmeal and sizable quantities of raw dough, should be thoroughly mixed with vegetable peelings and dry crusts for successful passage through the shredders.

The disposer is permanently lubricated and self-cleaning; drain-cleaning chemicals should not be put into the unit; they will tend to corrode the shredding mechanism. If anything causes the motor to stop during operation, shut off the switch and water and allow it to cool for 3 to 5 minutes, meanwhile investigating the cause of the trouble. The unit may be overloaded, or a piece of cutlery may have fallen accidentally into the open-top unit. In some models, a thermal relay protects against overload. The motor may be started again by a manually reset button.

Certain materials should never be put into the disposer; metal, china, plastics, paper, string, oyster and clam shells, and filtertipped cigarettes; in time the filters tend to mat over the cutting teeth and decrease their efficiency.

Waste disposers operate on 115 volt, alternating current. Although the motor current may vary from 5 to 7 amperes, the load conditions to which the appliance is subjected require a 15 to 20 ampere fuse. Because the load may vary, especially at the start, a thermal delay fuse is recommended. A separate circuit should serve the disposer. Waste disposer motors are usually ⅓ to ½ horsepower and may be capacitor start, induction run, or split phase. The larger motors provide extra thrust for the tough-to-grind loads. The capacitor start, which provides full power and lengthens motor life, draws less amperage from the household circuit for the motor starting process. The split-phase motor takes time to get up to speed and may cause lights to dim, since it draws large amounts of amperage from the circuit. These motors vary in revolutions per minute from 1700 to 10,000. Faster speeds of grinding may help to minimize the quantity of water required to dispose of the waste.

Special Features

Features that deserve special consideration and contribute to effective performance and satisfaction in use are the following:

1. A motor thermal overload protector switch, which will either reset automatically or can be manually reset. This overload button is normally located under the unit.
2. Automatic reversing switch. This switch, in reversing the direction of the motor, frees the jammed material so that normal operation may resume. When the disposer jams, it pauses for a brief time, then automatically reverses direction. A model may provide an off-on switch that continues to reverse the motor polarity until the threat of the jam is removed. The grinding action is equally efficient in either motor direction on a model of this type.
3. Rind positioner which aids fast, efficient grinding of grapefruit rinds and other large items.
4. Dishwasher drain guard baffle to prevent pump back of water or food wastes into dishwasher drain.

Here are features that are designed to absorb noise and vibration.

1. Extra thick, tough resilient polyurethane with stainless steel liner. Polyurethane does not rust and resists corrosion.
2. Fully insulated sound (shroud) shield; one layer of Fiberglas and two layers of asphalted fiber.
3. Impeller arm cushions.

4. Vibration insulating sink mounting. Tough, durable polyurethane isolates disposer from direct contact with sink mounting flanges.
5. Drain line cushion to prevent transfer of noise and vibration to plumbing line.

The following are features that contribute to durability.

1. Corrosive resistant drain chamber of Fiberglas, which is virtually indestructible, will not pit or corrode, and is unaffected by food wastes.
2. Positive pressure water seal. Prevents water leaks into motor area.
3. Manufactured of stainless steel: sink flange, shredding ring, fly wheel, impellers, drain chamber, and drain housing parts exposed to water.
4. Epoxy cast motor housing, used because of its ability to resist acids and alkalis.
5. Carboloy—steel cutting blades for fibrous wastes. Carboloy needs no sharpening.
6. Thermoplastic detergent shield—protects against caustic action of detergents.

Use with Septic Tank

A waste disposer doubles the septic tank load. Thus, one needs to consider the problem of adding the appliance to others using water in the kitchen. The disposer should not be used with a cesspool because there is not adequate bacteria to convert the solids to a liquid form.

TRASH COMPACTOR

The trash compactor is a relatively new appliance among those classified as ecology appliances. Trash pollution is a mounting problem in the United States. Conservative estimates are that, by 1980, the average person will dispose of around 10 pounds of trash daily or double the current rate. Thus, the compactor may benefit the consumer in several ways:

1. It provides a neater looking kitchen area by removing the trash container.
2. It reduces the number of trips to remove the trash to the outdoors.
3. It reduces the numbers of containers needed for waste.
4. It results in lower trash collection costs in those areas where such costs are based on the numbers of cans one has to be picked up by municipal systems.

The compactor compacts a week's accumulation for the average family of four in a disposable bag. At the end of the week the rectangular bag will contain approximately three 20-gallon trash cans or 30 pounds of trash. The bulk of the trash is reduced on a 4½:1 ratio. Four full garbage cans of trash would compact to the size of one.

Manufacturers of the trash compactors offer a number of different-sized units to choose from. Units vary in width from 12 inches to 17¾ inches and in height from 22 inches to 34½ inches, and all are less than 24 inches in depth. The compactor may be built in under the counter or used as a free-standing appliance.

All compactors convert electrical energy into mechanical force designed to reduce the volume of household trash. An electrically powered twin screw-driven ram moves down into the bin and compresses the trash under 2000 pounds of pressure. It automatically returns to the top to complete the cycle. Some models do not begin to compact or mash the trash until the drawer is filled to a specified level. Some of the important factors to consider in making a choice of a compactor are compacting pressure, compacting stroke, quiet operation, serviceability, and safety features. In determining the compacting pressure for satisfactory compaction, 2000 pounds is regarded as the minimum. Some models have up to 3000 pounds of pressure. The total weight of trash that a typical compactor bag will hold is about 25 pounds.

The maximum amount of compaction is obtained when the trash is compacted from the moment it is put into the unit. This property differs with brand, and in general the compacting function begins anywhere after a minimum of 2 up to 7 inches of trash is in the container.

The degree to which the compactor will be quiet will depend on the type of trash in the container. In general, the compactor is quieter than either the dishwasher or food-waste disposer, except in those instances where a loud noise is to be expected from the breaking of a bottle.

Serviceability of the compactor becomes most important if the consumer should decide to build in the unit. It is serviceable from the front or must it be worked on from the back? Installation of the compactor is essentially simple. The electrical circuit should be a 15-ampere circuit, with an adequately wired 115-volt outlet. No drains, vents, special wiring, or bolting is necessary. Thus, the need for servicing has been minimized in the basic design.

Compactors in general have more than a sufficient number of safety features. These are extremely important when children are in the home. All models have a key lock. Some key locks switch off the electrical power, and others lock the entire unit to prevent entry at any time. The electrical interruption switch prevents operation, if the drawer is open.

Specific quality features which should be considered in selection are the following: the unit should be durable; the critical parts should be resistant to moisture; the parts which come in contact with the trash should definitely be corrosion resistant and designed to take the expected abuse of broken bottles and crushed cans. Outside surfaces should be durable and easy to clean. Cracks and crevices which may collect dirt should be at a minimum. Paint should be of good-quality appliance-thickness enamel.

Intelligent use of the compactor should be exercised by the consumer. Items that cause excessive odor problems, are toxic, or are highly combustible, like paint or oil-filled rags, aerosol cans that contain explosive or toxic chemicals should not be put into the compactor. Aerosol cans containing nonhazardous materials such as shaving cream, whipped cream, or cheese spread may be compacted with little effort.

Experiments and Projects

1. Study the disposer design of different manufacturers to determine the advantages and disadvantages of each.
2. Examine a disposer to determine the location of the reset button.
3. Discuss the use of a disposer with a homemaker who owns one and uses it successfully.
4. Discuss the disposer with a serviceman to determine the kinds of problems consumers tend to have with the appliance.
5. Discuss the benefits of a waste disposer to consumers.
6. Study the plumbing design that is most effective for the best performance of a waste disposer.

CHAPTER 11

Bringing the Appliances Together in the Kitchen

The design and arrangement of space in the kitchen should be keyed to the patterns of living, the number of people to be served in the area, the size of the house, and the equipment and furnishings in the kitchen. The most important aspect to consider in planning a kitchen is the nature of the primary and secondary activities to be performed there.

ORGANIZATION WITHIN CENTERS

Space in the kitchen may be divided into centers for specific activities. When planning a center, determine (1) the activities, (2) the steps or processes in each activity, (3) the supplies and equipment needed for the performance of the process, (4) the working heights and space—storage, work counter, and floor space—necessary for the performance of the activity, and (5) the relationship of the specific center to other activity areas in the kitchen. Also, locate supplies and equipment to make the most efficient use of space, and locate the center in relation to other centers to fit into a logical flow of work.

Attention has been focused on four centers in the modern-day kitchen, the sink, mix, refrigerator, and range-serve centers. A fifth may be included if the oven is located apart from the surface units or burners.

Sink Center

The sink center serves primarily for food preparation and cleaning activities that require a supply of water and drainage. The need for water in the performance of all types of tasks in the kitchen requires that the sink be easily accessible from all other work areas. Each task requiring supplies and utensils which are first used with water should be stored in this area.

The sink designed with a work counter on each side fits well into the order of activities performed during food preparation and cleaning. This requirement is of less importance for a second sink included in the total kitchen design.

The amount of counter space needed should be determined by the amount of space in the house; the amount of food prepared on a regular basis; the meal service patterns—number of dishes, form of service; the work habits of the user—organization of activities; and the number of related activities performed at the same time. A general guide for the amount of counter space at each side of the sink is 3 feet. The recommended amount of base cabinet frontage space in the continuous connected centers of the U-shape kitchen is 4 feet 6 inches from the left side of the refrigerator to the right side of the sink, and 3 feet 6 inches from the left side of sink to the right side of the range. When counters accompany each foot of base storage space, this counter requirement then becomes 6 to 12 inches more than the general guide.

Counter space requirements may also be related to total number of square feet in the house. The recommended amount of linear counter frontage at the right of the sink as recommended by the Small Homes Council varies from 24 inches for a minimum house size of less than 1000 square feet to 36 inches for the liberal house size of over 1400 square feet. For the left of the sink, a variation of 18 to 30 inches is recommended for the specified sizes of houses.

Storage needs at the sink center should be determined by the kinds and numbers of items needed for food preparation and cleaning of food and dishes. Space of this type should also be provided for items such as foods needing washing or water for cooking, foods not requiring refrigeration, utensils used for foods which require water—for example, saucepans and a double boiler. Items which are frequently used should be stored at the place of their first use and where they are easy to see, reach, grasp, and replace.

Wall cabinets should be located above the counter in each center. For the best ventilation and light from the windows, wall cabinets should be positioned 10 to 12 inches from the window frame.

Appliances and equipment located in the sink center include the sink, dishwasher, garbage disposer, and the many small utensils and tools that are required for preparation and cleaning activities. The amount of space required for each appliance will vary with the different designs. The sink design may be a one-bowl, two-bowl, or three-bowl style. The three-bowl design permits a third small sink in the center which houses a garbage disposer and provides for the draining of liquids.

For most efficient use of space the sink should not be in a corner. At least 14 inches should be provided between the sink bowl and the turn of the counter for standing. This spacing also permits a more effective use of space in the corner for storage.

Mix Center

Major functions to be performed in the mix center are the preparation of salads, desserts, and baked products. A major appliance is not associated with this center, but many small appliances may be used here; the mixer, blender, can opener, juicer, meat grinder, and knife sharpener may well comprise a small equipment center. This would be a desirable location for the power center, a 300 watt multiple-speed motor mounted flush with the counter which may supply power for the small appliances.

Counter or work surface at the mix center should be more liberal than in any other center because of the number of activities that may be performed in this area, and the equipment that is needed. Counter frontage requirements vary from 36 inches, for the minimum and medium house designs of less than 1000 square feet, to 42 inches in the house design of over 1400 square feet as specified by the Small Homes Council. For effective work at this center, the counter height should be 5 to 7 inches below the elbow height of the user.

Storage should be provided for mixing equipment (bowls, spoons), measuring equipment (measuring spoons and cups), baking utensils, and small electrical appliances.

The specific needs related to the total design of the mix center should be carefully analyzed in relation to the kinds and the frequency of use of foods by the consumer. This need may vary widely depending on whether the homemaker uses basic ingredients, partially prepared foods, or totally prepared foods.

Range-Serve Center

The range-serve center is used for cooking and serving food. The major appliance in this center is the range which may be found in different arrangements. A conventional design with both oven and surface units or burners combined into one design will require a different arrangement of space than will the split-level or separate oven and units or burners.

Work counters should be present on each side of the free-standing range for the preparation and serving of food. Recommendations for counters beside the surface cooking units or burners vary from 15 to 24 inches. Two feet on each side of the free-standing range may serve as a general guide. When the oven is separate from the units or burners, 15 to 18 inches is recommended on one side of the oven. The counter at this location should be of a heat-resistant material.

The height of the appliances is important in this center. For effective leverage, the door of the built-in oven when fully opened should be between 1 and 7 inches below elbow height.

Storage in the range-serve center should be adapted for such items as skillets, sauce pans, measuring equipment, ready-to-eat foods, canned vegetables, small electrical appliances, and serving dishes for hot foods. An efficient ventilating hood will help to remove moisture, heat, smoke, and cooking odors from the kitchen.

Refrigerator Center

The refrigerator may be arranged in two ways in the kitchen—as a separate center with counter and storage space, or in combination with another center. Some of the functions to be served by this center are refrigeration of perishable foods, storage of leftovers, preparations of beverages, and storage of frozen foods. Each function requires counter space as well as storage. The amounts of each type of space should be determined by the types of activity.

Appliance design will influence the amount of space and placement within the refrigerator center. An undercounter refrigerator for a second center compared to the combination side-by-side refrigerator-freezer will have different space requirements. The refrigerator should be located where the flow of heat from other appliances in the kitchen will not be directed onto this appliance. Thus, the range, dishwasher, and separate oven should be located apart from the refrigerator.

ARRANGEMENT OF CENTERS

In arranging the centers in the total kitchen setup, one needs to consider the relationship between activities, distances to be traveled, separate centers compared to those which are continuous connected centers, and the types of floor plans (Fig. 11.1).

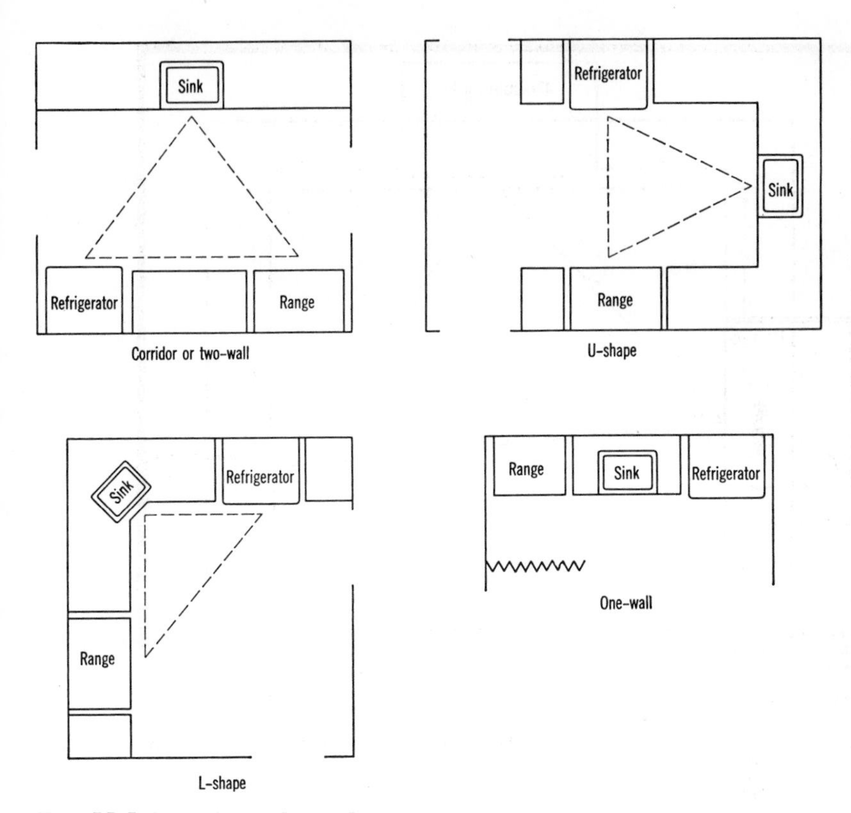

Fig. 11.1 Four basic floor plans.

For the most efficient use of space, the three most frequently used appliances—refrigerator, surface units or burners, and sink—should be connected at points of an equilateral triangle. The total of the distance around the triangle should be not more than 26 feet. The Small Homes Council of the University of Illinois recommends a distance of 4 to 6 feet between sink and range, measured from the center fronts of the appliances, 4 to 7 feet between sink and refrigerator, and 4 to 9 feet between refrigerator and range. These distances are based on the frequency of travel from one point to another in the kitchen with the greatest numbers of trips being made between the sink and the range. Other criteria by which to judge the effectiveness of this space design are: no major appliances are positioned in a corner—with the exception of the refrigerator with the door which opens onto the counter; continuous counter; adequate storage and

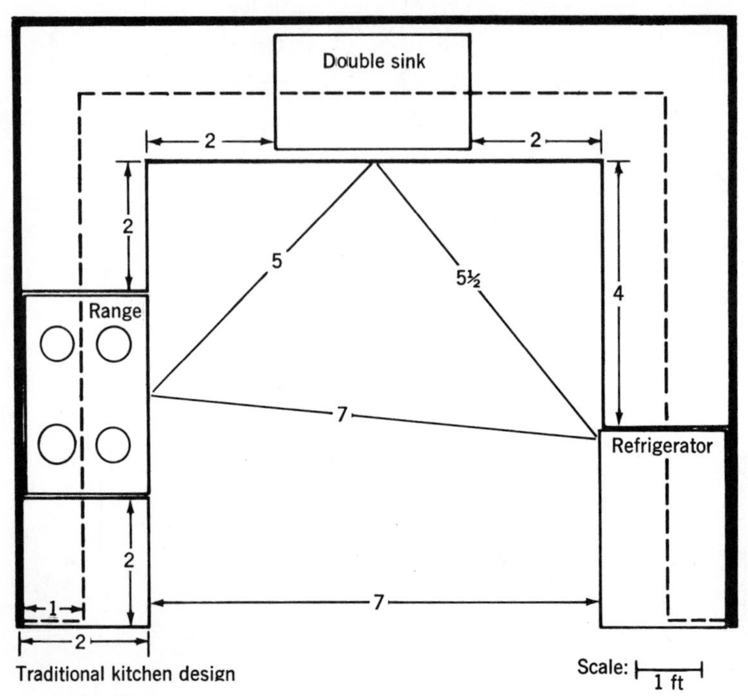

***Fig. 11.2** U-shape kitchen plan. Application of criteria for efficiency in design.*

counter space; effective clearances (5 feet or more) between applances or furnishings that are on opposite walls; and room entrances positioned so that traffic does not move through the work triangle (Fig. 11.2).

In the contemporary kitchen design where the centers are separate and often are accessible from all sides, more people may work in the kitchen and more activities may be performed at the same time. This arrangement may be important for those who wish to work with someone in the kitchen, to entertain in the kitchen, to prepare food for large numbers of people, and to serve food in the kitchen (Fig. 11.3).

Organization of Storage within a Center

The development of an organization plan for each center in the kitchen will contribute to the best use of storage space. As illustrated in Figure 11.4 and Tables 11.1 and 11.2, the plan is based on a care-

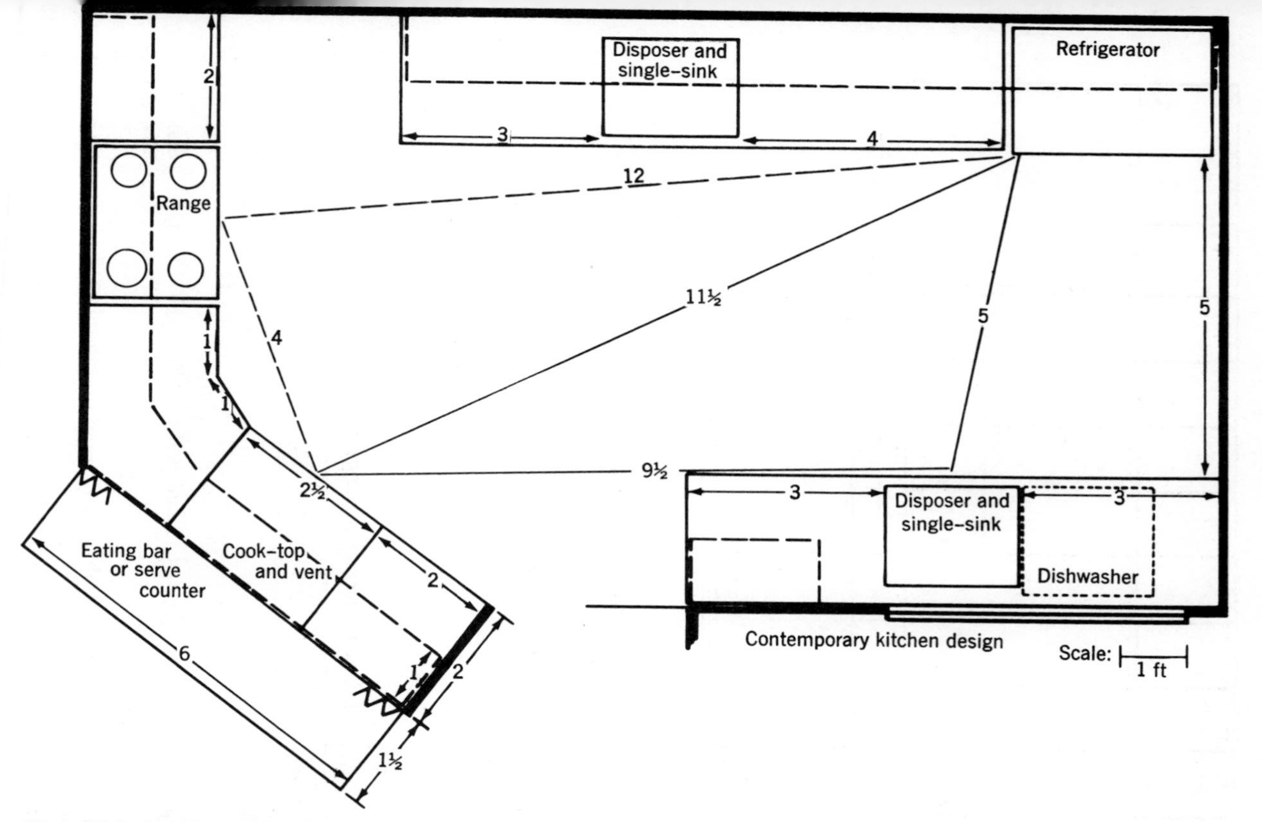

***Fig. 11.3** Contemporary, separate center kitchen design.*

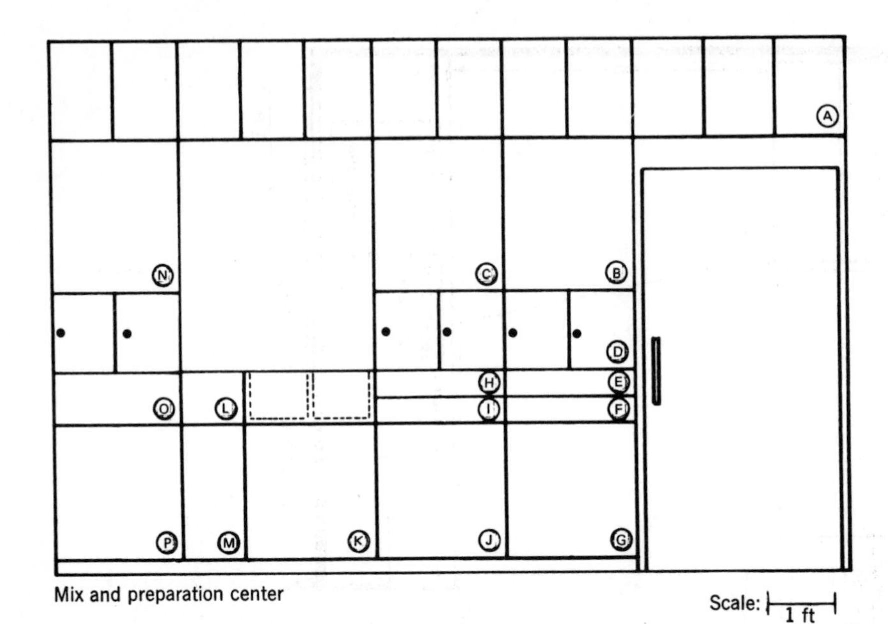

Fig. 11.4 Elevation drawings of mix and preparation center.

ful study of the activity patterns in each center, the processes involved in each activity, the equipment used for performing the activity, and the location within the center.

Table 11.1
Preparation and Mix Center Organization

Process	Equipment	Supplies	Space
Rinsing	Sink	Water, fruit, vegetables, fowl	Sink
Peeling	Sink, disposer, potato peeler, knife	Vegetables, fruit	Sink Mix counter
Scrubbing	Sink, vegetable brush	Vegetables, fruit	Sink
Chopping	Cutting board, knife, electric blender	Vegetables, fruit	Mix counter
Grating	Vegetable slicer, knife	Vegetables, cheese, meats	Mix counter

***Table 11.1* (con't.)**

Process	Equipment	Supplies	Space
Scraping of serve bowls in refrigerator container	Rubber scraper, refrigerator container, fork, spoon	Leftovers	Mix counter
Cutting food for storage	Knife, cutting board, spoon, fork, containers	Leftovers	Mix counter
Wrapping for storage	Foil, wax paper, freezer paper, fasteners, labels, marker	Leftovers	Mix counter
Tossing	Bowl, fork, spoon	Fruits, vegetables	Mix counter
Greasing	Baking pans, pastry brush	Shortening	Mix counter
Paring	Knife, potato peeler, disposer, sink	Vegetables, fruit	Mix counter Sink
Grinding	Grinder, pan, fork, blender, rubber scraper	Meat, vegetables, fruit	Mix counter
Straining	Strainer, sink, two bowls, spoon	Vegetables, fruit, macaroni, etc.	Sink
Spreading	Knife, spoon, spatula, scraper	Jelly, butter, etc.	Mix counter
Kneading	Pastry cloth	Flour and mix	Mix counter
Open cans	Can opener, wet dish cloth	Canned goods, bottles	Mix counter
Mixing	Bowls, spoon, electric mixer, electric blender, scraper	Ingredients	Mix counter
Measuring	Measuring spoons, bowl, liquid measuring cups, dry measuring cups, spatula or knife	Ingredients	Mix counter
Cutting	Cutting board, knife	Meats, vegetables, etc.	Mix counter
Icing	Knife, rubber scraper, spatula, bowl	Cake, etc.	Mix counter

Table 11.2
Location of Items To Be Stored

A. Adjustable shelves—Lightweight, seldom-used items
B. Adjustable shelves—Flour and sugar in canisters, sugar—brown and confectioners, shortening, vinegar, oil, syrups, chocolate, cocoa, cake mixes, frosting mix, pudding mix, jello
C. Adjustable shelves—Custard cups, casseroles, mixing bowls
D. Midway storage (6 inches)—Salt, pepper, spices, food colorings, extracts—vanilla and others, nuts, peanuts, nut grinder, chocolate chips, coconut, marshmallows
E. Drawer—Wax paper, aluminum foil, baggies and fasteners, labels, tape, marker
F. Drawer—strainer, vegetable slicer, funnel, can opener, grinder
G. Adjustable shelves—Electric mixer, electric blender, pull-out drawer—refrigerator containers
H. Drawer—Measuring spoons, pastry brush, mixing spoons, wooden spoon
I. Drawer—Rubber scrapers, cookie cutters, pastry blender
J. Adjustable shelves and pull-out trays—Sifter, measuring coups (dry and liquid), gelatin molds
K. Slide-out vegetable drawers, wastebasket
L. Drawer—Pastry cloth, rolling pin
M. Vertical dividers—Chopping board, trays
N. Adjustable shelves—Salad bowls, individual salad bowls, salad plates, relish trays
O. Drawer—Divided section for knives, potato peeler, vegetable brush
P. Vertical dividers—Cake pans, loaf pans, cookie sheet, pizza pan, muffin pan, pie pans

Lighting: Over sink—Fluorescent soffit above sink at the top of the cabinet cutout
Counter—Fluorescent fixtures under wall cabinets

Electrical requirements for this center:

1 circuit—General and task lighting	15A
1 circuit—Electric mixer and blender	15A
1 circuit—Refrigerator	15A

Total: 3 circuits in mix and preparation center

Clearance Space

Major appliances are customarily placed against the walls. Five to 6 feet of clearance space between the equipment on opposite walls is considered desirable. A distance of more than 6 feet requires unnecessary walking from one work center to another. If the clearance is 5 feet, two adults may work without interfering with each other and with a minimum of steps. A wider space is frequently needed when there are children and other family members desire to assist the mother in the kitchen tasks. Thirty inches is the least clearance advised for one adult, and this amount is scarcely adequate unless careful attention is given to the hinging of cabinet, refrigerator, and oven doors.

The amount of free space required depends on the breadth of the worker and on the type of perpendicular surfaces opposing each other. If the oven door is hinged at the bottom to form a shelf, if cabinet doors swing beyond the front edge of the work counter, if drawers pull out, and the opposite piece of equipment has a solid front, additional clearance space is needed. The worker should be able to move and stoop with ease. Other areas where the clearance is of importance are the door swings and space above counters. Doors should swing free of work areas and not against appliances where one may need to stand. Wall cabinets should be at least 15 inches above base cabinets. This clearance allows space for small appliances on the counter top without placing the top shelf beyond 72 inches from the floor.

Working Heights

Heights of work surfaces should be adapted to the height of the homemaker. Such operations as kneading, mixing, and stirring put a strain on the shoulder muscles. Work at too high a counter causes the elbow to bend and the shoulders to lift, resulting in increased tension in the small muscles and fatigue. Work at too low a surface causes a person to stoop, a position which tires the back muscles.

The practice in modern kitchen layouts of having all surfaces on a level, using the 36-inch height of the range as the unit of measure, places more emphasis on appearance than suitability. Different tasks performed in the kitchen frequently require work surfaces of different heights, and the present trend is to provide work areas at three levels, one for the sink and the counters on either side, a second for the mix center, and a third by the range. Counters of varying levels are also necessary if various members of the family help the homemaker in the kitchen. Alternate standing and sitting reduce fatigue and are possible if the height of the chair is properly adjusted to the height of the work surface. Correct sitting heights keep the arms in relatively the same position as in standing at work. Building in a sliding shelf or lap board makes it possible to sit for many, perhaps the majority, of the kitchen tasks. A bread board placed on the top of an opened drawer provides such a surface.

The height of built-in counters is not easily changed, but work tables that are too low may be raised by casters or blocks beneath the legs. When mixing or kneading at a low counter, a homemaker may place a utensil on a rack to bring it up to a convenient level.

Storage Cabinets

A sufficient number of storage cabinets conveniently placed and arranged makes it possible for the homemaker to have within easy reach everything with which to work. It is a good rule to store supplies and equipment at the place of first use, and to store together those things that will be used in one process. For example, vegetable brushes and paring knives belong at the sink; bowls, measuring cups and spoons, egg beaters, mixing spoons, and a spatula belong at the mix counter; skillets and saucepans with their covers and salt and pepper shakers belong at the range; serving dishes should be near the serving table.

Duplication of some equipment may be needed. Someone has suggested that knives, spoons, and measuring cups used in the various activity areas should have distinctive-colored handles so that they may always be returned to their proper location, a necessary procedure if the efficiency of the kitchen is to be maintained at its optimum level.

Kitchen planning specialists suggest that 6 square feet of shelf space be provided for each member of the family, but not less than 18 square feet total. In addition, 12 square feet of shelf space should be allowed for entertaining and the "accumulations" which occur in most households.

Cabinets are divided into three types: wall, base, and broom or floor-to-ceiling types. They may be separate units or be built into the kitchen as an integral part of the activity centers. If built in, the unit should extend from floor to ceiling. Space above a 7-foot height should be designed for things that are used infrequently. Many items in today's home are of this nature, and the demand for storage space has become one of the most serious challenges in the home.

Suitable heights for shelves may be found by measuring the height to which the housewife can reach with comfort; she should be able to reach to the middle of the shelf rather than merely to the front edge. Stretching over a work surface shortens the extent of reach—by 3 inches when the counter is 12 inches wide, and by 10 inches when it is 24, according to Wilson and Roberts, who carried on an investigation in Oregon and Washington. Some authorities place the top cabinet shelf 6 feet from the floor, a location which they have found to be the maximum accessible to the woman of average height. In fact frequently used items should be stored much lower; it requires four times the energy to reach 72 inches above the floor as to reach to 42 inches above, and twice as much to reach to 56 inches as to 42. Bending to bottom shelves of base cupboards also takes much energy, 19 times more to bend to three inches above the floor than that used at elbow height. When there is an upper section beyond comfortable reach, it may be used for the storage of utensils required only occasionally—during the canning season, or when roasting the Thanksgiving turkey. Sometimes top shelves have separate doors; such an arrangement keeps the equipment stored there more nearly free from dust.

Factory-built cabinets put out by a number of leading manufacturers in the field are constructed on 3-inch modules, commonly in widths of 12 to 48 inches. The standard base cabinet is usually 36 inches high, including the toe space at the bottom. Wall cabinets are approximately 30 inches high, with shorter ones, 15 to 21 inches high, for installation over refrigerator, range, or sink. When wall cabinets are built above the range, a 22-inch clearance should be allowed; above the sink, allow a 30-inch clearance. A 15- to 17-inch clearance should be allowed between counter and wall cabinet to accommodate small appliances.

Whatever the type of cabinet, shelves should be near enough together to prevent waste space, and of a depth to allow dishes or utensils to be stored in single rows easily accessible without bending or reaching. Shelves 12 to 13 inches deep and 8 to 12 inches apart provide storage for ordinary dishes and utensils, a wider space being left for larger kettles and saucepans. Adjustable shelves are recommended, so that the storage space can be adapted to changing needs. Adjustability is made possible by the use of slotted pilaster shelf standards, two on each cabinet wall, into which movable supports are fitted under each of the four corners of the shelf. A set of revolving shelves, commonly called a lazy Susan, in a corner cupboard brings all items within easy reach.

Shallow cabinets 6½ inches deep with a 6-inch clearance between shelves will take care of tumblers, cups and saucers, small pitchers, bread-and-butter plates, salts and peppers, and the numerous small items that clutter up the large cabinet and hinder its efficient arrangement. If a separate cabinet is not possible, narrow shelves can be built between the deeper ones to hold the smaller items. Sometimes, cutback shelves in the cabinet for single-line storage and racks on the inside of the cabinet doors solve the problem. The shelves should fit the articles to be stored.

A cabinet 4 or 5 inches deep is also effective for packaged foods. When cereals, cake flours, ready-mixes are stored broadside, their identity is readily recognized and they are easily accessible.

Spices, seasonings, extracts, and baking powder may occupy a narrow open shelf above the preparation counter or a series of step shelves 2 inches deep built onto a cabinet shelf. Manufacturers of modern cabinets design narrow closed cabinets with roll-up doors below the wall cabinets. These cabinets occupy 9 to 10 inches at the rear of the counter. These "midway" cabinets, as they are called, may be constructed locally. The most frequently used items may then be stored at the most accessible location.

As a result of extensive research, the Small Homes Council of the University of Illinois has suggested certain lengths for counter and cabinet frontage at the various kitchen centers, for the most satisfactory use of these areas.

Base cabinets may be fitted with drawers or shelves. Drawers are usually preferred, because they supply more accessible storage space. Shallow drawers are more suitable than deep ones for the storage of linens, small tools and appliances not easily hung from hooks. Partitioned drawers are needed for cutlery. Drawers should slide smoothly and never bind or sag. Ball bearings or nylon rollers are frequently used. In base cabinets it is an advantage to have pull-out shelves, often of a basket type, so that any piece of equipment may be plainly seen and easily removed, Some cabinet manufacturers are making these sliding shelves of open bars or heavy wire to increase visibility. Space under the built-in surface units or burners and below the built-in oven may provide effective base cabinet frontage if the storage space is at least 20 inches high. A kitchen should have base cabinets with at least nine drawers or pull-out shelves. One of the drawers should be designed with vertical pan dividers, and another should be not more than 3 inches deep.

A cabinet or drawer divided into sections offers the most satisfactory storage for trays, muffin tins, pie plates, and cooky sheets. The partitions should be removable for ease in cleaning. Whether all kitchen utensils are to be stored inside of cabinets or part of them to be hung near the place where they are used is a matter of choice. The use of a perforated hardboard panel permits the positioning of hooks which may be changed as the need arises to accommodate utensils of varying sizes. A pan storage cabinet near the range may be fitted with hooks for utensils and serving appliances. A rack on the door holds lids.

Dishwashing equipment and supplies are stored at the sink. A shelf below the sink for these articles or a wire rack on the door may be concealed by a lattice or louver which gives the needed ventilation. A towel rack in the cabinet beneath the sink improves the appearance of the kitchen, but here, too, adequate ventilation is essential.

The flush panel is preferable for cabinet doors, for it has an unbroken surface, easier to clean than the indented panel, although somewhat more expensive. All doors should be hinged in such a way as not to interfere with one another, with other pieces of equipment, or with the worker. Cabinet doors should not be so wide that they will swing out beyond the edge of the work counter. Two narrow doors occupy less space when open than one wide one. As a rule, single doors should open away from the direction from which they will be most frequently approached. Hinges should be partly or wholly concealed. All hardware used in cabinet construction should be non-tarnishing and of the highest quality. Occasionally manufacturers have replaced the metal handles with concealed slots at the base of the door panel and underneath the front of the drawers, into which the fingers fit. This makes for easier and faster cleaning of the panel surface. Magnetic catches are also used on doors. A slight pressure on the outside of the door causes it to open; when the door is closed, the catch automatically holds it shut. Toe space should be left at the floor to allow the worker to stand with comfort. Minimum dimensions of such a space, determined in the Washington-Oregon investigation, are 4 inches front to back and 3 inches vertically.

Cabinets or closets for cleaning equipment such as the electric cleaner, mops, brooms, dust cloths, and waxes and polishes are also usually 6 or 7 feet in height, with one or two doors and a combination of shelves and hooks. A study made by the Small Homes Council indicates that a space of 24 by 24 inches be allowed as a minimum for utility storage. Handles of mops, brooms, and electric cleaners require a free height of 60 inches. The cleaning cabinet may be built in or purchased as a separate unit. It should be placed in the back hallway, if space permits, or near the outside door to save steps when emptying dust containers and shaking mops and cloths out of doors.

Light and Ventilation

The kitchen may have doors and windows that affect the natural light available and ease of ventilating the room. The kitchen should, if possible, have at least two outside walls; windows placed on two adjoining walls, with an area equal to about one fourth or one fifth of the floor area, should then provide adequate light and cross ventilation.

There is some question as to the advisability of having windows directly over the sink. Light transmitted by high windows is reflected largely from the sky, and sky light is frequently more than twice as bright as light from the ground. Such windows facing the worker may, therefore, cause glare, which is to be avoided at all times. Windows built over the drainboards instead of over the sink or in the wall to the left of the sink will give a very satisfactory light. When windows are on only one side of the kitchen and an outside door is on the adjoining wall, light may be increased by having a pane in the door. Walls of glass bricks also improve the daytime lighting of the kitchen.

Artificial light is necessary morning and evening for several months of the year in all temperate climates, the amount depending upon the location of the home and the percentage of dark days. The kitchen is the homemaker's workshop and requires careful lighting. A fixture near the center of the ceiling with surrounding opal glass globe to diffuse the light evenly, or double fluorescent tubes will eliminate shadows and supply general illumination. In addition to the central fixture, separate lower lights over sink, work counters, and range increase efficiency by giving added illumination where it is needed. Standards determined by research by the Illuminating Engineering Society specify a level of 30 footcandles of light from the central fixture for general illumination. Standards for light from sources at the task areas specify 70 footcandles at the sink and 50 footcandles at the range and 50 to 150 at the work counter.

Doors and windows are commonly used for ventilation. Windows that extend almost to the ceiling may be opened at the top, permitting odors and steam to pass out. New homes frequently have shuttered ventilators through which a fan sucks the vitiated air; the opening usually closes automatically with the shutting off of the fan. A fan will occupy much less wall space than a window, since it may be placed above a cabinet or even in the ceiling.

Cooking produces heat, odors, moisture, and volatile substances, regardless of the type of fuel used; a fan removes these products. The fan itself is usually of the blade type, although the blower type is considered more efficient and is quieter in operation. The fan motor should be totally enclosed to prevent dirt from accumulating on the parts. The motor may need lubrication. Directions for the kind and amount of oil to be used and the frequency of application should be followed.

Since humidity hinders the evaporation of perspiration from the body, it is more enervating to the worker than heat. A hot, humid kitchen is suitable neither for the eating of meals nor as a playroom for children. Even a small, inexpensive electric fan that keeps the air circulating greatly reduces the fatigue of the worker. New gas and electric ovens are so well insulated that heat from baking is reduced to the minimum, and the warm vapors from the oven vent or from surface cookery may be prevented from spreading through the kitchen by building a flue-connected hood over the range; then the heat and volatile products are removed at the source.

A new hood that needs no outside venting is available for installation above a range. Grease, smoke, and cooking odors are removed electronically and are deposited on filter plates which act as magnets. The plates may be washed in warm water and a detergent. The filter never needs reactivating.

Eating Area in Kitchen

The majority of homemakers seem to desire some space for eating in the kitchen. In order that the routing may be the same as for service in the dining room, this eating place should be fairly close to the dining-room door. It should, however, be separate from the work area. The dining area with movable chairs and table has more of the atmosphere of a dining room but requires more space than when the furniture is built in.

An area 7½ feet square is recommended for the separate unit. When tables and benches are a part of the permanent construction, a space 6 feet square is usually sufficient. Seats should be 15 to 17 inches deep. The table will overlap the seats approximately 3 inches. A breakfast bar should be 18 inches deep. It may be used with stools or chairs.

Relation of Kitchen to Rest of House

Because of the nature of related activities, the kitchen should be located close to the dining room, living room, family room, utility room, childrens' play area, front entry, and to the intercom or telephone location. Outside areas which should be readily accessible from the kitchen are the service area, outdoor living area, childrens' play, and outdoor laundry drying.

Kitchen Safety

As a result of the number of accidents which occur in the kitchen, attention should be focused on the design features of this space that are potential hazards. Electrical safety should be given attention. Electric circuits, outlets, and switches should be located so that anyone working in the area will not come in contact with water and electricity-bearing surfaces at the same time. Step stools of sound structure and design should be provided if one is tempted to climb to reach inaccessible items. Floors should be finished with materials which are nonskid and easy to wipe clean of spilled liquids. Safety for children should be carefully evaluated with regard to chemicals, fire, electricity, tools, or utensils which may be sharp and cause cuts. Potential fire hazards should be given careful consideration.

MATERIALS

Materials for kitchen counters should be selected with much care and thought. Basically there are five materials from which to choose: laminated plastic, flexible vinyl, ceramic tile, glass ceramic, and laminated hardwood.

Laminated plastic is one of the most common materials in use today for kitchen counters. Some of these materials look and feel like genuine leather, slate, and butcher block. The surface is smooth, durable, hard, resists stains, and is easy to clean. The major disadvantages of this material are the ease with which it may be scorched by a cigarette or hot utensil and may be cut. Sharp knives cut the surface, and this type of mark does not heal on the laminated plastic. When struck by a heavy object the material will dent or crack. A high degree of care should be exercised in cleaning stains or soil as the gloss may be removed with certain types of cleaners.

Flexible vinyl has greater resiliency than the laminated plastic and involves less breakage with dishes. It resists stains, moisture, abrasion, and alcohol. Vinyl makes a smooth, quiet surface. However, some foods stain it, sharp knives cut it, and it is highly susceptible to heat damage. As a counter material one of the greatest advantages is the ability to cove the material from the counter top to the backsplash and thus eliminate the cracks at the back of the counter. The coving may also be done at the front of the counter. Vinyl comes in widths of 27, 36, and 45 inches and in many colors and patterns.

Ceramic tile is hard and durable. It is not affected by heat or fire. Sharp knives will not cut it. It is very stain resistant when properly finished on the surface, and the color is not affected by sunlight. It will crack when a heavy pan is dropped onto it. Ceramic tile is not resilient and must be used very carefully. It will dull knives and break dishes and glasses. The grout between the tiles will stain and become soiled.

Glass ceramic as a material is especially appropriate for the area where a heatproof surface is needed as next to the range cook top or oven. It is perfect as a landing spot for the hot utensils. This surface is impervious to stains and cleans easily.

Laminated hardwood is used for only a portion of the counter top. It makes an excellent cutting surface and is difficult to damage. When it becomes dented, stained, or burned it can be refinished if the thickness of the wood is adequate. Hardwood must be cared for and kept dry, or it will roughen and become difficult to clean. Sunlight will change the color of the wood. An occasional oiling will keep the surface in good condition for a long time.

An overview of the materials most frequently used for counters and their relative costs in order are:

Stainless steel.
Ceramic tile set in mortar.
Ceramic tile attached with adhesive to wood.
Hardwood—laminated.
Melamine—laminated with molded edges.
Melamine—laminated with stainless steel or cut melamine edge.
Vinyl
Laminated polyester.
Linoleum.
Tempered hardwood.

FLOORING

The type of flooring selected for the kitchen should depend on a number of factors. Three factors are universally desirable:

1. Comfort—with a high degree of resiliency the floor will be easy to stand on hour after hour.
2. Cleanability.
3. Color and design.

The choice among kitchen flooring materials is almost infinite if you consider wood, flagstone, area rugs, and others (Table 11.3), but the four materials which are used most frequently are carpet, sheet vinyl, vinyl asbestos, and ceramic tile.

Table 11.3
Comparative Analysis of Floor Surfaces

	Nonslip Characteristics										
	Dry	Wet	Polished	Resilience (comfort)	Ease of Cleaning	Resistance to Water	Resistance to Wear	Resistance to Residual Identations	Resistance to Marks	Warmth	Suitability for Underfloor Heating
Clay titles	G-VG	F	—	VP	G	VG	G-VG	VG	F-G	VP-P	Yes
Cork	VG	F-G	F-G	VG	F	F	G	VP-P	F	VG	Yes
Granolithic	G-VG	G	—	VP	G	VG	VG	VG	G	VP	No
Linoleum	G	P	F	F-G	G	P	G	P-F	F	G	Yes
PVC sheet & tile	G-VG	P	F	F-VG	G	F-VG	F-VG	F-G	F	F-G	Yes
Rubber sheet & tile	VG	P	P-F	VG	G	G	G-VG	F-G	P	G	No
Terrazzo	G	VP	VP	VP	G	VG	F-VG	VG	G	VP	Tiles: yes In situ: no
Thermoplastic	G	P	F	P-F	F-G	G-VG	F	P-F	P-F	F	Yes
Vinyl asbestos	G	P	F	P-F	G-VG	G	G	F	G	F	Yes
Softwood	G-VG	P-G	P-F	F	F-G	P-F	P-F	F-G	P-F	VG	No
Hardwood	G	F	VP-P	P-F	G	P	F-VG	F-VG	G	G	Yes
Carpet	VG	—	—	VG	F	VP	VP-P	P-F	VP	VG	Yes

VG—very good.
G—good.
F—fair.
P—poor.
VP—very poor.

Carpet has been widely used as a flooring material in the kitchen. This alternative has been less costly and has been very well accepted by the public for the aesthetic qualities.

The primary concern for the use of carpeting in the kitchen is related to the microbial survival within the material and the problems associated with soil removal. The survival of microorganisms within carpeting may be of public health significance. This material may serve as a reservoir of infection of young children as they crawl or play on the kitchen floor. Microorganisms might also be released from the carpeting to the air during periods of activity in the kitchen, serving as a source of respiratory infection or contaminating food.

In research conducted at Iowa State University in 1973, in homes of young families with children under the age of 18 months, the magnitude of the bacteria present in the carpet was highly significant and increased with time from one sampling period to another over a period of 35 weeks.

Vacuum-cleaner type was related to the magnitude of bacteria present in the carpeting. The upright beater-bar suction cleaner type was used on the carpeting which retained the lower quantities of bacteria. The higher quantities of bacteria were retained on the carpets which had been cleaned with the straight suction-type cleaner. In a study of the homemakers' responses as to their likes and dislikes regarding kitchen carpeting, the reasons given for liking carpet were resiliency, quietness, added warmth, added color and texture to the kitchen, less breakage, and less slippage. Those characteristics which the homemaker did not like were primarily related to the problems of cleaning: staining problems, spills that were difficult to wipe up, soils that showed up more easily, the carpet not feeling as if it was clean, and dirt tending to accumulate rapidly on the carpeting.

Experiments and Projects

1. Visit a house that is being built to observe kitchen cabinets being installed.
2. Ask a panel of homemakers to share their likes and dislikes in kitchen design.
3. Visit a kitchen cabinet distributor and study designs of cabinets.
4. Plan kitchens in houses with different amounts of floor space. Compare the advantages in each design.
5. Prepare meals in the mix center of the kitchen. Use foods in different forms—basic ingredients, partially prepared foods, and those

which require heating or warming only. Determine the counter and storage space required for the preparation of foods in different forms.

6. Plan a small appliance center in the kitchen.
7. Compare the space requirements of different models of ranges, the freestanding range as compared to the separate oven and units or burners.
8. Visit a flooring materials dealer—study carpeting and smooth floor coverings to determine the advantages and disadvantages of each when used in the kitchen.

CHAPTER 12

Appliances for Personal Care

Portable cooking appliances have been discussed in Chapter 5. The student may use a few of these in her dormitory. But the forecast indicates that the biggest advance in future sales of housewares will be in the line of personal care electric products rather than in those used in cooking and in house care. Personal care appliances are certainly increasingly popular with students in high school and college because they save time, which is always at a premium.

Among those most widely used are electric clocks with an alarm or other attachment, hair dryers, hair curlers, facial saunas, manicure sets, automatic toothbrushes, and electric shavers. To this list may be added the electric hairbrush, shoe polisher, clothes brush, crease presser, complexion brush, hair clipper, and the comfort appliances: the massager, heating pad, and electric blanket. A number of these products are cordless, battery-operated appliances. The future of cordless appliances depends on the availability of adequate batteries. The nickel-cadmium battery is used now and is rechargeable, but it may not supply enough power to operate certain personal care appliances; for example, the hair dryer must be used on the house circuit.

Operators safety seems to have special significance in the use of personal care appliances. There is no danger if the appliance has no current leakage, but precautions should be taken. In general it is safer to keep these appliances away from the lavatory and bathtub, since they are connected through the pipes to the ground. The homemaker should hesitate to wash dishes while drying her hair. Under such conditions, battery-operated appliances, if available, are a safer choice.

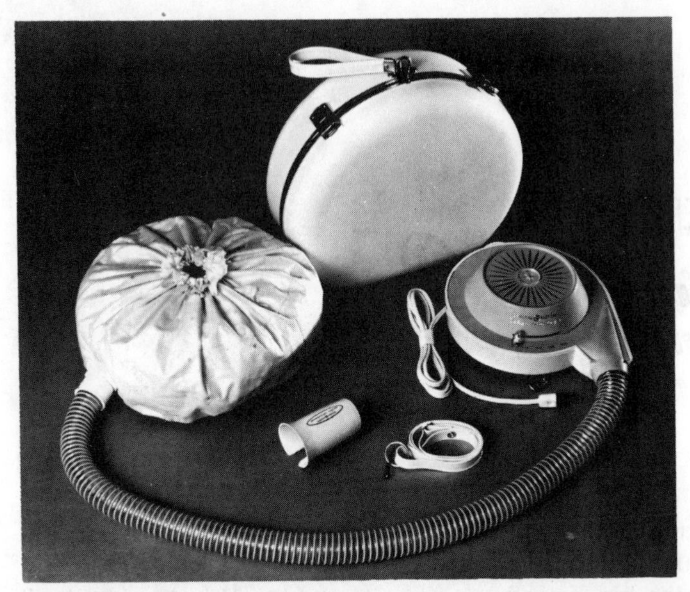

Fig. 12.1 Bonnet-type hair dryer with carrying case. A reach-in top allows user to check hair without disturbing set. Spot curl attachment allows touch-ups. Waist or shoulder strap gives freedom to move around while hair is drying. (General Electric)

Hair Dryers

Three types of hair dryers are marketed; the bonnet, most widely used, the rigid hood, like the professional dryer in beauty parlors, and the hand-held dryer, without a hood or bonnet (Fig. 12.1). The dryer is used on the 110/120 volt circuit and varies in wattage from 400 to 1000 watts. Wattage rating is not necessarily an indication of drying speed; it depends on the relationship of the heating element to the size of the fan motor. Drying time also depends on hood or bonnet design, which affects even distribution of air, and on the amount of air flow. Some bonnets feature many air jets that spread the heat evenly over the hair, others a stand-away ring that improves air circulation. Several manufacturers place the heating unit at the head of the drying tube next to the bonnet, a position that tends to speed drying time, but the person should be careful not to fall asleep during the drying period; there is a danger of the neck being burned unless the dryer is thermostatically controlled to prevent overheating. Three or four temperature settings are usually provided, high, medium, low, cool, but a few have only two, high and low, or high and cool, for summer comfort. Bonnets are of fabric in many colors of plain or floral designs, usually lined with plastic material and large enough to fit comfortably over the largest curlers. The bonnet may have a reach-in top or an opening on the top-back so that the user may check the dryness of the hair without removing the bonnet or disturbing the set of her coiffure.

One model has a wand that snaps on to the end of the hose for spot drying. Other models dry and at the same time style the hair with comb or brush attachments.

The hose is flexible and should be nonkinkable. It is 40 or 42 inches long about 1¼ inches in diameter. The cord length varies from 6 to 8 feet, the greater length allowing more flexibility in moving around to answer the telephone or doorbell. A waist and shoulder strap for carrying the power unit permits this movement. The homemaker can also iron if the iron is attached to a grounded outlet. She should not wash dishes or do laundry.

The professional rigid dryer may be on an adjustable pedestal or wall mounted, and some models have a solid state thermostatic heat control. Wattage varies from 750 to 1200 watts, with three or four heat settings provided. One type of control is a remote control, held in the hand, so that turning around is not necessary. Some models have mist attachments to aid in setting the hair between washings, and several have a perfume wick for scenting the hair. One dryer has a reading light attached to the center front of the rigid hood (Fig. 12.2).

High school and college girls like the soft bonnet type hair dryer because it lets them take a nap while their hair is drying; the rigid hood usually prevents this. In using the bonnet in this manner, care should be taken not to cover any air intake vents should the girl wrap herself in a blanket; otherwise the motor may overheat. A fire may even start if the student has papers scattered around her for studying and then falls asleep. Some models furnish automatic overheat protection by shutting off if airflow is restricted causing excessive heat to build up.

Various types of carrying cases are supplied with the dryer; baggage case, bandbox, vanity travel case, tote bag, and purse. One professional-type dryer has a double-walled hood that is made into a carrying case when the dryer is closed. Occasionally the case may have accessory space for manicure articles and hair brush and comb. Cases are compact and lightweight, 4 to 7 pounds, and sometimes as little as 2 or 3 pounds, and are often made of scratch-resistant vinyl.

Most dryers are advertised to operate quietly so that one may listen to radio or TV or talk over the telephone. The motor is commonly designed to reduce radio and TV interference.

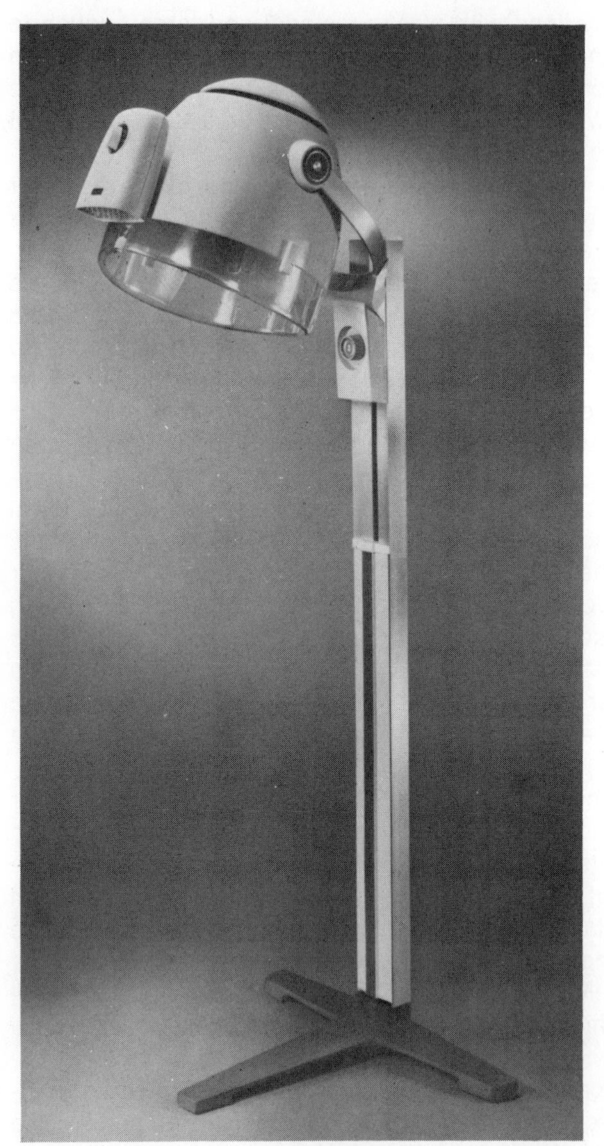

Fig. 12.2 A light attached to hood makes reading a pleasant way to pass the time. (Ronson)

Hair Curlers

Supplementing the work of hair dryers are hair curlers. They may be separate rollers, purchased by themselves, or be part of an appliance operated by electricity. Hair curlers vary in number, size, and weight of rollers, in method of holding the roller in the hair, in preheating time, and signal device for indicating when a roller is hot.

Some sets contain 12 to 20 curlers, frequently varying in size. One manufacturer indicates that his set of 18 contains 6 large, 8 medium, and 4 small rollers. Another manufacturer advertises 20 curlers, 6 jumbo, 8 large, 4 regular, and 2 small. Some emphasize tangle-free rollers and say that no water or lotion is needed; the hair may be rolled up dry, relying on a certain amount of natural moisture in the hair (Fig. 12.3).

In some types heated by electricity the metal core holds the heat, in others the plastic curlers may fit over metal spindles, but between the layers of heat-resistant plastic is a wax layer; when heat is conducted to the rollers, the wax is also heated and tends to retain the heat over a period of time.

***Fig. 12.3** Electric curler set of 20 rollers in four sizes. Dot on each roller indicates when it is ready to use. No water or lotion needed. Travel case has mirror for aid in setting and grooming. (Presto)*

Fig. 12.4 "Mist Set" curling wand, combines gentle heat and moisturizing mist to create firm, swirling curls in seconds. Provides curling action for waves, curls, marcelling, tendrils, and straightening—even use on wigs. Mist button provides up to 25 minutes of moisture for continuous use. Moisture gives hair extra body, tames cowlicks, and protects against split ends. (Oster)

One 12-curler set of solid polypropylene is heated by attachment to a thermostically controlled home or travel iron, and hence may be used anywhere. The curlers are slipped over tubular posts mounted on a metal plate, which is fastened to the soleplate of the iron. The heat passes through the posts to the curlers. They heat in 10 to 15 minutes, depending on the wattage of the iron; when the desired temperature is reached, red dots on the ends of the curlers turn black. Another type allows "touch-up" or "spot-curling" with an attachment that slides over the curler and is fastened to the dryer hose. Most curlers are held in place with bobby pins.

Separate curlers may be heated in boiling water or a curling iron may be used, now usually electrically operated. A modification of the curling iron is the "Mist Set" curling wand which combines gentle heat and moisturizing mist to create waves or curls in seconds. Moisture adds extra body to hair, smoothes down cowlicks, and protects against split ends. Thermostatic control supplies even heat; light signals when ready for use (Fig. 12.4).

A use-and-care booklet and a carrying case to hold curlers, clips, cord, and often a mirror are also supplied.

Electric Hairbrush

A final appliance for the care of the hair is the electric brush (Fig. 12.5). It offers real brush action, approximately 450 strokes per minute, and supplies stimulating scalp massage. One model has bristles that retract to prevent snags or pulls, and a switch that allows you to brush in any direction with either hand. Another model has a second brush that will remove tangles left in the hair when it is

***Fig. 12.5** A roto-stroke electric hairbrush will remove hair spray build-up and massage the scalp. (Ronson)*

washed—or at other times. The electric brush is useful in removing hair-spray build-up, and can be taken apart for easy cleaning.

Hair-Cutting Sets

Families with several children find hair-cutting sets a real economy. Even husbands discover that their wives are expert enough, except perhaps for a very special occasion, and women frequently learn to cut their own hair. A set features an electrical clipper, with an adjustable blade for cutting hair to different lengths, barber shears and tapered comb, attachments for "Butch" cuts and for trimming, combs for right and left tapering. An apron and duster brush may also be included along with an illustrated booklet of directions.

Electric Shavers

Men's electric shavers have been on the market for 35 years or longer. They may be used by women, too, but special models for women are now featured with separate cutting systems for underarms and legs (Fig. 12.6), as well as the conventional trimmer for the back of the neck. Men's shavers are provided with trimmers for sideburns and the mustache. The stainless steel cutter may have a very thin snap-on shaving screen to guide hair to the cutters. The shaver head may be flat with rotating or reciprocating cutters, or the head may be curved with reciprocating cutters or oscillating blades that move in an arc. One head is placed at an 11° angle and has scissorlike shaving action; another features hollow ground double-edged razor blades. Most manufacturers claim that their products will give extremely close shaves without causing irritation. A few models power-clean themselves. A built-in light, an on-and-off switch, and a detachable cord are conveniences. The use of the shaver should not interfere with TV and radio reception.

Cordless shavers, with or without rechargeable batteries, are also obtainable. Those with rechargeable batters usually store power for a week of shaves with a single charge. If constructed for use on a 120/240 volt circuit, they offer an advantage to the traveller in foreign countries. One company features a converter cord that adapts a converter shaver to any direct current 12-volt system, in addition to its use on the usual 110/120 volt AC/DC household current.

Lather Dispenser

For use with a shaver or as an aid in facial beauty care a lather dispenser is available. The dispenser contains a soap basket (any kind of soap may be used), and a water tank; it holds 8 ounces of

FOR SLEEK, CLOSE
SHAVING OF LEGS

FOR SAFE, GENTLE
UNDERARM GROOMING

***Fig. 12.6** Lady's shaver—one side of shaving head for close and gentle underarm grooming, the other side of shaving head for clean, sleek shaving of legs. (Oster)*

water and always should be at least half full during use. A heating element keeps the water warm. A rotating cylinder encircled by brushes on which the water drips rubs across the surface of the soap forming warm lather, its wetness depending on the adjustment of the water valve. A lever at the bottom of the appliance is held down to start the action; the more the dispenser is used, the faster it delivers the lather. If water is hard, the adjustment screw must be cleaned occasionally. This same company now features a new heated shavecream dispenser; it will hold any standard aerosol shave cream can and furnishes preregulated cream temperature, controlled by a heating system that shuts off automatically.

Manicure Sets

Certain hair drying models provide manicure discs of metal and emery board for shaping and manicuring nails. At least one manufacturer makes a complete manicure set with five attachments to trim, shape, and groom cuticles and buff nails. They include cuticle brush, nail shaper, buffer, cuticle pusher, and callus remover. Replacement emery discs and buffer pads are also supplied. This set may be used for pedicures, too. A case is provided with a built-in polish dryer under the front edge. The power cable, operated on 100/120 volt AC current and controlled by a push button on-off switch, coils inside the case.

Electric Brushes

Toothbrush. The electric toothbrush may be operated from a wall outlet, or it may be the cordless rechargeable type, which is frequently more popular because it eliminates any danger of shock from contact with water or the lavatory itself, connected as it is through the water pipes to the ground. A recent model uses a new motion, up-down and around at the same time (Fig. 12.7*a*). The base holds four or six brushes, and one as many as seven brushes, color coordinated for easy individual identification, the power handle, and occasionally the tube of toothpaste. One manufacturer advertises serrated bristles; another offers the contoured type as being safer for cleaning and gum massage because of the lack of sharp edges. Strokes may be as many as 11,000 per minute. Some toothbrushes are supplied with ball bearings to reduce friction.

Protected by a molded clear plastic cover, the model may stand on a shelf or counter, or be wall mounted on a bracket. The charger is frequently in the plug; it may be solid state controlled and often can be used on either 120 or 240 volts and, therefore, anywhere in the world. Some are supplied with a night light or a built-in electric outlet.

An automatic denture cleaner and an oral jet rinse are other cordless appliances that are available. In one oral hygiene pick, water in the upper reservoir is pumped to a jet tip, from which it is forced out in a series of pulsations. Pressure is controlled by a knob on the front of the unit (Fig. 12.7*b*). Four color-coded tips are included. In some other models the force may come from a pump that is built into the appliance, or it may come from natural water pressure from the faucet. The built-in pump is best; it allows more variety in pressure. Be careful not to direct the water under the tongue or into the throat. Some type of travel case is usually supplied.

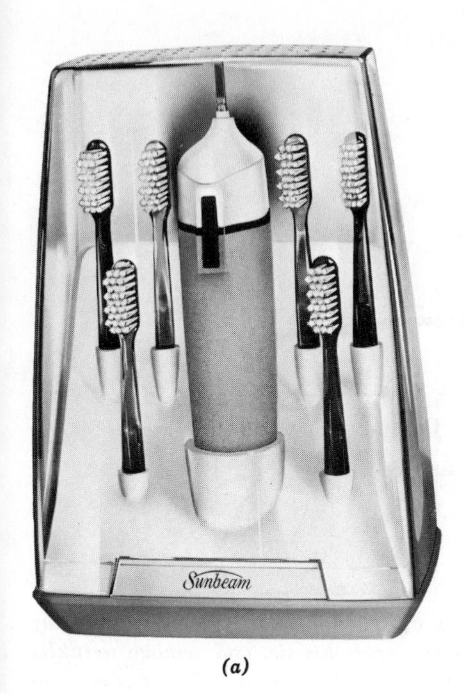

(a)

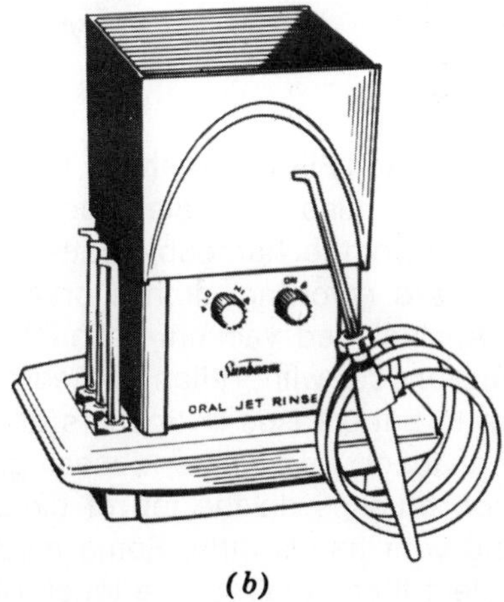

(b)

Fig. 12.7 (a) ***Cordless electric toothbrush with up-down and around motion, recharges automatically when stored in stand.*** (b) ***Oral-jet rinse. (Sunbeam)***

All of these aids to dental hygiene are recognized as desirable and are approved by the American Dental Association.

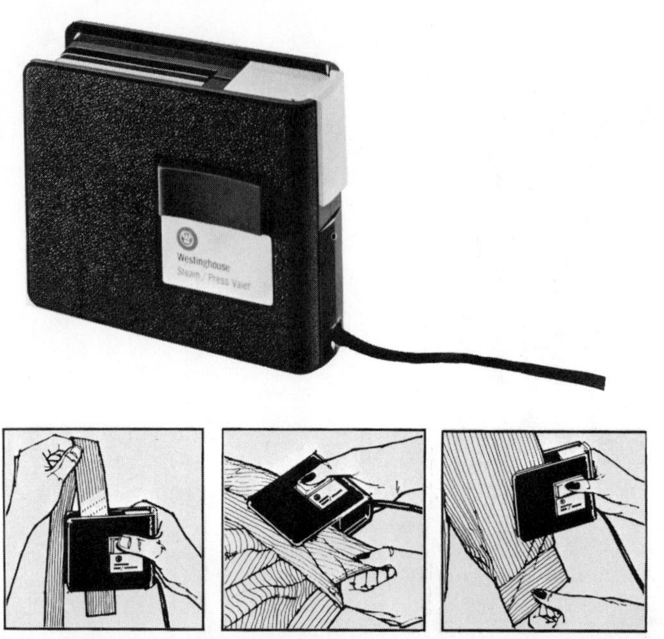

Fig. 12.8 This Steam/Press Valet was designed primarily for the traveling man. It will restore the crease in trousers, press his tie, and remove wrinkles from collars or sleeves of wash-and-wear shirts. Women also find it useful for a "touch-up" of blouses and dresses and to freshen pleats in skirts. (Westinghouse)

Clothes and Shoe Brushes. Other types of electrically operated brushes are also available, operated directly from the house wiring system or from rechargeable batteries, frequently solid state controlled.

One is a revolving clothes brush that removes dust and fluffs the nap. An included vacuum-action draws lint and dirt into a lint trap. Another model with Teflon-coated grids may be used with or without steam to put creases in trousers or pleats, smooth out hems and collars, or touch up wash-and-wear garments (Fig. 12.8).

Shoe brushes do the job of cleaning and polishing shoes without getting your hands dirty. Some models pick up brushes magnetically and eject them again at the touch of a button (Fig. 12.9*a*). One brush applies the wax to all areas of the shoe; another brush or lamb's wool buffing pad does the polishing. A two-speed motor has a low speed for applying wax without spatter and a high speed for polishing and buffing. Separate brushes, color coded for black and brown shoes, are supplied (Fig. 12.9*b*). One company places a convenient foot rest on top of the carrying case. Often, the fitted travel case has compart-

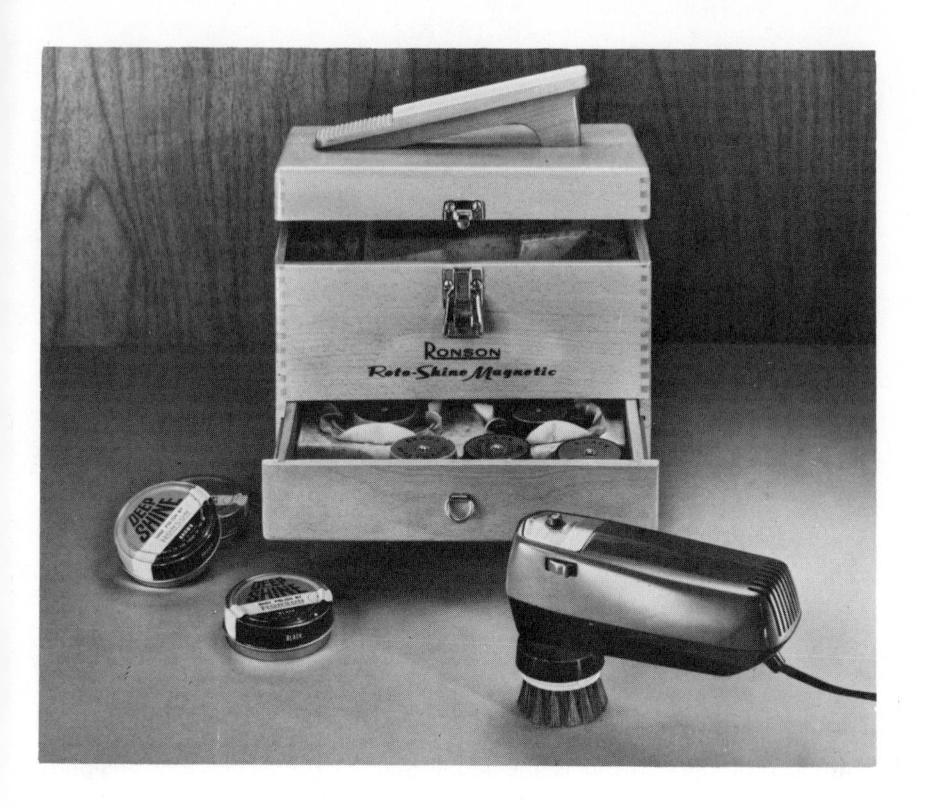

Fig. 12.9 (a) *This shoe brush picks up the brushes magnetically; a two-speed motor uses low speed for applying wax, high speed for polishing. (Ronson)* (b) *A shoe polishing outfit has a bristle brush and two buffing bonnets, one for black and one for brown shoes. (Regina)*

ments for cans of shoe polish, applicators, brushes, buffers, and extra flannels. A mud brush may be included.

Complexion Care Kit

At least one company markets a "facemaker," a complexion care kit, said to provide complete home beauty treatment. The power wand operates on rechargeable cells and is waterproof. It works only when the attachments are pressed against the face. The attachments include two complexion brushes, a general body brush, a facial massager, and a cold-hot pack. One brush with gentle circular motion removes old make-up and deep-down facial grime; the other brush is for cleaning when no make-up is on the face. The body brush can be used for cleaning hands, nails, elbows, arms, neck, or other areas. The massager stimulates the skin and effectively works in night creams and moisturizers. The cold-hot pack attachment has a special liquid center. Placed briefly in the freezer, the liquid becomes an icy cold pack; rubbed over the face it closes pores and adds a refreshing touch. Held under hot water it becomes a hot pack to prepare the face for brush use. A travel case provides a storage place for all of the parts (Fig. 12.10).

Facial Sauna

Another complexion aid is the facial sauna. The especially designed water container has an automatically controlled heater that converts a few ounces of water into a soothing vapor. The warm mist opens pores, providing, from the inside out, thorough facial cleansing from stale cosmetics and grime. It nourishes the skin, stimulates circulation, and relaxes tight, tense face tissues. The sauna also may relieve the stuffiness of head colds.

Sinusizer

An adaptation of the facial steam applicator, the sinusizer provides an inhalant in concentrated steam to help relieve sinus congestion and clogged respiratory passages. The lower section of a two-part housing contains the water and inhalant, which is vaporized at 115° to 120°F by heat from a two-post heating element attached to the cover of the upper section.

Make-up Mirrors

Special make-up mirrors are obtainable. One model is illuminated by diffused lights that are recessed on either edge of the mirror. It swivels to provide a magnifying mirror on the other side. By means

Fig. 12.10 Complexion care kit with five attachments. (Sunbeam)

of a gear system, another model rotates three colored plastic lights in front of two fluorescent lamps. The different colors allow you to apply make-up by the light in which you will be seen—daylight, office light, or evening light. This mirror also reverses from a normal to a magnifying surface. In either type the light is very bright and concentrated in a comparatively small area. To use the mirror successfully, it must be held within a few inches of the face.

COMFORT APPLIANCES

Blankets and Sheets

Electric blankets are made of a combination of rayon, cotton, and acrylic fiber in single or double bed sizes, and in king and queen sizes, too, the double sizes with single or dual controls. The control dial, marked for varying amounts of heat, should be illuminated or lighted to be read in the dark and, if possible, should be hung on the side of the bed. A rubber mounting is a convenience in preventing any marring of the wood. The thermostatically controlled heat should be uniform throughout the entire blanket, and all parts should be sealed against moisture. Do not tuck in the wired end of the blanket and do not use pins; they may damage the wires.

Some present-day electric blankets use a solid state control, which is so compact and small that it is placed right on the blanket itself, and automatically produces a condition of warmth that is based on the temperature under the blanket and also on the room temperature. The control has 27 click-stop settings, compared to 9 settings on the conventional blanket.

An electric blanket marketed by General Electric eliminates lumpy thermostats and tubes; it has a smooth wired surface that adjusts automatically to temperature changes and so provides protection throughout the entire blanket. The system is composed of two spiral wires separated by a nylon sheath. If the heating wires become too hot, the sheath together with the signal and heater wires turn off the heat. In other words the entire wired area is sensitive to excessive warmth, not merely at thermostatic locations (Fig. 12.11*a* and *b*).

A UL Seal of Approval should be provided and at least a one-year guarantee against defects in construction. Washing and summer storage instructions are desirable. Dry cleaning cannot be used; the cleaning fluid tends to cause deterioration of the insulation on the wires. Moth crystals should not be added during storage. The blanket may be washed in a washer and dried in the dryer at low heat. The lower the percentage of wool in the blanket, the less shrinkage in laundering. Most present-day electric blankets do not contain wool; they are of synthetic fibers, dacron or acrylic.

It is recommended that the electric blanket not be used as a bed covering over an infant, a helpless person, or one insensitive tc heat because of the danger of overheating.

To overcome this difficulty a blanket has been designed which contains, in place of the electric wires, a network of water-tight small plastic tubes, through which warm water flows. The same method is used to regulate the temperature in the astronauts' space suits. The control dial allows temperature selection, but in any case, the temperature cannot go above 120°F, so no scorching or burning is possible. A reservoir used with the blanket holds about 8 ounces of water, which are heated by a unit completely separated from the water supply. When warmed to the level determined by the dial setting, a pump forces the water through the tubes. It takes about 30 seconds for the water to circulate.

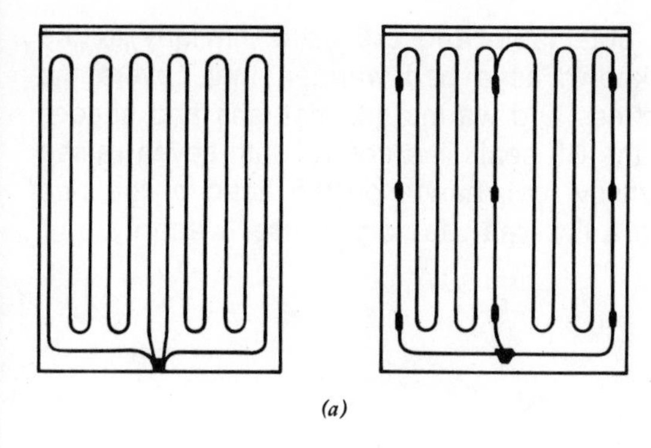

(a)

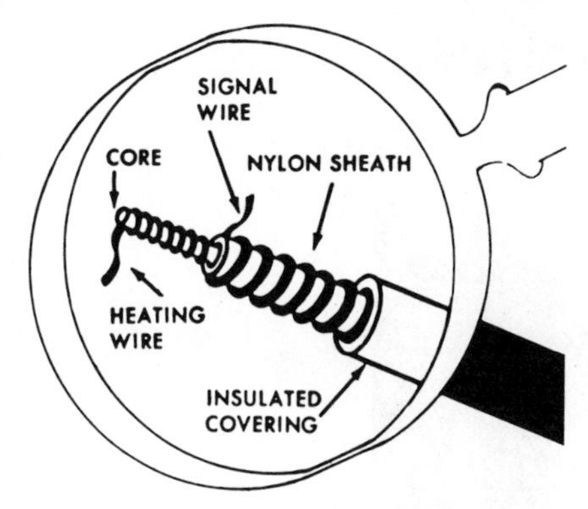

***Fig. 12.11** Feel the smooth surface on all General Electric blankets. The entire wired area is sensitive to excessive warmth.* (a) *Blankets with thermostats.* (b) *The Sleep-Guard® system is composed of two spiral wires separated by a sheath of nylon. Should the heating wire in your blanket become excessively warm the nylon sheath, in connection with the signal and heater wires, automatically turns off your blanket at the selector. The Sleep Guard is one continuous thermostat to provide security throughout the entire blanket. (General Electric)*

The reservoir needs refilling about once in every 30 to 50 days, depending on blanket use. If the reservoir runs out of water, the thermostat shuts off the control; after 8 minutes the control will cycle on and off in rapid succession until water is added.

The blanket of acrylic fiber should not be dry cleaned. It may be machine washed in lukewarm water for no longer than 3 minutes; no bleaches should be used. It can be spun dry, but not put through a wringer or twisted, for fear of damaging the tubes. An iron should not be used on it.

Electric sheets are also obtainable and are used similarly to the blanket. A type of sheet, known as a bed warmer, that covers the lower two-thirds of the mattress and warms the mattress and spaces beside the sleeper, carries the UL seal. The control with seven raised buttons, allowing adjustment by feel, hangs on the head of the bed within easy reach. It carries a five-year, factory-to-user warranty.

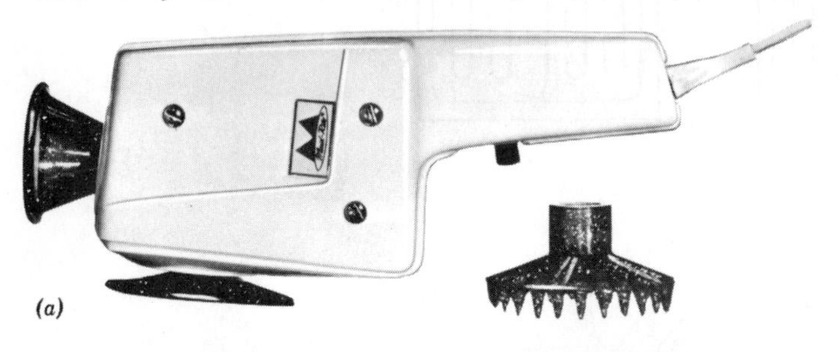

(a)

(b)

Fig. 12.12 Three types of massagers: **(a)** ***Two-way action, two-speed operation. Three applicators: smooth facial, suction cup for muscles, fingered disc for scalp. (A. F. Dormeyer Mfg. Co.)*** **(b)** ***Motor action delivers ratating, patting action (3600 pats per minute) to fingertips. (Oster)*** **(c)** ***For aching legs and feet a combination of motorized massage with water in motion. (Oster)***

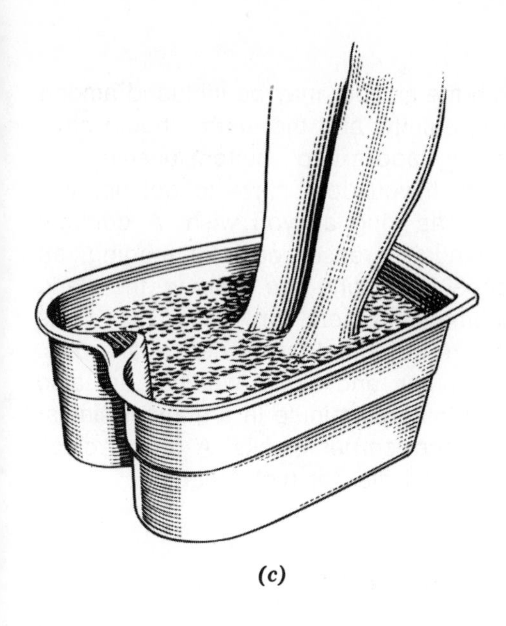

(c)

Heating Pads

Heating pads put the heat where the pain is. They usually have a quilted removable outer cover that may be washed or dry cleaned, and a sealed, 100% wetproof inner pad that permits damp applications. The pad may be contoured to fit around the pain. A lighted pushbutton control or slide-switch provides three separate heats. Some models are inflatable.

Massagers

Reference has been made already to the massager in the complexion care kit. Separate massagers are also available, often with both vibrating and pulsating actions, the first for gentle surface massage, the second for deep, penetrating results. Infra red heat may be provided as an additional benefit. Besides the general action, attachments include facial and scalp massagers. A foot massager allows you to sit in a chair with your feet on the appliance; the vibration refreshes and stimulates tired, aching feet (Fig. 12.12). Standing on it gives full body massage.

Another type is in the form of a pillow, one side of which provides high or low intensity pinpointed massage, the other side, full surface massage with or without heat. If desired, this model may be used as a heating pad.

Electric Clock

An electric alarm clock to awaken the sleeper may be included among comfort appliances. It should run quietly, and the alarm should have a pleasing tone. One model has an "add-a-nap" button; pushing will give you 10 extra minutes in bed. If you don't have to get up, you may continue to push the button as long as you wish. A contrast in color of dial numerals and hands increases ease of reading; an illuminated dial is a convenience for noting the time in the dark. Some clocks have radio attachments; others, an alarm that rings automatically every four or six hours to remind the owner to take pills.

At present digital clocks are popular, and digital watches are also available. They indicate the time minute by minute in a manner similar to the way time is indicated on street-corner clocks. A digital clock may have an alarm and an illuminated dial for night reading.

Experiments and Projects

1. Assemble representative personal appliances obtainable in the laboratory. If possible, add similar items brought by members of the class.
2. Make an outline of features to be observed for each type of appliance: model, wattage, electrically or battery operated, weight, type of unit, number and type of accessories, type of carrying case, etc. (Outlines of desirable information on all types of small equipment may be obtained from *Small Equipment Laboratory Manual* by Mary S. Pickett and Kathleen S. Schumacher, Family Environment Department, College of Home Economics, Iowa State University, Ames, Iowa 50010.)
3. Let individuals or groups practice the use of the different personal appliances and then demonstrate the use to other members of the class.
4. Evaluate the appliances used. Suggest desirable improvements in the form of the appliance or in the method of operation.

CHAPTER 13
Laundry Equipment and Aids to Satisfactory Operation

The basic principles of cleaning clothes by washing are the mechanical flexing and bending of fabrics in a detergent solution. The fabric should move freely through the water, and the water should move freely through the clothes. This process will contribute to wetting the soil and the fiber, loosening the soil from the fabric, and holding the soil in suspension for removal in the spin process of the cycle. Each washer design—automatic or nonautomatic—has been developed to make easier the task of adequately washing fabrics. Basic to the most effective operation of these designs is an understanding of the functions the specific washer can perform and how these may best serve to wash specific fabrics with certain care requirements.

WASHER DESIGN AND FUNCTION

Modern washers can be divided into two main groups—the automatic type, tumbler and agitator, and the conventional type, wringer and twin tub.

AUTOMATIC WASHERS

Agitator

The washing action in the top-loading washer depends on an agitator which moves the clothing through the water and the water through the clothing through an arc of approximately 180°. The agitator usually has vanes of aluminum, stainless steel, or plastic. The vanes are attached to a column that fits over a central shaft around which it oscillates. The vanes, blades, fins, or other methods for moving the clothes through the water vary in number and shape. Some vanes are short and thick, extending horizontally; others are thin and high; some are attached vertically to the central column; others curve in spiral around it; some are adjustable. Some agitators have a vertical motion as well as the horizontal component, and still others incline the plane of movement from the vertical, causing an undulating motion. The additional motions are designed to create greater turbulence in the water.

Two other designs for moving clothes through the water feature a pulsator with three flexible rings, which moves up and down on the shaft instead of around it, and an agitator with no vanes. The latter creates a wash action by the rolling and rocking of the blade-free mechanism as it bends up and down, sending high-speed waves in all directions at approximately 600 times a minute.

Tumbler

In the tumbler or front-loading washer, the clothes are placed in the washer through a door that opens to the front. Movement of the clothes occurs by lifting them out of the detergent solution by means of baffles on the rotating basket, and letting them fall back into the water of their own weight as they reach the top of the revolution.

During the wash and rinse period, the cylinder basket revolves slowly at approximately 50 rpm. At the end of this period, the water is drained from the machine. In the moisture extraction period, the basket revolves much faster to spin the clothes damp dry and ready for final drying. The speeds of these revolutions are approximately 600 rpm for the "normal action" and 300 rpm for "gentle action."

Wash Action

The speed of agitation in an agitator washer may vary from 50 to 100 rpm, depending on the cycle selection. An example of the speed of agitation as related to cycle choice may be as follows: gentle or reduced action, 50; delicate, 60; medium, 80; and normal, 100 rpm. Pulsator action may vary from 220 strokes per minute at reduced speed to 330 strokes per minute for regular or normal speed. The selection of wash action speed should be related to the type of construction of the garment, delicate or sturdy, and to those factors which require a specified level of flexing and bending necessary for effective removal of soil.

Controls

Automatic washers have two types of controls—selective and programmed. Selective controls permit the user to select and preset an appropriate function for specified types of clothing and textile items. The programmed controls are preset by the manufacturer and permit all variables to be controlled with a single setting. Some machines have both types of controls, incorporated in the same design (Fig. 13.1). With selective controls, the user has the responsibility for knowing the requirements of the fabrics and selecting each of these from the choices available. The programmed control requires that the user make only one or two decisions. Basically, the control selection should be related to the cleaning requirements of the items in the load, and the load volume to determine the water level—partial or full.

DESIGN FLEXIBILITY

Single-speed washers offer specific benefits to specific types of consumers: (1) first-time buyers on limited incomes, (2) buyers who prefer the basic machine, (3) families changing from wringer types to automatics who have a limited income, and (4) just-married couples or retired couples who have limited laundry needs or no children. The single-speed washer offers the essential elements of a good basic washer: adequate capacity and performance for most family wash loads, positive fill action for correct wash water level regardless of water pressure variations, and a clutch that makes the motor last longer.

The multispeed washers are beneficial to the consumer who represents the (1) average-sized families with young children, (2) families who have a wide variety of fabrics or large quantities of items in the weekly wash, and (3) large-sized families who have large capacity tubs. This washer design offers flexibility of washing and spinning speeds to match laundry loads, variable water levels for differing laundry needs, heavy-duty motor and transmission, balance and suspension system to prevent off-balance loads, and basic convenience features of filter and liquid bleach dispenser.

A deluxe or custom washer will be effective as a choice for the (1) working wives with schoolage children who need the utmost in

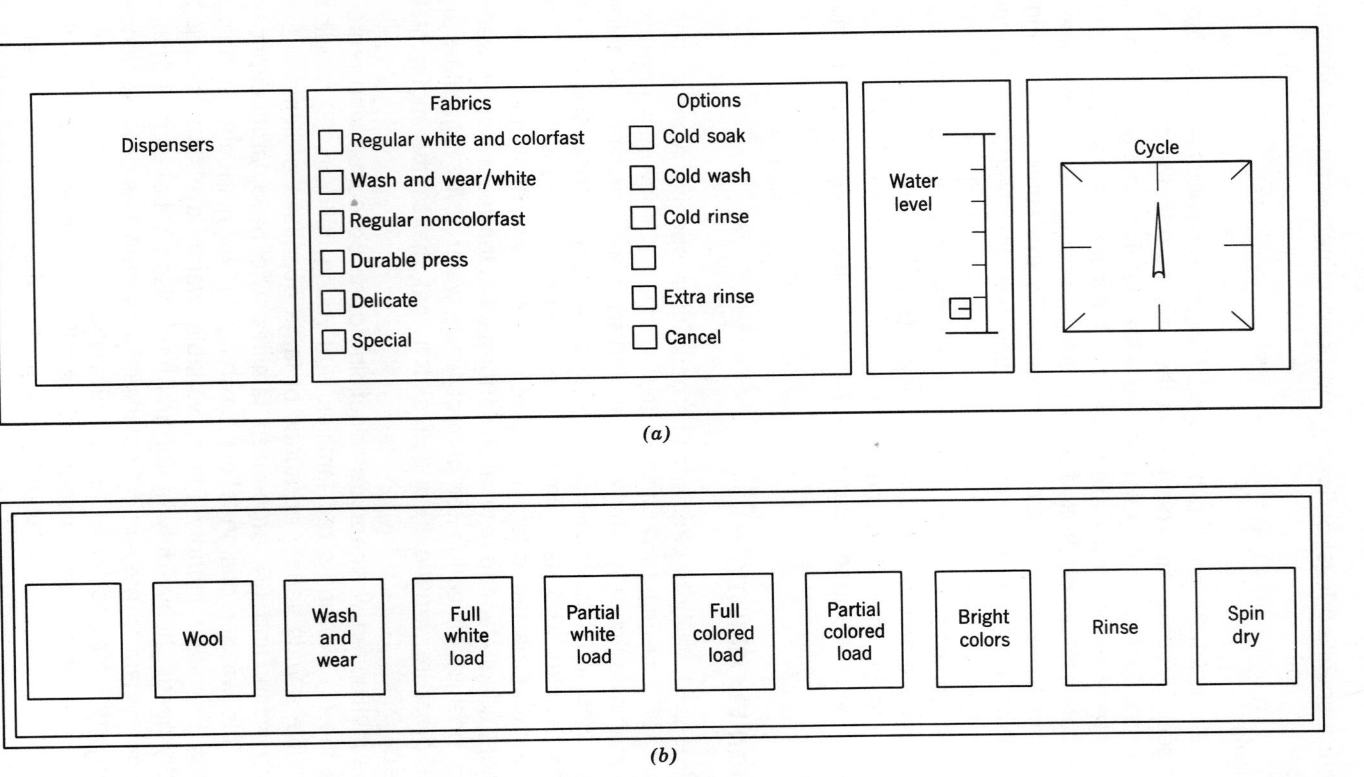

Fig. 13.1 Degrees of flexibility designed into washers at various levels of automation. (a) ***Selective controls.*** (b) ***Programmed controls.***

convenience and timesaving features, (2) sophisticated buyers who want the best, (3) consumers who want the greatest flexibility in controls, (4) buyers who want pushbutton ease or infinite control selections, or (5) families that have many varieties of washloads and want the maximum of cycles for every washday need. The deluxe model offers the ultimate in deluxe features and controls, infinite variety of cycles for each laundry problem, maximum in power and transmission, full line of convenience, and timesaving features.

Construction

The cabinet or outside frame of the washer serves to house, anchor, and separate the various parts of the washer design. In most washers the moving parts are found in the compartment at the bottom of the cabinet, separated from the other parts by the bottom of the tub. The positioning of the motor is related to the design of the washer type. In the tumbler type, the motor is mounted with its axis parallel to the floor with the drive belt to the cylinder basket pulley at the rear of the machine. In the agitator design, the motor is mounted with its axis at right angles to the floor, the belt pulleys traveling in a horizontal plane. The outside finish may be porcelain enamel. Some have a cabinet of acrylic enamel with porcelain enamel top and lid.

Each washer has two tubs. The inner tub may be made of porcelain enameled steel, aluminum, or stainless steel. This tub may be solid or have perforations in the sides or bottom which permit the passage of water in and out. Specific designs in the bottom of the tub provide for effective removal of dirt or residue and insure complete drainage. The tumbler-type washer basket is fitted with baffles placed at right angles to the circumference to carry the clothes to the top as the basket revolves. Baffles also help create turbulence, which loosens dirt more effectively. The outer tub serves to hold the water through which the cylinder revolves. This tub is generally made of porcelain enameled steel and has holes fitted with seals and sleeves for fastening the various hose connections. A large drain hose in the bottom of the tub may be provided with a baffle so as to prevent the air lock that may occur when swirling water passes down the drain.

The outside frame of the automatic washer is approximately square; that of the conventional washer may be square or round. The lid of the automatic is hinged to the washer body; the lid of the conventional washer, with occasional exceptions, is separate, but has a hook on the edge for attaching it over the rim of the tub where it will be out of the way but still easily accessible. The round tub of the conventional washer is supported by a leg frame or is on a

Table 13.1
Consumer-Care Guide for Apparel

	When Label Reads:	It Means:
MACHINE WASHABLE	Machine wash	Wash, bleach, dry and press by any customary method including commercial laundering and dry cleaning
	Home launder only	Same as above but do not use commercial laundering
	No chlorine bleach	Do not use chlorine bleach Oxygen bleaches may be used
	No bleach	Do not use any type of bleach
	Cold wash Cold rinse	Use cold water from tap or cold washing machine setting
	Warm wash Warm rinse	Use warm water or warm washing machine setting
	Hot wash	Use hot water or hot washing machine setting
	No spin	Remove wash load before final machine spin cycle
	Delicate cycle Gentle cycle	Use appropriate machine setting; otherwise wash by hand
	Durable-press cycle Permanent-press cycle	Use appropriate machine setting; otherwise use warm wash, cold rinse and short spin cycle
	Wash separately	Wash alone or with like colors

This care guide was produced by the Consumer Affairs Committee, American Apparel Manufacturers Association and is based on the

	When Label Reads:	It Means:
Nonmachine Washing	Hand wash	Launder only by hand in luke warm (hand comfortable) water. May be bleached. May be drycleaned
	Hand wash only	Same as above, but *do not* dryclean
	Hand wash separately	Hand wash alone or with like colors
	No bleach	Do not use bleach
	Damp wipe	Surface clean with damp cloth or sponge
Home Drying	Tumble dry	Dry in tumble dryer at specified setting—high, medium, low or no heat
	Tumble dry Remove promptly	Same as above, but in absence of cool-down cycle remove at once when tumbling stops
	Drip dry	Hang wet and allow to dry with hand shaping only
	Line dry	Hang damp and allow to dry
	No wring No twist	Hang dry, drip dry or dry flat only. Handle to prevent wrinkles and distortion
	Dry flat	Lay garment on flat surface
	Block to dry	Maintain original size and shape while drying
Ironing or Pressing	Cool iron	Set iron at lowest setting
	Warm iron	Set iron at medium setting
	Hot iron	Set iron at hot setting
	Do not iron	Do not iron or press with heat
	Steam iron	Iron or press with steam
	Iron damp	Dampen garment before ironing
Miscellaneous	Dryclean only	Garment should be drycleaned only, including self-service
	Professionally dryclean only	*Do not* use self-service drycleaning
	No dryclean	Use recommended care instructions. No drycleaning materials to be used.

Voluntary Guide of the Textile Industry Advisory Committee for Consumer Interests.

pedestal base that carries the legs. The closer the legs extend to the top of the tub, the better braced the tub is and the more adequate is the support. A rubber gasket between tub and base cushions the tub and reduces vibration and noise. One washer has a square tub with rounded corners. A few models have double-walled tubs with air spaces in between to insulate the tubs and prevent the wash water from cooling too rapidly. In this construction the tub rim rests on the edge of the steel frame and bolts through the tub are avoided, leaving the interior entirely smooth. All pedestal-based washers have bolts extending into the tub.

Sometimes the bottom of the conventional tub is rounded like a bowl, with a sediment trap beneath the agitator, an arrangement that helps prevent dirt from circulating back into the clothes. If the sediment is drawn off through the drain pipe from time to time and fresh hot water is added to the tub, the number of complete changes of water may be reduced.

Legs fitted with large casters and of adjustable height are desirable on the conventional type for ease in moving it from place to place. Automatic washers usually have leveling screws for adaptation to uneven floors. Cushioned on springs or on rubber shock absorbers, the motor is placed beneath the tub to be shielded from water, and is insulated from the metal chassis. All moving parts should be enclosed, and the washer grounded. The cord is rubber covered.

Electrical System

The automatic and conventional washer may be operated on a 115-volt, 60-cycle AC power supply. Electricity is used in the automatic washer to power the drive motor, timer motor, and the electrical components used in controlling the flow of water in and out of the machine. The motor used in most automatic washers is a ⅓ to ½ HP, 115-volt, 60-cycle, capacitor-start motor. It is equipped with thermal overload protectors which will disconnect the motor from the circuit if an overload condition occurs. The combination washer-dryer is an all-electric design which requires 220 volts to operate efficiently the heating element for the drying circuit.

For safety, the power supply cord of the washer should be equipped with a three-pronged grounding plug. It should be plugged into a matching grounded three-pronged type wall receptacle.

Some of the specific electrical components which are fundamental to the operation of the automatic washer are the timer, fill switches, lid switches, solenoids, and drive motor. The timer is the heart of the control for all functions. The most recent advance in design of electrical components has been in the use of solid state controls. This control takes the form of a thermistor for sensing heat in specific portions of the cycle. In a washer, the thermistor is used to sense and regulate the heat of the incoming water. In some models of the dryer, a series of thermistors has been used to register the temperature of air at the duct outlet, replacing bi-metal disc thermostats. Thermistors, because of their infinite setting potential, are more efficient in regulating temperatures than the off-on thermostatically controlled switch.

A second important electrical component, less well understood, is the solenoid. This is a means for converting electrical energy into mechanical energy to trip switches, open and close valves, and actuate mechanical linkages. The solenoid consists of a coil of wire in which a magnetic field is developed when current passes through it. This powerful magnetic force is able to move a metal core or plunger inside the coil. The movement of the plunger is resisted either by a spring or by gravity, so that when the energizing current is switched off, the plunger returns to its normal position. The solenoids do not generate enough force to engage gears or operate a heavy mechanism. Instead, they start a train of events by setting free more powerful forces.

Operating Characteristics of the Washer

The process of selecting a combination of conditions appropriate for the specific type of laundry load does no more than close circuits to the drive and timer motors. The drive motor is combined with the agitator or, in the tumbler washer, the cylinder basket, and the water pump. A second circuit in the electrical system supplies current to a timer motor which controls all the operations of the machine during the entire wash and spin cycles. When the various selector dials are energized for the type of washing desired, a set of cams is aligned, presetting a switch and adjusting a pair of water-inlet valves. When the machine offers both normal and gentle action, a different set of cam wheels is brought into action for the selection. These cam wheels are revolved slowly by the timer motor. As they revolve, their surfaces make contact with switches at various points in their travel, opening and closing circuits to energize solenoids, and segments of circuits to activate or stop the various components in the system.

Water Temperature

Water temperature should be considered in relation to the following factors: fiber, fabric, finish, type and degree of soiling, color of textile, and effectiveness of specified laundry aids at certain temperatures, and desired level of sanitation. The provision for varying water temperature on automatic washers ranges from hot to cold. The rinse temperatures are generally limited to two—warm and cold. Many washers provide a warm rinse automatically with the hot wash temperature and a cold rinse with a warm or cold wash temperature.

The temperature of the hot water will be determined by the supply and distribution of hot water in the home, thus the need for concern for those factors which may cause temperature loss as water is heated at the water heater and delivered to the washer. The setting at the water heater should be appropriate for the cycle choices. Plumbing designs should be such that losses in water temperature will be minimized—straight runs of pipe, well-insulated pipes, pipes located in areas where temperature differential is minimized, and water heater located as close to the washer as possible. Another factor that may cause a loss of temperature in the washer is the temperature of the clothes and of the appliance. Cold water temperature will be the same as that for the cold water system in the house (Fig. 13.2).

Warm water temperatures are regulated by a mixing valve. This water temperature may be used for both washing and rinsing. It is designed to combine hot and cold water in specific proportions. The warm water temperature will depend primarily on the temperature of the hot water.

Water Fill

The two factors that are used to control the fill are time and pressure.

Time fill is determined by an electric timer that controls the time the water inlet valves are held open during the fill period for washing and rinsing. The time set for fill will provide the proper level of water when the pressure at the inlet valve remains within fixed limits, usually between 20 and 120 pounds per square inch. If the water pressure varies outside this range, it is possible that the washing or rinsing action will begin without the proper amount of water for the load. This may result in ineffective washing conditions or damage to fabrics.

The second method of fill—pressure or metered—is controlled by a switch which insures that the electrical circuits will not begin operation until the washer has filled to the proper level. Water is permitted to enter the tub until a certain weight or volume exerts a pressure on the fill switch, closing the circuit to the agitator control solenoid, and the machine begins operation.

***Fig. 13.2a** Automatic washers now have a separate cycle with cool-down, designed especially for laundering permanent press. Providing up to 9 minutes of wash time, the cycle allows wash and rinse water temperatures to be adjusted according to the laundry load, has an automatic cool-down rinse before spinning, and a special regular-speed short spin. A washer that represents the ultimate in flexibility, it has been made even more versatile with the addition of agitation and spin-speed control buttons. (Maytag)*

A recent design engineered into a new automatic washer is a jet circle spray fill-rinse system. The new pressurized fill-and-rinse system sprays jets of water down into the tub through 12 openings in a circular tube positioned around the top of the tub. The multientry water system gets clothes under water much faster and reduces the billowing and air pockets often associated with the single-inlet fill methods. Other advantages of this system are associated with the loosening of imbedded dirt and the fast cool-down provided before deep rinsing (Fig. 13.2*b*).

The amount of water desired for a load of laundry may be predetermined in both the time fill and pressure fill designs. In the time fill, the controls may be set to allow shorter intervals of time for the fill period, thus reducing the initial fill within certain limits. In general, the rise fill is always for a full tub, unless the controls are reset manually at the proper time in the cycle. In the pressure fill design, controls may be provided which will vary the amount of pressure required to activate the pressure switch, and thus control the height

***Fig. 13.2b** Pressurized fill-and-rinse system sprays jets of water down into tub through 12 openings in circular tube positioned around top of tub. (Frigidaire)*

of water level in the tub. Many washers with the pressure fill provide controls that permit preselection of varying degrees of partial fill for both the wash and rinse phases.

The total amount of water, hot and cold, used in the laundering of a load of clothes in the automatic washer may range from 25 to 57 gallons. Hot water use, on hot wash, warm rinse cycles, may range from 18 to 53 gallons. Many washers have a water level selector in order that the tub fill may be controlled as the homemaker chooses. These models may have "small," "medium," and "large" fill settings. Other models may have an infinite control.

Rinse

The effective rinse process will perform three functions: remove suspended soil, remove suspended laundry aids, and prevent minerals in the rinse water from combining with residual laundry aids to create a film.

Three types of rinses are found on automatic washers—spray,

deep, and overflow. The spray rinse introduces water into the wash basket during the spin phase, creating a flushing action that is very effective for removing soil and detergent residue, and flushing away suds. The number of spray rinses will vary with washer design but all will serve the same purpose.

A deep rinse fills the wash basket with clean water and there is a brief period of agitation or tumbling to flush the water through the fabric and to further remove the soil and detergent. The two laundry aids which are to be added in this phase of the cycle are the fabric softener and water conditioner.

An overflow rinse is incorporated in the laundering cycle of the washer with a solid tub (nonperforated). This rinse action consists of filling the washbasket to overflowing to float soil over the top of the basket. The solid tub may have small holes below the agitator to facilitate the removal of heavy, sandy soil.

Water Extraction

Water is extracted from clothing by the centrifugal force that is exerted by a spinning washbasket. Three methods have been employed to extract the moisture from the clothes—spinning alone, bottom draining followed by spinning, and a combination of bottom draining and spinning at the same time. With each method, a certain type of soil is being removed from the clothing. Among the factors that may influence the amount of moisture retained in the clothing are the following: the basic design of the washer, perforated or solid tub; the spin speed or revolutions per minute per cycle choice; the size of the washbasket; the length of the spin period; the volume of clothing in the load; the composition of the load; the amount of water in the washbasket; the type of fabric—fiber, weave, and weight; and garment construction.

The spin speed may vary from 300 rpm for delicate fabrics to 1010 rpm for regular white and colorfast materials.

Cycles

In recent years the trend has been in the direction of the use of fewer cycles. It is recognized that each new man-made fiber does not require a separate and unique washing and drying cycle. All washable fabrics can be properly washed if only three cycles are available. Separate washing procedures are needed for:

1. "Regular" loads of sturdy fabrics which do not have permanent-press characteristics; variations within the cycle should be possible to adjust for size of load and degree of soil.
2. "Permanent Press" loads—these fabrics need the extra cleaning power

of warm or hot wash water combined with a "cool-down" rinse to avoid the creation of additional wrinkles during water extraction.
3. "Delicates" need a special gentle washing procedure to protect against possible harm to sheer fabrics, to lace-trimmed garments, to knits, and to washable woolens.

The permanent-press article is no longer a special item in the laundry. It is in daily use and gets soiled just as all other fabrics. This finish may make certain soils more difficult to remove, and since some man-made fibers have an affinity for certain soils, it is necessary to use methods for removing heavy soil when laundering permanent-press loads. On more recent models of washers, the permanent press cycle offers the alternate water temperatures of hot, warm, or cold water temperatures to permit handling various types and degree of soiling. Combined with the added flexibility in the wash phase of the cycle, there must be a "cool-down" procedure before complete water extraction to reduce the formation of wrinkles from the spin. A soaking procedure is important to loosening soil and for spot and stain removal on permanent-press fabrics.

In designing the gentle wash action, there are two ways that this may be accomplished in an automatic washer. The most common method in the past was the use of the two-speed motor to obtain a slower oscillation rate of the agitator which would produce less vigorous water currents. A shorter wash time might be recommended as well as slower agitation speed.

A second method of accomplishing a gentle wash action is to simply alternate brief periods of agitation at regular speed with periods of soak. Relatively long periods of soak in a detergent solution may provide better soil removal with less "wear" than customary shorter periods of "gentle agitation."

The cycles which provide gentle wash action have been changed to the use of an alternating series of agitation periods and soak periods to provide a delicate procedure on a single speed washer. This enables the usually lower priced single-speed washer to provide for the proper laundering of delicates, washable woolens, knits, and double knits. The shortened periods of regular speed agitation, followed by relatively longer periods of soak, will remove dirt as effectively and with as little effect upon the fabric as the more delicate cycle using reduced agitation and spin speeds.

Knit Cycle

The most recent development in laundry equipment is the special knit cycle. This cycle is similar to the gentle or delicate cycle which has been on many models for years.

The laundering procedures recommended for knits depends on the fiber content, stability of the fabric, and the degree of soil. In general, hot water washing is appropriate for white or heavily soiled garments, warm water removes normal soil on medium colors, and cold water prevents the fading of bright or dark colors. Regular laundry detergent may be used, but liquid or predissolved detergents give the best results in cold water.

Color and fiber content should be given careful consideration before using a bleach. Chlorine bleaches will yellow wool knits but may be used on most cotton and synthetic fabrics.

Fabric softeners are recommended for all synthetic knits to prevent the buildup of static cling and to minimize wrinkling. Wool will not require a fabric softener because the fiber is inherently soft and absorbent.

Special consideration should be given to the fact that knits are more likely to stretch or become distorted during laundering than woven fabrics. A "full" or at least normal water level should be selected to allow garments to circulate freely with minimum abrasion and strain. Smaller loads in both the washer and dryer will also help minimize wrinkling.

Fabric type determines the cycle choice in the knit. Cotton single or double knits may be laundered as any other cotton garment if the fabric has been properly finished to resist shrinkage. Cotton knits without a stablizing finish should be blocked to shape and air dried on a flat surface. Synthetic knits are usually machine washable and dryable. Because the fibers are heat sensitive, they should be treated as permanent press. Most automatic washers have a permanent press cycle which provides special cool-down rinses and controlled spin speed. Delicate knits include loosely knit specialty knits, sheer single knits used in lingerie, or other garments labelled "hand washable." There are two methods of achieving a hand wash or delicate cycle on the automatic washer. One is to use slow or gentle agitation for a short period of time. The second method alternates brief periods of agitation with soaking time.

Wool knits may require dry cleaning, or they may be washable. Wool double knits should be dry cleaned, but wool single knits may be washed in an automatic washer. Since wool should not be exposed to much agitation, a gentle, delicate, or special wool cycle should be used in the laundering. Unless the manufacturer clearly indicates the need for drying the garment in the dryer, this agitation should be avoided, and the garment should be blocked and air dried.

The main enemy of synthetic knits in the dryer is overdrying. Such a condition leads to garment shrinkage and distortion that arises from manufacturing methods which stretch the garment or fabric. Static cling and fabric harshness are two other end results of overdrying. To avoid overdrying, three basic changes have been made in the design of the dryer—reduced energy input, limit of air temperatures, and assignment of specific areas on the control dials for the knit cycle.

A separate soak cycle has come into prominence in recent years. This in combination with the soak procedure alternating with a regular agitation for the laundering of delicates has proven to be of value especially for those consumers who have made the change to low-phosphate or nonphosphate detergents. To compensate partially for the loss in washability and less effective laundering results, a period of soaking preceding the normal wash has proven to be valuable in cleaning. Various kinds of soak cycles may be found on modern washers. Some are separate, some are integrated into the total washing procedure. The separate soak requires the user to return to the washer, as the soak cycle ends, to add laundry aids and then reset the controls in order to run through the regular wash cycle. In the integrated design, the total setting is made with a single setting of the controls for a soak and a normal wash cycle.

Among the automatic washer designs, ranging from the simplest to the most complex, the numbers of cycles may vary from two to eight or even more. With each cycle there will be a preset combination of wash time, wash and rinse water temperature, agitation and spin speeds, and a special sequence of operations to meet particular laundering needs.

An illustration of the degree of flexibility and automation designed into the automatic washer may be seen in the following typical cycle designs.

1. **Regular.** White and colorfast items, work clothes, cottons, and linens. Hot wash, warm rinse, fast agitation, and spin speed. Ten- to 15-minute wash period.
2. **Colored cottons and linens.** For bright or dark colored items. Colored cottons and linens. Warm wash and rinse. Otherwise, same as regular cycle.
3. **Delicate.** For lingerie, some wools, less sturdy wash and wear. Sheers, curtains, lace trimmed articles, loosely woven fabrics. Warm wash, cold rinse, slow agitation and spin. About 5-minute wash period.
4. **Permanent press, wash and wear, sturdy.** Man-made fibers. White or colorfast man-made fibers or blends such as: Acrilan, Arnel, Dacron, Dynel, and Orlon. Resin finished white or colorfast cottons. Wash time —4 to 10 minutes depending on the soil. Wash temperature—hot or warm. Rinse temperature—cold. Fast or medium agitation, slow spin. The permanent press cycles include a "cool down" between the wash

and first spin to prevent wrinkles; the wash water is partially replaced by cold water before the spin. Some washers also offer a delicate permanent press cycle with slow agitation and spin speed.

5. **Special items.** Wool, extra-delicate items. Warm or cold wash and rinse, extra slow agitation, fast spin, 3- to 5-minute wash period. Some of these cycles include soak period during the wash.
6. **Cold water wash.** For heat-sensitive colors and fabrics such as knits, prints, and deep colors, man-made fiber fabrics and blends. Cold wash and rinse, fast or medium agitation, slow spin, 5- to 8-minute wash period. Many washers without a separate cold water wash cycle will permit the selection of cold water on any of the other cycles.
7. **Extra rinse cycle.** Useful for diapers, heavy soil, and heavy loads where extra detergent has been added. Usually an option that can be selected with any of the other cycles. This repeats the cycle's rinse.
8. **Soak cycle.** A prewash for heavily soiled items. Water temperatures are selected, a brief agitation followed by a soak period, usually 10 minutes, then a spin. On some machines, the controls are then reset for the wash cycle; others automatically proceed to a regular wash.

Hand Wash Cycle

In the hand washable cycle the washer will fill with 7½ gallons and give low speed agitation, 34 strokes per minute. The time of the wash must be set to the minutes desired for the clothes.

Additional Features

Other features to be found on the automatic washers are as follows.

Dispensers. The **bleach dispenser** may be two types—timed or drained by gravity. Both kinds add the bleach into the circulation system where it is mixed with the wash water before contacting the clothes. The gravity drain dispenser automatically adds the bleach at a specified time after agitation has begun. This timing gives the optical brightener in the detergent time to become affixed to the fabrics before the bleach takes effect. Other dispensers are designed to dispense detergent and fabric softener. The fabric softener is added to the deep rinse water.

Counter-Balanced Lid. This lid has been balanced to close easily, come down slowly, and be easier on the fingers. This is accomplished by the addition of spring-mounted hinges.

Suds Saver System. The hot sudsy water may be pumped into a storage tub, and later returned to the washer for a second use. This water should have added hot water and detergent to perform effectively.

Pump Guard. This guard may be located under the tub just ahead of the drain pump. When the machine drains, the water must pass through the pump guard before it enters the pump. A special baffle prevents any foreign objects such as bobby pins, hair pins, needles, key chains, and others from causing damage or clogging.

Lint Removal System. Lint may be removed in one of two ways. The first method is a design to circulate wash and rinse water through a filter or screen where the lint will collect. This device may be built into the agitator; some are built into the recirculation system within the inner tub; and still others are separate attachments or accessories which must be positioned in a specific location within the inner tub and then removed to load or unload the washer. The homemaker must keep the lint removed from this filter for the most effective performance.

A second type of lint removal system is a scientifically positioned channel in which the lint collects and is then flushed down the drain with the force of the water being expelled from the wash or rinse cycle. This system is automatic.

Self-Leveling Rear Legs. The self-leveling device may compensate for a difference in floor level up to 1½ inches.

Large Capacity

The concept of large capacity has been offered by the manufacturers of many washer designs. This capacity is appropriate when washing large loads of fabrics which may properly be washed together. Capacity can be most misleading as a guide in choice of washer unless the basic principles of effective laundering are given careful consideration. Some manufacturers provide automatic water level control to properly and economically wash smaller loads even in the large tub while others offer a small, removable tub to accommodate the small loads.

Height

The washer and dryer may be elevated higher than normal counter heights in the current models. This feature needs to be questioned for those who may have used the top of the appliance as a work surface, especially the dryer for folding clothes. The homemaker may feel that this is compensated for by the convenience of loading and unloading from the front without stooping or bending.

Sanitizer

To keep the porcelain-enamel perforated wash tub clean and sanitary, a simple sanitizing procedure has been recommended by one manufacturer of laundry appliances. The consumer may add ½ cup of liquid chlorine bleach to the empty tub, select warm rinse water and minimum water, set timer to sanitize setting, and start the machine. The washer will fill, agitate, and spin the water out, insuring tub cleanliness.

Safety Features

Among the many safeguard and protective devices common to many automatic washers are the following:

1. Out-of-balance safety switch; stops operation in case an extreme out-of-balance load condition occurs.
2. Washer lid switch: stops all washer operation if lid is raised.
3. Safety brake: stops the spinning tub at the end of the cycle and in case the lid is opened during the spin.
4. Push-to-stop timer: washer can be stopped by a simple push of the timer knob.
5. Torque limiting clutch: limits acceleration of the spinning tub.
6. Internal motor protection fuse: stops washer operation if the motor should overheat.
7. Vacuum break in the water fill system: protects the fresh water supply.
8. An external grounding kit: insures a grounded appliance.
9. Electrical service cord with a third wire for grounding through the electrical service supply outlet.
10. Underwriters' Laboratories (UL) approval: appliances meet UL specifications.

Each manufacturer offers a number of models from which to choose. These washers vary in number of cycles, variety of speeds of agitation and spin, number of water temperatures, absence or addition of soak cycle, types and number of rinses in cycle, presence or absence of dispensers for washing aids, lint filter, clothes guard, type of washtub, and color of equipment. The washers also vary in levels of automaticity, flexibility, and convenience. The homemaker should acquaint herself with the various features and decide what characteristics most satisfactorily meet her needs before making a choice.

NONAUTOMATIC WASHER

Among the reasons for selecting a conventional washer are the following: lower initial cost; no installation cost; portability which

Fig. 13.3 Nonautomatic washer. (Speed Queen)

eliminates need for permanent installation—the washer may be stored in a place other than near the source of water and drain; limited service needed; and less water may be needed for total wash (Fig. 13.3). Points to consider in selection are discussed below.

Water Extraction

Water is extracted from clothes by the pressure between the rotating rolls of a wringer or by centrifugal force, which spins the water from the fabric as the cylinder or tub revolves at a fairly high rate of speed. Wringer rolls are made of soft or semisoft rubber, or the wringer may have one hard one and one soft roll. Although it is not possible to obtain the degree of compression with soft rubber rolls that is obtainable with hard rubber, there is a larger area of contact between soft rolls, which may compensate, partially at least, for the lower pressure. Soft rolls cause less injury to the clothing; buttons are not as easily removed or broken, for the roll adapts itself to irregularities in the fabric; creasing is less evident, and ironing, therefore, is easier. It is true that clothing seems to cling more readily to the soft rubber.

A wringer may have a pressure selector that permits the choice of the correct pressure for any fabric from a sheer handkerchief to a heavy blanket, or the tension on the rolls may adjust automatically. The wringer usually swings and locks into at least four and sometimes as many as 28 different positions. The tub should be well balanced so that there is no possibility of its tipping when the wringer is swung into a position away from the tub. The direction of the rotation of the rolls and of the tilt of the drainboard is reversible. The drainboard, of aluminum or steel, should have rounded corners to prevent the catching or snagging of clothing, and should be wide enough to direct the passage of the material.

The wringer post that contains the connections to the motor should be integral with the body of the washer to provide a smooth, easily cleaned surface.

The centrifugal extractor or spinner on the conventional washer is a perforated basket or cylinder which revolves in its own tub within the same housing as the washtub. The spinner may whirl while a second tub of clothes is washed, and rinsing may be done in the same basket. A spinner does not remove buttons and is a satisfactory way to extract the water from filled comforters, pillows, and blankets, leaving a soft, fluffy product. If desired, the whirling may be continued until a large percentage of the water is removed. On the other hand, the clothing must be entirely lifted from the tub into the basket and, if the material is very bulky and heavy, this is difficult; a wringer is helpful here. Clothes must be packed evenly into the basket, or considerable vibration will occur. There is no danger of catching the fingers in the spinner.

Washing Mechanism

Most nonautomatic washers have an agitator-type wash action, and may be operated at more than one speed. In this design there is a choice between normal and gentle wash action, gentle for delicate or slightly soiled articles, normal for a more vigorous action for sturdier and more heavily soiled items. The agitator should be sturdy, light, easy to remove, and of a material which resists rusting, pitting, and roughening.

Controls

The lever for controlling the agitator of the conventional washer should be located in an easy-to-reach position and clearly indicate variations in time or speed. It is an advantage to have the lever independent of the switch that starts the motor. Controls should be located so that the clothing of the user will not catch on them. Some washers have an automatic device that may be set for any selected washing time from 8 to 20 minutes and will stop the motor when the period is completed. The control may also be set at a special position for continuous action. It is an added advantage if, along with the time, the dial indicates the various materials commonly laundered, such as rayons, linens, and heavy work clothes.

Wringer controls should be within comfortable reach from any position at the washer. A centrally located safety release, easily manipulated and instantaneous in action, is essential.

The Underwriters' Laboratories have adopted new safety standards that require a virtually automatic release of the wringer rollers should the hand or clothing get caught. The standard requires that power to the wringer be controlled by the operator or that the rollers stop revolving if force is applied opposite to the direction of infeed. The open top wringer makes it easy to see and free tangles when necessary.

Motor Overload Protector

A special control should automatically protect all types of washers from overload by shutting off the motor the instant the overload occurs and so prevent blowing a fuse or damaging the mechanism.

Motor

The motor should be ⅓ HP capacity, permanently lubricated, and equipped with an on-off switch.

Wash Process

The conventional washer is manually filled to the water line, indicated on the agitator or side of the tub, with water of the correct temperature for fabrics in the load. The tub is filled with the least expenditure of energy by a Y-shaped hose attached to the mixing faucet; if the hot- and cold-water faucets are separate, use a mixing hose with one arm of the hose fastened to each faucet. If the wash is large, it may be necessary to change the water between loads to prevent redepositing of the soil on the materials.

Utilities

The amount of water the conventional washer uses is important if water supply is limited, difficult to pump, expensive to buy, and if conditioning or heating water is inconvenient or expensive. Soft water can be conditioned and heated manually. The washer requires a 115-volt, 60-cycle circuit, fused with a 15-ampere fuse. Provisions are required for draining the water. Draining is accomplished by gravity (water flows from drain hose into floor drain) or by means of a water pump. The pump operates as long as the motor is turned on, by opening a valve in the pump.

Lint Filter

A filter collects lint from the water as it passes through. It should be easy to remove in order to rinse clean.

SPECIAL WASHER DESIGNS

Combination Washer-Dryers

The features of a separate washer and dryer are combined in a single unit in this washer design. A study of the features of both the washer and dryer will enable the user better to understand the merits of this design. Two additional features are unique to this combination. It takes less space than the two separate appliances and uses less water. If space is a concern, a combination washer-dryer is a desirable alternative.

The convenience aspects of this design are also worthy of special consideration. It is more convenient because it washes and dries in a continuous operation, eliminating the need for transferring wet clothing from a washer into a dryer. The main disadvantage is that only one process can be accomplished at one time.

Portable-Permanent Convertible Washer

This design may be used either as a portable or as a permanent installation machine (Fig. 13.4). Connected as a permanent machine it is an automatic equipped with controls for selecting from five water temperatures, three water levels, two washing speeds, and four washing cycles. When the hose is connected as a portable automatic washer, it provides every feature of the permanent installation except that wash and rinse water temperatures are manually controlled at the faucet to which the washer is connected. It is equipped with a convenient handle for moving: hose holders, power cord holder, and single hose, snap-on faucet attachements.

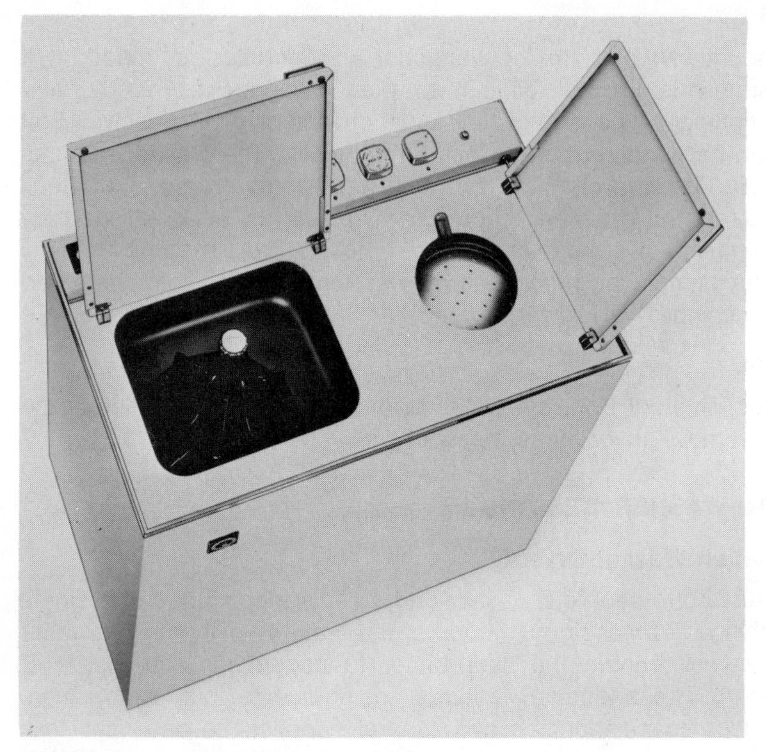

Fig. 13.4 Portable washer. (Speed Queen)

Compact Laundry

As the population in this country moves more and more to living environments of the design of the apartment, condominium, and mobile home, the manufacturer of laundry equipment has become more highly sensitized to the need for smaller and smaller laundry appliances. Indicative of the nature of the shift in housing starts is the decrease in percent of single family starts and the increase in percent of multifamily and mobile homes.

In designing to meet this need, two basic approaches have been used by the manufacturers. First the twin tub washers and matching dryers. This design has two compartments: one for washing and rinsing and one for the spinout. Controls are provided for each function. The washing process is basically to wash the clothes in one compartment, transfer the clothes into the second chamber for spinning, drain out the wash water, transfer the clothes back into the wash area for rinsing, and then a final transfer for final spinout.

Some of the special features to be found on the portable washers are settings which control time of wash for delicate, durable press, normal, and heavy soil, and a water and suds saver. In one design the pump has an off-on control for recirculation of wash water or, by connecting a reversed fill hose, draining of water; the pump is driven by transmission pulley. The timer control must be on for the pump to operate. The washer has a nylon impeller and a recirculation capacity of approximately 3.5 gallons per minute.

The second type is the automatic washer and matching dryer. These appliances operate in much the same way as the full size automatic. They are normally smaller in capacity and physical dimensions (21 to 24 inches wide). The washing process involves the loading of the clothes into the single wash basket, where they remain for the normal cycle of wash, spin, rinse, and spin. As in the case of the twin tubs, they normally come equipped with casters and unicouples.

AIDS IN THE LAUNDERING PROCESS

WATER QUALITY AND QUANTITY

In the conditioning of water for textile use the removal of impurities is important in order to: decrease corrosion of materials used to form the interiors of water-conditioning equipment, softener and heater, washers, and plumbing pipes; eliminate color, odor, and suspended substances; and improve the ability of water to interact in the wash solution.

In addition to the hardness minerals in the water there are those brought into the wash water by the soil on dirty clothes. A load of clothes may add several grains per gallon to the wash water; water of one grain hardness could be increased to three to seven grains per gallon. The result may be that no load of dirty clothes is washed in the absence of hardness minerals.

More specifically, water characteristics that are pertinent to laundering are total hardness, iron and manganese content, turbidity, color, total dissolved solids, alkalinity, and temperature.

Hardness

The hardness of water will vary from one area to another and from one season to another. The types of water will depend on the geology, biology, and climate in the geographic areas. According to many who are concerned with the quality of water found in the natural state, the following general ideas have been advanced:

1. In areas of the country where water is naturally low in hardness minerals, the water is also low in alkalinity, silica, dissolved solids, conductivity, and pH. This type of water may be found in the New England States, and the Middle and South Atlantic seaboard.
2. Other areas in the United States have much water that may be characterized as high in hardness minerals, alkalinity, dissolved solids, and pH.

The hardness of the water is proportional to its content of calcium and magnesium salts. These salts in order of their relative average abundance in various supplies are: bicarbonates, sulfates, chlorides, and nitrates. Calcium and magnesium soaps are formed when hard water is combined with soap or synthetic detergent. These precipitates are gummy, gelatinous, and sticky substances and are capable of entrapping loosened soil. Once this has become deposited on the fabric, it is very difficult to remove.

The relationship between water hardness and laundry performance is: calcium and magnesium interfere in the overall washing process when using detergents or soaps; action of hardness minerals affects soil removal, soil redeposition, and rinsability; condensed polyphosphate builders in sufficient concentration can effectively remove minerals by the sequestering action.

Water may be classified into the following degrees of hardness.

Grains per U.S. Gallon	*Degree of Hardness*
0 to 3	Soft
3 to 6	Moderately hard
6 to 12	Hard
12 to 30	Very hard

Measurement is by weight and is sometimes expressed in parts per million (ppm). One grain per gallon equals 17.1 ppm.

Hard water tends to damage the fibers and make clothing harsh to the touch, and its continuous use may shorten the life of clothes as much as 40%; therefore it should be softened either mechanically or with packaged softener.

If possible, soften the water in the entire household system, thereby eliminating deposits of scale in the pipes and in the hotwater tank. A common method of system softening is based on the principle of ion exchange, in which a tank, similar in appearance to a small hot-water tank, is used. The tank contains a bed of zeolite, synthetic resin beads, or a combination of the two, and one or more beds of filtering gravel. Hard water contains salts of calcium and magnesium, usually in the form of sulphates. The incoming water is sprayed over the top of the ion-exchange material and filters down through it, replacing the calcium and magnesium with sodium from the sodium silicate of the zeolite; the calcium and magnesium precipitate out, leaving the water soft.

When the sodium is exhausted, a fresh supply of common salt (sodium chloride) is added and the zeolite is regenerated. After the system is installed, the only operating cost is that of the sodium chloride, which is relatively inexpensive.

Automatic regenerating of the tank is now possible. An additional tank for dry salt is provided beside the ion-exchange tank, or the salt tank may be incorporated within the same container as the softener. Regeneration is electrically controlled by a time switch that determines the frequency of regeneration needed.

In certain communities, water softening service is available on a monthly basis. The water-softening tank is replaced each month with a new one, or more or less frequently, depending on the hardness of the water and on the water consumption habits of the family. This arrangement allows a small tank, occupying very little space, to be used. The homemaker pays for the service monthly, just as she pays for her other utilities, and is relieved of all maintenance responsibility. The household system is recommended in areas where water hardness exceeds 8 grams. It is an advantage to soften both hot and cold water lines, but the softening system should include a by-pass for water used in toilets and for the outside faucets to which hoses are connected for lawn and garden sprinkling.

Packaged softeners may also be used. They are of two kinds; those that form a precipitate, and those that soften by sequestration, that is, "tie up" the hard-water minerals. The precipitating softeners are less expensive, but the water is cloudy and, unless the precipitate is removed, either by skimming it from the surface of the water or by drawing it off from the bottom of the tub, it may deposit in the meshes of the fabric and cause grayness. The sequestering softeners, sometimes called water normalizers, of which Calgon is an example, are preferable.

If the rinse water is hard, the nonprecipitating water softener should also be added to it to aid in more effective rinsing. The softener will remove previously formed washing film from fabrics, prevent minerals introduced in rinse water from forming a second film, and aid in removing washing ingredients and suspended soil, reducing alkalinity, and removing lint.

They may also be effectively used to remove lime-soap deposits from fabrics that have been washed in hard water. This process is defined as reconditioning. The clothes are washed in a full tub of water on the regular cycle with one cup of a nonprecipitating water softener. No other laundry aids are added to the water. The conditioning process should be repeated until suds no longer form.

In areas where the water is extremely hard, the nonprecipitating softeners may fail to give satisfactory results; in fact, a graduate student reported that home water in one area was so hard that no softener was effective.

Iron and Manganese Content

These two minerals are objectionable in the laundry solution because of their staining tendencies. Continual washing in water containing 0.3 ppm of iron will produce staining of white goods. Even at a lower concentration of 0.1 ppm, iron may cause poor color or poor whiteness maintenance in laundered clothes. The presence of iron in the water should be given careful consideration, since the iron may combine with additives or finishes on fabrics to produce adverse effects. Chlorine bleaches should not be used in water containing iron.

Iron in the water may be derived from two sources. It may be in the water itself as it comes in contact with iron-bearing solids in the ground. When this water containing dissolved iron is pumped out of a well or from a city pumping station, the water appears to be free of any iron and is clear. The iron exists in a completely soluble, inorganic, or dissolved ferrous form. After the water stands in the open for awhile, the iron combines with oxygen from the air and becomes a ferric or insoluble form and settles out as a yellowish or reddish brown product. This same process occurs when the water is chlorinated. Thus the addition of chlorine bleach to such water causes oxidation which, in turn, produces yellow water and staining of fabrics in the wash. Heat will also accelerate this change.

Iron may also be found in water in an organic state. This means that the water actually contains organic compounds and the iron is linked in some way within the organic molecular structure. Iron in combination with the proper concentration of soap, water softeners, and detergents may be held in suspension during the wash, but settle out onto the clothes in the rinse when the concentration of these laundry aids is lowered, another reason for the yellowing of clothes.

Turbidity

Turbidity in water is a result of the presence of clay, silt, and finally divided organic matter. Excessive quantities of these materials in the water will increase the soil load and may produce graying effects similar to those produced by soil redeposition.

Color

Color is normally the result of the extraction of decaying vegetable matter—roots, stems, and leaves. Washing in highly colored water will give rise to staining. These stains will usually respond to bleaching, but the problem of whiteness maintenance is more difficult in a highly colored water.

Total Dissolved Solids

High quantities of dissolved mineral matter in water increase the tendency for soil redeposition during washing and can give effects similar to hardness constituents. Also, waters of high mineral content usually have high electrical conductivity. This may cause problems of corrosion because of galvanic action when dissimilar metals are brought into contact.

Alkalinity

The alkalinity of a water supply refers to its ability to neutralize acids. Alkalinity in natural water is due to the concentration of bicarbonate and carbonate ions. Alkalinity of water is generally of little significance in the wash water, since the alkalinity of the detergent is usually much higher. Where the amount of detergent is very low, rinse water of excessively high alkalinity can cause problems by increasing the tendency of cotton and other cellulosic fabrics to yellow during tumble drying or ironing.

pH

A pH determination can be expressed either in terms of hydrogen ion concentration (pH) or in terms of hydroxyl ion (pOH) concentration. A very low pH indicates a high concentration of hydrogen ions and considerable acid intensity. A very high pH indicates a substantial concentration of hydroxyl ions and considerable alkaline intensity. In either case the question is how much of either is required to neutralize the other.

DETERGENTS

Many processes have been used in cleaning of fabrics. The most common of these, detergency, aids in the separating of the soil from the fabric by wetting the soil, wetting the fabric, separating the soil from the fabric, and holding the soil in suspension for removal by extraction and rinsing.

The detergent further reduces the surface tension of the water, allowing it to spread out in a film and penetrate the fabric. The detergent solution removes the dirt in the following way. Inorganic soils, such as dust and mud, are removed by adsorption, the adhesion of the dirt particles to the surface of the foam. Greasy materials mixed with dirt are emulsified by the solution, which holds the dirt in suspension until it is removed by the mechanical agitation of the washer. Dust and mud are usually fairly easy to remove. Grease, other food spots and stains, body excretions of oil or waste epithelial cells, and bacteria are organic soils and often need special prewashing treatment.

Soap is the product resulting from the action of an alkali on animal or vegetable fats. A single fat is rarely used but several are combined, the kinds varying with their availability and the type of soap to be produced. Soaps are classified as light duty or heavy duty. The light-duty soap is pure, with no free alkali, mild in action, recommended for washing of sheer, delicate fabrics. Heavy-duty soaps have builders added. Builders are alkaline substances which aid in softening hard water and promote the cleaning action by combining with the grease on badly soiled clothes. Some investigators have found that the use of soap with soft water is the most effective means of removing soil from materials.

Made at first for use in the textile industry, synthetic detergents or syndets—also frequently referred to simply as detergents—are now finding almost universal application in the home. They are readily soluble in water of low temperature, even in most hard waters, unless extremely hard, and form a neutral or slightly acid solution with no precipitate or scum.

For the most part, synthetic detergents are sulphated fatty alcohols and esters, although the exact formula is usually not made public. They are crystalline bodies, commonly marketed in the form of flakes, beads, powders, and now often in liquid form.

Detergents are classified similarly to soaps as light duty, used for washing delicate fabrics and animal fibers; heavy duty, containing builders for increased cleaning ability. The heavy-duty syndets may be further divided into high- and low-sudsing products. The low-sudsing types are recommended for use in the tumbler models.

Composition

Present-day synthetic detergents have been chemically tailored to meet a wide variety of cleaning needs. Among the ingredients combined to perform this function are the following.

1. **Surfactant:** functions as a wetting agent, loosens soil, makes soil more soluble, and aids in the suspension of soil.
2. **Builder:** softens water by sequestering hardness minerals, helps to disperse particulate soil, aids in surfactant efficiency, helps to suspend soil. Silicates or mildly alkaline complex phosphates.
3. **Fluorescent fabric brighteners:** absorbed by fabric, converts invisible ultraviolet light to visible blue light. Additional blue light counters yellowness and makes fabric appear whiter.
4. **Suds stabilizer or suppressor:** provides stability to foam. Stabilizer makes foam last longer, foam aids sudsing property of surfactant.
5. **Anti-redeposition agent:** helps to hold soil in suspension and prevents the soil from redepositing on the fabric. CMC (carboxymethyl cellulose).
6. **Corrosion and tarnish inhibitor:** protects metals.
7. **Bleaches:** oxygen type slower reacting but effective for all kinds of fabrics.

Other ingredients that may be found in some of the detergents are fabric softener, perfumes, dyes for coloring, antibacterial agents, enzymes, and water softeners.

Among the many factors which should be considered in the selection of a specific detergent are the fiber and fabric, the amount and nature of the soil, the composition of the detergent, water quality, and the type of washer. The amount of detergent to be used in a load of laundry should be related to the quality and quantity of water, quantity of clothes, amount and kind of soil, type of detergent, and washer design (Table 13.2).

One of the components of the synthetic detergent which has commanded major interest in recent years has been the phosphates. An understanding of the contribution made by the phosphate to the detergent may be seen from the functional point of view:

1. They increase the efficiency of the surface active agent.
2. They reduce redeposition of dirt by keeping the dirt particles in suspension.

Table 13.2
Laundry Product as Related to Water Hardness (Soap and Detergent Association)

Washing Product	Soft Water (0.0–3.5 GPG) Amount[a]	Medium-Hard Water (3.6–9.5 GPG) Amount[a]	Hard Water (9.6 GPG and over) Amount[a]
All-purpose detergent	1–1¼ cups	1¼–1¾ cups	1¾–2½ cups
Normal suds All-purpose soap	1–1½ cups	1½–2½ cups	2½–3½ cups

[a]Recommended amount of laundry product for 6–8 pound load of clothes of average soil in top-loading automatic or wringer washers (17 gallons water).
Increased water volume, heavy soil, or additional loads in wringer washers would require more product than the amounts shown above for both detergent and soap.

3. They furnish necessary alkalinity to insure proper cleaning.
4. They provide resistance against changes in alkalinity due to water supply or wash load which impair cleaning performance.
5. They emulsify oily and greasy soil.
6. They soften water by sequestering (tying up objectionable minerals).
7. They contribute to the reduction of the level of germs on clothes, thereby reducing cross infection.

From a water quality point of view, phosphates

1. Break down satisfactorily through hydrolysis in sewage treatment plants and surface waters and thus lose their sequestering ability.
2. Do not interfere with other aspects of waste treatment operations.
3. Can be effectively removed in waste treatment plants.
4. Are a material which is part of the natural environment and with which there are years of experience.

From the safety point of view, phosphates are

1. Nontoxic.
2. Safe for colors.
3. Safe for all fibers and fabrics—particularly important today with many synthetic fibers available.
4. Safe for use in washers—they are noncorrosive and nonflammable.

One or more of the following ingredients may be found in the synthetic detergent or soap:

Oxygen bleach to aid in the removal of stains and soil.
Borax to aid in the removal of stains and to impart a freshness.
Bacteriostats to inhibit the growth of some bacteria.
Bluings to impart a blue tint on fabrics and thus tend to counteract yellowing.
Dyes to impart color to the product.
Enzymes to aid in the removal of stains and oils—especially proteins.

Nonphosphate Detergents

The nonphosphate detergent will require some special consideration on the part of the consumer if the same cleaning results are to be achieved as with the phosphate type. The manufacturers' instructions should be closely followed as the package will be marked differently than the phosphate product. The basic guide of dissolving the detergent in the water before adding the clothing is one way to protect the fiber.

Spot removal is perhaps the most difficult problem resulting from the inability to use phosphates, and no easy answer has been found in those areas where the consumers have been using these products. Though time consuming, as well as expensive, some success may be realized with the use of water softeners, vinegar, bleaches, and commercial spot removers.

The heavy use of bleach should be avoided as a means of compensating for the less effective cleaning action of the nonphosphate detergent. Fabric life will be shortened and produce a substantial increase in lint which, if pumped into drains, can contribute to plumbing problems. Fabric stiffness may be improved by the use of vinegar, fabric softener, and water softener.

Flame retardance may be inhibited as the residue from the nonphosphate detergent becomes deposited on the fiber. This property may be reactivated, after laundering, by soaking the item in a vinegar solution or liquid detergent for about one hour. Such soaks are also effective for removal of the fatty deposits that may be left on the fiber from soaps. Color fastness should be carefully considered, since vinegar may cause streaking.

Vinegar rinses may be effective in inhibiting residue buildup on the washer resulting from the use of nonphosphate detergents. The vinegar may be added to the machine and run through a wash cycle without a wash load. Caution should be exercised, however, because vinegar will attack the porcelain and should not be used in large amounts. Washers most susceptible to damage from heavy vinegar use are older machines since many may not be as acid resistant as newer models.

As the proportion of phosphates is lowered in the detergents, the consumer should recognize the contributions made to these products by this ingredient. The phosphate has usually been replaced by sodium carbonate, silicate or carbonate-silicate mixtures in powdered detergents and by sodium citrate in liquid detergents. All of these ingredients—phosphate, sodium carbonate, silicate, and citrate—serve to soften water, either by tieing up the hardness minerals or by precipitation of these minerals. In hard-water areas of the country, problems may occur from a residue of hardness minerals which precipitate out on the fabrics and in washers when the carbonate-based no-phosphate powder is used in the laundry. This residue may stiffen fabrics, cause flame-retardant finishes to become ineffective, and can cause appliance problems.

Some guides which may be used to compensate for the less effective cleaning resulting from the removal of phosphates from the detergents are to maximize the efforts in pretreating, soaking for longer periods of time, delay the adding of the clothes until the detergent is combined with the wash water, and use the most appropriate water temperature for the fabrics being washed (white and colorfast: 140 to 160°; noncolorfast: 110 to 120°; permanent press: hot or warm wash and cold rinse). Hot water minimizes hardness residue particles.

Enzymes

To assist further in the removal of soils from fibers, enzymes or proteins have been introduced as a cleaning aid. Dirt and stains normally found on soiled clothing consist of a complex mixture of organic and inorganic matters bonded to fabric by small amounts of body proteins and perspiration. The nature of the soil will differ depending on the geographical area, climate, and occupation of the wearer. In general, however, the binding agent in the soil appears to be of protein origin and, therefore, is susceptible to the action of an enzyme.

Soil that is commonly found on clothing and household textiles is typically protein-bound soil. The inorganic components of clothing soil are usually colored and generally based on clay from airborne earth-derived soil. Clay alone is readily removed by effective laundering. The combination of the inorganic soil with the protein binder produces a more stubborn soil that is not readily removed in laundering. The strong adherence of this combination is further aggravated by the constant abrasive action of normal body movement, especially at the collar and cuffs.

Fabrics also acquire stain. If the protein-containing stains can be broken down into components of a much smaller size, their function as binding agents may be eliminated and the dirt may be more easily removed by the water-detergent solution. Examples of some protein-containing stains are blood, gravy, excrements, fruit juices, and other food residues. Enzymes in general do not have an effect on stains such as those produced by urine, tea, and coffee unless such stains are bound to the fiber by a protein, as in the case of coffee or tea with milk.

As the enzymes are dissolved in the water-detergent solution, the enzyme molecules attach themselves to the protein of the stain or soil. An enzyme-protein complex forms, allowing reactive groups of atoms in the enzyme molecule to catalyze or speed up the breakdown of the protein molecule to smaller particles. These are then in a form to be more easily removed from the clothing by the detergent-water solution.

The nature of the enzyme reaction may be seen in the following equation:

Enzyme (E) + (substrate) soil or stain (S) acted on by enzyme =
enzyme − substrate complex (ES) =
product (P) + enzyme (E)

The enzymes are considered to be highly specific in catalytic action and in the laundering process; their effectiveness is specifically directed to the stains and soils, not toward cellulose, nylon, polyester, wool, or other fiber types.

Among the conditions which affect the action of the enzyme are:

1. Concentration of enzyme in the solution.
2. Concentration of soil or stain on the fabric.
3. pH of the solution.
4. Temperature of the solution.

Generally speaking, there is an optimum relationship between the concentration of the enzyme and the substrate-soil or stain for maximum activity. Likewise, enzymes function optimally at a particular pH and temperature. Each of these factors must be considered as the enzyme is combined with the other ingredients in the detergent or as the enzyme is used alone in a presoak product. All ingredients in the detergent or soak solution must be compatible with the enzyme for effective action.

Active chlorine compounds inhibit the action of enzymes and should not be used in laundry aids containing them. If chlorine bleach is to be used, this should be used after the completion of the enzyme detergent wash cycle. In contrast to this response to chlorine bleach, the enzymes have been shown to be quite compatible with the sodium perborate type bleaches.

Bleaches

Bleaching is the process of improving whiteness of textiles by oxidation or reduction of the coloring matter. The factors affecting selection of a decolorant are:

1. Color of the textile, degree of color removal desired, permanency of the bleaching effect, requirements for deodorizing, use as disinfectant, and economy.
2. Amount and kind of finishes on fabrics.
3. Form of textile as fiber, yarn, or fabric.
4. Catalyst or materials to be combined with bleach to speed chemical reaction.
5. Temperature of the solution.
6. Duration—time bleach may be in contact with fabric.
7. Reaction between fiber and bleach solution.
8. Rates of decoloration and depolymerization of fiber.
9. Amount of bleach to be used in the solution.
10. Oxidation potential.

Bleaching is rarely necessary if clothes are always washed in soft water as hot as 160°F. Of course, there may be certain disadvantages in so hot a home water supply. In hard-water areas scale will form in the water heater and connecting pipes and may clog them; in some soft-water regions this hot water may cause more rapid corrosion of pipes and heater. And there is always the danger of children turning on a faucet and being burned.

Since soft, hot water is not always available, bleaching is widely practiced. Bleaches tend to break up the dirt by chemical action, making it more easily removed.

Bleaches commonly used in the home are of two types: chlorine and oxygen. Both work through an oxidizing process whereby the chemical release of oxygen activates the bleaching process. The two oxygen producing products are sodium perborate and hydrogen peroxide. Chlorine bleach may be found in both the liquid and dry form.

The amount of bleach to be used should be determined by the amount of water. One tablespoon of bleach for each gallon of water has been used as a guide. An increased amount will be necessary for the large capacity washer and less when a low level of water is used for a small load.

Chlorine bleach should be used in measured amounts, in dilute form, and may be added before or after the machine is filled with clothes. If the chlorine is put into the washer first, the bleach should be diluted in at least one quart of water and added after the machine has started to agitate. Chlorine should never be applied directly to fabrics. This procedure will cause color change and fabric damage which may not be evident until several washings later. Holes, tears, and discoloration are indications of possible bleach damage.

Better laundering results from the addition of the bleach about half way through the wash period. The delay enables the fluorescent dyes to work with maximum efficiency and allows enough time for the bleach to be effective. The suds formed will also serve as a buffer and protect fibers. This method is also recommended when an enzyme detergent is used since chlorine bleach inactivates the enzyme. Delayed bleaching gives the enzyme time to work before the bleach is added to the solution.

The washer design which has the automatic bleach dispenser provides both automatic dilution and delayed addition of the bleach. In addition, there is an assurance of greater safety for the fabrics.

The quality of the water should be given careful consideration when using chlorine bleach. Chlorine should not be added to water which has a high quantity of iron. Brown spots will appear on the clothing when this has been done.

In the liquid form, chlorine bleach can be used on white and most colorfast fabrics as well as all man-made fibers with the exception of spandex. Permanent press fabrics and soil release finishes, white and color, are also compatible with chlorine bleach. Fibers which cannot be bleached with liquid chlorine include wool, mohair, and silk as well as the spandex. Some flame retardant fabrics are also included in this category.

Chlorine bleach in the dry form performs equally as well as the liquid form. Dry bleach products may be added directly to the wash water at the beginning of the cycle. This is possible since this form contains its own brighteners and delayed bleaching action, which begins as the powder is dissolved.

The hypochlorite bleach should be used in the soak cycle with a detergent, or in the wash cycle to allow for adequate rinsing. Do not use chlorine bleaches in cold water or without a detergent which protects the fabrics by its buffering action.

Sodium or calcium perborate is a powder, easy to handle, and comparatively safe to use on colorfast materials, on wool, rayon, and nylon, and on certain fabrics with special water-repellent and crease-resistant finishes. Excessive quantities are not likely to cause damage, since the bleaching action depends entirely on oxygen freed from the perborate. The liquid, hydrogen peroxide, is a good bleach for wools and silks, but has a tendency to deteriorate with time.

Oxygen type bleaches are produced in liquid and powder form. These are light duty bleaches and may be used safely on all washable fibers and most colors. The oxygen type bleaches are most effective on fresh stains and slightly soiled clothing. Hotter water or a soak period gives better results.

Fabric Softeners

With the appearance of synthetic fibers, a new problem of fabric care arose for the consumer. The synthetic fiber has little natural moisture and thus needs extra moisture to prevent a build-up of static electricity. The fabric softener has been designed to serve this purpose.

Composition and Function. The basic ingredient of the fabric softener is a dialkyl quaternary ammonium salt in an alcohol and water solution. Dyes and perfumes are also added for aesthetic purposes. The surface of each fiber and each yarn is coated with a fatty derivative of a quaternary. These fatty materials are excellent lubricants and permit the closely woven fabrics to slip easily over each other. The soft, wrinkle-free clothes obtained by the use of a fabric softener are the results of this type of internal lubrication.

As an antistatic material, fabric softeners are most effective. When two materials, such as wool and nylon, are rubbed together, a charge of static electricity is built up on the contacting surfaces. Natural fibers are water loving and always contain a certain amount of moisture. This moisture forms a path on which the static charge can travel. Thus, on natural fibers the static electricity is rapidly carried away and dissipated. This is not the case with the synthetic fibers.

Benefits of the System. In addition to preventing a build-up of static electricity on synthetic fibers, other benefits which may be attributed to the fabric softener are:

1. Imparts soft quality and antiwrinkling properties.
2. Reduces friction and abrasion, thus lengthening the life of the fabric.
3. Improves tear strength.
4. Increases ability of fabric to resist redeposition of soil in wash, thus assisting in whiteness retention.
5. Prevents matting of woolens, pile fabrics, and knits.
6. Prevents snagging or picking of items such as hose.
7. Decreases drying time.

Factors to Consider in Use. The fabric softener is described as substantive, that is, it is attracted to the fabric and stays on it. This attraction is furthered, since the softener has a positive charge and the fabric a negative one. Thus the amount of softener for each load should be considered in relation to the volume of fabric in the washer. Resin finishes on fabric have been found to be compatible with the fabric softener.

Fabric softeners have been found to be incompatible with some of the other laundry aids, especially the synthetic detergents. Thus the fabric softener should be used in the final rinse after other laundry aids (synthetic detergent and water softener) have been removed from the fabric. For those who use the fabric softener dispenser on the automatic washer, this may present a problem. Should the fabric softener become dispensed during the time that detergent or water softener is in the water, dark stains may appear on the fabrics. The soil in the path of the softener may become attached to the fabric just as the softener does.

Another disadvantage in the use of the fabric softener occurs when too much softener has been used. The absorbency of the softener is reduced, which can be particularly undesirable for babies' diapers and towels. To handle this problem, the softener should not be used for several following washes to remove the excess. One recommendation has been to use the softener every fourth wash for best results with diapers or towels.

Other adverse effects on fabrics may be a general yellowing and dulled colors. Softeners may become lumpy during storage and should be strained before use in an automatic dispenser. The small lumps may clog the dispenser lines to the tub.

When a separate bluing is used, the fabric softener should be used after the bluing rinse. If starch is used, the fabric softener should be used before starching. The softener may be used effectively in waters of various hardness and temperatures.

The most uniform results from the fabric softener are realized when the softener is well distributed in the final rinse water. The softener should not be poured directly on to the fabrics.

PROBLEM AREAS IN CARE OF CLOTHING

Yellowing

Causes. *Incomplete soil removal.* Poor removal of oily soils. Insufficient detergent, low wash temperature, lack of bleach. May result in yellowing or scorching in the dryer due to higher than normal sensitivity to heat.

Hard water yellowing. Incomplete soil removal. Calcium in hardwater combines with substances in detergent and in the soil to form insoluble greasy curds. Causes retention of soil and body oil and acts to prevent removal. Fabric greasy, nonabsorptive, disagreeable odor.

Minerals in water. Iron or manganese content of water. Concentration of detergent will aggravate degree of yellowing.

Chlorine retention. Resin finishes for cotton and rayons retain chlorine when bleach is used. Remove with reducing agents as a dye remover.

Synthetic fabric yellowing. Pick-up of body oils and dye. Nylon and Dacron particularly susceptible to this yellowing.

Fabric softener yellowing. Undiluted fabric softener contacts garment being laundered. Normal rinse will not remove such high concentration. When garment is dried, especially in tumble dryer, yellowed area may result.

Swamp water yellowing. Decaying vegetation—highly colored due to organic compounds.

Brown line effect. Chemical damage to cotton fabric when a water surface remains at one point on the cotton fabric for extended period of time. May occur when load of wet clothes is allowed to stand in pile for some time. Susceptible to scorching or heat yellowing.

Tendency of certain fibers to yellow with age. Some manmade fibers, notably nylon and polyester, will tend to yellow with age, particularly if not laundered properly for removal of oily soils.

Graying

Causes. *Overloading the washer.* When the washer is too full to permit free circulation of the load, some articles may not receive proper flexing action, and washing efficiency will be impaired.

Insufficient hot water. If water temperature is too low, efficiency of the detergent will be lessened and full cleaning power will not be available.

Improper soaking. Soil released from a fabric in soaking will be redeposited in a chemically altered form unless it is held in suspension by sufficient detergent. This can occur either when no detergent is used or when too little is used.

Excessive wash time. Detergent is used up in several ways—by combatting water hardness, taking soil into suspension, and important to this problem, holding it there. If the wash time is too long, the detergent is unable to hold soil in suspension and consequently redeposition occurs.

Insufficient detergent. Far more laundering problems are created by the use of too little detergent than too much. It is the chief source of tattle-tale gray.

Improper use of suds-return system. The cleaning power of the washwater is reduced with each load. May need to add additional hot water and detergent before reuse.

Inadequate rinse.

Excessive Lint

Causes. *Overloading.* In addition to causing undue abrasion of the fabrics, overloading inhibits free movement in the washbasket and the flushing away of natural lint from fabrics.

Improper sorting. When items giving up lint are laundered together with items that naturally attract lint, the latter will become unsightly from lint.

Insufficient detergent. Use of too little detergent can reduce wetting action and lint will not readily be flushed from the fabric surface.

Excessive Wearing or Tearing

Causes. *Improper use of liquid chlorine bleach.* If concentrated bleach, just as it comes from the bottle, is allowed to come in contact with cotton and other fabrics, the fiber will be weakened seriously and any pressure will result in tearing.

Failure to mend rips and tears before laundering. Unrepaired rips can be made larger by the flexing action during the wash cycle.

Snags. Clothes chutes, clothes baskets, or fasteners on other garments can sometimes snag garments. Inherent weakness of fabric, with lower thread count, is bound to have a shorter wear life. Treating fabrics with a resin to provide wash-and-wear characteristics also shortens wear life.

Shrinkage

Causes. *Shrinkage of wool and hair fibers.* These fabrics, when untreated, have a natural tendency to felt or mat when wet, causing shrinkage. During laundering, the problem is aggravated by excess agitation and heat.

Shrinkage of other fabrics. Those garments not treated during manufacture to be shrink resistant may lose their shape and fit during laundering.

High agitation and spin speeds for delicate fabrics or weak construction.

DRYERS

To dry today's fabrics effectively clothes dryers have increasingly complicated designs. The principle of drying employs heat, motion, and air. As dry air from the room is drawn into the dryer cabinet through the various vents, a blower system forces the air through a heat duct arrangement. At this point the air is heated by either a gas burner assembly or an electric heating element. The heated air then passes into the drum interior. As the air passes over and through the clothing it absorbs moisture and soon approaches saturation. This action is called evaporation. The more heat applied, the faster the rate of evaporation and the faster the drying action. The amount of heat applied should be considered with regard to the types of fabrics.

The blower, which operates continuously while the drum is revolving, forces the moisture-laden air out of the dryer through the lint screen and vent. The principle of operation of the dryer involves two basic physical principles. The first is that heated air holds more moisture than cool air. Air at 32°F can hold only about 2 grains of moisture per cubic foot, while air at 70°F can hold about 8 grains of moisture. Air at 160°F holds 18.6 grains of moisture per cubic foot.

A second principle concerns the movement of air. Air in motion can absorb moisture at a more rapid rate than still air. By heating the air and then putting it in motion, average temperature 165°F and moving at 125 to 175 cfm, it is possible to hasten the process of drying to the point where an efficient unit can remove approximately 10 pounds of water (over 1 gallon) from a load of clothes in one hour.

A third basic principle in the drying of clothes is the motion provided by the movement of the drum. Three baffles, found on the interior walls of the drum, lift, turn, and tumble the clothing as the drum rotates and provides even distribution of heated air in the drum. This action constantly exposes the many surfaces of the clothing to the drying action of the heated air.

Process of Drying

Automatic drying is based on the principle that temperature rises more quickly as the rate of evaporation decreases. When a dryer starts to operate, the temperature of the exhaust air rises very rapidly. At the point where the refrigeration effect of evaporating moisture from the clothes balances the input of heat, the temperature levels off. As the clothes are dried the temperature of the exhaust air rises abruptly. At this point a temperature-sensing device is activated and the source of heat is cut off and the temperature subsides. The blower motor continues to run until a low point in temperature is reached where the temperature-sensing device again closes the circuit to the heater and causes the temperature to start rising. Each succeeding heat-on cycle will be shorter and shorter. The dryer continues to cycle and, again, the heat in the exhaust air flow begins to climb, each time more rapidly as the clothes have less and less moisture. This process continues in clock-timed dryers until the time reaches the end of the set period, or until the machine is shut off manually.

Among the many factors which may have an effect on drying time are operating voltage; wattage of heating element; load type—size, composition, and fabric; moisture content of clothing; cycle choice; and presence of lint in the screen.

For the most efficient use of the dryer, clothes of a similar weight and texture should be dried in the same load. When the weight and texture vary among the items in the load, the lighter weight items will be overdried in the time required for drying the heaviest items. The problems of overdrying need to be given consideration when loading the dryer. Under normal conditions, all fabrics contain some moisture; approximately 2% to 5% of the total weight is normal for all fabrics. This condition results in a softer fabric.

Cycles

The automatic cycles provide the correct temperature and timing or moisture-level control for the type of load to be dried. Controls may be set for the load type—heavy fabric, delicate, damp-dry, permanent press, or wash and wear. The setting for wash and wear or permanent press finished fabrics is somewhat different from the other settings. These fabrics need to be raised to a predetermined temperature to relax the fiber. The dryer will operate until the fabrics are dry and have reached the temperature range necessary for the fabric to become relaxed and wrinkle free. The temperature range recommended by the manufacturer is generally between 160°F and 180°F. When the fabrics have reached this temperature, there is a conditioning or cool-down period during which the fabrics are stabilized in the smooth form.

CONTROLS

The three general types of dryers are: electric timer control, autodry, and electronic. The electric timer control permits the consumer to set an electric timer for a specified number of minutes. When the preset time has elapsed, the dryer will stop. There usually is a cool-off period, during which time the heat is turned off and the tumbling and air flow will continue.

The automatic dryer type has a dial control which may be set for degree of dryness. Different sectors on the control dial are provided for regular loads and for permanent press loads. The operation of the control involves a timer and a thermostat, though the consumer does not set a specific time. There is a timed cool-down period before the dryer stops.

The electronic control is an electronic moisture-sensing device which continually measures the degree of moisture in the load and stops the load when it is properly dried. The consumer selects the cycle as related to a type of load. The operation of the control is not dependent on time or temperature but only upon degree of moisture in the fabrics being dried (Fig. 13.5).

The dryers that employ the temperature and time controls operate in essentially the same way. This approach to automatic drying is based on the fact that temperature varies predictably with the rate of evaporation. When the heat source is turned off and the clothes are still wet, evaporation continues, causing the temperature within the dryer to drop rather quickly. Therefore the period during which the heat is turned off is relatively brief. As the clothes get drier, there is less evaporation. Thus it takes longer for the temperature within the dryer to cool down to recycle, causing the timer to run longer than before. In adjusting the timer for the type of load, less time is required for the lightweight fabrics than for the heavy loads. The temperature control will be indicated by fabric type: cottons and linen, synthetics, or delicates. The time control will be indicated by the size of the load: light, medium, or heavy.

The automatic dryness control which directly senses the amount of moisture in the load is usually mounted on baffles which rotate inside the drum. A continuous electrical signal is sent through the sensing elements. Since water is a good electrical conductor, the wet clothes offer relatively little resistance to this electrical signal.

As the clothes become dryer, the resistance increases and causes the electrical signal to change. Finally (depending on the setting) the change triggers a relay which is part of the electronic control circuit and fuel ceases to flow or shifts control to another device such as the thermostat.

The important decision to be made with this control is the desired level of moisture to be retained in the clothing. For the regular dry setting the dryer will operate until the sensing device determines that the moisture in the clothing has reached a very low level—approximately 2%. The dryer will then continue to operate without heat for a cool-down or conditioning period.

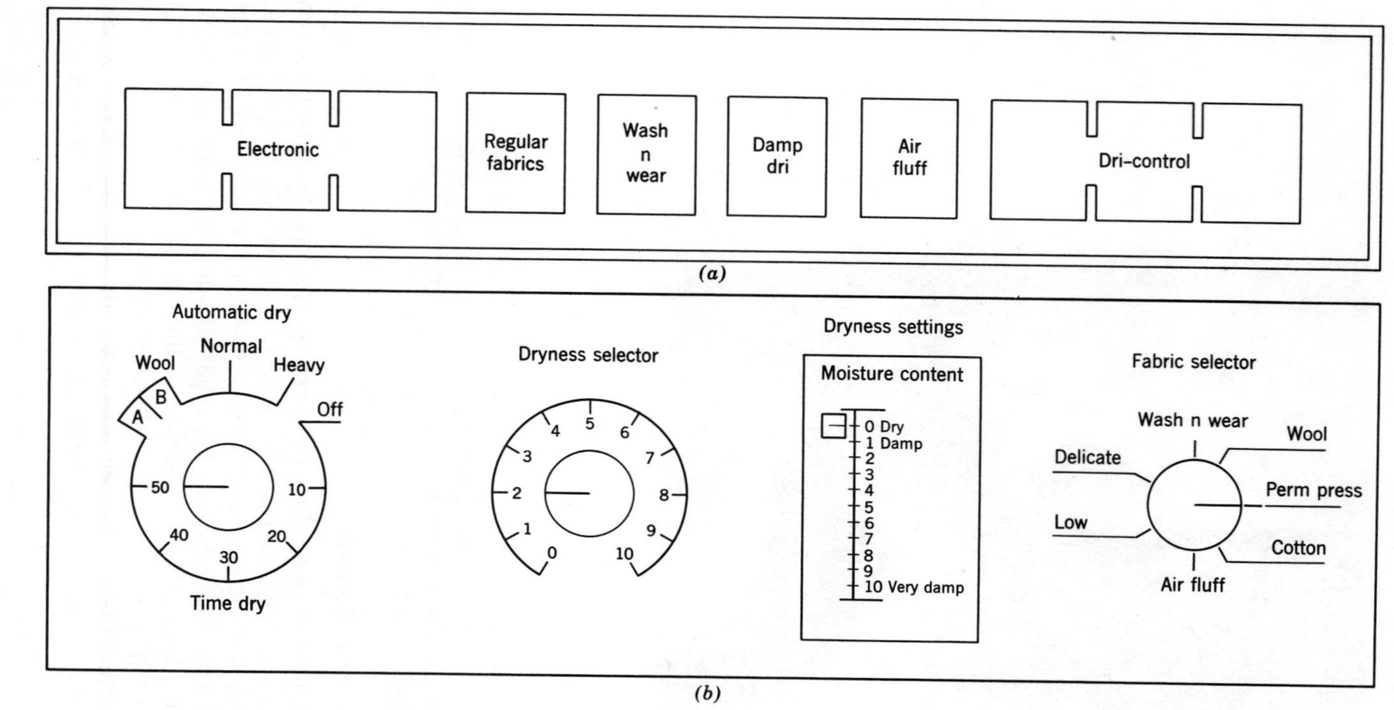

Fig. 13.5 *Degrees of flexibility designed into dryers at various levels of automation.* (a) *Selective controls.* (b) *Programmed controls. (Maytag)*

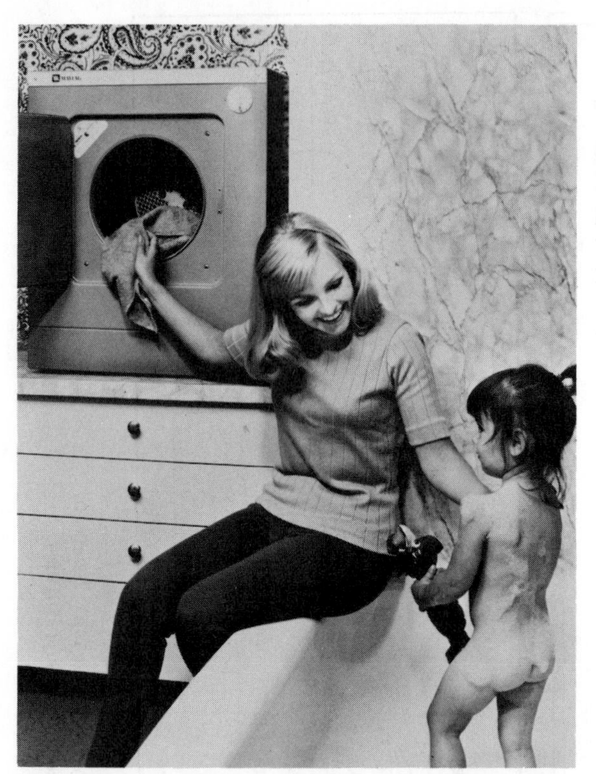

Fig. 13.6 Portable dryer designed to operate on 115-volt circuit. May be set on a counter, hung on a wall, or rolled around on casters. (Maytag)

The damp dry setting on the electronically controlled dryer shuts off the dryer when the moisture has been reduced to approximately 15 to 30%. The amount of moisture should be the desired level of dampness for ironing.

Temperature Control

Temperature is controlled by thermostats. The operating thermostat controls the drum temperature. This may be an adjustable thermostat providing a range of drying temperatures or a multiple of fixed setting thermostats, depending on the dryer design. The maximum allowable operating temperature depends on the fabrics in the load and may vary from one to as many as seven temperature settings.

The "automatic" model may also employ an auxiliary control thermostat working in series with the regular cycling thermostat. These thermostats sense heat increase and decrease only—they do not react to moisture control directly. Both thermostats are placed in the path of the air exhausted from the dryer.

A high limit thermostat is provided on both the electric and gas dryer, whether timed or automatic. In gas models the relay circuit is opened when the temperature in the burner duct reaches approximately 275°F and closed when the temperature drops to 195°F. In electric models the high limit switch is opened when the temperature in the heater housing reaches approximately 210°F and closed when the temperature drops to 160°F. The high limit switch will be opened when there is a restriction in air flow caused by a clogged lint trap, blocked duct, failure of either the auxiliary or cycling thermostat, or failure of the blower. The basic problem related to temperature control is the effect on fabrics. Excessive temperatures are likely to shrink fabrics, cause them to become more wrinkled, and hasten deterioration. The stretch fabrics loose their elastic characteristics when overheated.

Timer Controls

The timer controls all electrical circuitry to the operating parts. The number of cycles and circuits controlled depends on the model. The timer assembly on the dryer is a switch blade and contact device that consists of a timer motor and switch box assembly.

Blower Assembly

The blower assembly consists of a fan, bearing hub assembly, pulley, and two mounting flanges. The assembly has a dual function. It draws hot dry air into the dryer drum and exhausts moisture saturated air through the lint screen and out the dryer exhaust.

Some models of dryers use two fan speeds, normal and high speed selections, which operate in conjunction with the two-level heat input systems. The capacity of these blowers varies considerably, ranging from 70 cubic feet per minute for low speed to 220 cubic feet per minute for high speed.

Multiple Speed Drying

Multiple speed drying is appropriate for the drying of large and bulky items. This is accomplished by speeding up the drum above normal speed to provide better tumbling of the heavier loads. The faster speed reduces the "balling" which occurs with large or bulky items.

A slower than normal drum speed is designed for small loads and articles composed of the lighter weight synthetic fibers. The slower speed prevents sticking to the drum of small loads and lightweight articles. Recent changes in the design of drum baffles or vanes provide a variation in tumbling of the clothes load to eliminate hot spots and uneven drying.

Motor

The main drive motor is a split phase motor of ¼ to ⅓ HP capacity. It provides power for the fan or blower and the drum through a mechanical belt drive system. The built-in overload switch will turn off the heating element when the motor stops.

Electrical Requirements

A three-wire, 120/240 volt, 60 cycle, AC standard 30-ampere fuse branch circuit is desirable for the electric dryer. The electric dryer may have incorporated in the design the possibility of operation on either the 240- or 120-volt installation. When the dryer is converted to operate on 120 volts, a 20-ampere separate circuit is desired. The wattage will range from 1000 watts to 1350. This provision requires at least twice the time and possibly more to dry the same load of clothes as when it is installed to be operated on 240 volts.

The gas dryer requires a regular 115-volt circuit.

Heat Input

Heating elements are located so that the heated air flows into the drum immediately after passing over the source of heat, thus reducing heat loss and improving operating efficiency. The heat input for the electric unit may range from 4600 watts at the gentle speed to 5600 watts at the super speed. Gas input ratings to the gas burner vary widely among dryer designs, ranging from 18,000 to 37,000 Btu per hour.

Modern fan designs increase the rate of air flow, shorten drying time, and permit effective drying at lower temperatures.

Lint Screen

The lint screen is designed to collect loose lint that has been removed from the clothes, and helps to keep the interior of the machine lint free. A heavy load of lint will reduce the drying efficiency of the machine. A recent dryer design has an indicator to show when the lint screen should be cleaned.

Moisture Control

The moisture laden air may be directed to the outdoors or be collected in a filtrator. The filtrator causes the moisture to condense and drop into a disposal pan.

Venting the Dryer

Automatic dryers should be vented to the outside air. Basic to effective drying is a rapid movement of air, the removal of moisture-laden air to the outside. Factors to consider in the installation of a good venting system are the length and design of the run of pipe, the manner in which the tubing is joined, the location of the termination of the vent outlet, and the position of the vent exhaust hood on the outside of the house.

The most efficient vent duct system is straight, as short as possible, and concealed in an inconspicuous place. The system should be limited to two elbows, with a maximum of 6 feet of straight pipe and the vent hood. The system should never be permitted to run more than 30 feet, with 4 feet subtracted for each elbow in the run. A long vent run tends to reduce the efficiency of the dryer's airflow system.

For the best movement of lint through the run of pipe, all joints should be made so the exhaust pipe of one length of pipe is inside the intake end of the next pipe. A severe angle in the pipe can also inhibit the movement of lint and can mean a potential lint build-up. For adequate air circulation, the vent exhaust hood should be at least 12 inches above ground level. The vent outlet should terminate where the resultant lint cannot create a fire hazard.

An outside exhaust is recommended for proper combustion for the gas dryer. A sufficient amount of air must be available at the location where the dryer is installed. When normal infiltration of air does not meet the air requirements, outside air must be made available.

The maximum run of no more than 15 feet is recommended for the gas dryer exhaust. Any system exceeding this length will have reduced drying efficiency. The ducting should be as short as possible for easy removal to facilitate periodic cleaning of lint.

Cabinet

Acrylic enamel on chemically treated steel is used for the outside cabinet. The top is often finished with rust- and stain-resistant porcelain enamel.

Drum

The dryer drum has baffles that lift and tumble clothes during the drying process. Either the front wall, rear wall, or peripheral wall is perforated with holes through which warm dry air may be brought into the drum and moisture-ladened air may be exhausted by the fan. These perforations are important in the gas dryer design for the passage of combustion products.

The interior of the drum may be enameled, galvanized, of stainless steel, or Teflon coated. The material of the drum should be resistant to rust and corrosion because of its continual exposure to high levels of humidity.

Variations in dryer drum speed will permit proper care for fabrics; gentle for delicates and super speed for bulky or highly absorbent loads. One design permits 33 rpm for a light load and 49 rpm for a heavy load.

The speed of the drum is standard, since faster speed would tend to hold the clothes to the periphery of the drum by centrifugal force causing loss of tumbling action and a longer drying time.

Load Capacity

The proper load size varies with the size of the articles and the type of fabrics. The practical load depends on the volume of articles to be dried rather than their total weight. A full load of synthetic garments by volume would not weigh the same as an equal volume of cotton.

For best drying, the drum should not be overloaded. If the articles are seen to be revolving in one mass at the same speed as the drum, the load is too large.

Dryer Additives

Three approaches have been used to provide for fabric softening within the dryer. These are (1) an aerosol spray, (2) a foam spray, and (3) an impregnated fiber sheet. These products should be used with care. The aerosol spray type should not be directed onto the fabrics because some fabrics will be spotted or stained. This stain or spot may be removed after a few washes although a faint watermark may remain.

The foam product should be used with care to prevent a buildup which combines with lint, thus reducing drying efficiency and making lint removal difficult. This becomes difficult as the position of the user may not enable one to evenly distribute the product within the drum.

The impregnated sheet should be used carefully, especially with the small load. There is a chance that it may be attracted to the air outlet and reduce air flow. To prevent this from happening, a couple of dry towels may be mixed with the small load. The sheet may also be pinned to the corner of a hand towel.

SAFETY FEATURES

Consumer and appliance safety features to be found on recent designs of the dryers are:

1. Safety start button: dryer will not start until time is set, door is closed, and start button is pressed.
2. Safety door switch: stops all operation when the door is opened.
3. Friction door latch: allows door to be opened from the inside under even gentle pressure.
4. Limiter control: stops heat source should the appliance overheat.
5. Motor switch: insures that no heat is on if blower is not at operating speed.
6. Motor protector: stops the dryer if motor overheats.
7. Grounding screw provided to add fourth wire ground to electric models.

For safe and efficient operation, three factors that are related to lint should be considered: the lint filter must be cleaned before each use; the rear air intake screen must be cleaned frequently; and the dryer must not be overloaded. If a dryer is used repeatedly with a clogged lint filter or air intake screen, lint can accumulate inside the door above the filter or behind the drum near the heating coils. In time, these accumulations may become excessive and cause a fire.

Other Special Features

1. A feature called a "finish guard" provides automatic tumbling of clothes for 10 seconds every 5 minutes for an extended period of time after drying completion to protect against wrinkling. (20 to 120 minutes)
2. Cycles designed to dry without tumbling. A drying rack will be provided with this design feature.
3. Tumble press control for restoring press in clean permanent press fabrics that are wrinkled from short wearing or storing.
4. Automatic interior light.
5. Two-way door pulls down or opens from the side.
6. Adjustable tone end-of-cycle reminder buzzer sounds to signal when drying stops.
7. Larger capacity dryers.
8. Automatic gas igniter-recrystallized silicon carbide glows to red hot intensity in fulfilling the function of automatically igniting the gas flame. Temperatures near 3000 are reached in 10 to 15 seconds.

9. Lint indicators to advise when lint filter is full.
10. Light on back panel to illuminate panel.
11. Built-in filter to clean incoming air.
12. Automatic sprinkler.
13. Ultra-violet lamp creates ozone rays to give clothes freshness without outdoor fading.

SERVICING

Recent developments in the design of washers and dryers have resulted in features which require less service. Among the features which make servicing easier and less expensive are part standardization, reliable design simplicity in nonstandardized parts, removable control consoles, lift-off tops and front panels, coupled with the use of electronic analyzers to quickly diagnose and correct problems. Another important feature is an improved design of components to reduce the number of moving parts and thus reduce service potential. Many of these features require less technical knowledge on the part of the serviceman.

Internal components are accessible from the front so that appliances generally need not be moved. Cabinet tops may be easily removed for access to the tub, outer seals, and the spray fill-and-rinse system. The front panel may be removed to give access to the motor, wiring, transmission, and other major parts.

IRONS

The modern iron has been designed to meet the many needs of finishing of clothing after home laundering, self-service dry cleaning, or the simple touch-up after hanging for a time in the closet. The design may incorporate both the dry and steam features. The steam features may range from the spray for light-duty steaming to the concentrated spray of steam for heavy-duty ironing tasks.

The soleplate, of aluminum, stainless steel, cast iron, steel plated with chromium, or Teflon coated, should be without flaw, smooth and flat. A good iron has tapering sides with beveled edges and a narrow point to get around buttons and into gathers. Some have special indentations on the sides of the soles for ironing under button edges. If the iron is well balanced, the point will not damage fabric.

The iron handle should be of a size and shape convenient to the hand, of heat-resistant material, and far enough from the body of the iron to eliminate any danger of burns. A properly constructed iron should be insulated so that the heat will be concentrated in

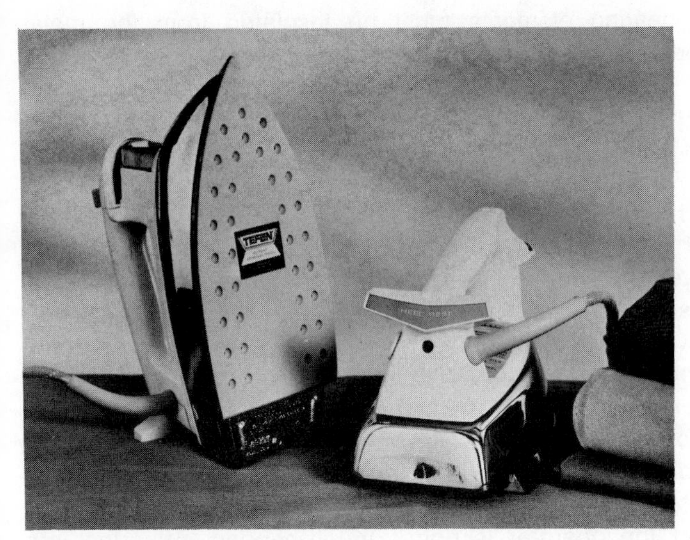

Fig. 13.7 Heel rest; designed to prevent accidental tipping. (Sunbeam)

the soleplate and the upper surface will remain cool. In one iron, the upper surface is separated from the main body of the iron by a narrow space through which air circulates to cool the top. In another the handle is open at the front end. A thumb rest may be molded into one side of the handle and occasionally into both sides.

With the exception of certain irons used entirely for steaming that are placed on a stand, the electric iron has an attached heel rest that holds the iron in a well-balanced, upright position between periods of use. A deluxe feature found on one iron is a wide angle heel rest designed to guard against accidental tipping (Fig. 13.7).

An automatic iron has a heat control connected to a thermostat in the soleplate, which controls the flow of electricity. Most controls have a range of heats suitable for different types of materials. In general, the scale will list the names of the fabrics such as rayon, silk, wool, cotton, and linen, and will sometimes indicate a range of temperatures for each fabric. The control is frequently located at the front of the handle where it is easy to see and accessible for manipulation. A control reduces the kilowatthour consumption by supplying only the heat needed, thus eliminating the fire hazard, since the iron cannot reach an excessively high temperature. This control also greatly lessens the wear on the heating element.

The point is often somewhat hotter than the rest of the iron, with the heat gradually lessening toward the heel, which is coolest. In all irons, the heating elements must be insulated from the metal body to prevent the operator from receiving a shock.

Linens and cottons require a fairly high temperature; silk and woolen materials, a comparatively low one. Viscose rayons and unweighted silks use a heat suitable for wool, but man-made fabrics, especially the acetate rayons, require even lower temperatures and may be melted by a hot iron (Table 13.3).

Cords used with irons may have an insulation of asbestos beneath the fabric covering. They are flexible and sometimes tend to kink. The cord is usually permanently attached to the iron, on the side or at the back. A fairly stiff, molded-rubber collar, several inches long, around the cord at its junction with the iron, holds the cord away from the heat. There are also attachments for the end of the board to keep the cord out of contact with the clothing. One type of cord coils and uncoils automatically as the iron is pushed back and forth. Some cords are reversible and may be turned to either side, depending on whether the operator is right- or left-handed. After the iron has cooled completely, the cord should be wound very loosely around the iron for storage. The standard automatic household iron weighs 2¾ to 3½ pounds and has a rating of 1100 watts.

When the lightweight iron is used in ironing heavy materials, the fabric should be properly dampened or the user will need to exert a good deal of pressure to remove creases. Very damp fabrics usually require a somewhat longer ironing period, although the high heat tends to speed up the process. In extensive tests made by Potter at Virginia Polytechnic Institute, best ironing results were obtained when a material was dampened to the optimum moisture content, which was found to average about 35%.

Always connect the iron to a wall outlet. When the cord is separate, connect the iron plug to the iron first, and then connect the wall plug into the wall outlet; when disconnecting, reverse the procedure.

An iron should be kept clean, the soleplate free from rough places, stains, and rust. To avoid scratching the soleplate, do not iron over hooks and zippers.

The spray-steam-dry iron is designed for flexibility and may serve many of the functions desired in the finishing of fabrics. Important in this design is the ease with which the operations may be changed from one to another. A convenience feature is the built-in water window which indicates water level and when to refill.

Vents on iron soleplates should provide steam without sputtering or spattering. There should be enough steam to penetrate the fabric, but not so much as to cause wetting or spotting. The number of

Table 13.3
Safe Iron Temperatures for Modern Fabrics

Fabric	Safe Ironing Temperature	Fabric	Safe Ironing Temperature
Acetate	350°F	Kodel (Polyester)	350°F for Kodel II 350°F for Kodel IV
Acrilan (Acrylic)	325°F		
Arnel (triacetate)	400°F		
Anril (High wet modulus Rayon fiber)	375°F	Lycra (Spandex fiber)	300–325°F
		Nylon	325° F for type 66 300°F for type 6
Avron (High tenocity Rayon staple fiber)	375°F	Orlon (Acrylic)	300°F
Benberg (Cuprammonium Rayon)	375°F	Rayon (Viscose)	375°F
		Saran	150°F
Caprolan (Nylon-type 6)	300°F	Silk	300°F
		Verel (Modacrylic)	275°F
Cotton	425°F		
Creclac (Acrylic)	300°F	Vycron (Polyester)	300 to 325°F
Dacron (Polyester)	325°F	Wool	300°F
		Zantrel (High Wet Modulus Rayon)	375°F
Dynel (Modacrylic)	225°F		
Flax	450°F	Zefron (Acrylic staple fiber)	350°F
Fortrel (Polyester)	325°F		

Source: Burlington Mills.

vents beyond a dozen does not affect the steam penetration when ironing. When pressing, there is an advantage to added vents.

The push-button spray offers on-the-spot dampening for lightweight fabrics where slight amounts of moisture are needed to combine with heat for a better finish. The pattern and fineness of spray

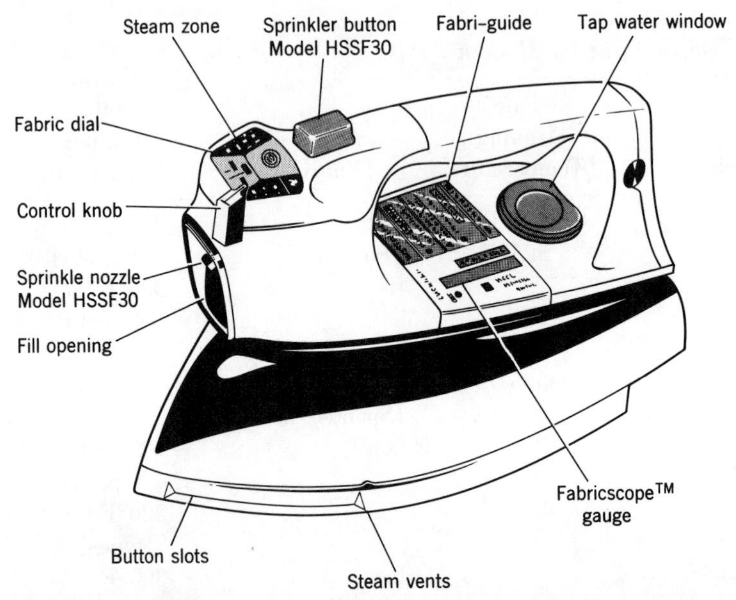

Fig. 13.8 Steam-spray-dry features designed to handle all types of ironing. Gauge tells when iron is in proper temperature area. Single knob control dial permits steam and dry ironing with settings for all types of fabrics. (Westinghouse)

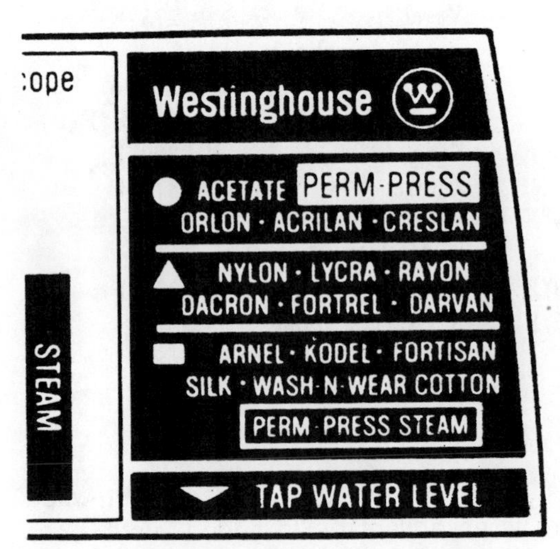

Fig. 13.9 Fabriguide. Symbol (circle, bar, or triangle) on guide corresponds to fabric to be ironed and control indicated should be selected accordingly. (Westinghouse)

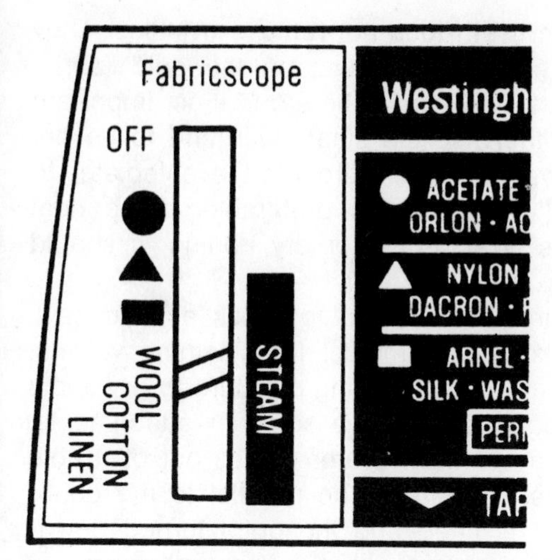

***Fig. 13.10** Fabriscope. Indicator of fabriscope gauge will tell when iron temperature has reached general recommended range of temperature selected on dial and signals time to iron. (Westinghouse)*

is important in the effectiveness of this feature. (Figs. 13.8, 13.9, 13.10).

The concentrated spray of the steam-feature will provide increased steam for heavy duty ironing jobs. Important to the successful use of this feature is a basic understanding of the fabric content of the garment to be sure it will not be damaged by the high quantity of moisture. The jobs which are best accomplished by this feature are: making stubborn wrinkles disappear, setting professional-type creases in trousers or pleated skirts, touching up permanent press fabrics, or pressing newly sewn seams flat.

This feature replaces the spray in a conventional steam-spray iron, but may be used while dry or steam ironing. The concentrated extra steam comes from the soleplate vents, rather than a spray of water from the front of the iron. This quantity of steam will penetrate a heavy pressing cloth when one is used. The iron delivers the extra, deep penetrating quantity of steam at the touch of a button on top of the iron. This button is located in the same position as the spray button on steam-spray irons.

A special steam feature built into a portable steam presser may be another important appliance for touch-up finishing of a garment just cleaned at the self-service dry cleaner or laundered at home. This presser may be held to the spot and the garment finished as desired.

Riedel[1] points out that steam itself does not remove wrinkles. They are removed only when the steam condenses on the cold fabric, followed by the application of heat from the iron. It is important not to have the iron too hot, otherwise the steam will tend to be dry instead of moist. A medium-warm temperature for the soleplate, to produce steam at about 225° or 230°F, is preferable. If the heat control for the different fabrics is accurate for dry ironing, it should be satisfactory when steam is used.

The steam iron is a help in allowing you to press as you sew. it may be used for blocking knitted garments, for steaming velvets, suede purses, and shoes, and always for adding moisture to dry spots that develop during ironing. Some irons have sprinkler attachments. Just press a button, and a spray of warm water comes out of a hole on the front of the handle. The iron should be parallel to the board when the sprayer is used; if it is lifted and the point turned downward toward the fabric, a stream of water may come forth.

Soft water, either rain water or distilled water, is important for filling the iron. Hard water forms a lime deposit that gradually cakes the sides of the cavity and clogs the steam vents. Some irons may be taken apart and cleaned, but this involves unnecessary work and the possibility of doing some damage to the parts, especially to the silicone seal between the cover and the soleplate.

Care of the Iron

Distilled Water. Distilled water is preferable for use in the steam iron because it contains no minerals, dust, or alkali particles. These particles in tap water will eventually clog the steam iron and reduce the steam flow.

Seasoning. When distilled water is used, periodic seasoning will improve the steaming performance of the iron. Tap water can be used for seasoning, since it contains minerals which will coat the steam channels. If the iron is used as a steam iron frequently during the month, it may be necessary to reseason the iron every two or three months.

Flush the Iron. In order to avoid a build-up of lint in the steam vents, flushing the iron steam vents is recommended by some manufacturers. This is advised especially if the iron is used as a dry iron more frequently than as a steam iron. Scorched lint may be mistaken for

[1]Mary R. Riedel, ***An Analysis of the Steam Iron,*** Proctor Information Center, Proctor Electric Company, New York.

rust. Should brown spots or residue appear on the garment while ironing, it is time to flush the iron.

This feature may now be accomplished on one brand of iron with the operation of a button which flushes steam from the iron's water tank out through the steam vents. This water and steam carries away loose minerals, sediment, and lint from the iron.

Use of the Iron

Some guides for the successful use of the iron in finishing follow.

1. Sprinkle large pieces such as table linens, organdy curtains, bedspreads, and heavily starched wearing apparel which need a high quantity of moisture to iron smoothly.
2. Spray lightly while steam pressing seams, darts, hems, and pleats in sewing.
3. Experiment with the iron to determine the proper heat setting for use with each fabric. If you are not sure of the fabric, start with the lowest setting. If this setting is too cool, then proceed to the next higher one. Test the settings on an inside facing or seam of the garment. If the iron sticks even slightly, return to the lower setting.
4. When spray ironing, test with a piece of the fabric first. Some fabrics may water spot.
5. When spray ironing, start spraying at the lower edge of the garments such as skirts, blouses, shirts as you move on the forward stroke. Some garments need to be touched up with the dry iron after spraying for a smooth finish.
6. Use long, slow, smooth strokes in ironing. This gives the moisture an opportunity to penetrate and the iron time to dry out the fabric.
7. After use, turn off all controls on the iron. Remove the cord from the outlet by grasping the plug rather than pulling on the cord.
8. Empty the water from the iron after each use, emptying while still hot. The heat will dry the reservoir.
9. Let the iron cool completely before storing. Loop the cord loosely around the iron when storing.
10. Always store the iron upright on its heel. Do not store in cartons as any residue of moisture may cause the soleplate to discolor.
11. Avoid ironing over snags, zippers, and rivets, since they scratch the soleplate.

LAUNDRY CENTER

In contemporary house planning the laundry center may be located in the bathroom, bedroom, hall, or utility room. Here natural lighting is better, and frequently the ventilation, so necessary for removing heat and steam, is improved. Because of the kinds of bacteria present on household textiles, the laundry area should not be in the same room as the food preparation and service area. Research conducted in the equipment laboratories at Iowa State University indicates a high degree of survival bacteria on household linens even after washing and drying.

Each work center should be provided with counters for sorting clothing to be washed and for folding dry and ironed garments, and also with cupboard space for storage of detergents, starch, bluing, stain removers, measuring cups, and pretreating brushes. Bins to hold soiled articles may be located below the counters, or baskets on wheels may serve even more satisfactorily, since they eliminate stooping and the transferring of dry and damp clothes by hand. A water supply separate from the connection to the washer is desirable, and also a hot plate to heat water for removal of certain types of stains and for the making of cooked starches.

The walls of the laundry should be light in color and moisture-proof; the floor nonabsorbent, resilient without being slippery, and easily cleaned. When a cement or wood floor must be used, put rubber or linoleum mats in front of the appliances. Artificial lights should be provided over the washing and ironing centers. Blue daylight lamps are especially recommended for the laundry room because they help the homemaker to see spots of dirt and stains on the clothing. They should be 150- or 200-watt lamps to provide sufficient light. Fluorescent tubes of 40-watt ratings are frequently used and give satisfactory light.

With electric outlets in every room, ironing may be done wherever it is most convenient. A cool, well-ventilated location is desirable. When you sit to iron, use a posture chair to eliminate unnecessary strain and expenditure of energy. If you prefer to stand, be sure a resilient mat is beneath the feet.

Energy Conservation

The purpose of applying energy-saving techniques is to reduce the energy needed to heat water and to operate the washer, dryer, and iron. Washers with adjustable water level controls are capable of significant energy conservation. Select the water level to match the size of the load. Soak heavily soiled garments instead of washing them twice. Select the wash time to match load and soil levels. Washing longer than necessary wastes energy. Plan laundry schedules so you do not have to use the washer and dryer more than is needed. When possible wait until there is a full load. Do not overload the machine.

Much of the energy used in the laundering process is used in the heating of water. Locate the washer as close to the water heater as possible to minimize heat loss in the piping. Use warm and cold water whenever possible.

Use the correct amounts of cleaning products. Recommendations as to amounts are the results of testing in order to get the best performance and satisfaction from a product or appliance.

Care and maintenance can result in an optimum use of energy for laundering. Lint filters in both the washer and dryer should be cleaned after each use. Be sure the dryer is vented. Avoid using extension cords for major appliances and do not overload circuits.

In drying clothes, the permanent press and fluff cycles should be used whenever possible rather than the hot cycle. This is especially true for the synthetic fiber garments. Select the drying temperature and time appropriate for the purpose. Overdrying wastes energy and results in a wrinkling of fabrics that is difficult to remove. By removing the garments promptly when the dryer has stopped, much ironing can be eliminated. Iron large amounts at one time rather than several small loads. Use the lowest setting on an iron possible to do the job. Press items that need a cool or warm iron first as heat is building up and again at the end to utilize heat after the iron is turned off.

Experiments and Projects

1. Compare lingerie washed and dried by hand to some done in the automatic washer and dryer to show improved washability in washer and dryer.
2. Observe the seam construction of a garment after washing and drying automatically to point out the need for sturdy seam construction.
3. Discuss the sorting of clothes and the mixing of colorfast and white cottons and linens.
4. List the man-made fibers, blends, and finishes used to make permanent press garments.
5. Discuss hang tags and the type of information that they give.
6. Make a collection of hang tags from garments and analyze them.
7. By use of a display, discuss the stains commonly found in baby clothes and solutions needed for stain removal.
8. Teach proper selection, use, and care of the washer and dryer by using the owner's manual, equipment texts, and other references.

9. Take a field trip to an appliance store to observe and compare features of laundry appliances.
10. Wash and dry a load of clothes and explain the features used on the washer and dryer.
11. Discuss how a washer and dryer can be used to keep laundry from accumulating by washing as clothes become soiled.
12. Have students survey friends and neighbors to learn of items people wash and dry in the automatic washer and dryer.
13. Discuss the cost of the operation of a washer and dryer including water heating costs and dryer operation costs.
14. List points to consider in buying clothing and linens if one is going to wash automatically.
15. Plan loads of clothes for large, medium, small, and very small loads.
16. Plan a clothes budget for a family with two to four children using an automatic washer and dryer, and one using a wringer washer and line drying the clothes. Estimate the length of time clothes will wear in both cases.
17. Estimate the time required to do a week's laundry by an operator when an automatic washer and dryer are used and when a wringer washer and line drying are used.
18. Determine the size of a water heater needed for use of an automatic washer with a given size family.
19. Set up a plan for the removal of stains from table linens and tea towels.
20. Set up an instruction guide for the use and care of the washer and dryer using the owner's user's guide.
21. List the laundry aids needed; group them into types and determine the method of using each.
22. Discuss the possible need of service on a washer and dryer, how to reduce service needs, and how to recognize improper operation.
23. Develop a guide for selecting a washer and dryer.
24. Make a floor plan for a utility room including laundry appliances.
25. Make a plan of the basic features of washers and dryers and then include the added features that are available. Justify the use of plus features.
26. Compare the advanges and disadvantages of a separate washer and dryer and a combination washer-dryer.
27. Discuss packaged and commercial water softeners. Make a field trip to a water softener company, if possible. Demonstrate the effect of water softeners when using soap and detergents.

28. Teach safety in the use of electrical appliances which are connected with a water supply.
29. List safety features on washers and dryers. Study the purpose of each.
30. Relate the importance of using hot water in the prevention of the spread of communicable diseases and in removing heavy soil.
31. Set up a method of laundering curtains, draperies, bedspreads, blankets, and rugs.
32. Discuss fabrics which should not be laundered automatically.
33. Recondition regular family laundry. Wash clothing on regular cycle. Add 1 cup of nonprecipitating water softener, such as Calgon, to a full tub of water. Observe the nature of the wash solution.
34. Compare water extraction from a preweighed load of clothes washed in different cycles of the same washer.
35. Compare the water extraction from cycles of different washers.
36. Study detergents in local stores. Determine the quality of information available on packages. How do they compare in price? What purpose are they designed to serve?
37. Compare the wetting ability of water. Fill two containers with equal qantities of hard water (½ cup water). Place ½ teaspoon detergent in one cup. Do not use an additive in the second cup. Cut two 1-inch squares of wool. Place one sample on the surface of each container of water. Observe the wetting action.
38. Study the effects of water conditioners when detergent is used in hard water. Repeat the study using soap instead of detergent.
39. Bring in naturally soiled clothes from families. Sort and study soil and stain problems. Determine procedures for the removal of soils and stains.
40. Study the effects of chlorine bleach when it is used correctly as compared to an incorrect method of use. Use white cotton fabric samples—soiled with chocolate, lipstick, mustard, oils.
41. Study energy conservation as related to the use of the washer, dryer, and iron.

CHAPTER 14
The Sewing Machine

Today's sewing machine designs provide a wide range of choices from which to make a selection. The decision should be based on the nature of the fabrics, and the kind and amount of sewing to be done. The kind of sewing for which the machine is to be used may be the most important consideration. Many homemakers use the machine for occasional plain sewing, mending, patching, and for the construction of simple garments. For these needs, where machine-made decorative effects are of little value, the straight-stitch machine is suitable.

The more flexible design zig-zag machine with the swing needle features a wider range of stitch designs and provides for needs such as specialty sewing requires. This flexibility is comparable to the function served by the attachments provided for the straight-stitch machine. A critical matter at this point is—how many of the attachments are ever used with the straight-stitch machine?

Other matters to consider in the choice of a machine are associated with the cabinet or portable design and round bobbin type—oscillating or full rotary.

The portable machine may be light or mediumweight and performs more effectively with lightweight fabrics. The heavyweight or cabinet design is used for average sewing and medium to heavyweight fabrics.

The oscillating bobbin design should have a removable shuttle and shuttle race is thread locks are to be easily removed. The rotary bobbin is less subject to thread lock and operates more smoothly than the oscillator type.

PRINCIPLES OF OPERATION

The straight-stitch machine employs the motion of the needle moving up and down in a straight line, making stitches that are locked into the fabric. In the machine described as the zig-zag, the needle may move from side to side as well as up and down. A basic circular or semicircular motion is combined with the movement of the needle to produce the back stitch.

Higher levels of automation in the sewing machine are described as semiautomatic and fully automatic. In the semiautomatic, the same motions performed by the straight-stitch and zig-zag machines are combined with some provisions for decorative stitching. The fully automatic machine performs a triple action-control of the needle, movement of the fabric, and position of the needle. With this high level of automation, the fully automatic machine may perform all of the motions performed by the other three designs.

SELECTION

Criteria which may guide selection are associated with ease of use, ease of making adjustments, ease of care, and those which affect durability.

Criteria to Determine Ease of Use

1. Is the upper tension setting shown by markings that are easy to see?
2. Does the lamp direct light where you need it?
3. Is the lamp placed so that it will not burn you during normal use of the machine, for instance, when raising the presser foot?
4. Is the stitch-length control scale easy to read?
5. Will the machine stitch backwards?
6. Are there adjustable stop positions for the forward and reverse stitching control?
7. Is the upper thread tension released when the presser foot is raised?
8. Can the pressure on the presser foot be released for darning and embroidery?
9. Is there a footrest on the electric foot control?
10. Is the machine quiet and free from objectionable noise and vibration?
11. Does the machine run smoothly at all speeds?
12. Is the knee or foot electric motor control comfortable for you to use?
13. Does the motor start smoothly and provide easy starting as well as slow running?
14. Is the machine easy for you to thread?
15. Is the bobbin easy to take out and put back?

16. Is the bobbin easy for you to thread?
17. Try the machine on some of your own materials, both straight and curved seams. Is it easy to guide when stitching curved seams?
18. Notice whether the material has a tendency to drift to right or left, whether one layer of material tends to creep over the other during sewing. Does the machine satisfy you in this respect?
19. Is the bobbin winder easy to use and does it fill the bobbin evenly?

Criteria to Judge Ease of Making Adjustments

1. Is the bottom tension conveniently located and easy to adjust?
2. Are the tension adjustments clearly explained in the instruction book?
3. Is the stitch-length control easy to use?
4. Can the toothed feed dog be dropped out of the way for darning and embroidering?
5. If the feed dog can be dropped, is the adjustment easy to get to?

Criteria to Determine Ease of Care

1. Are the cover plates easily removable and all parts readily accessible for cleaning, oiling, and greasing?
2. Is the wiring protected against oil drip?
3. Is the light bulb easy to replace?
4. Is the machine easy to dust and wipe clean?

Criteria to Determine Durability

1. Is the wiring located where it is protected from wear?
2. Are adjustments provided for wear between moving parts?
3. Is the cabinet well constructed; are hinges sturdy, legs well braced, and is the leaf well supported and level when opened?
4. Are the service parts carried in stock?

Other special features to consider are as follows:

1. Elastic stitching.
2. Overlock stitch—sews straight seam and overcasts at the same time (for durable knits and jerseys or adding elastic to tricot).
3. Special stitch for sewing soft bulky knits.
4. Low gear—permits sewing at extra slow speed, with complete control and increased motor power.
5. Hinged presser foot for sewing over pins.
6. Double-needle sewing for specified stitches.
7. Forward-reverse stitch for hemstitching.
8. Automatic bobbin winder.
9. Pop-up pressure release regulator for different fabrics.
10. Front tension control for easy threading.

11. Darning possible without a hoop.
12. Feeding quality over areas of heavy fabrics.
13. Removable cover for effective cleaning and oiling—all moving parts accessible for cleaning.
14. Jam-free designs of hooks.
15. Dial drop feed.
16. Darning release.
17. Special attachments built into the machine design—machine baste, reverse stitch, gather, mend, narrow hemmer, binder, zipper foot, tucker, buttonholer, patching.

ADJUSTING THE MACHINE

Regulating the Sewing Machine

To achieve the desired results, the machine should be adjusted properly for different fabrics. High quality performance is related to the proper needle and thread, correct tension, and correct pressure.

The size of the thread and the weight of the fabric will call for a specific size of needle. Needles are made in different lengths for different brands and styles of sewing machines. The distance from the top of the eye to the top of the needle must be correct for the machine. The eye of the needle must be large enough for the thread to pass through it freely. If the needle is too small, the thread will fray and break, and the needle may bend or break. The needle must be strong enough so that it does not bend when it goes through the fabric, especially when stitching across seams. If the needle is too coarse for the fabric, it will snag the fabric or break the weave. This needle will also make holes in the fabric that are too large for the thread to fill. The correct thread, needle, and stitch length for specific fabrics may be obtained from Table 14.1.

Regulating the Stitch Length

The stitch length selector controls the number of stitches per inch and determines whether the sewing direction will be forward or in reverse. The stitch length control can be a lever or a thumbscrew. The stitch length lever has numbers on each side which represent the number of stitches per inch; the higher the number, the shorter the stitch. Generally, shorter stitches are best for lightweight fabric, longer ones for heavyweight.

The stitch length thumbscrew can be turned to the right to lengthen the stitch and to the left to shorten it.

Table 14.1
Fabric, Thread, Needle, and Stitch Length

Fabrics	Thread Sizes	Needle Sizes	Stitch Length Setting
Delicate—tulle, chiffon, fine lace, silk, organdy, fine tricot	Fine mercerized 100 to 150 cotton synthetic thread	9	15 to 20
Lightweight—batiste, organdy, jersey, voile, taffeta, silk, crepe, chiffon velvet, plastic film	50 mercerized 80 to 100 cotton "A" Silk synthetic thread	11	12 to 15 (8 to 10 for plastic)
Medium weight—gingham, percale, pique, linen, chintz, faille, satin, fine corduroy, velvet, suitings, stretch fabric	50 mercerized 60 to 80 cotton "A" Silk synthetic thread	14	12 to 15
Medium heavy—gabardine, tweed, sailcloth, denim, coatings, drapery fabrics	Heavy duty mercerized 40 to 60 cotton	16	10 to 12
Heavy—overcoatings, dungaree, upholstery fabrics, canvas	Heavy duty mercerized 20 to 40 cotton	18	6 to 10

Regulating Thread Tensions

Tension is necessary to control the threads that interlock to form a stitch. A perfectly locked stitch is one in which the two threads are drawn into the fabric to the same degree. This is accomplished by two tensions, upper and lower. The upper tension controls the needle thread and the lower tension controls the thread from the bobbin (Fig. 14.2).

To regulate tension on the needle thread the pressure foot must be down. The upper or needle thread tension is exerted by the closely fitted discs between which the upper thread passes. If the needle thread breaks, the upper tension is tighter than the bobbin tension. When the bobbin thread breaks, the upper tension is looser than the bobbin tension. To balance the tensions, change the upper tension setting by either tightening or loosening until both threads break. When both threads break, the needle thread tension has been balanced with the bobbin thread tension (Fig. 14.2).

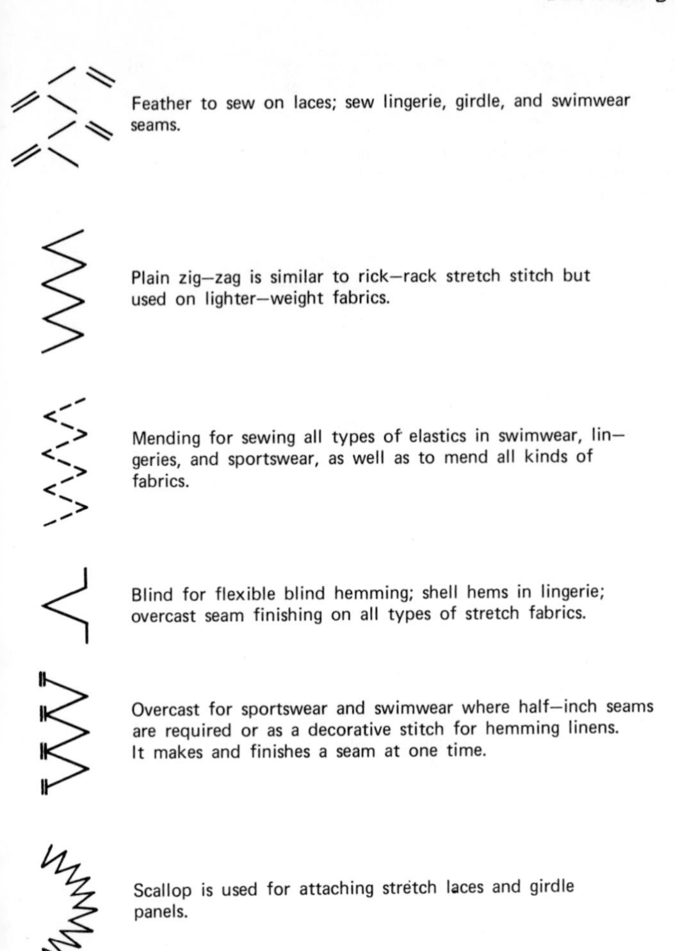

***Fig. 14.1** Stitches and their purposes.*

The tension on the bobbin thread may be regulated by the small screw on the bobbin case or shuttle (Fig. 14.3). A slight turn of the screw on the bobbin will make a big tension change. Some models of sewing machines do not require bobbin tension adjustment. On such models, the bobbin thread is not drawn under a bobbin tension spring; tension is placed on the bobbin itself.

The bobbin tension for best flexibility varies with the machine. Here is a guideline to follow: normal bobbin tension is the one at which your machine will stitch a 12-stitch length with size 50 Mercerized thread on two layers of muslin without puckers, and without thread loops on the underside (Fig. 14.4).

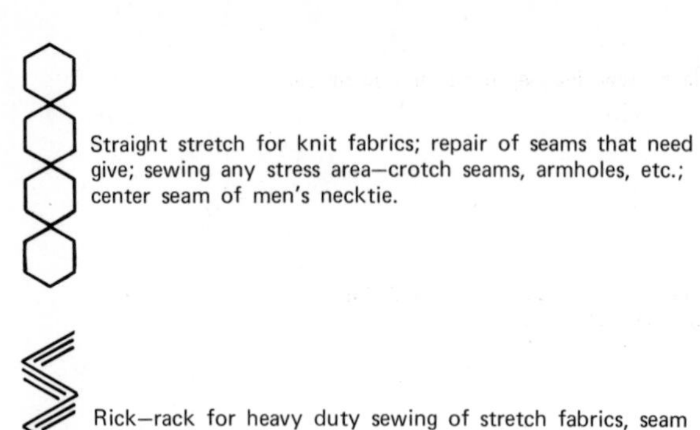

Straight stretch for knit fabrics; repair of seams that need give; sewing any stress area—crotch seams, armholes, etc.; center seam of men's necktie.

Rick—rack for heavy duty sewing of stretch fabrics, seam finishing, edge finishing on stretch fabrics.

Smocking used on yoke, insert or band on baby clothes, little girls' dresses, or on ligerie.

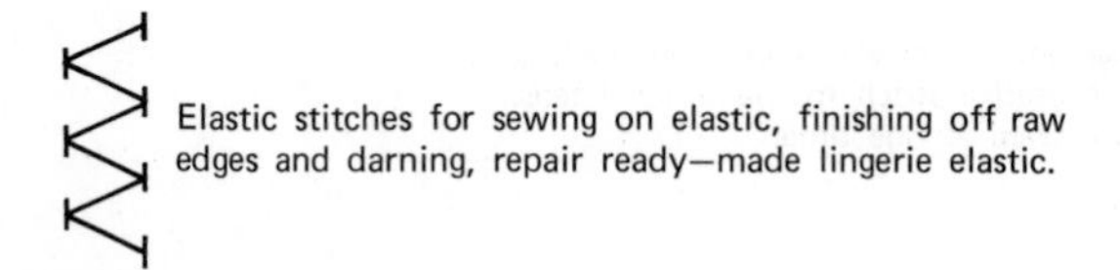

Elastic stitches for sewing on elastic, finishing off raw edges and darning, repair ready—made lingerie elastic.

Serging or pine leaf for overcasting fabrics that ravel easily; formation of swimwear or girdle seams that need a great deal of stretch.

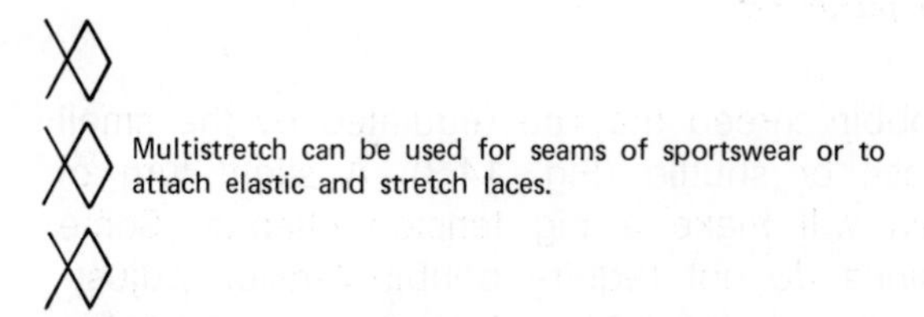

Multistretch can be used for seams of sportswear or to attach elastic and stretch laces.

***Fig. 14.1** (Continued)*

Regulating Pressure

Correct pressure is important if the fabric is to feed smoothly, evenly, and without being marred over the feed-dog. An understanding of the fabric weight will enable one to make the correct adjustment.

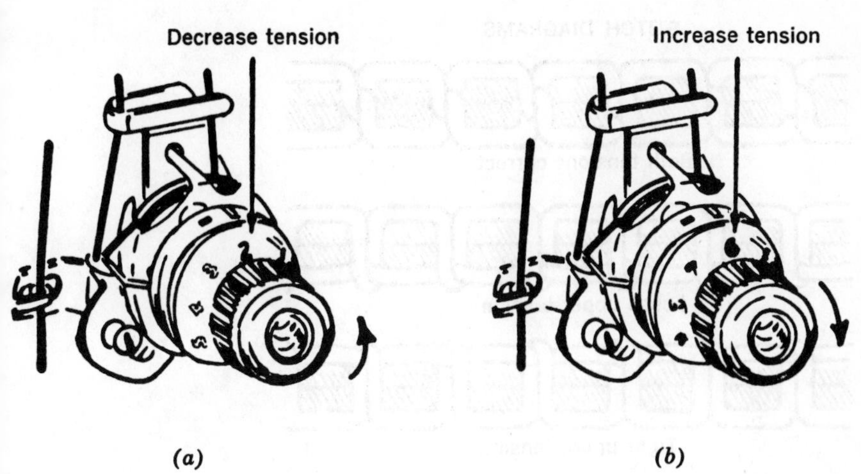

Fig. 14.2 (a) *Regulates needle thread tension. To decrease tension, turn the thumb nut to the left (counterclockwise) until the required tension is obtained. The lower the number, the lower the tension.* (b) *To increase tension, turn the thumb nut to the right (clockwise) until required tension is obtained. The higher the number, the higher the tension. (Singer Sewing Machine Co.)*

The pressure adjustment increases or decreases the force exerted by the presser foot on the fabric (Fig. 14.5). For example, heavy fabrics require a high number setting; lightweight fabrics, a low number. Soft, sheer fabrics require less pressure than crisp fabrics to feed properly. Spongy, thick fabrics or those with a pile require less pressure to prevent crushing. Heavy, dense fabrics require heavy pressure.

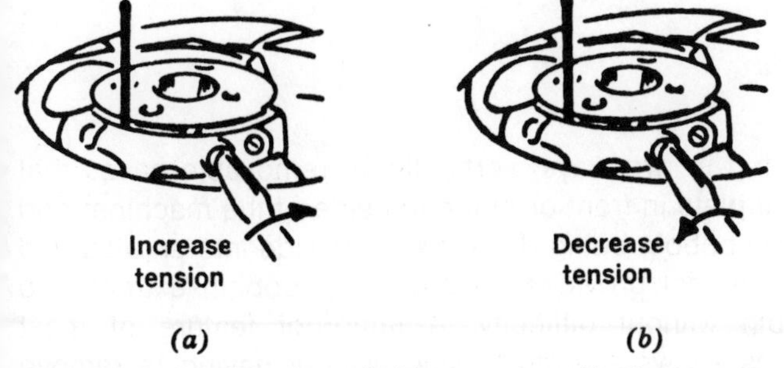

Fig. 14.3 (a) *Regulates bobbin tension. To increase tension, turn screw to right.* (b) *To decrease tension, turn screw to left. (Singer)*

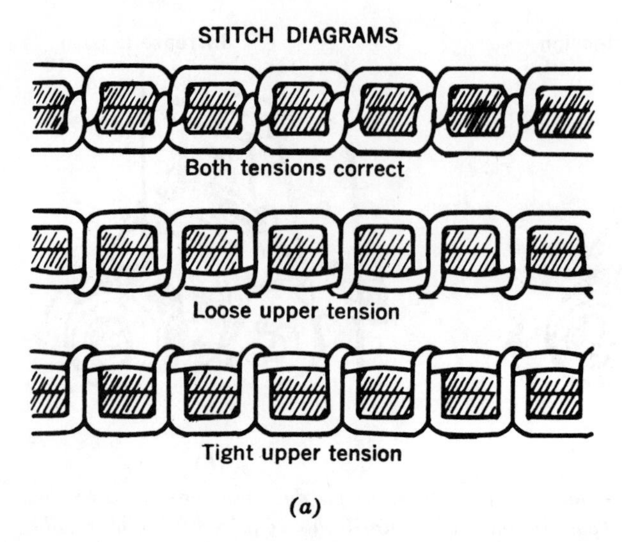

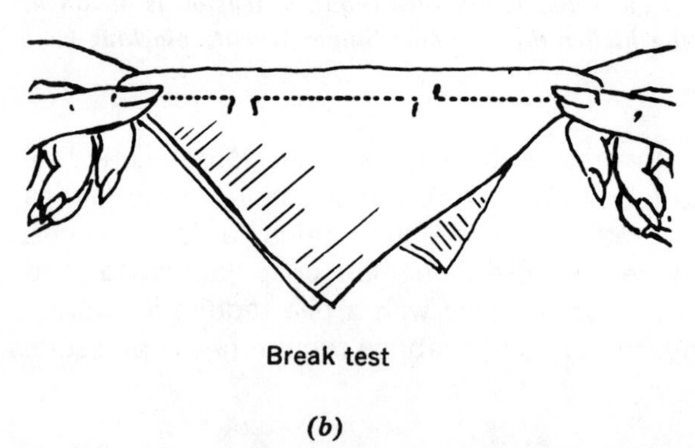

(b)

Fig. 14.4 (a) *Stitch diagrams.* (b) *Guides to effective tension adjustment. (Singer)*

Bobbins

Bobbins are inserted in two ways: vertically, in removable cases that go under the throat plate in front or at the left side of the machine; and horizontally with the bobbin being dropped on its side into an attached bobbin case. One model provides an automatic bobbin extractor to remove the bobbin without difficulty. A jamproof feature of most bobbin mechanisms allows smooth sewing without having to remove any thread that may tangle around the bobbin hook. A transparent bobbin will permit one to see the amount of thread remaining on the bobbin. To wind the bobbin, the thread should pass through a tension regulator for the bobbin to wind evenly. One model has an adjustable tension regulator as a further assistance to uniform winding.

While winding the bobbin, the needle action must be stopped on most sewing machines. The winding process stops automatically when the bobbin is full. One model has a self-winding bobbin, which is activated by moving a lever under the throat plate; press the foot control and the bobbin winds from the top spool. To resume sewing, release the lever to the original position and close the throat plate.

Stitches

An important difference between machines is the variety and number of stitches they can perform. All of the deluxe machines are described as automatic zig-zag machines, and all of them do regular stitching; zig-zag stitching (a side-to-side movement) in varying widths; and a blind hem stitch, which also can make a shell edging, plus a variety of decorative stitches (Fig. 14.1).

A special feature to prevent skipped stitches and separate and penetrate the fibers of closely linked knits may be found on a deluxe model of sewing machine. The presser foot-needle combination set is most successful when sewing on polyester knits, nylon tricot, and elasticized fabrics as these fabrics are so closely linked that a standard sewing machine needle experiences difficulty when penetrating the fibers.

When the needle encounters resistance, the thread is pulled taut preventing a loop from forming in the shuttle area. The result is a skipped stitch. The set is designed to be used with zig-zag stretch stitches where skipping most commonly occurs.

The threading process is similar on most machines. Some have special features that make threading easier—features such as slots (rather than holes for the thread to go through) or a white plate located behind the needle to make it easier to see the eye. Some models have automatic threaders, either attached to the machine or as an optional accessory.

Presser feet are commonly screwed into the presser bar. This feature has been simplified on some models by introducing "soles" which slip onto a master bar, or presser feet that will easily clip into place. The presser-bar-lifter attachment which may be found on some models fits on the lower front of the machine and makes it possible to raise or lower the presser foot with your knee, leaving the hands free to move fabric when doing intricate work.

Foot power control may be more easily accomplished by a pedal which permits one to place the whole foot flat on the pedal, which is a more natural position than holding the foot at an angle and provides better control of the speed.

Free arm models have a narrower sewing surface that is helpful when inserting sleeves, topstitching cuffs, and for working on other small-area or tubular sewing. Additional sewing space may be provided by an extension-table attachment which may be clipped into place.

An additional special feature to be found on many new machines is a lever to adjust the feed-dog position. The lever may be used to lower the feed-dog slightly for sewing on fine or delicate fabrics; or to lower completely for darning or embroidery so fabrics may be moved in any direction.

Features which make sewing easier are those which permit an extra-long basting stitch; one machine does this by using the blind stitch and a special needle with two eyes. Marking with tailor tacks is made easy with a special presser foot and a very loose top tension.

Care of the sewing machine has been simplified by the self-oiling and self-cleaning features. The self-oiling feature is achieved by the release of oil from the porous steel bearings, as oil is needed. On another design, the shuttle area in the bobbin is self-cleaning.

Safety features on newer models include an off-on power switch so the motor may be stopped while the machine is plugged in, and a means of locking the exposed hand wheel (or the needle) so the machine may be moved without damage.

CARE

The sewing machine is a highly designed, yet sturdy precision instrument, and may be cared for in a few simple steps.

Regular Care

1. Remove lint as it accumulates. The rate will depend on how much you sew and the nature of the fabric.
2. With a soft cloth, clean the tension discs, bobbin case, take-up lever, and thread guides. Use the lint brush to clean the feed-dog and hook or thread-handling parts under the slide and thread plate. Remove the bobbin case while cleaning and oiling this area (Fig. 14.6).

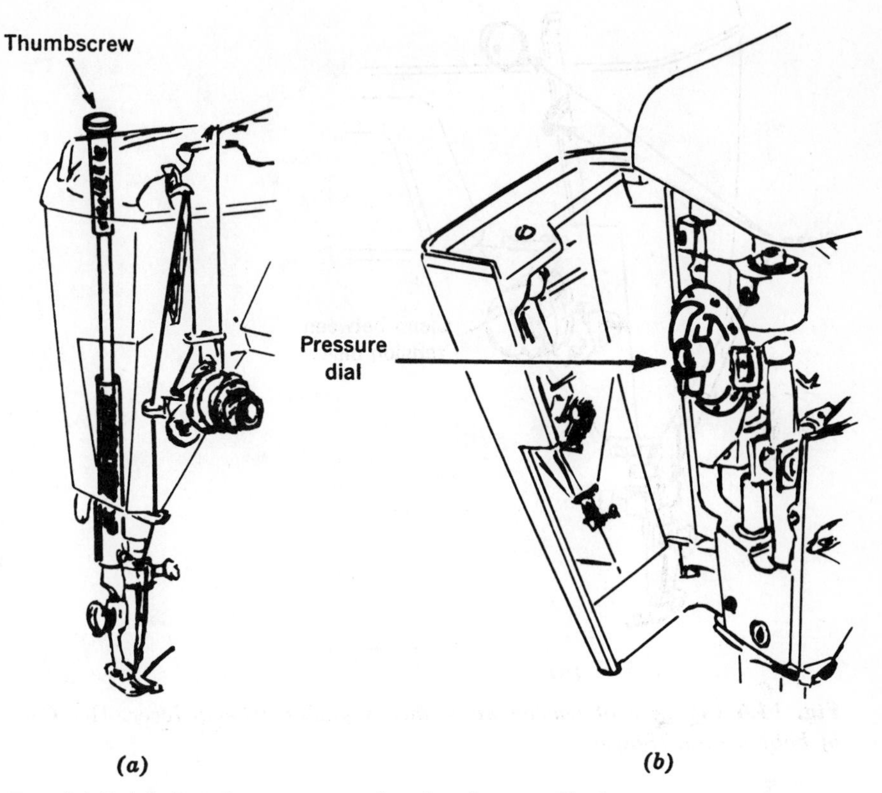

Fig. 14.5 (a) *Regulates pressure by thumbscrew. To increase pressure, turn thumbscrew to right (clockwise). To decrease pressure, turn thumbscrew to left (counterclockwise).* (b) *Regulates pressure by pressure dial. Lower presser foot before turning dial. To increase pressure, turn dial to higher number. To decrease pressure, turn dial to lower number. For darning, turn dial to specified position which releases pressure and permits darning without embroidery hoop. (Singer)*

3. Apply the correct type of oil to moving parts under the throat and slide plate. Slowly work moving parts to enable the oil to cover each part.
4. Remove excess oil by stitching on a scrap of fabric.

Periodic Care

1. The face plate. When the design permits the face plate to be removed, remove the lint and add a drop of oil to moving parts. Oil should be applied to each point where metal moves against metal (Fig. 14.7).
2. Top of the machine. Follow the manufacturer's directions in determining

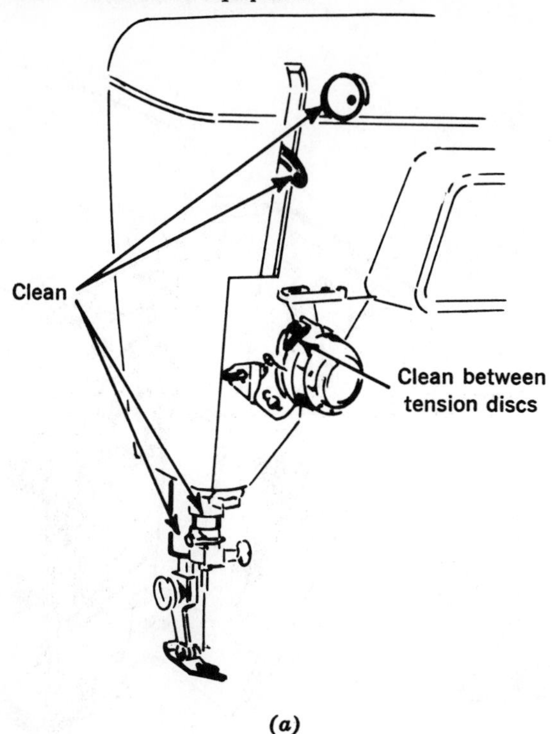

Fig. 14.6 (a) *Care of tension discs, thread guides, take-up lever.* (b) *Care of bobbin case. (Singer)*

whether to use a lightweight oil or a heavyweight lubricant. Take special care to oil wherever one moving part comes in contact with another. Because a lubricant is heavier and longer lasting than oil, it is often recommended for gears, or points of greatest contact of metal on metal. Remove the top cover and oil or lubricate the moving parts (Fig. 14.8).

3. Bottom of the machine. Tilt the machine back and remove the bottom cover. Turn the hand wheel and oil where metal moves against metal. Turn the hand wheel to find this area. Oil or lubricate as recommended by the manufacturer (Fig. 14.9).
4. When oiling is completed, operate the machine with a scrap of fabric under the presser foot to distribute the oil to all moving parts and to remove excess oil.

Storage

Place a small piece of fabric under the presser foot with the needle down through it when the machine is not in use for a long period. The fabric will protect the feet and foot. With the presser foot down,

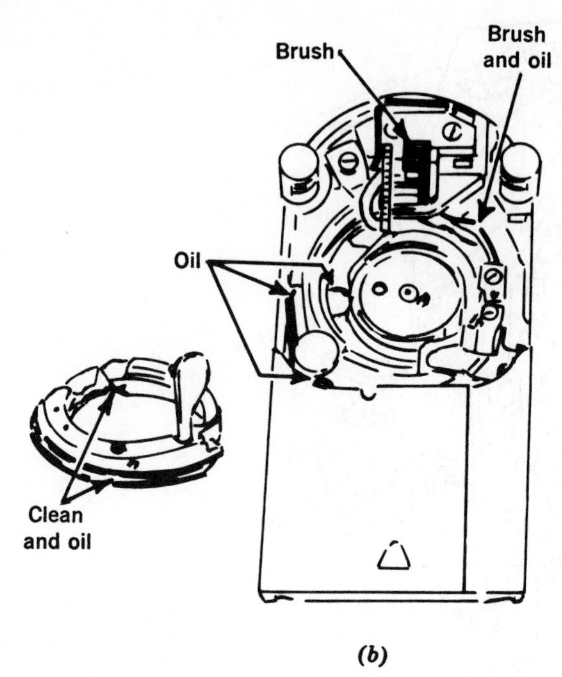

(b)

the pressure spring is relaxed and life is prolonged. Excess oil will be carried onto the fabric by the needle.

Experiments and Projects

1. Study a manufacturer's instruction booklet and identify the parts of the machine.
2. Clean and oil the machine. Bring yours to class.
3. With fabrics of different weights, sew a straight seam. Determine the correct thread and needle for the task.
4. Use each attachment with the machine to determine:
   (a) Its function.
   (b) The skill required to do an effective job.
   (c) The correct procedure for use.
   (d) If instructions are clear and understandable.
5. Invite a sewing machine dealer to bring several brands of machines to class. Ask him to demonstrate them.
6. Determine the pressure necessary to move fabrics of different

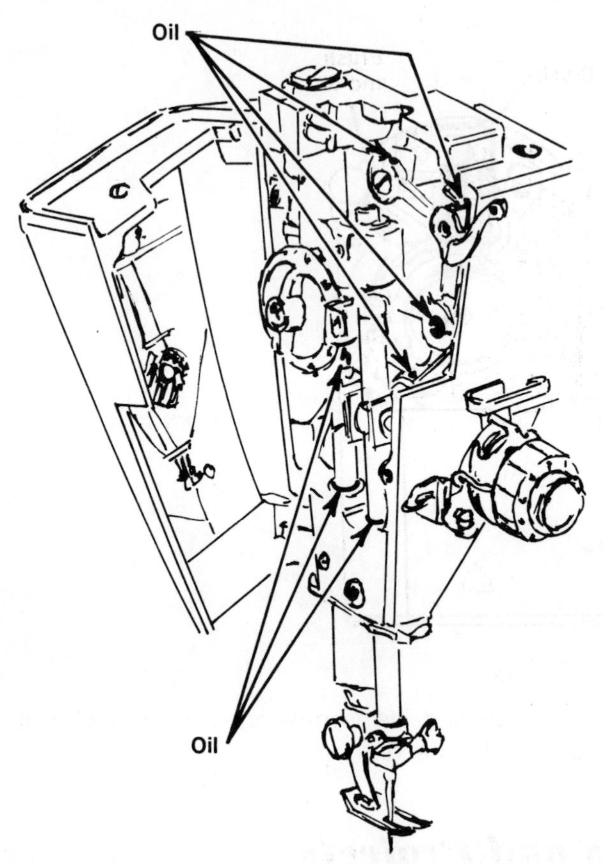

Fig. 14.7 Care of face plate. Clean and oil behind face plate. (Singer)

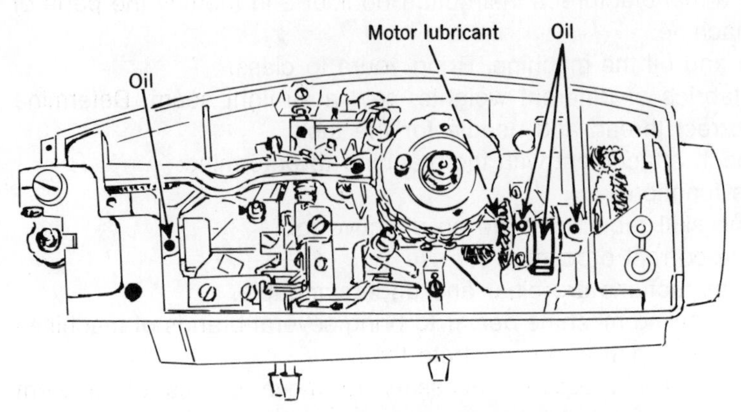

Fig. 14.8 Care of top of machine. Oil and lubricate top of machine. (Singer)

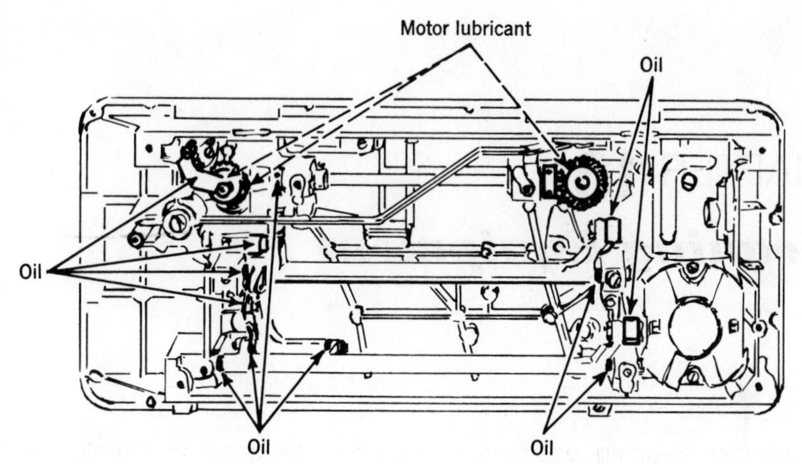

***Fig. 14.9** Care of bottom of machine. Oil and lubricate bottom of machine. (Singer)*

weights smoothly, so that seam edges and ends feed evenly, and no feed marks show on the underside of the fabric.
(a) Soft, sheer fabric.
(b) Crisp, sheer fabric.
(c) Heavy, dense fabric.
(d) Soft, bulky fabric.
(e) Spongy velvet.

CHAPTER 15
Cleaning Equipment

Modern-day cleaning equipment has been designed to handle the cleaning needs that accompany the many developments in floors and floor coverings. Many improvements have been made over the years to keep pace with the cleaning needs of the home; cleaners are lighter in weight; motors are smaller in size, have more power and suction, and have more filters which insure that only clean air will leave the cleaner.

SELECTING CLEANING EQUIPMENT

Various types of cleaning equipment are available today. Proper selection becomes a matter of knowing the composition, texture, or finish of the surfaces to be cleaned; the types of soil that collect in and on them; and the type of cleaning equipment that operates most effectively in removing dirt from each surface.

In selecting an efficient electric cleaner, consideration must be given to the design, since this affects performance, and to the quality of the materials and workmanship used in its construction, since these determine durability. The efficiency of an electric cleaner is dependent on such factors as the size and shape of the nozzle; the size, shape, and design of fan and fan chamber; the type and design of the rotating part; and the speed and efficiency of the motor. The quality of the various parts and the over-all cleaning effectiveness of an electric cleaner can be determined only by carefully controlled tests. In general, however, cleaners combining sweeping and carpet vibration with suction remove more embedded dirt than cleaners employing suction alone.

Equipment that can be used for several jobs should be selected whenever possible, since it reduces the initial investment, simplifies the problem of storage and care, and necessitates carrying fewer tools from room to room during the cleaning process.

ANALYSIS OF THE CLEANING PROBLEM

Although the thorough cleaning of all home furnishings is important, studies indicate that 85% (sometimes as much as 97%) of all the dirt that accumulates in a room is in the carpet. Cleaning carpets and rugs is, therefore, of prime importance in modern homes. To understand the problem it is helpful to have some basic knowledge of carpet structure.

Carpet Structure

Carpet is really a two-part fabric. The interweaving of cotton warp yarns with weft yarns of cotton, jute, or Kraft-cord produces a thick, firm back. Tufts of fibers such as wool, cotton, rayon, nylon, or blends of fibers are woven or tufted into the back in loops and may be cut or left uncut. The method of attaching the pile yarns to the back determines the type of carpet. In the different types and grades of carpet the number of rows of pile tufts per inch lengthwise varies from 4 to 14, and the pile height ⅛ to ½ inch or more.

Between the rows of pile tufts, and between individual tufts, there is much open space where dirt may collect. Owing, then, to these structural characteristics carpet has an enormous capacity for storing dirt which other home-furnishing fabrics do not have.

Carpet Dirt

The properties of the complex substances that accumulate in and on carpet also add to the carpet-cleaning problem. Physically carpet dirt may be separated into three groups according to particle size: surface litter, which includes thread, hair, lint, paper scraps, and sewing-room litter; fine dirt and dust, which, being light in weight, collect in the upper part of the pile tufts; and sand and grit, which, because of their weight, readily become embedded in the base of the pile tufts, in the open spaces between the pile rows, and even in the tiny pockets formed by the interweaving of the backing yarns (Figs. 15.1 and 15.2).

Surface litter is unsightly but less harmful than the other two types of carpet dirt. It is usually more easily removed. Greasy, sticky organic substances, which combine with the finely divided particles just below the carpet surface, cause that portion of the dirt to cling

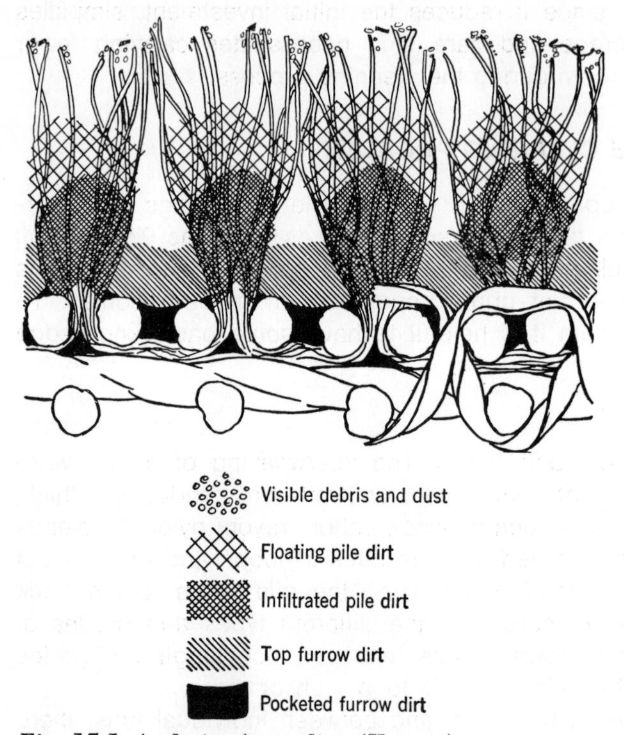

Fig. 15.1 Analysis of rug dirt. (Hoover)

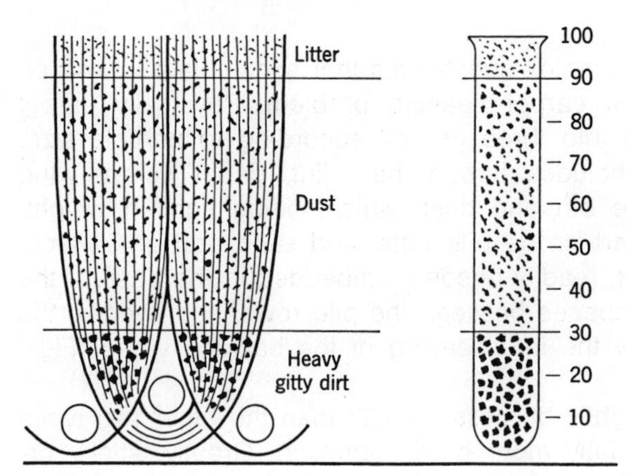

Fig. 15.2 Test tube shows proportion of dirt by weight in average carpet. (Hoover)

tenaciously to the pile yarns. This, and the more deeply embedded dirt and grit, are the most difficult to remove; if the embedded dirt has sharp cutting edges that press against the soft pile yarns under the friction of use, the result may be a loss of pile tufts and a worn appearance.

Carpet-Cleaning Equipment

The primary function of an electric vacuum cleaner is to remove surface dirt and embedded dirt and grit from floor coverings. Most surface litter can be removed readily by surface cleaning methods: the sweeping action of a carpet sweeper or the suction and sweeping action of the vacuum cleaner. To remove both light and heavy dirt and grit embedded below the pile surface requires carpet-vibrating action. This loosens the embedded dirt and bounces it to the carpet surface where suction and sweeping action can carry it into the dirt receptacle.

ELECTRIC CLEANERS

Types and Identifying Characteristics

Electric cleaners are usually classified, according to the principles of cleaning action employed, into straight-suction, motor-driven or air-driven brush, and motor-driven agitator. Straight-suction cleaners are currently made in both tank and canister models. Motor-driven brush machines are available in upright and canister models, and air-driven brush machines come only in canister models. The motor-driven agitator machine is available as an upright model. Each type of cleaner has definite design characteristics by which it may be identified (Figs. 15.3 and 15.4).

Those characteristics that contribute to effective construction, materials, and workmanship for all vacuum cleaners are:

1. A signal to indicate that the bag is full, or needs to be emptied or replaced.
2. Dust bag that can be easily removed or replaced.
3. Dust bag that will do a thorough job of filtering air from dust.
4. Effective adjustment to height of pile of carpet.
5. A control for regulating amount of suction from high to low for different fabrics, plus above the floor cleaning.
6. Tools and hose that lock together securely while in use.
7. Tools that are well designed for the task of cleaning.
8. A bumper around tools to protect baseboards and furniture.
9. A quiet motor.

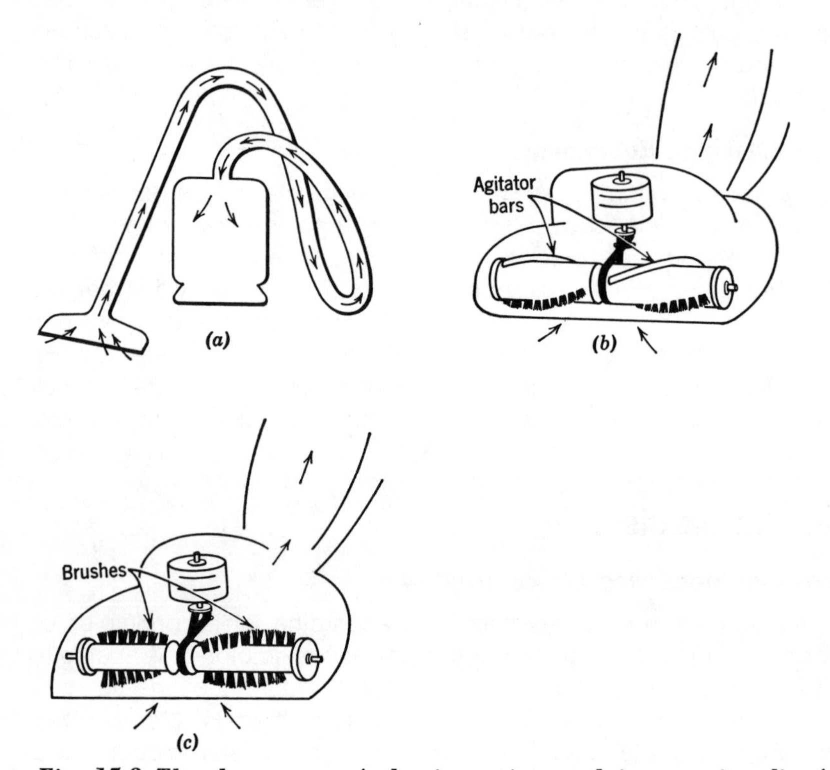

Fig. 15.3 The three types of cleaning action used in removing dirt from carpets and rugs. (a) *Straight suction.* (b) *Motor-driven agitator.* (c) *Motor-driven brush. (Hoover)*

10. Radio and TV filter to prevent interference.
11. Light on front of cleaner.
12. Desirable weight and size to handle easily.

The operating performance of a vacuum cleaner is dependent on suction and air flow, which together may be called the suction air power. The cleaner develops suction by means of an electrical motor-driven fan or blower inside the machine that forces air in one direction and thus creates a vacuum or suction at the opposite or intake end of the machine. New developments and designs of motors and fans have made possible machines with greatly increased suction-air power. The greater the suction-air power at the contact points of the cleaning tools, the greater the cleaning ability of the cleaner.

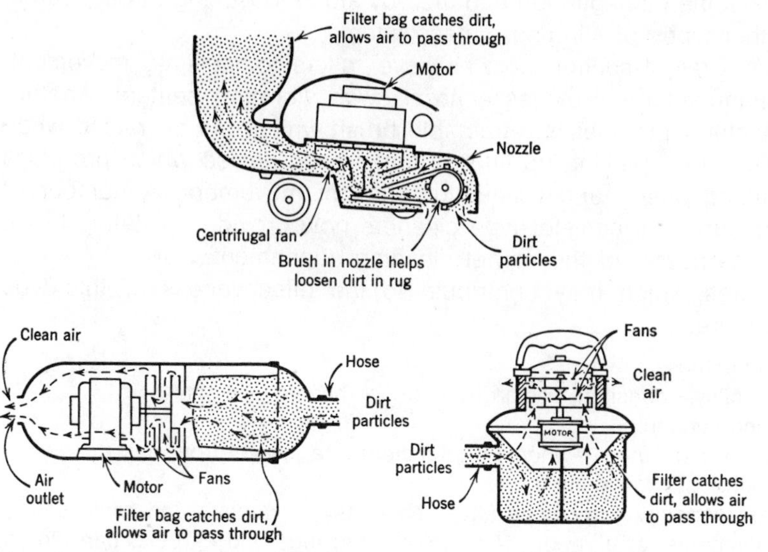

***Fig. 15.4** Principal parts of the upright, tank, and canister cleaners, showing path of dirt as it passes through each part. (Vacuum Cleaners Manufacturers Association)*

In the upright vacuum cleaner, the action is supplemented with mechanical brushing and beating.

The continued cleaning efficiency of the machine depends on how long it can maintain high suction power. As soon as dust and dirt begin to collect in the bag, both suction and air flow are decreased. Thus the suction-air power is doubly decreased.

Straight-Sunction Cleaners, Canister or Tank. This type of electric cleaner may be identified by the slitlike nozzle opening and usually by the absence of any rotating part in the nozzle. Suction is the principal cleaning action, and straight-suction cleaners usually have a motor of ½ to 1⅓ horsepower in order to provide the high suction required. One exception to this is the electric broom, which uses a ⅙-horsepower motor. This is an upright suction cleaner, light in weight and primarily designed to take care of the short daily cleaning usually done with a broom or carpet sweeper. Straight-suction cleaners may also add sweeping action by means of a stationary brush mounted outside the nozzle lips. Such brushes have one or two rows of bristles, and they may be detachable, permanently attached but adjustable, or floating. They combine with suction to provide some pile agitation and thereby aid in loosening thread, yarns, and other types of clinging surface litter.

Other straight-suction nozzles have rollers to facilitate movement. One manufacturer provides a nozzle with vibrating beaters. Another manufacturer provides a retractable brush which can be raised when not needed for picking up litter, a firm plastic device which produces a combing effect, and rollers for ease of movement. A number of manufacturers of canister-type cleaners now provide a rotating brush within the nozzle of the carpet cleaning attachment.

Features which may contribute to the effectiveness of this type cleaner are:

1. Compactness.
2. Portability—consider design, comfort of handle, weight.
3. Strong suction.
4. Ease of movement—wheels of sufficient size, swivel front caster.
5. Effective suction control.
6. Large capacity disposable filter bag—easy to remove and replace.
7. Attachments for a variety of cleaning jobs: rugs and carpets, bare floors, upholstery and fabrics, dusting, crevices.
8. Positive method of locking hose, wands, and tools.

Motor- or Air-Driven Brush Cleaners. Suction is also used in motor- or air-driven brush cleaners, but sweeping and carpet vibration are of greater importance in this type cleaner. Carpet vibration is limited, however, since it must be accomplished by bristle tufts that, by their nature, are flexible and bend or give when they touch the carpet. The rotating brush is the identifying feature of this type of cleaner and usually consists of either one or two helical rows of fairly stiff bristle tufts. A few cleaners use one row of stiff bristle tufts and one of soft bristles. In the motor-driven brush machine, power is transmitted from the motor to this rotating brush by means of a belt passing around the pulley on the end of the motor shaft and around another pulley near the center of the brush roll.

In most motor-driven brush cleaners the nozzle is supported by two wheels placed immediately back of the nozzle opening. This permits the carpet to be lifted to the nozzle lips for the entire length of the nozzle. Since the motor-driven brush is rotated at high speed by the motor, regardless of the speed at which the operator pushes the cleaner, it sweeps the carpet more thoroughly than is possible with the stationary brush on a straight-suction cleaner.

One present-day upright cleaner features an "automatic carpet compensator" that acts, in combination with a four-way selector device, to maintain the correct degree of suction for various heights

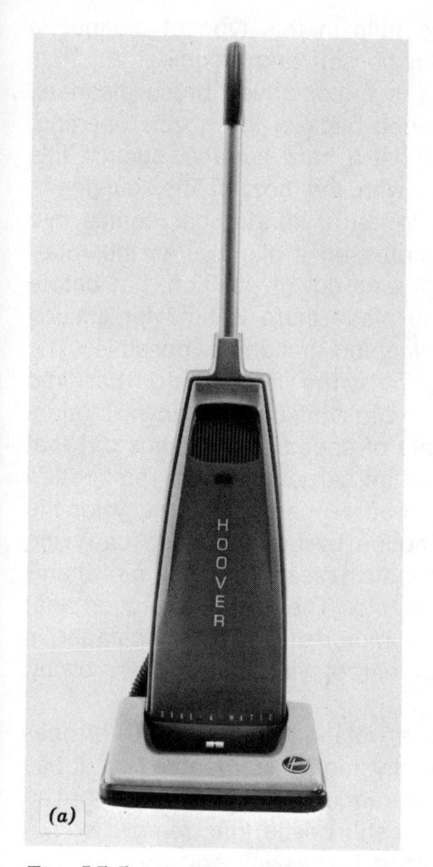

Fig. 15.5 (a) *Motor-driven brush-agitator cleaner. Dial-a-matic. Note bag indicator which shows when disposable bag should be changed. (Hoover)*

of carpet, from low to extra high. Models of other manufacturers also provide dials for selecting the proper suction and the correct position of nozzle for different heights of nap, from flat to shag (Fig. 15.5*a*).

Motor-Driven Agitator Cleaner. This type of electric cleaner also employs three cleaning principles, suction, sweeping, and carpet vibration. Carpet vibration is accomplished, however, by a rotating clinder which differs in design from the brush roll in that it is equipped with two helical vibrator bars. The rigidity of the smooth vibrator bars on this rotating cylinder produces a more intense vibrating action than is possible with flexible bristle tufts, and, since they do not wear even after years of use, their effectiveness remains constant. The function of the bristle tufts in this type of cleaner is largely sweeping; therefore, they can be soft and flexible.

In this type of electric cleaner, as in motor-driven brush cleaners, the nozzle-supporting wheels are placed back of the nozzle opening. Since carpet cannot be vibrated against a hard surface, suction lifts the carpet off the floor into contact with the nozzle lips, curving it slightly up into the nozzle opening. When a vibrator bar comes into contact with the uplifted carpet it depressed it slightly. As the rotating cylinder turns, the vibrator bar passes out of position, but before the row of bristle tufts is rotated into place there is time for suction to lift the carpet again so that sweeping action is possible. The agitator continues to turn and again brings the vibrator bar into contact with the uplifted carpet. This constant up and down motion, or vibration, takes place at a high rate of speed and loosens dirt that has become deeply embedded within the carpet. With this embedded dirt loosened and lifted to the surface of the carpet, it is easy for the air stream, which has been set in motion by fan action, to carry the loose dirt into the nozzle, through the air passage into the fan chamber, and thence into the dirt receptacle. The cleaning action that takes place under the nozzle of the motor-driven brush cleaner is similar except that the amplitude of carpet vibration is less, owing to the flexibility of the bristle tufts.

Electric cleaners equipped with a rotating agitator are erroneously credited with removing a slightly higher percentage of loose pile yarns from carpet than a cleaner equipped with a motor-driven brush. Motor-driven brushes that have stiff bristle tufts remove somewhat more loose carpet pile yarns than those with soft bristles. The removal of loose yarns from pile tufts is, however, unimportant in the overall life of a carpet or rug. Any fibers that can be removed from well-made carpets are only shearings and fibers that were too short to be caught and held by the weft yarns of the carpet backing. Dirt and grit, if allowed to remain in a carpet by ineffective cleaning methods, are undoubtedly more harmful.

Several cleaners, among them the agitator type, are constructed with the bag enclosed within a metal housing. In at least one model the housing carries a dial control that changes the suction, according to need, from use in carpet cleaning to the manipulation of tools. The dials allow the operator to choose high, medium, or low suction for any task. One model has a "Time-to-Empty" signal to let the homemaker know when the bag is so full of dirt that suction has been reduced (Fig. 15.5*b*).

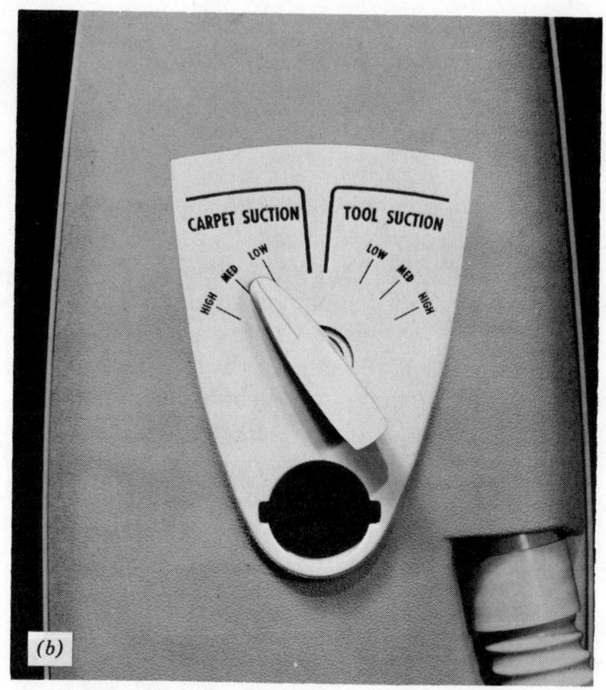

Fig. 15.5 **(b)** ***Suction control. Medium-to-low, correct suction for most rug thicknesses. Low: setting for deep pile carpets, shag rugs, cotton rugs. High: for low pile, sculptured or multilevel rugs. Cleaning tools: high for most cleaning, low for cleaning draperies and delicate fabrics. (Hoover)***

In addition to the usual suction adapter some cleaners are equipped with a blower attachment which can be connected at the exhaust outlet. With the flexible hose and nozzle brushes attached at this point, the cleaner may serve as a hair dryer or for spraying light liquids. Some cleaners include a disinfecting attachment, and others a floor-waxing or polishing brush.

Features which will contribute to efficiency of the motor-driven agitator design are:

1. Two-speed motor: (a) one for carpet and rug cleaning; (b) one for utilizing cleaning attachments.
2. Rotating brush and bar.
3. Ease of movement.
4. Comfortable handle.
5. Positive handle positions.

6. Effective suction.
7. Suction control.
8. Convenient on-off switch.
9. Large capacity disposable filter bag.
10. Release pedal.
11. Ease of connecting attachments.

Small Hand Cleaners. Several companies manufacture small hand cleaners that may be used, in some instances, instead of attachable cleaning tools. One manufacturer makes a basic power unit which can be used either as a hand cleaner or an upright cleaner. Hand cleaners are of two types: straight-suction and motor-driven brush. They usually weigh from 3 to 6 pounds, have a handle for ease of manipulation, and are suitable for cleaning automobiles, upholstered furniture, mattresses, and other surfaces on which the larger cleaner is inconvenient.

Features which contribute to efficiency and convenience are:

1. Lightweight.
2. Compact design.
3. Easy to carry.
4. On-off switch easy to reach.
5. Filter bag easy to remove and replace.
6. Easy to maneuver.

Principal Parts

An electric cleaner has many parts, but the principal ones are nozzle, fan and fan chamber, motor, and filter or dirt receptacle. Attachable cleaning tools are also important and are available with most cleaners.

Nozzle. In motor- or air-driven brush and motor-driven agitator cleaners the nozzle is the part that houses the rotating brush roll or cylinder. The nozzle of the straight-suction cleaner is similar but somewhat narrower and usually contains no rotating part. It is the perimeter of the nozzle, or the nozzle lips, that makes direct contact with the floor covering or other surface to be cleaned. Design of these lips is of considerable importance. In upright, motor-driven brush and agitator cleaners, the nozzle is an integral part of the main frame or casting. In tank and canister cleaners a flexible tube and one or two metal tubes form the connection between the tank or canister and the nozzle.

Nozzles of most cleaners are equipped with bumpers of plastic or rubber to prevent marring of furniture (Fig. 15.6).

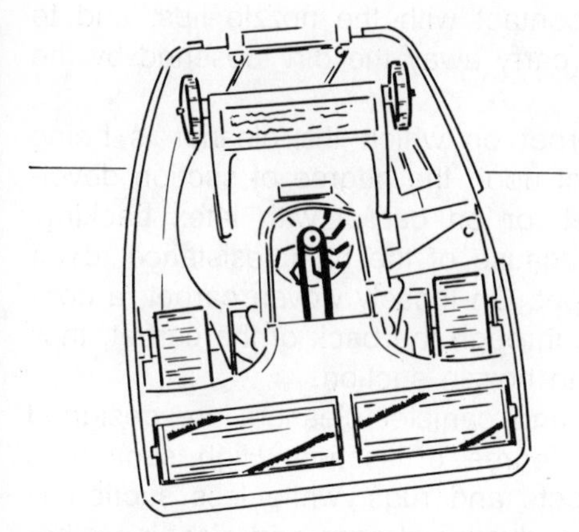

***Fig. 15.6** Nozzle of motor-driven agitator cleaner. Note placement of wheels and pulley. (Hoover)*

Fan and Fan Chamber. The efficiency of the fan in producing suction depends upon the number and design of the fan blades, the size of the fan, the speed of operation, and the design of the fan chamber in which it rotates. The fan in an upright cleaner usually consists of one set of blades arranged about a hub or fan eye. A fan mounted at the end of the motor shaft rotates at the same speed as the motor. It is located in front of the motor, if mounted vertically, or below the motor, if mounted horizontally. A two-stage fan, consisting of a double set of curved blades, is used in tank- and canister-type cleaners and is housed with the motor and dirt receptacle.

The rapid rotation of the fan in an electric cleaner throws the air outward by centrifugal force and creates a partial vacuum at the hub. Outside air at atmospheric pressure rushes in through the nozzle opening to fill this vacuum. This inrushing air reaches its highest velocity and becomes most effective in dirt removal immediately under the nozzle lips. Dirt that is encountered at this point on the carpet surface can, therefore, be removed by this high-velocity air stream and carried into the cleaner and eventually into the dirt receptacle.

Straight-suction cleaners depend largely upon suction for their cleaning ability, and high suction will generally be found in this type of cleaner. In motor-driven brush and agitator cleaners a larger percentage of the cleaning effectiveness is due to sweeping and carpet vibration. The primary function of suction in these cleaners is to lift carpet off the floor into contact with the nozzle lips, and to provide a current of air that will carry away the dirt loosened by the sweeping and vibrating action.

The characteristics of the carpet on which the cleaner is being operated have considerable effect upon the degree of suction developed. On densely woven carpet, or on carpet with latex backing, nozzle suction is usually high because of the high resistance to air flow through the back of the carpet. On loosely woven carpet, a considerable amount of air will pass through the back of the carpet, thus increasing air flow but decreasing nozzle suction.

Current models of most tank and canister cleaners are designed to provide for more than one degree of suction. High suction is needed for cleaning heavy carpets and rugs, while less suction is required for cleaning scatter rugs, dusting drapes and cleaning other furnishings. Some cleaners have the suction control in the cleaner, others in one of the metal parts of the hose or in the wand. Each position has its advantage.

Motor. Universal motors of the type used on electric cleaners vary from ¼ to 1⅓ horsepower. One horsepower is the equivalent of 746 watts, and the output wattage of a ¼-horsepower motor will be, therefore, ¼ of 746 or approximately 187 watts. The efficiency of a motor is the ratio of the wattage output to the wattage input, and the efficiency of universal motors is usually about 50%. The significance of this is that the input wattage must be approximately twice the output wattage. It is the input wattage of an electric cleaner motor that is marked on the name plate, and from this figure the cost of operation can be determined. A universal motor may be operated on either direct or alternating current, but should not be applied to any voltage more than 10% above or below the voltage stated on the motor name plate.

Motors used on electric cleaners vary in speed. Motor speed of tank and canister cleaners is usually higher than that of upright cleaners. High speed in a motor does not necessarily mean high nozzle suction. The amount of suction depends primarily upon size and design of fan and fan chamber and nozzle and the resistance of air passages and dirt container. Exceptionally high speed in an electric motor may tend to shorten motor life. Motors used on electric cleaners usually have a small ventilating fan to assist in maintaining a low operating temperature.

In most electric cleaners the motor rotates on either ball bearings or self-aligning, self-lubricating sleeve bearings, neither of which require lubricating by the user, although they should receive periodic attention at a factory-operated service station.

Dirt Receptacle. The types of dirt receptacles used on electric cleaners vary. The first dirt container used was a bag of cotton cloth, designed to retain the dirt and allow air to pass through the tiny interstices in the cloth. Much research has been done to develop cotton cloth that will serve continuously as a filter and not lose its effectiveness after a short period of use. To meet these requirements one manufacturer has set up the following specifications: the number of looms per weaver must be limited in order to guarantee cloth without defects; the humidity of the spinning rooms must be controlled to reduce variations in yarn tensile strength; the cloth must be tested frequently during the weaving to make sure that it meets the predetermined filter standard.

Next came disposable bags of paper or paperlike materials which are now used as filters in many cleaners. One such filter used on a canister cleaner is cone shaped and fits over a perforated steel motor guard that projects downward into a metal dirt receptacle. Another cleaner uses a centrifugal separator which is mounted above the surface of a pan of water. The separator revolves at a high speed and churns up a cascade of water. The sprays of water wash the dirt from the air. The dirt drops into the water below, and the air passes out through an outlet at the top. One cleaner of the agitator type has a dirt bag of fiber felt, a paperlike material having excellent filtering ability; it is durable and easy to empty because of its smooth inner surface. This bag is protected by a zippered cloth cover.

The size of the dirt container used on an electric cleaner is of great importance and is related to the volume of dirt that the cleaner handles during the usual cleaning period. The required area depends upon the efficiency of the filter, particularly its resistance to clogging.

Attachable Cleaning Tools. Most upright cleaners have attachable cleaning tools, consisting generally of an adapter or connector, a flexible tube, two metal extension tubes, and several nozzle brushes suitable for cleaning various surfaces other than carpet. There is usually a 5-inch brush for cleaning draperies, cushions, screens, mattresses, high moldings, etc.; an 8-inch brush for cleaning bare floors, under low furniture, etc.; a round brush with long soft bristles for light dusting; and a flat nozzle called a crevice tool for cleaning upholstery tufting, the spaces between the arms and seat of upholstered furniture, and other places difficult to reach (Fig. 15.7). These

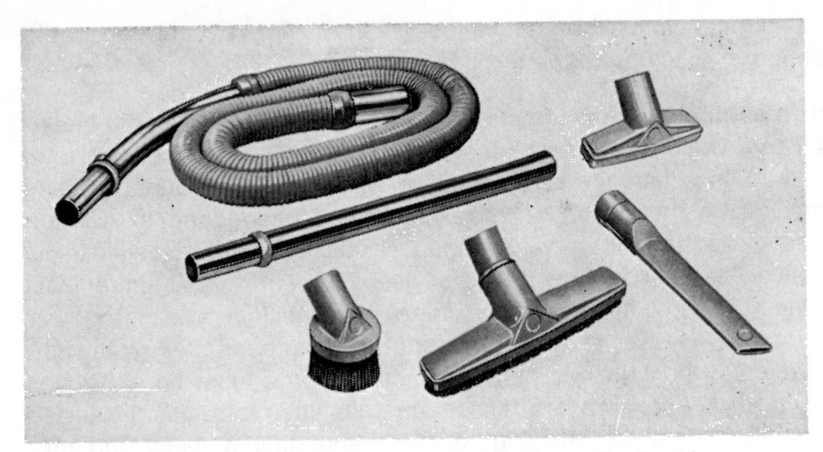

Fig. 15.7 A set of attachable cleaning tools. (Sears, Roebuck)

nozzles are similar to the nozzles furnished with tank- and canister-type cleaners.

Cleaning action by attachable cleaning tools depends on suction and pile disturbance. Such tools are intended largely for dislodging fibrous clinging substances and removing surface dust which is not embedded. A brush nozzle should, however, be so constructed that the maximum air flow is at the bristle tips. This is accomplished by means of a hard-rubber or plastic inner nozzle slightly shorter than the bristle tufts. Without such an inner nozzle or shroud the air flow cannot be concentrated at the point of contact between the bristle tips and the surface to be cleaned. The shroud also prevents bending and collapsing of the brush.

USE AND CARE OF ELECTRIC CLEANERS

Nozzle Height

Unless the carpet is cemented to the floor, all motor- or air-driven brush, or motor-driven agitator cleaners lift the carpet off the floor during the cleaning process. They are, therefore, designed with nozzle-supporting wheels to hold the nozzle lips at the correct height above the carpet. Since carpet thickness varies in different homes and even in different rooms it is necessary to provide some means of adjusting the nozzle. In many machines this is done by a screw, cam, or a foot lever which readily lowers either the front or rear of the machine. An adjustment approximately ⅛ inch above the carpet surface gives the most effective cleaning. If the nozzle is too high, suction cannot lift the carpet; if too low, too much effort is required to move the cleaner. Such manual devices, whatever the type, should be clearly marked to indicate the correct nozzle height.

On the agitator cleaner, and on several motor-driven brush cleaners, the nozzles are self-adjusting for most types of rugs and carpets. This is accomplished by using two wide wheels immediately back of the nozzle and two narrow wheels at the rear of the machine. For cleaning particularly heavy or soft pile carpet these cleaners usually have a mechanical device for additional adjustments of the nozzle.

With the nozzle of a tank or canister cleaner, the contact between the carpet and nozzle lips depends largely on the pressure exerted by the user, since there are no nozzle-supporting wheels. Suction also affects the pressure with which carpet is pressed against the nozzle. For best performance with this type of cleaner it is necessary that the entire perimeter of the nozzle be held firmly against the carpet during the cleaning operation. Maintaining this seal between nozzle and carpet is sometimes difficult particularly when cleaning under furniture. To eliminate breaking the seal some manufacturers use a "hinge type" connection between nozzle and wand. This permits the use of the nozzle under low furniture and beds without tilting of the nozzle.

Frequency of Use

Practically all authorities agree on the importance of cleaning carpets and rugs frequently. In fact, the Carpet Institute recommends that "an electric cleaner be used for a short time every day on each carpet or rug" and "at least once a week all carpet areas be given a **thorough** cleaning with an electric cleaner." The reasons in back of this recommendation are sound. With the exception of certain fibrous materials, such as cat and dog hair, an electric cleaner can quickly remove surface litter and the dust and dirt in the upper part of the carpet pile. It is the deeply embedded dirt and grit that are more difficult to remove and require long cleaning periods. Daily cleaning removes the heavy gritty dirt while it is still on or near the surface and before it has time to become embedded deeply into the pile tufts. Daily cleaning, therefore, requires less total cleaning time per week to keep floor coverings in a sanitary condition.

The total amount of time required weekly to keep carpets and rugs in a home satisfactorily clean depends on the type of cleaner, the type and area of the carpet, the density of the carpet dirt in and around the pile tufts, and the resistance to air flow through the dirt receptacle.

Tests in a manufacturer's laboratory have indicated that several short cleaning periods each week are more effective in carpet dirt removal than one longer cleaning period of equal total length. In tests conducted at the State College of Washington, Roberts found that, for equal dirt removal, straight-suction cleaners must be used for a longer period than other types.

Speed of Operation

A survey conducted by a vacuum-cleaner manufacturer showed 1¾ feet per second to be the average speed at which women operated their cleaners, but tests in the Iowa State University household equipment laboratories indicated that more dirt was removed when motor-driven brush and agitator cleaners were operated at a speed of 1¼ feet per second than at a higher speed. Energy expenditure was also less. With straight-suction cleaners, however, cleaning efficiency increases with increased speed, but not in proportion to the increase in energy required to operate the cleaner.

Cleaning the Dirt Receptacle

The stream of air set in motion by the fan in an upright cleaner starts at the nozzle lips, flows through the nozzle opening, into and around the fan chamber, and through the exhaust outlet into the dirt receptacle. In a tank or a canister cleaner, the dirt is filtered from the air before the air enters the fan chamber and is exhausted. In a cloth or a paper filter, the openings through which air can pass are small. If these openings become clogged with dirt the bag pressure, or resistance to air flow, becomes great enough to reduce nozzle suction and consequently the dirt-removing ability of the cleaner.

Although nozzle suction is dependent upon fan efficiency and the

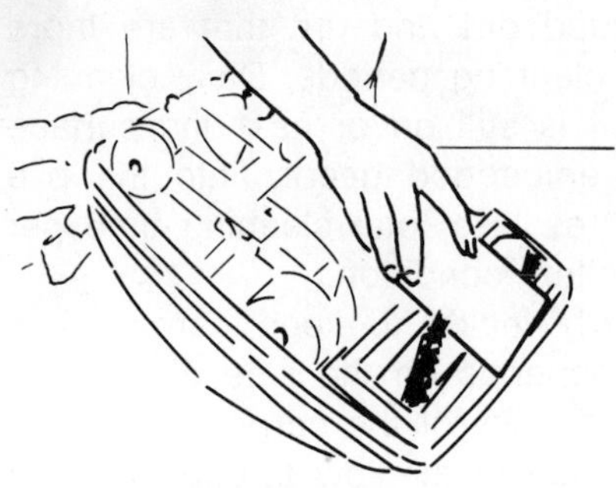

Fig. 15.8 Method of determining correct bristle length. (Hoover)

seal between the nozzle lips and the floor covering, the effective suction is really the static pressure developed by the fan system minus the system losses, which include the back pressure caused by the resistance to air flow in a clogged dirt receptacle. If the filter is allowed to become clogged the back pressure steadily increases and the cleaning efficiency of the cleaner decreases. To maintain high cleaning efficiency, fabric dirt containers should be emptied and cleaned frequently. The water filter should be emptied at least after each cleaning period. Disposable paper bags should be changed before they become full.

Bristle Length

Since the effectiveness of the motor-driven brush cleaner depends largely upon the sweeping and vibrating action of the rotating brush it is necessary that the brush bristles at all times be of sufficient length to touch the carpet as the brush rotates. Recognizing that bristles of all types become worn after a period of use, manufacturers of motor-driven brush cleaners provide some means of adjusting the brush to successively lower positions in the nozzle. The number of adjustments varies on the different makes of cleaners. The method of determining whether the bristles are the correct length consists merely in placing a straightedge across the nozzle opening (Fig. 15.8). If the bristle tufts touch this straightedge they are sufficiently long to touch the carpet as the brush roll rotates. When, however, the bristles are worn so short that, even though the brush roll has been adjusted to the lowest possible position, the bristle tufts do not touch the carpet, then suction becomes the only cleaning action employed by the cleaner although the brush roll is being rotated by the motor. A new brush roll should then replace the worn one.

This same method may be used to measure bristle length on the motor-driven agitator cleaner.

Belt Tension

Electric cleaners that have a rotating brush roll or agitator in the nozzle must be provided with some means of transmitting power from the motor to the rotating part. Rubber belts are quite generally used and have been found to be most practical. The belt passes around a pulley on the end of the motor shaft and around another pulley on the brush roll or agitator and transmits power from the motor to the rotating part. If the belt breaks or becomes inoperative for any reason, so that the rotating part fails to turn, or turns at a speed too low to be effective, the efficiency of the cleaner is reduced.

The belt is, therefore, an important part of electric cleaners of these types.

Belts differ greatly in the total period of time over which they will give satisfactory service by retaining sufficient tension to drive the rotating part at the correct speed. When used in what might be called an average home, a well-made belt should give good service for several months, perhaps a year, barring such accidents as cutting by sharp objects which the cleaner may pick up. It is advisable for the user of an electric cleaner to check the condition of the belt frequently.

Lubrication

The need to oil or lubricate an electric cleaner has now been eliminated by most manufacturers by either self-lubricating sleeve bearings or ball bearings. When an electric cleaner must be oiled by the operator, the manufacturer's directions should be followed accurately.

Electric Floor Scrubbers and Polishers

In kitchens, workrooms, and, occasionally, in other rooms of the house, hard floor coverings are used such as linoleum, cork, vinyl, or asphalt tile. These coverings require occasional scrubbing and waxing to be kept attractive. A machine recently developed for this purpose is the electric scrubber (Fig. 15.9). This machine automatically dispenses the washing solution, scrubs the floor, and then takes up the dirty water. One manufacturer provides, in addition to the scrubbing action, a rinse of clean water to remove any detergent film. This machine can also be converted into a straight-suction cleaner by changing the nozzle.

The electric floor machine used for waxing, polishing, and rug shampooing is similar to the vacuum cleaner. A ¼- or a ½-horsepower motor is connected to a revolving brush by a belt or a system of gears; a handle attached to the frame guides the machine; a cord makes connection with the convenience outlet; and a switch turns the current on and off.

Some floor machines have only one brush; others have two. The brushes rotate vertically on a horizontal shaft, or horizontally around a vertical pivot. Brushes are made of different kinds of fibers, depending upon the use to which they are to be put. Stiff-fibered palmetto brushes are frequently used for scrubbing floors with water and scouring powder; the softer Tampico brushes are used for polishing. Provision for applying wax differs with different machines. On some, the wax is applied to the floor from a container on the machine

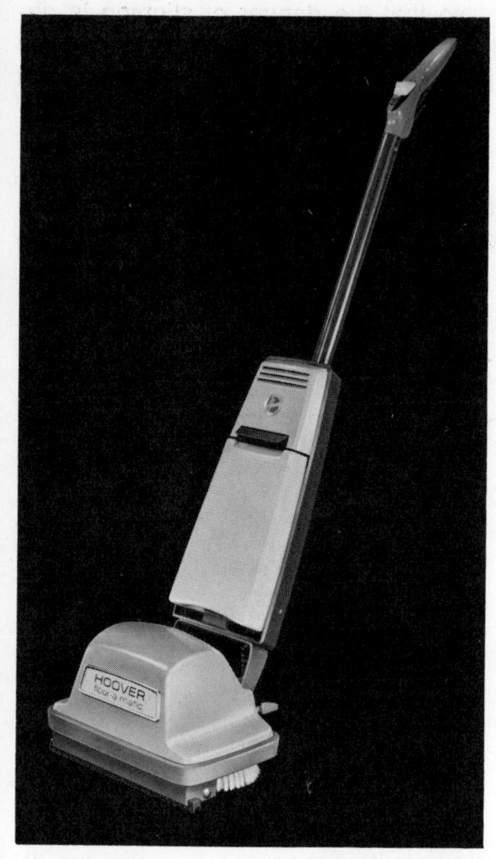

Fig. 15.9 Rug and floor conditioner, Floor-a-matic. Scrub, wax, polish, shampoo, wet pick-up features. (Hoover)

either by an electric pump or by dripping out. With others, it is applied by another brush attachment, by a special wax mop, or by hand with a cloth.

Preparatory to refinishing a floor, it is often necessary to remove accumulations of dirt, or old varnish and paint, with the aid of a steel-wire brush. The wire brush is used with a pad of steel wool at least ½ inch thick. After removal of the varnish and dirt, the floor may be smoothed with a sandpaper disk before applying the new coat of paint, varnish, or shellac.

The Tampico brushes usually give a satisfactory polish, but a higher luster may be obtained with the polishing pad. By these processes the wax is thoroughly worked into the wood and any excess wax is absorbed by the brush, so that the danger of slipping is decreased or entirely eliminated. Wax should be allowed to dry on the floor for 30 minutes before polishing. Never attempt to polish wet wax.

Built-In Vacuum Cleaner

The central vacuum cleaner system consists of a power unit with a dirt-collecting container located in a part of the house separate from

***Fig. 15.10** A built-in cleaning system supplies constant maximum suction power, quiet operation, with no recirculation of dirt or need to empty the dust bag. All you have in the various areas of your home is the slender wand with its numerous attachments for a variety of cleaning jobs; and the flexible, lightweight 25-foot vinyl hose, easily moved from outlet to outlet throughout the house. Dirt is drawn through the hose to the central power unit. It is so quiet that you can hear the doorbell, telephone, radio, TV, and conversation. (Whirlpool)*

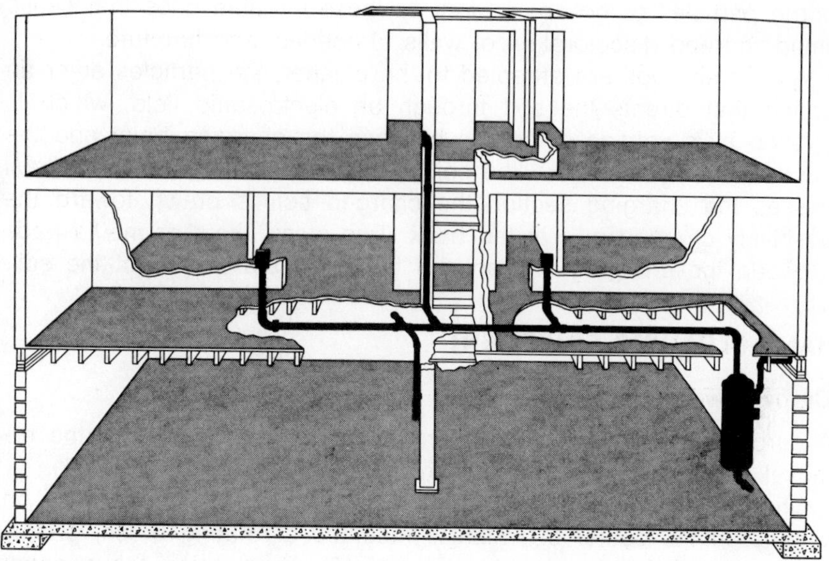

***Fig. 15.11** A typical layout for a two-story house. The power unit may be placed in the basement (as here), in a closet, or in the utility room. The tube-conveying system is connected to one or two outlets on each floor, providing total cleaning coverage. Three or four outlet jacks are sufficient for most homes. (Whirlpool)*

the living areas. The container is connected to dirt-carrying tubes which open into the walls in various areas of the house. To these outlet valves, which have touch-plate covers, is fastened the flexible hose with its wands and tools. Even in a large house, only one, or at most two openings are needed on each floor, because the flexible hose is very long, 25 feet or more and easily bends around corners and between furnishings. The system may be controlled from a master switch or from the touch-plate switch at the wall valve (Figs. 15.10 and 15.11). A central system has definite advantages over the conventional portable cleaner: maximum cleaning at all times, quiet operation, no recirculation of dust, and no machine to push and pull around.

Electrostatic Air Cleaner

A cleaner that employs the electrostatic principle that opposites attract, has been designed to clean the air of dust particles. It is used with central residential heating and cooling systems and may be added to an existing system. This cleaner will remove visible airborne particles or pollen, as well as those invisible ones that cause smudging and discoloration of walls, draperies, and furniture.

As dirt and soil are attracted to the cleaner, the particles enter an orifice that directs the soil through an electrostatic field, which is created by a voltage difference between the charging point and the edge of the orifice. Here the soil is negatively charged. As the soil passes the charging section the charged soil is drawn toward the positively charged grid at the back. Two glass filters or media pads between the charging section and the grid help to collect the soil.

HAND CLEANING EQUIPMENT

Carpet Sweepers

A carpet sweeper is a hand-operated machine designed for the removal of surface dirt from rugs and carpets.

The cleaner consists of a long-handled metal or wooden case on wheels, in the center of which is mounted a rotating brush, preferably made of Chinese bristles (Fig. 15.12). By pushing the sweeper

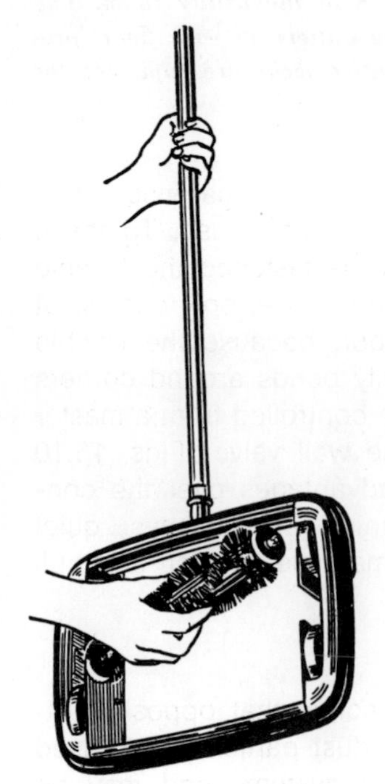

Fig. 15.12 Carpet-sweeper case and rotating brush. (Landers, Frary, and Clark)

back and forth the brush rotates and picks up dust, ravelings, etc. The bottom of the sweeper consists of two pans, one in front and one in back of the brush. As the dirt is picked up by the rotating brush it is carried into the case and dropped into one of the pans. For best cleaning results, a sweeper should be pushed with as little pressure as will do the work and with smooth, even strokes. A rubber bumper around the edge of the sweeper case protects furniture from being marred.

Frequently a metal rod across the top of the sweeper is lifted to open both pans at the same time. Thread, hair, or string that is wound around the axle of the brush should be removed at regular intervals so that the brush may rotate easily and efficiently. The brush should be kept clean, and dust pans should be emptied after each use.

Brooms

One of the most commonly used pieces of cleaning equipment is the broom. Brooms come in a variety of sizes and weights. They are made of broom corn and other fibers, most commonly Chinese palm and Tampico, although some of the newer types are made of nylon.

A good-quality broom-corn broom is made of a mixture of curly and rough fibers stitched together at the top. These fibers should have few split ends, and the splits should be short.

Broom handles are made of smooth-finished maple, pine, and birch. Shellacked, varnished, lacquered, or enameled maple is used on the best brooms, and pine, untreated but smooth finished, or birch is usually used for cheaper ones. Pine wears well, but birch breaks easily.

Broom-corn brooms should not be put in water, as water causes deterioration of the fiber. They should be stored by hanging free of the floor or by resting on the end of the handle.

Brooms of Chinese palm do not flip up the dust and they wear down evenly. As they are not affected by water they are especially useful for cleaning porches, walks, or rough floors such as basements.

To clean Chinese palm or Tampico brooms dip them up and down in clean water. If badly soiled they should be washed in lukewarm suds, rinsed thoroughly in clear water, and hung up to dry. Brooms should never be stored with the fiber resting on the floor.

Brushes

A variety of brushes is available for such tasks as cleaning venetian blinds, draperies, upholstery, and radiators, and for scrubbing floors and toilet bowls; in other words, to meet the requirements of almost every household cleaning job. Some are designed for one specific task, others are for dual use. They tend to gather and hold dust instead of scattering it.

Brushes are made from different kinds of fibers; horse and goat hair, pig bristles, palmetto, Tampico, nylon, and plastics. The bristles are twisted into wire, which should be of the nonrusting, galvanized type, or stapled or cemented into wood blocks. The type to be selected depends upon the household furnishing to be cleaned. Where a stiff brush is desired the bristle brushes are usually the most desirable and satisfactory. Brushes of hair are the softest. Hair and bristle brushes hold dust better than fiber brushes.

Proper care of brushes is important: All brushes should be washed periodically in warm suds, rinsed thoroughly in clean water, shaken to straighten bristles or fibers, and preferably hung to dry. Brushes in which the fibers are set in wooden blocks should be dried with the bristles down if they are not hung to dry. All brushes should be stored hanging on hooks.

Many jobs may be done more easily and with less expenditure of energy by the attachments of the electric cleaner, and they should be used whenever possible.

***Fig. 15.13** Well-designed general storage cabinet for cleaning equipment and supplies. (Hoover)*

CLEANING EQUIPMENT STORAGE

Vacuum cleaners and their tools, brushes and brooms, dust cloths and mitts, cleaning powders and solutions, and waxes should be stored in an easily accessible, centrally located place (Fig. 15.13). A tray, box, or basket with a handle helps in carrying small supplies. If mounted on wheels, it will prevent stooping and conserve time and energy. Shelves for storage should be rounded and edged with a strip of moulding to keep items from falling off. Hooks in a section of pegboard are useful for hanging brooms and brushes.

Experiments and Projects

1. Use soils of various types on carpets to compare effectiveness of soil removal by upright cleaner with that removed by canister and tank types. Oatmeal and cornmeal are examples of soils that will penetrate the pile. Paper scraps and cornflakes will provide surface litter.
2. Select carpet samples of different fibers, pile height, and weave. Determine the type of cleaner that is most appropriate for the care of each sample. Samples of carpets may be obtained from rug and carpet dealers.
3. Study vacuum cleaner information from various manufacturers with regard to:
   (a) Design of basic cleaner.
   (b) Purpose of each attachment.
   (c) Cleaning principle involved.
4. Use scrubber and waxer on samples of smooth floor coverings. A tile of smooth floor covering may be mounted on plywood to provide test squares of a size appropriate for this experiment.
5. Study information on the convertible scrubber-vacuum cleaner.
6. Compare the use of attachments for above-the-floor cleaning.

CHAPTER 16
Lighting the Home

Light as an element of design should be used for visual comfort and to achieve desirable emotional responses. Factors related to visual comfort are those which enable one to see easily, accurately, and in comfort.

Emotional responses are influenced by the manner in which light has been used to communicate ideas about color, texture, shape, form, and line. Structural aspects of the interior may be accented as light is used to interpret visual elements which define space, to denote which surfaces shall be lighted and which shall remain dark, and to convey means by which patterns of brightness may be merged with the structural patterns of the house design. The nature of a space is greatly dependent on distribution and patterns of illumination. As light is used as an integral part of the total environmental design, it means giving the house an atmosphere in which one may respond in the most favorable way to other members of the family and friends.

To further understand the benefits of effective lighting in the home, there is need for an understanding of how man responds to light. Human seeing responses improve as lighting levels are increased. These responses bear directly on human performance and productivity. Mental and physical responses are slower and less precise when lighting is not suited to the seeing tasks being performed. As seeing conditions are improved the following responses are more positive: seeing becomes more reliable and takes less time; texture, color, and fine detail are seen more clearly; human energy is conserved; and productive seeing is prolonged.

Eyes and light work together to provide man with sight—through which approximately 85% of the responses to the environment are experienced. Each aspect of man's growth, development, and performance may be influenced by the luminous aspect of his environment. More specifically, the physical, psychological, psychophysical, and aesthetic responses of man are associated with this environmental factor. The physiological aspects of man's response to light are the following: the structure of the eye, the seeing process, perception, elements of seeing, factors involved in the seeing task, recommended levels for seeing for task performance, neural and muscular reactions to environmental stimuli, and energy. Each factor involves an understanding of man and his physical response to light in the environment. Diagrammatically, the physical-emotional relationships involved in home lighting have been presented by Kirk. These relationships have been demonstrated in many studies (Fig. 16.1).

The elements of seeing are the eye, the source of light, and the object to be seen. Each element must be given careful consideration in terms of the part to be played in effective seeing. The age of the person, his health, and his activity may well determine the quality of the response of the eye to the environment.

Based on research reported in 1969, Mrs. Blackwell[1] determined that the need for illumination on a given viewing task increases as the age of the observer increases. The average performance (contrast discrimination) of those 62 to 66 years of age required seven times more light than the average performance of the 17 to 29 year old.

The source of light will influence the quantity of light available for seeing and the color of the light. The nature of the object or surface to be seen combined with the color of the light source will determine the color to be seen, and the many other factors related to perception of the object or surface to be seen.

The psychological responses of man are related to the quality and quantity of light in the seeing environment. The specific responses which have been found to be related to the lighting in the environment are stimulation, comprehension, motivation, and distraction. Psychophysical responses related to the lighting are visibility, visual performance, and how man evaluates visual responses in terms of physical characteristics.

Many aesthetic aspects of the environment are influenced by the quantity and quality of light. Among those of significance are: spatial effects (bright or dismal—restful or exciting), scale and definition of space (long or small), color and color rendition, texture, line,

[1]Institute for Research in Vision, Ohio State University, 1968.

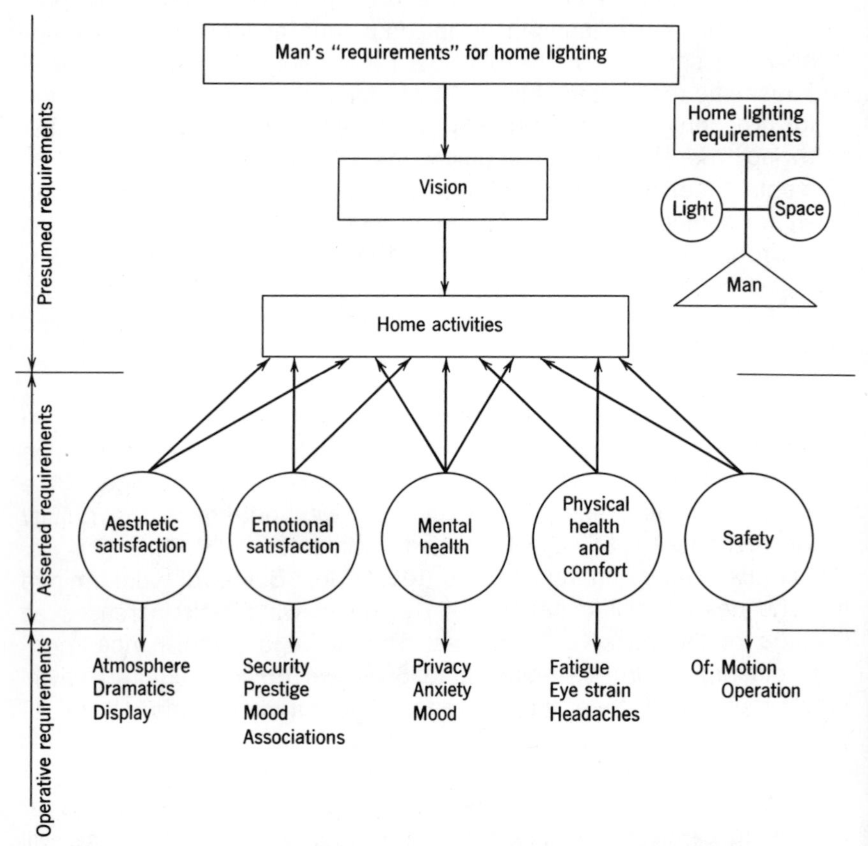

*Fig. 16.1 Physical-emotional relationships involved in home lighting. (*Source: *Alexander Kira, "The Requirements of Lighting in the Home: A Definition of the Problems." Ithaca, N.Y., Cornell University Housing Research Center, March 1958, mimeo.)*

and structural features. Each aspect of the residential interior becomes more significant to those living in the area, as the lighting is integrated into the design.

A basic understanding of the language of light enables one to use light to achieve many objectives.

Physics of Light

Even in very early times the thoughtful observer must have noted that light traveled in straight lines, since it was not possible to see around a corner, but it was not known that light had velocity.

In 1676, Romer, a Danish astronomer, studying the eclipses of the moons of Jupiter, found unexpected variations in the time eclipses occurred, depending upon the position of the earth in its orbit. Romer suggested that these discrepancies might be due to the fact that light had velocity and, by determining the differences in time when the eclipses took place, calculated the velocity at 186,000 miles per second. In 1924, Michelson, employing stationary and rotating mirrors to reflect light between two mountain tops in California, found the velocity of light to be 186,300 miles per second.

The eye sees almost entirely by means of light reflected from an object or surface. Light itself is invisible; it is the agent through which objects are perceived, a form of radiant energy capable of producing sight. There are two theories as to the manner in which it is transmitted. The corpuscular theory, the first one suggested, assumed that the lighted body shot off particles at high speed and in straight lines. This theory fits in with many of the laws of light transmission, but does not agree with others.

The other theory for the propagation of light is known as the wave theory, that light waves are probably a form of electric waves. Light waves show a transverse motion, the waves traveling in a direction perpendicular to the path of the light. This theory of light propagation presupposes a medium. Like an ocean wave, the light wave has a series of crests and troughs, which travel forward, but the medium itself is not carried along. After the discovery of electrons it was suggested that light waves were formed by electrons oscillating in the source of the light.

Measurement of Light

Just as a unit is necessary in measuring volume or distance, so a unit is essential for the measurement of light. The oldest standard light unit was a sperm candle of specified size and rate of burning. The illumination given by a horizontal beam from the standard candle was known as candlepower. The light from any candle flame, 1 inch in diameter, is equivalent to approximately 1 candella. The present light standards are incandescent lamps, although the term candella has been retained.

Footcandle. It is a basic law of light that the intensity of illumination on any surface varies inversely as the square of the distance from the source of the light, the surface being at right angles to the beam. This law is strictly true only when the source is so small that it approaches a point in size. The relationship may be expressed in an equation:

$$\text{Intensity of illumination} = \frac{\text{Candlepower of lamp}}{\text{Square of distance from lamp}} = \frac{\text{cp}}{d^2}$$

Since, in the equation, cp = ft-c × d^2, intensity of illumination is measured in footcandles. The illumination on any surface or the distribution of light around a source may be determined by a photometer. The footcandle meter is a portable photometer. By means of a light-sensitive cell, which is independent of the personal error of the investigator, the density of light rays from any direction is converted directly into electric current, the amount of which is indicated on a scale, calibrated in foot-candles in terms of tungsten-filament standard lamps. The sensitivity of the cell must be equivalent to that of the eyes if it is to evaluate the intensity of illumination in accord with terms of luminous sensation.

A small footcandle meter, adapted to home use, has a scale, calibrated from 0 to 75 or 100 foot-candles. Captions above the scale indicate what part of the range is suitable for various tasks. A perforated disk may be placed over the cell for reading illumination values beyond the scale, the footcandles read being multiplied by 10 (Fig. 16.2).

Lumen. Light flux or flow, the quantity which may be used, is measured in lumens. If light from a standard candle falls upon a screen 1 square foot in area, every point of which is 1 foot distant from

Fig. 16.2 Footcandle meter for determining the level of illumination at the place where light is used. (General Electric)

the candle, the screen is said to receive a lumen of light. Incandescent lamps commonly used in the home give from 10 to 20 lumens for each watt of energy consumed, and fluorescent lamps, 60 lumens or more. A photometer is used to measure the lumen output, in establishing the rating of any lamp.

Foot Lambert. The lambert is the unit of brightness. Brightness of the light source or of a reflecting surface within the line of vision and the contrast between the brightness of general and task illumination have a very important bearing on visual acuity and eye strain. Brightness should, therefore, be determined.

Relationship among light terms. It is customary to express light terms by letter symbols. Equations may then be written to show the relationship among the different terms. I is the symbol for candella; E for illumination in footcandles; F, for lumens; and B, for brightness. $I = Ed^2$; since E, footcandles, is easily obtained from the meter reading and the distance, d, may be measured, I, candlepower, can be determined. $F = 4\pi I$; lumens are measured over a spherical surface, so the multiplication factor 4π is used. The area of a sphere is $4\pi r^2$. In this case the equivalent of r^2 is d^2 contained in the equation for I. $B = E \times$ Reflection factor. The reflection factor for any surface is obtained by dividing the footcandle reading of light reflected from the surface by the footcandle reading for the light incident on the same surface. To avoid error in results, the surface for which the reflection factor is to be determined should be at least 12 inches square.

The footcandle meter may also be used to determine the transmission factor of any translucent material by dividing the footcandles of light transmitted through the material by the footcandles of incident light. The brightness B of the translucent material may be obtained by multiplying the footcandle reading of transmitted light, when the cell of the footcandle meter is held against the translucent material, by a constant 1.25. The constant 1.25 is used because the light, which is incident on the cell almost equally from all angles, is diffused by the material.

For greatest eye comfort the brightness of the shade on a desk lamp should not exceed 50, on floor and table lamps not over 200 when the background is light, and not over 100 when the background is moderately dark.

Reflection of Light

Man sees primarily by reflected light, thus the importance of the ability of each surface to reflect light. Light colors for room surfaces and furnishings are important, and often essential, in achieving desirable relationships between areas in the view. Dark colors will need higher levels of light for effective viewing in the room (Table 16.1).

Table 16.1
Recommended Reflectances and Munsell Values for Residences

Surface	Reflectance (Percent)	Approximate Munsell Value
Ceiling	60 to 90	8 and above
Fabric treatment on large wall areas	45 to 85	7 to 9.5
Walls	35 to 60[a]	6.5 to 8
Floor	15 to 35[a]	4 to 6.5

Source: Lighting Design Criteria For Interior Living Spaces, Illuminating Engineering Society.
[a]In areas where lighting for specific visual tasks takes precedence over lighting for environment, minimum reflectance should be 40% for walls, 25% for floors.

The reflectance of these major surfaces and the amount of light they receive serve to form the background against which most seeing takes place. They are always within our field of view. As a result of our visual comfort, mental attitudes and emotional moods are influenced by the relationship which exists between the sources of illumination, the surfaces and objects to be seen, and the backgrounds against which they are viewed.

Glare

When light is transmitted or reflected in such a way that an unpleasant sensation affects the eye, fatigue is increased and attention distracted, the cause is usually glare. Glare is light wrongly directed. Direct glare comes from unshaded or insufficiently shaded lamps shining directly in the line of vision, reflected glare from a bright light falling on a polished or glossy surface from which it is reflected into the eye. Very shiny paper in a book or magazine may cause reflected glare. Spotty light gives another kind of glare, glare by contrast. This happens when one or two direct-type floor or table lamps make bright spots of illumination in a room and the rest of the room is in comparative darkness. The eye must constantly readjust itself to the brightness and the darkness, and discomfort of disabling effects result. The level of general illumination should not be less than ⅓ of the footcandles supplied by task lighting sources. A ratio of 1 to 3 is desirable.

Other causes of glare may be associated with the nature of the source—luminance, size, and location; number of sources in the area; brightness or luminance in the field of view; individual sensitivity to glare; and the nature of the visual task.

Daylight in the Home

Natural light may be provided by windows and skylights in the modern home. How adequate this light is would seem to be determined by the number of windows and skylights. Glass, however, may absorb or reflect as much as 35% of the light that falls upon it. The thickness of the glass does not seem to have much effect, but the smoothness does, a rough surface absorbing radiant energy more readily than a smooth. Fifteen to 20% of the total light may be absorbed by dirt on the glass, and in very smoky localities this may amount to as high as 25 or even 50%—sufficient argument for keeping the windows clean. Glass has, nevertheless, an important part to play in diffusing the light, and although it reduces the available light, it is not an advantage to have the windows open, for the glass distributes the light and so gives more illumination at the farther side of the room than would be obtained if no glass were there.

Other conditions, such as the use of shades and draperies, also greatly influence the amount of daylight that is available in a room. Randall and Martin have studied these conditions and report the following rather startling facts. Drawing the roller shade so that the upper half of the window is covered, cuts off 60% of the daylight, but only 14% is lost when the shade is restricted to the upper fifth. This great variation is due to the difference in the light transmitted by the upper and lower halves of the window. Except in congested districts of a city, the light coming through the upper half is largely reflected from the sky. By contrast, the light transmitted through the lower panes is reflected from the ground, other buildings, and shrubbery, all of which are so much darker than the sky that they may absorb more light than they reflect. Windows may themselves be a source of glare if the walls on either side are dark in color—again glare by contrast. To avoid this, paint walls and ceiling light. An off-white is a very satisfactory finish.

Opaque shades may cut off 98% of the light, whereas translucent shades usually transmit as much as 17.6%. A screen covering the whole window reduces the available daylight 50%, but not more than 15% when the screen is only over the lower half. Painting the screens diminishes the light further, 9% on an entire screen, 3% on a half; and when the painting is repeated year after year to prevent rusting, the transmitted light is reduced to a still greater extent. For these reasons bronze or copper screens which do not require painting are preferable.

Windows are regarded as one of the most satisfactory means of decorating a room. Curtains and draperies, however, may cut off as much as 75% of the light. Removing the valance may double the available daylight at the farther side of the room. Moreover, many heavy draperies and curtains increase the shadows and tend to produce a spotty condition which may result in glare, whereas a clean window and very thin curtain materials diffuse the light and soften the shadows. If draperies are desired for purposes of decoration, the windows must be more numerous and larger to make up for the loss in light. A window area equivalent to one-fourth of the floor area is desirable. The use of glass blocks for wall construction has improved daytime lighting of the home.

In the solar house, the exterior walls on southern exposures are almost entirely of glass. This large number of windows increases the daylight illumination of a room to a high level. Since excessive contrasts between bright and dark areas of the room are consequently eliminated, glare is avoided and ease of seeing greatly improved.

Further experiments proved that the illumination on a vertical plane parallel to a window was twice that on a horizontal plane at the same place. A simple trial will convince anyone of the truth of this statement and show the need for holding the book or magazine at the proper angle when reading, instead of allowing it to lie on a table.

Electric Light

Fixtures are classified as follows: the direct which throws all the light downward on the surface to be illuminated; the indirect which directs light on the ceiling and the upper walls from where it is redistributed in well-diffused light throughout the room; and semi-indirect which may direct approximately 60% of the light toward the ceiling and send 40% downward on a limited area, or vice versa, the semi-direct which distributes 40% over the ceiling and 60% downward. Another type, the multidirectional, frequently an enclosing globe of opal glass, diffuses light evenly in all directions.

There are two sources of artificial light, the incandescent bulb and the fluorescent tube. They are installed in ceiling fixtures, floor and table lamps, and structural light designs that direct the light to where it is to be used. Light is needed for general illumination and for task and decorative purposes.

Whatever the type, wherever the location, all lamps should be shaded. If indirect luminaries are used, the ceiling and upper walls must be light in color to reflect the light. Dark colors absorb a large portion of the light rays.

Figure 16.3 illustrates the different types of fixtures and indicates the distribution of light below and above them.

It will be observed that the greatest quantity of light is obtained

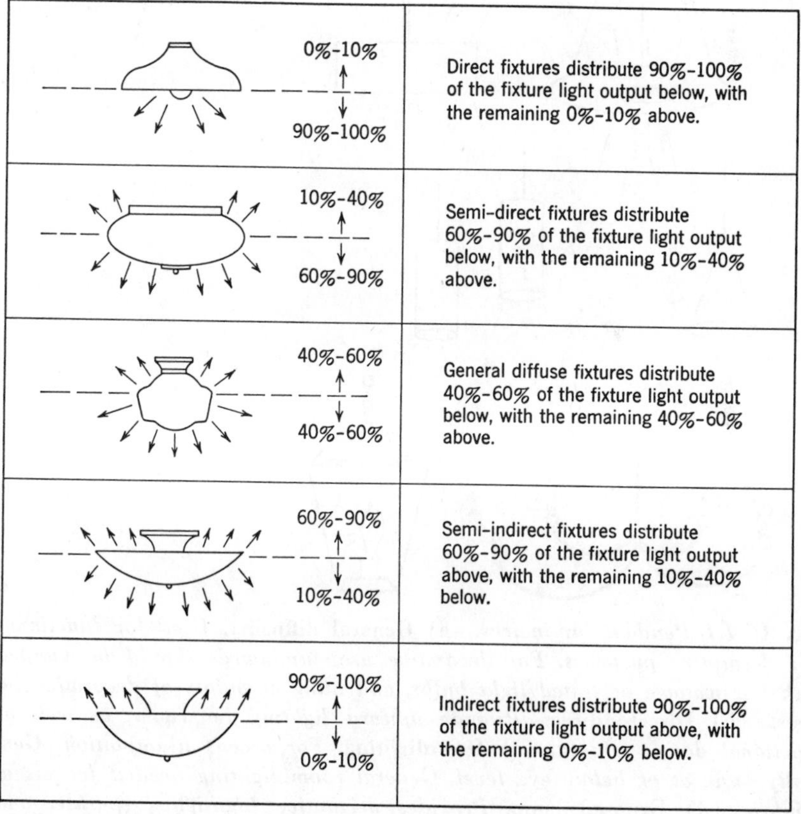

Fig. 16.3** All lighting fixtures are classified according to the way they distribute their light. If an imaginary horizontal line is drawn on a level with the source of the light, the amount of light directed above and below the line determines the fixture classification. (A Method for Predicting General Lighting Levels in the Home**, *Home Lighting Section, Westinghouse Electric Corp., Bloomfield, N.J.)*

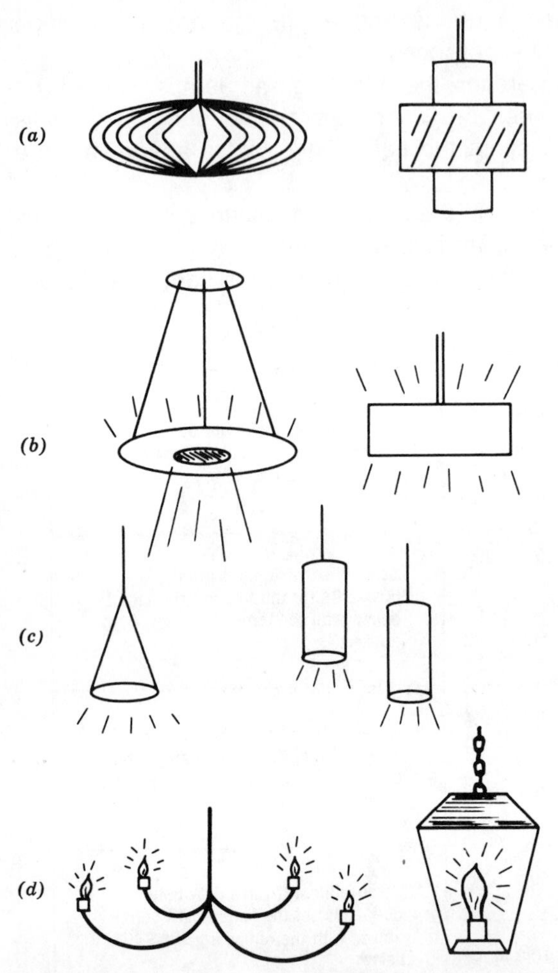

Fig. 16.4.1 Pendant luminaires. **(a) General diffusing.** ***Used for functional and decorative purposes. For decorative use, luminaires should be lamped with low wattage or tinted light bulbs, and hung at points of decorative interest.*** **(b) Direct-indirect.** ***Provide upward lighting for room, as well as functional down lighting.*** **(c) Downlighting.** ***For accent illumination. Generally hung at or below eye level. General room lighting needed for visual comfort.*** **(d) Exposed lamp.** ***Provides decorative highlights, sparkle, and accent.***

by the direct luminaire. It concentrates the light at the desired surface and the minimum amount is lost in transmission. The quality, however, is inferior, for the space covered by the light is comparatively small and forms an objectionable contrast with the darker

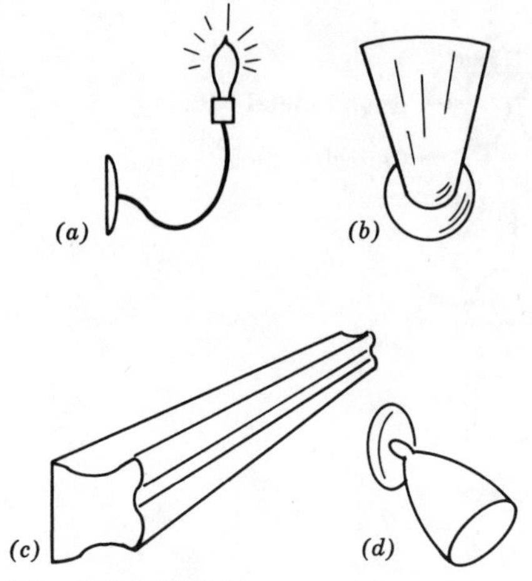

***Fig. 16.4.2** Wall-mounted luminaires.* (a) *Exposed lamp.* (b) *Small diffuser.* (c) *Linear.* (d) *Directional.*

surroundings so that glare is produced. The totally indirect light loses most in transmission and absorption, on account of the length of path the rays traverse, but the light is highest in quality. The almost perfect diffusion practically eliminates shadows. The semi-indirect fixtures share the advantages and disadvantages of the other two types according to the proportion of light which is directed to ceiling or floor (Fig. 16.4).

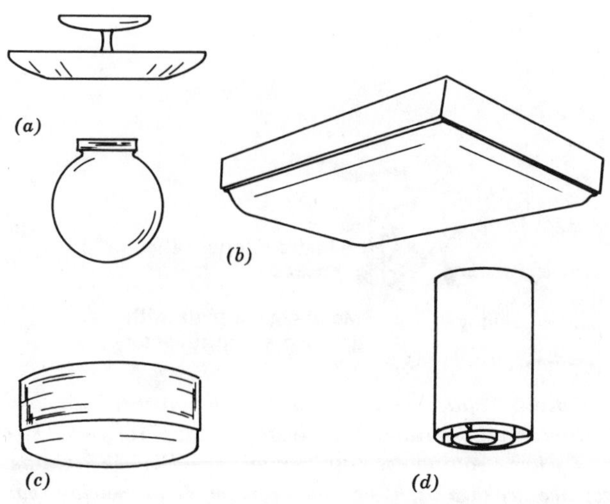

***Fig. 16.4.3** Ceiling-mounted luminaires.* (a) *and* (b) *General diffusion.* (c) *Diffuse downlighting.* (d) *Downlighting.*

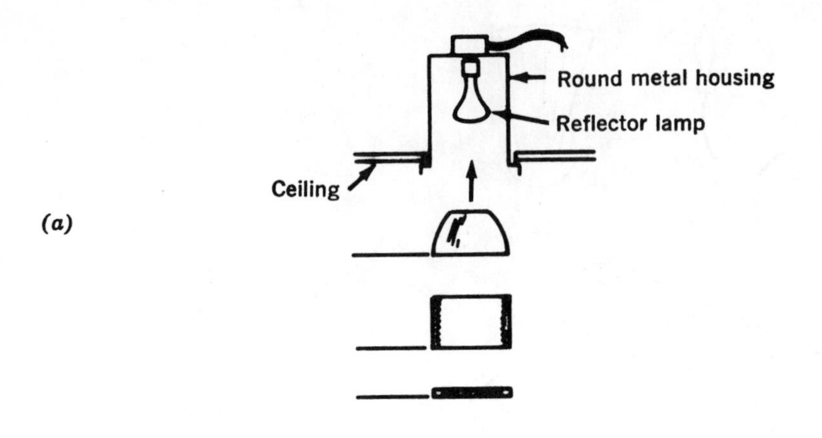

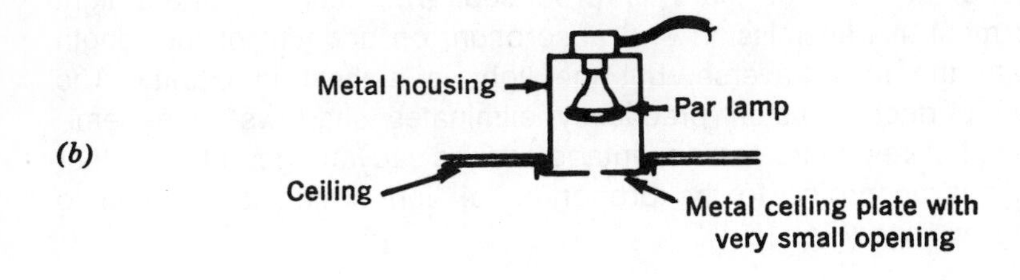

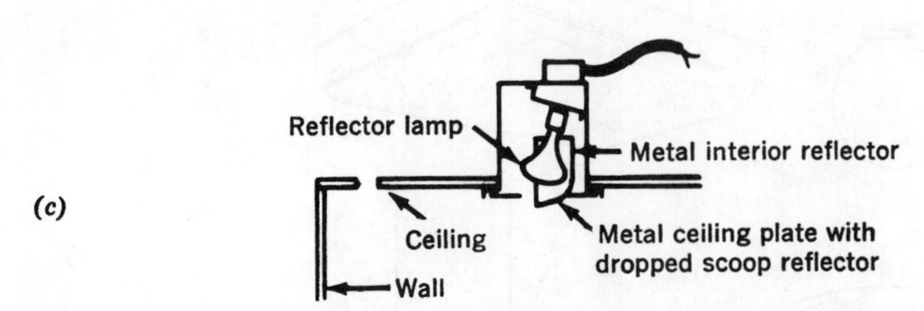

Fig. 16.4.4 *Recessed lighting equipment. Accent lighting over plants and tables.* (a) *Accent lighting, wall lighting, task lighting, stair lighting.* (b) *Low-level accent lighting where equipment cost is vital.* (c) *Picture lighting, sculpture lighting, fireplace surface lighting, piano music and sewing machine lighting. (*Source: *Lighting Design Criteria for Interior Lighting Spaces. Residence Lighting Committee of the Illuminating Engineering Society, 1969.)*

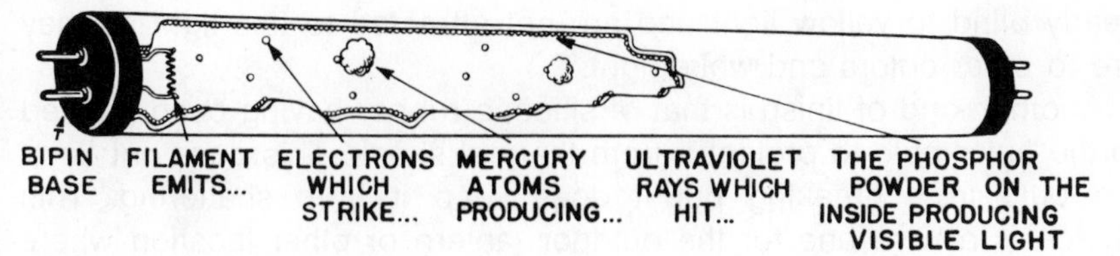

***Fig. 16.4.5** The fluorescent tube. (Westinghouse)*

Electric Light Sources

Incandescent Light Sources. The incandescent light source is the most popular form of lighting in the residential setting. Light from this source is produced by heating a tungsten filament to a white-hot stage. Due to the manner in which light is produced, this source also produces high quantities of heat.

Light source manufacturers provide information on the bulb sleeve or wrapper about the wattage, average light output in lumens, and average hours of life. The basic relations of wattage, lumen output, life, voltage, and efficiency become significant when selecting these sources. As the voltage is increased, the current is increased, lumen output is increased for a given wattage, but life is decreased. The reverse is true when the voltage is decreased.

Incandescent light bulbs come in a variety of rated lives. In the case of the 60-watt bulb and under, the rated life is approximately 1000 hours; the 75 watt and over are generally rated at 750 hours. An exception to this is where longer life is desired. The extended life service lamp has a rated life of 2500 hours. Some manufacturers use different shapes, finishes, or names to identify longer life ratings.

The finish on the incandescent light bulb is important to the quality of light. White-light sources may be finished with inside frost and soft white. The inside frost lamp gives off an identical quantity of light to that of the clear lamp, however, the soft white finish does absorb approximately 2% of the light that normally is generated within the source.

Another finish deals with color. Color is produced with incandescent lamp bulbs through the application of enamel or ceramics on the bulb wall. These finishes produce certain colors by absorbing all colors other than the one desired and therefore transmit the selected color. Colored light becomes less efficient in terms of light output but this should not be a basic concern as these sources are more appropriate for decorative or festive purposes.

Some colored sources are selected for very specific purposes. The yellow bulb is useful for controlling insects. Night-flying insects are nearly blind to yellow light and are not attracted to this light as they are to other colors and white light.

Another kind of finish is that of silicone rubber having been applied to the outer bulb to protect it from thermal shock. This does not keep the bulb from breaking but it does keep it from shattering. This choice is a wise one for the outdoor lantern or other location where the rain and snow might affect the bulb.

An applied finish which has been designed for specific purposes is the dichromatic finish. This finish is a microscopic thickness of vacuumized salts applied as an interference filter. In one case, this finish has been applied to a reflector surface which projects as much light out as possible yet transmits the heat out the back of the lamp. A second example of the use of this technique is with the dichro color application. The dichro color has the layers built up on the inside of the lens. This permits an interference filter to filter out the unwanted colors and allows only the desired color to pass through. The method that is used for introducing the dichromatic finish influences the wattage of the bulb as well as the degree of color transmission and density. These sources also are found in the flood and spotlight distribution which permits greater flexibility than the other type of lamps.

A type of process lamp which is used very frequently in the home is the reflector type or R lamp. The bulb comes in a variety of sizes ranging from R14 to R20, R30, and R40. These are available in both the floodlight and spotlight distribution, in color as well as white. The entire set of lamps cannot withstand exposure to water or thermal shock. The colored bulbs come in six colors which are very suitable for residential applications. The blue-white enhances complexions and red toned objects as well as enhancing blue colors while the pink is very suitable for application on red brick walls and other areas where a warmer tone might be most suitable. The colors are designed for decorative purposes.

Other lamps in the decorative category are often used as bare or exposed bulbs. These include the glow lamps, flicker lamps, wobble filaments, opaque lamps and many things that are designed to be seen instead of seeing. The consumer needs to recognize that these sources should be used in low wattages to avoid the excessive brightnesses of the high wattage bulbs.

FLUORESCENT LIGHT SOURCES

The **fluorescent lamp** is tubular with a small filament, known as an electrode, sealed into each end and a small amount of mercury in the tube. When a switch is thrown, the electricity first heats the electrodes and vaporizes the mercury, then arcs across the tube from one electrode to the other (Fig. 16.5). Electrons given off from the arc collide with the mercury atoms, producing ultraviolet rays. These invisible rays impinge on the phosphor powder with which the tube is coated, changing the invisible ultraviolet rays into visible light.

There are four fluorescent tubes commonly recommended for home use: warm white, cool white, deluxe warm white, and deluxe cool white. Both warm whites blend well with incandescent light; the deluxe warm white especially enhances the complexion and warm tones of furniture and decorations. The deluxe cool gives very accurate color rendition and is recommended if cool greens and blues predominate. In the deluxe tubes, there is a loss of about 25% of the light output.

A fluorescent tube has an average life of 7500 hours on a cycle of 3 hours burning per start. The frequent starting of a tube may affect the depletion of the emission coating on the electrodes, causing a blackening at both ends of the tube; when the coating is exhausted, the light will tend to flash off and on.

Fluorescent tubes, as a general rule, operate most successfully at an ambient temperature of 80°F, but some low-temperature tubes have been developed, with special glowswitch starters, to start at O°F. This has made possible the use of fluorescents out of doors. Rapid-Start tubes of 30 and 40 watts, with intensity selectors, may be dimmed from full brightness to nearly full blackout.

Most tubes give approximately 60 lumens per watt. They are comparatively cool to the touch while in operation, but do produce enough heat to affect air conditioning. The prices of the fixtures are high in relation to the price of the current they use; the cost of current represents only about 30% of the total cost of the light.

PORTABLE LAMPS

A study of the design characteristics of portable lamps will enable the consumer to better determine the function that may best be served by the lamp. Each lamp provides some level of general light. Some may provide a decorative emphasis in the interior design. Others may act as a portable lamp for close-seeing tasks, provide general light, and be a very satisfying aesthetic design as well.

The portable lamp that is best designed to fulfill the lighting requirement for close-seeing tasks has some specific characteristics that may serve as guides in selection. These features are associated with specific parts of the lamp: the height of the lamp to the lower edge of shade; the size, color, and density of the shade material;

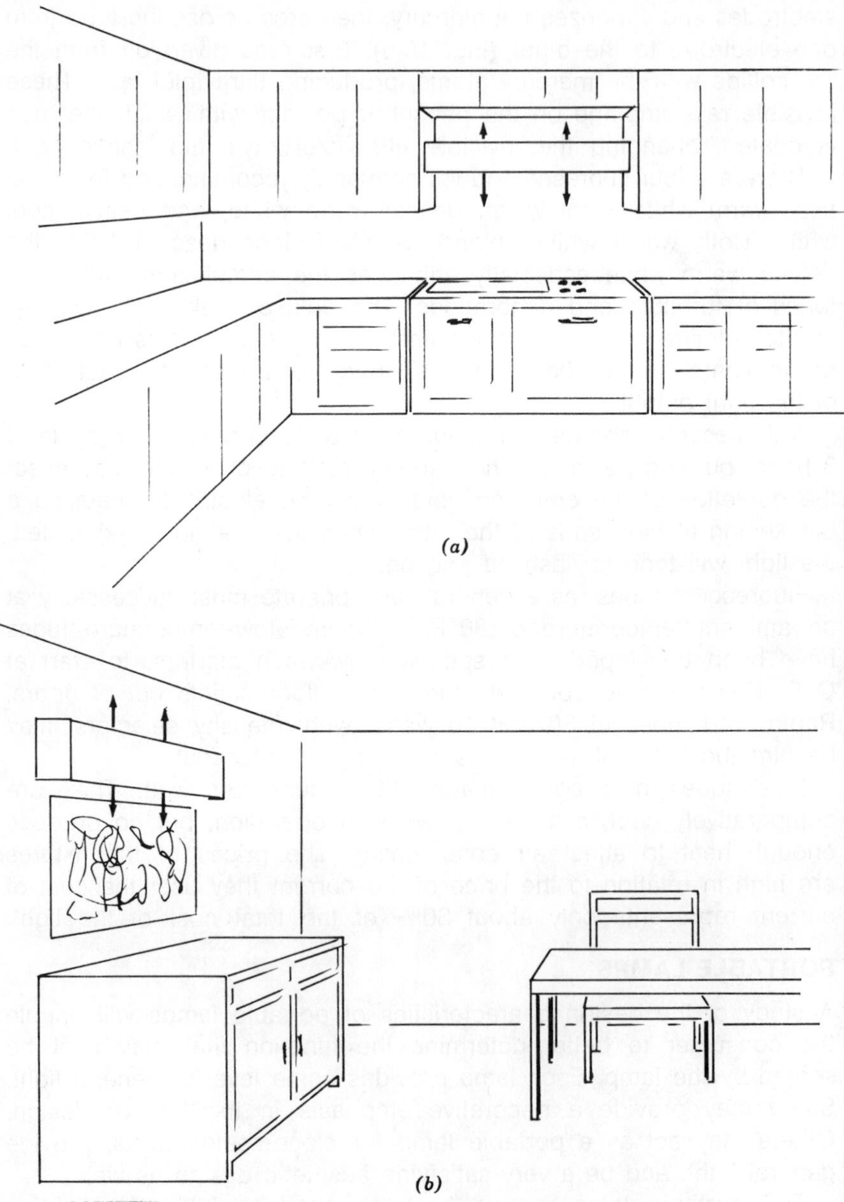

(a)

(b)

Fig. 16.5a and b Wall brackets (high type). High wall brackets provide both up and down light for general room lighting. Used on interior walls to balance window valances both architecturally and in lighting distribution. Mounting height determined by window or door height.

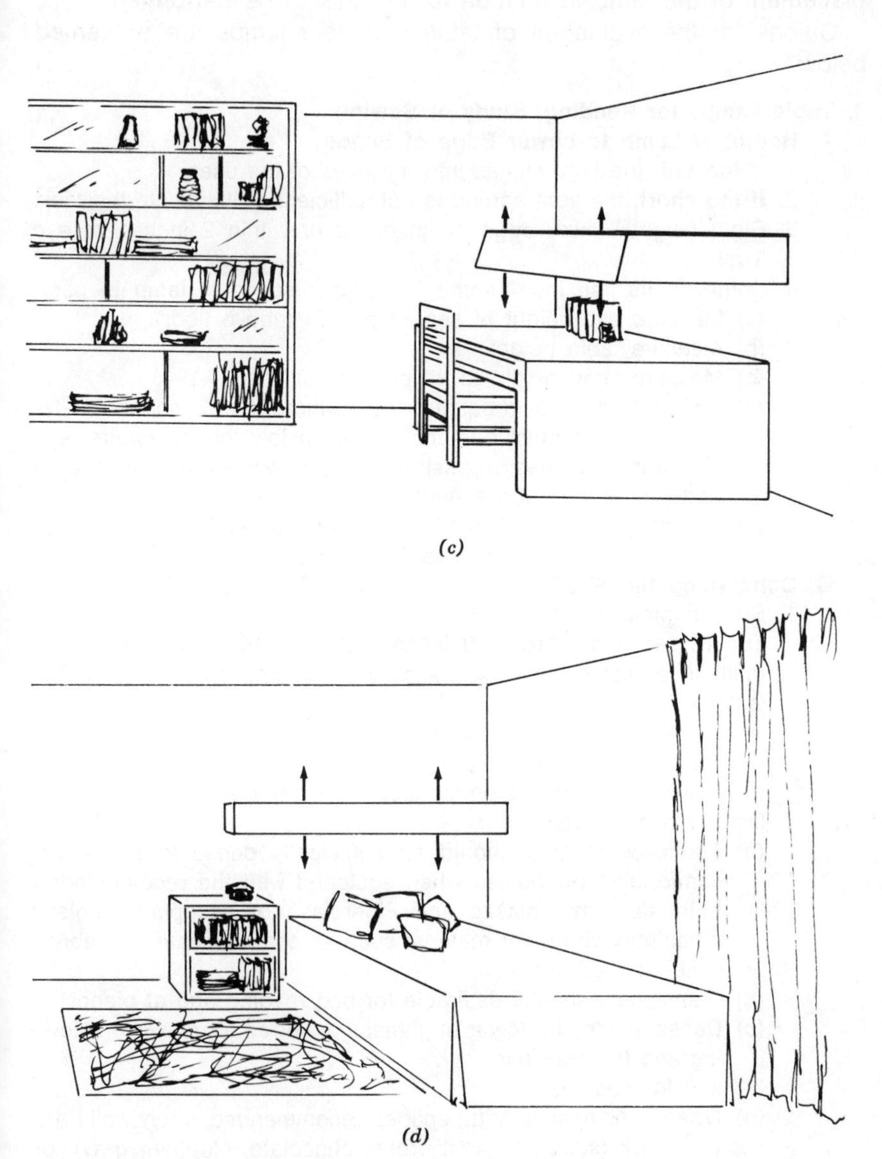

Fig. 16.5c and d Wall brackets (low type). Low brackets are used for special wall emphasis or for lighting specific tasks such as sink, range, or for reading in bed. Mounting height is determined by eye height of users, from both seating and standing positions. Length should relate to nearby furniture groupings and room scale.

the position, diffusion properties, and wattage of the bulb; and the placement of the lamp in relation to the task to be performed.

Guides for the evaluation of table and floor lamps are presented below.[2]

I. **Table Lamps for Reading, Study or Sewing**
   A. **Height of Lamp to Lower Edge of Shade**
      1. **If too tall,** the light shines into the eyes of the user.
      2. **If too short,** the light spread is not sufficient to illuminate the task.
      3. Right height—lower edge of shade at or within 2 inches of eye level.
      4. Some chairs "sit low," some "sit high"; so, to fit lamp to user:
         (a) Measure eye height of seated person (above floor).
         (b) Measure table height.
         (c) Measure lamp height to lower edge of shade.
         (d) "b" plus "c" should equal or be no more than 2 inches greater than "a." For most persons seated in low lounge chairs, eye height is 40 inches; in desk or straight chair—44–46 inches; in bed—20 inches above mattress.
         (e) For study desks, "b" is 29–30 inches, "a" is 44 inches; therefore, "c" should always be 15–16 inches.
   B. **Concerning the Shade**
      1. Size (minimum).
         Top 8½ inches Depth 10 inches Bottom 16 inches, except in multiple socket or shallow shades with top shielding.
      2. Color.
         (a) Inside—white or near-white.
         (b) Outside—preferably natural or blended with wall color.
         (c) Dark shades not recommended for study desks.
      3. Density or translucency.
         (a) Translucent types should be sufficiently dense to avoid uncomfortable brightness when equipped with the recommended bulb size (thin plastic and Fiberglas not acceptable unless combined with other material such as opal laminates or fabric-covered plastic).
         (b) Translucent shades desirable for bed reading and at pianos.
         (c) Dense shades for lower brightness preferred for television viewing and for desk use.
         and for desk use.
         (d) Opaque or near-opaque shades recommended when walls are very dark (such as forest green, chocolate, elephant gray), or when walls are a very brilliant color.
   C. **Under the Shade**
      1. Position of bulb in shade—best lighting distribution when bottom of socket is even with (or slightly lower than) bottom of shade.

[2]**Source.** General Electric Company.

2. Diffusion or "softening" of light for reduction of "shine" from glossy surfaces.
3. Wattage—minimum of 150 watts in any single-socket lamp for reading, sewing, or study.
   (a) Use of three-way 50/100/150 allows a low level for relaxing or conversational atmosphere.
   (b) 50/200/250 used in 8-inch diffusing bowl gives nearly 80% more light.
   (c) Some large-scale lamps are equipped with mogul socket and diffusion bowl and use the 100/200/300 three-way.
   (d) In lamps with two or more sockets, the total should add to between 120 and 180 watts (the latter essential when sockets are in base-down position).

D. **Placement of Lamps**
1. Place lamp for best quality and quantity of light for the task.
2. When styling calls for much taller lamps, it is desirable to place them in floor lamp position to assure the comfort of the user.

II. **Floor Lamps for Reading, Study and Sewing**

A. **Height of Lamp to Lower Edge of Shade**
1. Placement slightly to rear side of user makes possible heights greater than 42 inches to lower edge of shade.
2. If height to shade bottom is 40–42 inches, lamp may be placed in table lamp position.
3. Lamps 47–49 inches to shade bottom usually have mogul sockets for 100/200/300 three-way lamps. Lamps 43–47 inches to shade bottom have medium sockets for 50/100/150 or 50/200/250 three-way or single 150 watt.

B. **Concerning the Shade**
1. Size (minimum) in inches.

	Top	Depth	Bottom
(a) Senior	10	10	18
(b) Swing-arm	10	10	16
(c) Junior	10	9	16
(d) Swing-arm	8–10	9	13–16

2. Color and density (see **Table Lamps**).

C. **Under the Shade**
1. Diffusing bowls give greatest amount of diffusion.
   Bowl size—10-inch diameter for 300-watt bulb and 100/200/300 (mogul socket); 8-inch diameter for 50/100/150 or 50/200/250 (medium socket)
2. Wattage—minimum of 100 watts for casual reading, 200 or 300 watts for prolonged reading or sewing.
3. Position of bulb in shade.
   Husk, or socket covering, should be 1½ inches below shade.

D. **Placement of Floor Lamps**
Place lamp for best quality and quantity of light for task.

The most common forms of portable lamps are the table lamp, floor lamp, wall-hung lamp, dresser lamp, and the pole or tree lamp. Each form may be used to satisfy the lighting need for close-seeing activities. As research progresses and the lighting needs of the human being are better understood, the quantities of light for a seeing task increase and the design of portable lamps should be very critically evaluated.

Special consideration needs to be directed to the lamp designed for study at the desk. In recent years, much attention has been directed to the design of the lamp that will provide 70 footcandles of high quality light on the desk. The design of the lamp has varied with each manufacturer, yet the basic design criteria fulfill the 70 footcandle requirement established by the Illuminating Engineering Society.

Structural Lighting Elements

Structural lighting may be used for general, task, or decorative purposes in the home. These forms become integrated into the basic design of the house structure. Figures 16.5 and 16.6–16.10 illustrate the various designs and their purposes.

***Fig. 16.6** Lighted soffits. Bathroom or dressing room soffits are designed to light the user's face. They are used almost always with large mirrors and countertop sinks. Length is usually tied to size of mirror. Add luxury touch with an attractively decorated bottom diffuser.*

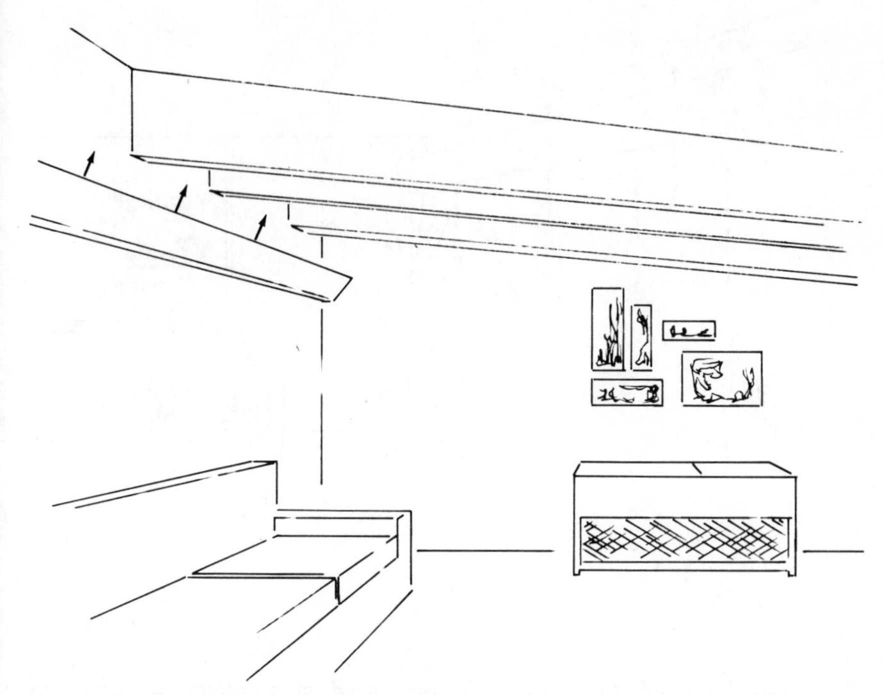

Fig. 16.7 ***Coves. Coves direct all light to the ceiling; should be used only with white or near-white ceilings. Cove lighting is soft and uniform, but lacks punch or emphasis. Best used to supplement other lighting. Suitable for high-ceilinged rooms and for places where ceiling heights abruptly change.***

USE OF ELECTRIC LIGHT IN THE HOME

Various factors affect the amount of artificial light in any area: the color of ceiling, of walls, and, to a lesser extent, of floors and floor coverings, the color of furnishings, of draperies, and of decorative accessories. Dark colors tend to absorb light, light colors, to reflect. If dark colors are preferred, more light must be provided to give a high enough level of illumination to prevent eye strain. Usually, light ceilings and light upper walls are recommended.

At the **entrance** a fixture may be placed at either side of the door or above it to direct light on the steps and on the faces of visitors. When the fixture does not light all of the steps adequately, additional lights should be provided on the edges of the steps or by means of a post light. An illuminated house number is a convenience.

The **entrance hallway** should be well lighted by a ceiling fixture

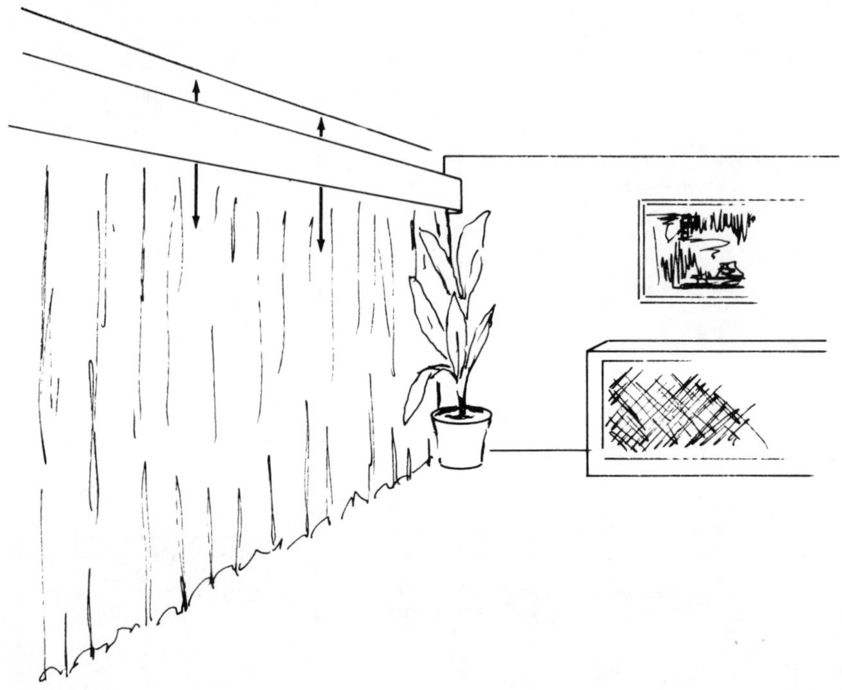

***Fig. 16.8** Valances. Valances are always used at windows, usually with draperies. They provide up-light which reflects off the ceiling for general room lighting and down-light for drapery accent. When closer to ceiling than 10 inches, use closed top to eliminate annoying ceiling brightness.*

and other supplementary light as needed. The fixture in the lower and upper halls must be located so that no shadow will fall upon the stairs. Lights should be placed at both the top and bottom of the stairs so the light will always be ahead of the person in ascending or descending (Figs. 16.11*a*, 16.11*b*).

The **living room,** used for so many different kinds of activities requires a variety of lighting techniques. In the modern home, ceiling fixtures are frequently omitted. Semi-direct floor and table lamps supply some general illumination, which may be increased by cove lighting concealed in troughs around the walls, by valance lighting over windows or by cornice lighting placed close to the ceiling along one or more walls of the room (Fig. 16.12). Emphasis is directed toward raising the level of the overall illumination in a room to such an extent that local lighting from floor and table lamps will no longer cause spotty areas against a somewhat dim background. Portable lamps will supply the additional light needed beside easy chairs, at the piano, or at the desk.

Fig. 16.9 *Lighted cornices. Cornices direct all their light downward to give dramatic interest to wall coverings, draperies, murals, etc. May be used also over windows where space above window does not permit valance lighting. Good for low-ceilinged rooms.*

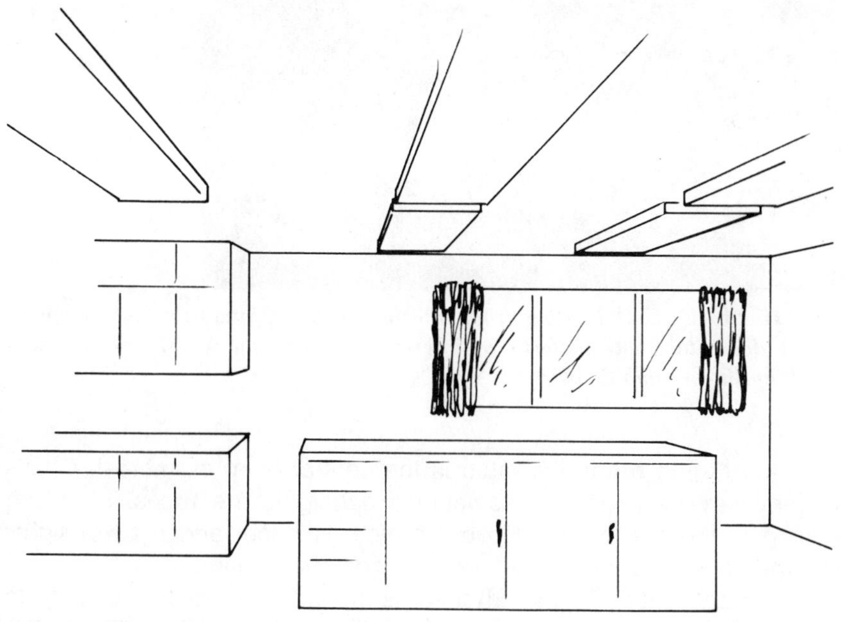

Fig. 16.10 *Luminous wall panels. Luminous wall panels create pleasant vistas, are comfortable background for seeing tasks, and add luxury touch in dining areas, family rooms, and as room dividers. Wide variety of decorative materials available for different covers.*

Fig. 16.11 (a) ***Hall lighting from interior of closet filters through louvered closet doors to create patterns of interest on floor and walls. Low level of light outlines bamboo planting.***

In the **dining room,** the table is the central point of interest. Chandeliers may be used and sometimes spotlights are recessed in the ceiling in the same area. Cove, cornice, valance, and bracket lighting may supply general, task, and decorative lighting.

A central fixture in the **kitchen** is hung close to the ceiling to illuminate all parts of the room equally. Additional lights over the several work areas—sink, range and mix centers—are needed to provide light for critical seeing tasks. When lights are installed below

Fig. 16.11 **(b)** ***General hall lighting for safety; task lighting for cleaning. Light for appreciation of beauty of floor and wall finishes. Emphasis on bamboo planting. (General Electric)***

wall cabinets, they should be placed below the front edge of the cabinet to direct the light away from the face (Fig. 16.13).

Bedrooms are used for many purposes besides sleeping: sewing and studying, reading and letter writing, dressing, and applying make-up. A central fixture is essential for general illumination. Floor or table lamps at special locations and boudoir lamps meet other needs (Fig. 16.14). Dressing-table lights should have sufficient height that the shade center will be located in line with the center of the

***Fig. 16.12** Living room. Beauty of walls accented with cornices employing both incandescent and fluorescent sources. Luminaires mounted on ceiling beam accent picture and floral arrangement. Portable lamps contribute light for general, task, and decorative purposes. (General Electric)*

cheek. The lamps should throw the light on the face rather than on the mirror. The light fixture for reading in bed should be placed on the wall above the head of the bed, high enough to throw a wide circle of light, not just on the book but also on the surrounding area. For this purpose, a tall lamp on a bed-side table may be used. A spotlight fixture, that allows one person to read without disturbing a second person in the same room is not desirable because of the great contrast between the brightness of the book and the rest of the room.

For a night lamp, a small louvered light may be built in above the baseboard, or a bipost lamp with an opaque shade attached above it may be plugged into a convenience outlet.

Closets should be lighted; the bulb may be operated by a pull-chain or by an automatic switch in the hinge of the door.

Lights in the **bathroom** should be placed on either side of the mirror and in the ceiling above the head to illuminate all of the face

***Fig. 16.13** Kitchen ceiling luminaire provides general light. Task lighting over counters, at the sink, and at the range. Combined, these sources of light accent finishes on walls, floor, and cabinets.*

***Fig. 16.14** In bedroom ceiling mounted luminaires provide general light. Valance light accents beauty of drapery fabric and supplies some general lighting reflected from the ceiling. Wall mounted luminaires provide light for grooming at mirror. (General Electric)*

equally. Large bathrooms should also have a central fixture. If the shower compartment is completely enclosed, a moistureproof fixture should be installed in the ceiling of the shower stall, but controlled by a switch outside the shower space. A night light in the bathroom is a convenience.

The sun porch is usually a part of the living rooms of the house and is lighted in a similar manner. An unsheltered terrace requires entrance lights like those at the other outside doors of the house.

A recreation or family room is lighted in the same way as the living room (Fig. 16.15). Adequate general lighting is especially desirable, and fixtures should probably be protected by wire guards if table tennis or other games are played in which balls or rings are tossed around.

To prevent accidents the laundry area should have the necessary number of fixtures to allow the homemaker to work without shadows falling on the several appliances. Basement stairs should be as carefully lighted as those in the front hall.

It is not always possible to purchase new lighting fixtures for a house already equipped with fixtures, but often the old equipment may be improved at comparatively little expense. Bare lamps may be shaded, dark shades replaced with light-colored ones.

Several different types of fixtures that screw into lamp sockets are obtainable. One has a louver, delivering a high level of direct illumination. Another has a silvered bowl, giving a shadowless, glareless light, largely indirect. Plastic bowls or diffusers, often quite shallow in design, are made to clip on to a bare light bulb. They are comparatively inexpensive.

Eyesight Conservation

Light is valuable, as it has a bearing on the protection and care of eyesight. Sight is the only one of the senses that is dependent upon an outside agency for its functioning, and it is, therefore, doubly important that this agent should be adequate in doing its part.

Primitive man lived out of doors, did few tasks that required close vision, and went to rest at nightfall. The level of his illumination was high, several hundred times, perhaps at noon even several thousand times, the amount of light that modern man receives from artificial sources. The fact that sunlight is high in foot-candles does not prove that artificial illumination should be. Sunlight comes from a distance and is adequately diffused by sky and atmosphere; the source of artificial light is near at hand. Small concentrated sources often produce sharp shadows and harsh effects, and consequently cause eyestrain and, indirectly, mental and nervous strain. The important consideration is to have the light properly directed and distributed.

This objective is probably attained when seeing is performed with assurance, ease, and speed. No hard and fast rule can be stated; eyes differ, and light requirements vary with age. Comfort is not

***Fig. 16.15** Portable lamps with three-light sources provide high levels of light at chair and couch for tasks such as casual reading and sewing. Lower levels are appropriate for conversation or TV viewing. Ceiling luminaires provide high quality and quantity of general light and accent light on the fireplace brick wall and other decorative objects. (General Electric)*

always a safe criterion, for it is frequently a matter of individual preference and depends on habit, which changes with time. Real eye discomfort is associated with aching, redness, burning, scratchiness, and a sensation of fatigue. Older people need more light than younger, since the size of the pupil decreases with age and the eye tissues are less translucent.

The need for superior artificial light is realized when it is known that about 20% of elementary school children have faulty eyesight. Insufficient and wrongly directed light causes the child to hold his book too near his eyes, squint, and soon lose interest in the learning activity. If a person is not able to read easily with the printed page 12 to 15 inches from the eye, something is wrong with either the sight or the light. Children's eyes should be carefully protected from unshaded and "spotty" light, where there is too great contrast between the illuminated and unilluminated areas of a room. The eye cannot adjust to brightness and darkness at the same time. When the children have grown to college age the percentage has doubled and 40% do not see normally. This condition seems doubly distressing when it is learned that over 70% of all muscular energy needed for work and recreation is dependent upon seeing, that 80% of the impressions received from the surroundings come through the eyes.

The eye becomes defective through use as well as abuse. The pupils gradually dilate under work requiring continuous or great visual effort, and only partially recover during the night—all too short at times—or over a weekend. Good lighting will prevent or reduce defective vision, which results in the unnecessary waste of human resources. Eyestrain frequently results not so much from defective eyes as from the effort of good eyes to see when there is not enough light for sight.

The ability to see with certainty and ease is increasingly important because of the growing number of complicated occupations that demand keen vision and the almost universal driving of automobiles. Seeing is definitely tied up with human reactions.

Luckiesh[3] lists the following factors as significant in influencing visibility: size of object and details—dependent in part upon distance away; brightness, and contrast in brightness with surroundings; color contrast; precision of task; speed of seeing required; normal or subnormal vision; pupil size and retinal adaptation, affecting visual acuity; fatigue of eye muscles; distractions.

With the present emphasis on a higher level of illumination, bright-

[3]Matthew Luckiesh, formerly Director, Lighting Research Laboratory, General Electric Company, Nela Park, Cleveland.

ness and brightness contrast have demanded special consideration. Surface brightness should not exceed ½ candlepower (cp) per square inch when in direct view. In marginal view it may be as high as 3 candlepower per square inch. Out-of-doors contrast in brightness is usually 1 to 18–25 candlepower; in a luminaire with polished surfaces it may be 1 to 2000 or 3000 candlepower. Working surroundings should not be so bright as to emphasize insignificant details and distract attention. It has been found that the sensitivity of the visual sense is generally greatest when brightness of task and surroundings is approximately the same. On the other hand, if the brightness or contrast is low, more light will increase seeing facility. For example, compare the ease of reading the usual book or magazine with a newspaper or telephone directory. There is no question as to which requires more illumination. A moderate amount of brightness and color is desirable. They stimulate and hold the attention of the worker. Shadows, too, are valuable if they make seeing easier.

Defective eyes need two to three times as much light as normal eyes. Interesting experiments have shown that light-eyed people require a higher level of light than dark-eyed for the same task. Light eyes are also very much more sensitive to glare, perhaps because the light iris, being more transparent, transmits more light. Negroes were found to see two to four times better at night than white persons.

Such information is valuable in determining which individuals should exercise special care in night driving. Glare apparently bleaches the visual purple, which is essential for sight. Vitamin A aids in regeneration of visual purple. Dairy products, green vegetables, and carrots especially are a source of this valuable vitamin. Carotene, a vitamin-A concentrate, is taken by color-matching inspectors to relieve eyestrain and consequent fatigue.

Low Voltage Lighting

Low voltage lighting may be used in the residence for general, task, and aesthetic purposes. As task lighting, the portable high intensity lamp may be combined with general light sources to provide the needed light for sewing. An even brightness at a low level may be achieved with a down-lighting design. Beautiful portraits may be framed for effective accent on a wall. Interesting lighting effects may be produced on books and art objects People may be flattered beautifully with light reflected from walls. The lighting of a conversation area to create a quiet atmosphere may result from a combination of regular voltage and low voltage systems.

Low voltage lighting is produced by a filament of short, thick, tightly coiled wire. The filament is of a larger diameter than that used for regular voltage sources. This factor is very important if the filament wire is to effectively carry the greater current. At 12 volts, a filament of a certain wattage must carry 10 times the current of a 120 volt filament. The short, thick wire has less surface area per watt, therefore, less heat is lost by convection of the inert gases in the lamps. All of this means greater efficiency.

Because the filament is short and stubby, it is more of a point source. Therefore, greater control of the light is possible either through external or built-in reflectors.

An obvious advantage in outdoor lighting is the safety factor. The shock hazard is greatly reduced when using low voltage rather than regular 120-volt wiring systems. Because of this, many codes permit simplified wiring systems, including open wiring when properly installed.

Lighting for Television. With the extensive and increasing use of television sets, the problem of comfortable viewing without harm to the eyes must be considered.

Commery suggests certain general rules to follow: no reflected light from screen; the brightness of surroundings of set at least one-tenth of the brightness of the screen itself, eliminating any sharp contrasts; no light directly in the line of vision.

Good viewing habits should be established by the family. Sit as directly in front of the screen as possible but at a distance. This distance will then provide the best focus, decrease fuzziness in the picture, and make the image have depth and appear lifelike. Look away occasionally to rest the eyes. Children should not sit on the floor, and their viewing habits should be checked frequently to prevent damage to young eyes.

Care

Keep the lamps and fixtures clean. Dust accumulating over months will absorb as much as 50% of the light otherwise available. The deposit of dust is so gradual that the consumer is not conscious of the dimming until it becomes very bad, but all the time he is paying for light he does not get. Clean fixtures and lamps frequently and at regular intervals. Inverted fixtures should be wiped out every two weeks and the pendent types washed at least once a month. Avoid installing luminaries that catch dust easily and are cleaned with difficulty. Bulbs may be wiped off with a damp cloth; they should not be immersed in water. The metal supports of the fixtures are often finished with a protective coating of lacquer, which will be destroyed by acids. Wiping off the dust with a damp cloth and then rubbing with a dry one is usually sufficient to keep the metal in condition. Ceilings should also be kept clean if they are to reflect light adequately. Discard blackened lamps. They may reduce the footcandles as much as 60%.

Energy Conservation and Lighting

Among the many ways to optimize the use of electricity in lighting are the following:

1. Design lighting for expected activity with less light in the surrounding nonworking areas. When the activity is of a critical nature, design for the needed amount for effective seeing.
2. Design with more effective sources or luminaires or fixtures. Luminaire effectiveness depends upon how well the light provided enhances the visibility of visual tasks. Luminaires that consume equal wattage and provide equal illumination levels may not provide equal visibility of seeing tasks.
3. Use efficient light bulbs or lamps. When determining the efficiency of the bulb, consider color, life, physical size, and the light output per wattage input. For incandescent or fluorescent lamps, the higher the wattage, the more efficient the lamp. For overall design, prime consideration should be given to more efficient sources as the fluorescent.
4. Use thermal-controlled fixtures. Select fixtures from which the heat may be directed out the back of the fixture.
5. Use lighter-colored finishes on ceilings, walls, floors, and furnishings. Lighter colored surfaces will reflect higher quantities of light.
6. Use efficient incandescent bulbs. One 100-watt bulb produces more light than two 60-watt bulbs (1740 versus 1720).
7. Turn off lights when not needed. This practice is one of the most effective management practices in the conservation of electricity.
8. Use efficient fluorescent bulbs. The fluorescent light source is more efficient than the incandescent bulb. Higher wattage, long length tubes are more efficient than lower wattage shorter length tubes.
9. Control window brightnesses. A skillful use of daylighting may add much to the total amount of light available for effective seeing.
10. Keep lighting equipment clean and replace sources which have become darkened. Older, dark bulbs may be used in areas where lower levels of light are appropriate.

Terminology

Candlepower. Luminous intensity.

Fluorescent lamp (tube). A low pressure mercury electric-discharge lamp in which a fluorescing coating (phosphor) transforms ultraviolet energy into visible light.

Footcandle. A quantitative unit for measuring illumination: the illumination on a surface one foot square on which there is a uniformly distributed flux of one lumen.

Footlambert. A quantitative unit for measuring luminance. The footcandles striking a diffuse reflecting surface, times the reflectance of the surface equals the luminance in footlamberts.

General lighting. The lighting designed to provide a substantially uniform level of illumination throughout an area, exclusive of any provision for special local requirements.

Glare. Any brightness or brightness relationship that annoys, distracts, or reduces visibility.

Incandescent filament lamp (bulb). A lamp in which light is produced by a filament heated to incandescence by an electric current.

Lumen. The unit of luminous flux, light output at source.

Luminaire. A complete lighting unit consisting of a lamp (bulb) or lamps, together with parts designed to distribute the light, to position and protect the lamps, and to connect the lamps to the power supply.

Quality of lighting. The distribution of brightness and color rendition in a visual environment. The term is used in a positive sense and implies that these attributes contribute favorably to visual performance, visual comfort, ease of seeing, safety, and aesthetics for the specific visual tasks involved.

Task plane (work plane). The plane at which work is done and at which illumination is specified and measured. Unless otherwise specified, this is assumed to be a horizontal plane at the level of the task.

Experiments and Projects

1. Make a list of activities to be performed in each room of the home. Determine how the light may be used to achieve good general, task, and decorative lighting.
2. Determine the reflection values of colors of samples of fabrics for carpeting, upholstering, wallpapers. Combine the colors with effective light in a room arrangement.
3. Study the effect of incandescent and fluorescent light sources on different colors; red, green, blue, blue-green, orange, yellow, lavender, and purple. How would you apply this knowledge in the home?
4. Plan the task lighting for the various areas in the kitchen.
5. Plan the task lighting in the bathroom.

6. Plan task lighting for the bedroom of a teenager, where the rest area is combined with study, grooming at a dresser, and sewing activities.
7. Study the various sources of lighting that may be used as colored sources (dichro filters, etc.).

CHAPTER 17
Water in the Home

Water is essential for life. From prehistoric times the need for a reliable and adequate supply of water has determined man's choice of a place to live. With an increase in population, attention also had to be given to sanitary means for the disposal of waste. Town and city governments assume the responsibility for providing the needed water and the care of sewage for the people dwelling within their corporate limits, but those who live in rural areas must themselves obtain a source of usable water and prevent its contamination by animal or human waste. The availability of a sufficient quantity of cold and hot water at the turn of a faucet, of a well-fitted bathroom, and of modern automatic laundry equipment, all needed for cleanliness and health, are marks of the present-day civilization and cultural development of a people. Installation of a house plumbing system has become a professional job, requiring the employment of trained personnel to assure the household satisfactory results.

SOURCES OF WATER SUPPLY

In most homes there are three sources of water supply: the kitchen sink, the bathroom facilities, and the laundry, all making available both hot and cold water. Water may be heated by various methods, but gas and electric heaters are commonly used.

WATER HEATER

The water heater is an essential appliance in the modern home. The quantities of hot water needed by the family will differ depending on the habits of the family members, the geographical location, and the methods of washing clothes or dishes. Even the leisure time uses of water have contributed to an increased need for hot water. As indicated by Tricomi, a tub bath requires about 7 to 8 gallons of water, and a quick shower may consume only 2 or 3 gallons. A dishwasher uses approximately 9 gallons of hot water, whereas washing dishes by hand may require from 2 to 8 gallons.

Selection

When choosing a water heater, three factors—tank size, input rating, and recovery capacity—which determine the hot water producing ability of the heater are important. The size of the tank will determine the numbers of gallons it can hold at any one time, while the recovery rate refers to the ability of the water heater to replace depleted supplies of hot water. The recovery capacity is usually stated in terms of the number of gallons of water that can be raised 100° in temperature in one hour's time; this may be directly related to the input rating of the heater. For a given tank size, the higher the input rating the higher the recovery rate.

Other factors that may influence the selection of a water heater are: water heater design, tank material, water needed for automatic washer and dishwasher, temperature of the inlet water, temperature of outlet water required, number of people living in the household, and number of bathrooms in a home. Many formulas have been devised to determine the minimum quantity of hot water used daily in the American home.

An analysis of the hot water needs by size of family may serve as a guide in the selection of the correct size heater. An undersized heater, which cannot supply the amounts of hot water needed by the family for the performance of many activities at the same time, requires careful management for each to be adequately supplied with

Table 17.1

Family Size	Gallons Consumed, Per Month
2	700
3	950
4	1200
5	1450
6	1700
7	1950

Source: Ernest Tricomi, *How to Repair Major Appliances.*

water of the correct temperature. The problem of doing several loads of wash and personal bathing at the same time is an illustration of a peak demand on the hot water supply. Table 17.1 indicates an average monthly hot water requirement for different sized families.

To determine the capacity of the water heater, the monthly usage may be divided by 30; for a family of four, 1200 divided by 30 indicates the need for a 40-gallon capacity heater. An understanding of the capacity needed coupled with the recovery rate of the heater will enable the consumer to arrive at a wise selection of a heater.

Water heaters with a fast recovery rate employ about 9000 watts, making a complete recovery in about 2 hours. Other heaters may be slower, using a lower wattage, depending on the needs of the family.

Location of Water Heater

The location of the water heater is of prime importance if the consumer is to make the most efficient use of this resource. The heater should be located as close to the points of use as possible. Each time the hot water faucet is turned on, cool water is delivered first which has once been heated by the water heater at some expense. The amount of water standing in the pipes plus that which might be lost through a leak may result in a significant waste. Thus, long runs of pipes should be avoided in the design of the water supply system. The number of running feet of pipe has become such a critical factor in hot water economy that many manufacturers now offer small capacity heaters for location in the kitchen near the dishwasher and kitchen sink. Another may be located in the utility room near the washer.

Construction

Most water heaters have certain characteristics in common, regardless of the source of heat. They are made of galvanized iron or steel, copper-bearing steel, copper, stainless steel, or glass-lined steel. Galvanized tanks are subject to corrosion in certain water regions. To prevent this, a rod of magnesium or magnesium alloy is inserted into the tank. Since magnesium is a more active metal than the inside zinc coating of a galvanized tank; the hot water reacts with the magnesium, corroding it instead of the zinc. Replacement of the rod is occasionally necessary. If water is very soft the tank should be of copper and the connecting pipes of brass. Brass pipes should not be used with a galvanized tank, however, because of electrolytic action between the brass and the zinc, with resultant corrosion. Glass linings on metal keep the hot water free from rust and other impurities. Tanks are commonly finished on the outside with synthetic enamel.

A blanket of rock wool or Fiberglas, 1 to 3 inches thick, is used for insulation; the greater the thickness, the less loss of heat from the tank. A heat trap, in the form of an inverted L-shaped pipe at the top of the tank, prevents circulation of hot water to the house plumbing system by means of convection currents when hot water is not being drawn. Temperature and pressure relief valves should be installed, so that if the burner in the nonautomatic types is accidently left ignited or if anything goes wrong with the automatic appliance and steam is formed, it will blow off into the drain without danger of explosion. The relief valve is installed in the cold-water line adjacent to the heater, with no shutoff valve between it and the tank. A baffle above the cold-water inlet spreads the incoming water over the bottom of the tank and minimizes the mixing of cold water with the water that has already been heated.

The outlet pipe for the hot water is usually at the top of the tank and the inlet for cold at the bottom, but occasionally the manufacturer places the cold-water inlet on the side. Insulating the hot-water pipes is recommended. One investigator found that an inch of insulation on the pipes resulted in a 12% decrease in heat loss when water consumption was 40 gallons a day and, consequently, the thermostat could be set for a lower temperature. He also advised using small-sized pipes. His research proved that a 26% saving in heat loss could be obtained with smaller pipes. Hot water was more quickly available at the faucet with less waste of water and more satisfactory service.

ELECTRIC WATER HEATERS

The two main ways of heating water electrically are the induction and immersion methods. The induction method is accomplished by strapping the electrical heating elements to the outside of the tank and covering them with insulation. The heat from the elements passes through the walls of the tank to the water.

In the immersion method, electric elements are passed through the walls of the tank directly into the water (Fig. 17.1). The tubular units may vary in wattage from 1000 to 2250 watts each. The arrangement of the heating elements may vary as follows.

1. **Upper part of tank.** None, 1, 2, or 3 units; 1000 or higher wattage each; zero to a maximum of 4500 watts.
2. **Lower part of tank.** Two or 3 units; 1000 or higher wattage each; maximum 4500 watts.
3. **Total.** Maximum 4500 watts.

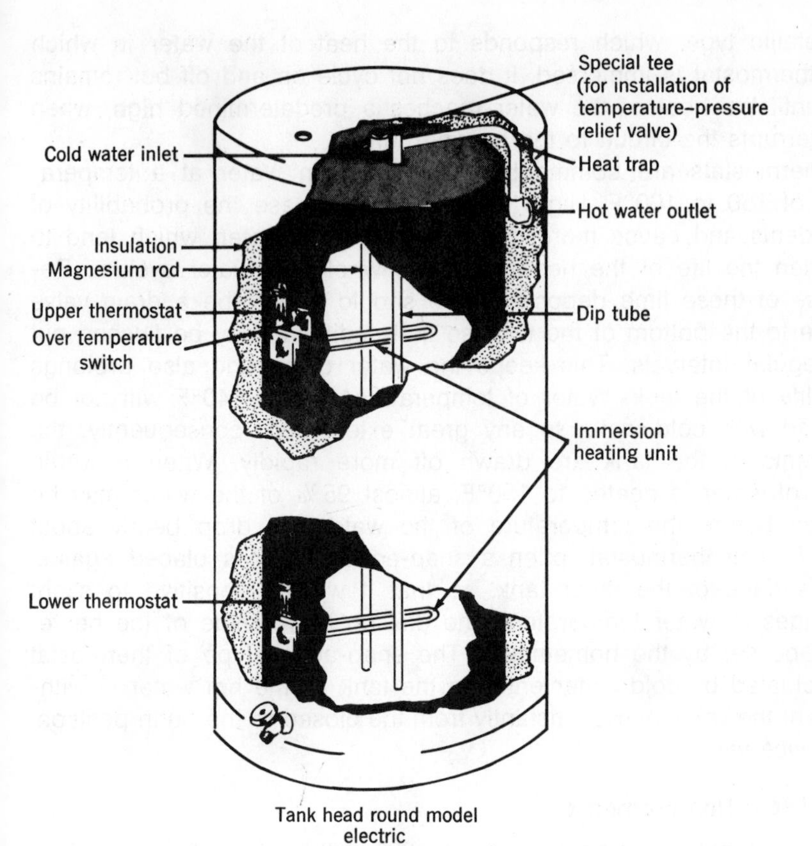

Fig. 17.1 Electric water heater. Glass lined to prevent rust and corrosion; calrod immersion units to achieve desired temperature by direct contact with heat source; over-temperature switch cuts off electrical power to heater when water reaches 190°F. (Hotpoint)

Principle of Operation

The principle that warm water rises is applied in the water heater design. Cold water enters the tank at the bottom and hot water leaves the tank at the top. Water heaters which have a top and bottom heating element of one or more tubular units each are controlled by two thermostats, or a single thermostat having a double-throw action.

Control of Temperature

The thermostat is the device designed to control the temperature of the water. The thermostat usually used in the water heater is a bimetallic type, which responds to the heat of the water in which the thermostat is immersed. It does not cycle on and off but remains on until the heat of the water reaches a predetermined high, when it interrupts the circuit to the heating unit.

Thermostats are commonly set to maintain water at a temperature of 150 to 160°F. High temperatures increase the probability of accidents and cause more deposits from hard water, which tend to shorten the life of the heater and the whole hot-water system. Because of these lime deposits, there should always be a drain valve close to the bottom of the tank so that sediment may be flushed out at regular intervals. This keeps the water clean and also prolongs the life of the tank. Water of temperatures below 140°F will not be diluted with cold water to any great extent and, consequently, the contents of the tank are drawn off more rapidly. When a whole tank of water is heated to 150°F, almost 95% of the water may be drawn before the temperature of the water will drop below about 120°F. The thermostat, often a snap-action type, is placed against the surface of the inner tank, so that it will be sensitive to slight changes in water temperature; the dial on the outside of the heater may be set by the homemaker. The snap-action type of thermostat is actuated by cold water entering the tank as the hot water is withdrawn; the valve passes instantly from the closed to the open position, and vice versa.

Electrical Requirements

The electrical water heater requires 230 volts unless it is especially designed for special applications. When supplied with less than 230 volts, the heater will supply hot water at a slower rate than on full voltage. To achieve the recovery rate required by the need for hot water when the voltage is lower than 230 volts, a water heater of higher wattage rating than normal is necessary. The higher wattage will lower the recovery time.

Safety Features

Pressure Relief Valve. As noted a pressure relief valve is provided to guard against dangerous pressure build-up in the event of a failure of the heat control. The pressure relief valve is set to relieve pressure from 25 to 35 psi above normal water pressure, to a maximum allowable setting of about 125 psi.

Thermal High-Limit Switch. A thermal high-limit switch is designed to cut off all power to the heater when the upper limit is reached.

This switch acts in the same manner as an overload protector of an electric motor, except that it is actuated by the heat of the tank wall or the water itself. The thermal-limit switch is set at the factory at a high-limit cutoff setting of about 195°F.

Installing an electric heater is a fairly simple operation since it needs no flue hookup to a chimney.

GAS WATER HEATER

The automatic storage gas water heater consists of a vertical tank, insulation, gas burner, and automatic controls which are combined into a single compact unit. The input rating may be up to but not including 50,000 Btu per hour.

Gas enters the heater through the thermostatic control, is ignited by the automatic pilot, and burned in the combustion chamber below the storage tank. The heat from the gas flame is transferred directly to the water in the tank through the concave tank bottom. Flue products are vented from the tank through a vertical tube up the center of the tank. A baffle inserted in the flueway slows down the flow of flue products and creates greater turbulence, thereby forcing the products to come into closer contact with the surface of the tube and increase the heat transfer. This flueway also increases the surface that is in contact with water, and additional heat may be transferred to the water. The warm water from the bottom of the tank rises along the hot surfaces to the top of the tank. Colder water circulates down along the outer tank walls to be heated. In this manner, all water is heated to the same temperature throughout the tank (Fig. 17.2).

Two variations of this design are offset flue or ducted flue, where the vertical flue is placed off center in the tank, and the external flue design. In the external flue water heater, the flue products are vented up the outside walls of the heater, pass into a collector at the top of the tank, and are then vented through a draft hood to the chimney (Fig. 17.3).

For installation and plumbing arrangements, both the cold water inlet and the hot water outlet are located in the top of the gas water heater. To prevent the stored hot water from becoming mixed with the cold water, a dip tube extends to the lower part of the storage tank, thus discharging cold water into the tank close to the thermostat. As hot water is drawn from the storage tank, cold water which enters the tank near the thermostat level causes the thermostat to open, permitting gas to flow to the burner and thereby heating the

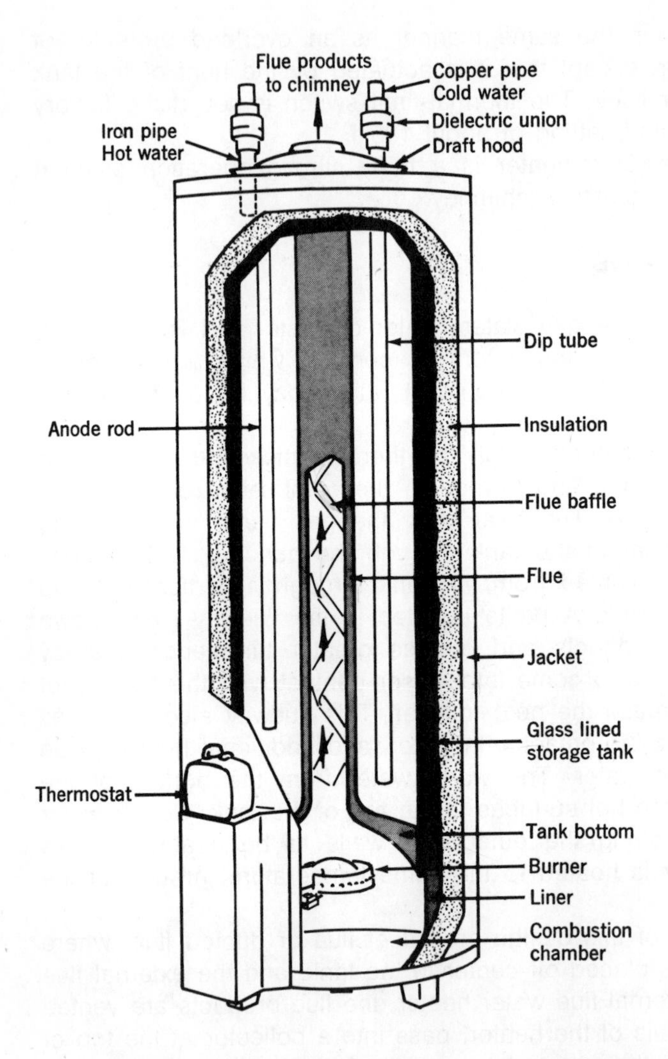

***Fig. 17.2** Gas water heater-center flue. Flue products vented through tank by means of a vertical tube up the center of the tank. Spiral flue baffle inserted in flueway slows velocity of flue products causing turbulence of flue gases. This forces them to scrub or come in contact with surface of tube increasing heat transfer. (*Home Service Manual, *A. G. A. Laboratories)*

cold water to storage temperature. To prevent heat loss from the stored water the storage vessel is covered with a generous layer of insulation. This is then covered with a sheet metal outer jacket.

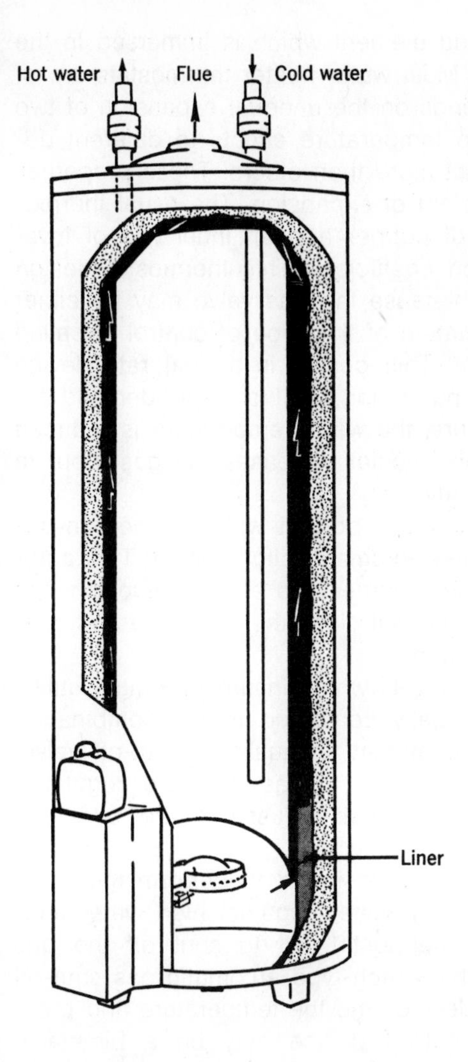

***Fig. 17.3** Gas water heater-external flue. Flue products vented up and around outside surfaces of storage vessel. Can be designed to have higher recovery capacity because of large vent transfer area in contact with hot flue gases.* (Home Service Reference Manual, *A.G.A. Laboratories)*

Controls and Safety Devices

The two types of control systems designed into the gas water heater are for temperature and control of flow of fuel. The thermostat is provided for the control of water temperature. A water heater thermostat consists of a heat sensing element which is immersed in the water and controls a gas valve. Most water heater thermostats are of a rod and tube type which depends on the unequal expansion of two metals. Different metals react to temperature effects to different degrees; some expand and contract more than others. These properties can be expressed as the coefficient of expansion. The usual thermostat design has an outer tube of copper and an inner rod of invar which has a very low expansion coefficient. This thermostat design is called a snap-action device because the gas valve may be either fully open or fully closed. A variation of this type of control is called a demand augmentation control. This design is a dual rate device which operates at a normal input rating until a large demand for hot water arises. When this occurs, the water temperature is reduced below a certain level and the thermostat increases the gas input to the burner to reheat the water quickly.

A second system which becomes important with the modern-day design of gas water heaters is the automatic pilot system. This automatic pilot serves to light the main burner and also provides for the automatic pilot safety device, which will close the gas valve and supply if the pilot flame is extinguished.

Because of the demand for controls which insure maximum utility, the automatic pilot device is usually combined into a combination control system containing the thermostat, manual gas shut-off valve, automatic pilot, and in some designs, a gas pressure regulator. Mechanisms may also be provided which adjust both the pilot and the main burner input rating.

Two other types of safety devices designed to perform the function of protecting the water heating system against high water temperature should the operating thermostat fail to shut off the gas supply to the main burner are the switch-type automatic gas shut-off device or energy cut-off safety device, and the temperature and pressure relief valve. The energy cut-off device may be a bimetallic thermostat mounted on the exterior surface of the water heater, or it may be a rod and tube sensor immersed in the tank itself, where it can sense the water temperature within the top 6 inches of the tank and limit it to a predetermined temperature. The control operates a cut-off switch that opens the electric circuit causing the safety gas valve to interrupt the gas flow to the pilot and main burner, should the temperature in the tank exceed the preset temperature. This is also a solid state device, another example of electrical controls being used more and more to increase the safety of gas appliances.

The temperature-pressure relief device is a combination of two separate safety controls. A pressure relief valve is set to open at a preselected pressure to allow water, or probably steam, to escape, thereby reducing the pressure in the tank. The temperature relief valve is actuated by a hydraulic thermal element or a rod and tube. The valve is set to open at a predetermined temperature, allowing excessively hot water to escape.

Softening Water Supply

The value of soft water is discussed in Chapter 13. Both hot and cold water may be softened, but in hard water areas, if only one is softened, it should be the hot supply. Heating hard water tends to cause a deposit of lime in the water heater and pipes, which gradually will clog the pipes and hinder water flow. It is usually recommended that the softener be by-passed in providing water for the toilet and for sprinkling the lawn. Some people may also prefer unsoftened water for drinking.

Some investigators report that persons who drink soft water habitually are more subject to the hardening of the arteries than those who drink water containing minerals.

BATHROOM FIXTURES

The bathroom with the sink and tub for cleansing the body, and the toilet for the sanitary disposal of human waste matter, is an essential part of the home. These fixtures are made in a variety of colors to harmonize with any chosen plan of decoration.

Sink

The sink may be rectangular, round, or oval in shape and should be of ample basin area. It is supported on legs or fastened to the wall, and has an overhanging lip to prevent splashing and a concealed overflow outlet to the drain. Sinks may have depressions for soap at either side of the bowl; certain new models feature single or twin soap receptacles located below the bowl rim in a recessed nook in the side of the bowl, with a drain beneath, preventing the frequently messy, slippery counter surface or rim (Fig. 17.4). This receptacle may also serve as the overflow to the drain. A shelflike back ledge above the bowl holds a glass or toilet articles. There may be towel bars on either side as part of the supporting frame, but at present basins are often built into a counter as part of a dressing table, with drawers and shelves below for the storage of towels, face clothes,

Fig. 17.4 The bowl of the sink is reflected in the mirror above it, showing the position of the soap receptacle below the bowl rim. (Crane)

and toilet items. This counter construction conceals the connecting pipes.

The counter tops may be of tile, plastic laminates, and in de luxe bathrooms, even of marble. The sinks themselves are of vitreous china or porcelain enameled iron or steel and are obtainable in a number of different colors, sometimes decorated with flowers or geometric designs. Occasionally the bowl has a self-rimming feature that eliminates metal bands, or it may be a drop-in type installed below the counter surface. The wall above the lavatory counter often displays a wide mirror, or one or two shallow cabinets with mirrored doors for the storage of toothpaste, make-up materials, and frequently used small medicine bottles and pills.

The faucets, individual or connected to a mixing spout, may be on the back rim of the bowl, on a slanting panel below the shelf that backs the bowl, or on the counter top itself. Many faucets are single control, either the push-pull or the dial-operated types. The open end of the water spout should be at least one inch higher than the water level in the basin to prevent any possibility of the spout being submerged in dirty water that could get into the house supply and cause contamination. When two faucets are used, the one for hot water is placed on the left. A rubber stopper or pop-up cylinder closes the waste pipe, into which the overflow also opens.

Tub

The tub is rectangular or square in shape and occasionally contoured to provide wide corner areas for sitting or for holding toiletries. The rectangular tub usually has a flat bottom and may have a sloping end and a wide rim on the room side. To prevent slipping when getting in or out, one manufacturer features a slip resistant surface that covers the entire bottom of the tub. It has 1-inch diameter circles fused to the cast-iron body of the tub; the surface is comfortable to sit and stand on and is easy to clean. Sometimes the tub is sunk into the floor for stepping-down access; at other times it is partly recessed, providing sides only 14 inches high, for easier entrance and exit. The square tub has an integral corner seat for sit-down showers or for foot bathing. With grab rods at either side of the seat, it provides an especially comfortable bathing arrangement for elderly people. The recessed niche above the side of the tub for holding the soap usually is also provided with a grab bar; other bars may be installed to help in getting into and out of the tub, and near the toilet seat as an aid to older or partially disabled persons. A mixing faucet is installed above the end or side of the tub, and there is a rubber stopper or pop-up waste closure and an overflow opening to the drain. Piping is usually concealed in the wall, but should be accessible for servicing.

Plumbing authorities recommend the selection of porcelain enameled cast iron bathtubs rather than porcelain enameled steel tubs. The cast iron tubs are more rugged and longer lasting. If tapped the cast iron gives a solid sound, in contrast to the somewhat metallic sound from the steel.

Shower Bath

The shower bath, now frequently called the shower receptor by manufacturers of plumbing fixtures, is increasingly popular with men of the family, and with older people of both sexes. The shower may be a part of the tub installation or may be in a seperate compartment, finished in tile or marble, made of glass, or sometimes of bonderized steel coated with synthetic enamel in a variety of colors.

Various types of shower heads are available which provide a wide spray pattern or a more concentrated one; the spray may be fine, normal, or one for drenching by the turn of the volume regulator. A swivel ball-joint connection allows one to direct the spray to any position. Some spray faces are made of delrin that resists the build-up of corrosive minerals and, therefore, helps prevent clogging. New for use in homes is a flow-control feature that reduces water consumption by approximately one-half, especially desirable where hot water is in restricted supply or is expensive to provide. The flow control on a shower head limits the flow to about three gallons per minute instead of the usual six gallons a minute. Used for some time in hotel, motel, apartment building, school, and hospital bathrooms, it is now being installed in homes.

It is recommended that the shower mixing valve be controlled by a temperature regulator, especially if the water heater is maintained at a high temperature. Otherwise, when cold water is used in other areas of the home, especially when a toilet is flushed, a sudden surge of very hot water in the shower may cause a serious accident.

Toilet

The toilet is always made of vitreous china which is resistant to the rather strong alkaline agents usually employed in cleaning it. The important features to look for are the water area of the bowl, the flushing action, and the size of the passageway through which the excreta are ejected into the sewer pipe. The size should be at least 2 inches in diameter. There are several types of flushing action: siphon jet action, reverse trap, and washdown, to list them in their descending order of efficiency and sanitation. The siphon jet has more positive action and is quieter than the others (Fig. 17.5). The trap is at the front of the washdown bowl and in the rear of the siphon jet closet, an easy way of distinguishing between them. An elongated bowl is also marketed. About 2 inches longer than the conventional bowl, it holds a larger volume of water. The seat over the bowl is often shaped to promote correct posture.

The tank that holds the water is usually connected to the bowl by a common housing, and may be placed at various heights above the seat. In one model the tank is very low, not more than 2 or 3 inches above the seat, and the bowl and tank become a one-piece fixture, most attractive and with exceptionally quiet flushing action. When flushed, water flows in around the rim to remove any clinging soil and then discharges rapidly into the siphon passageway. The

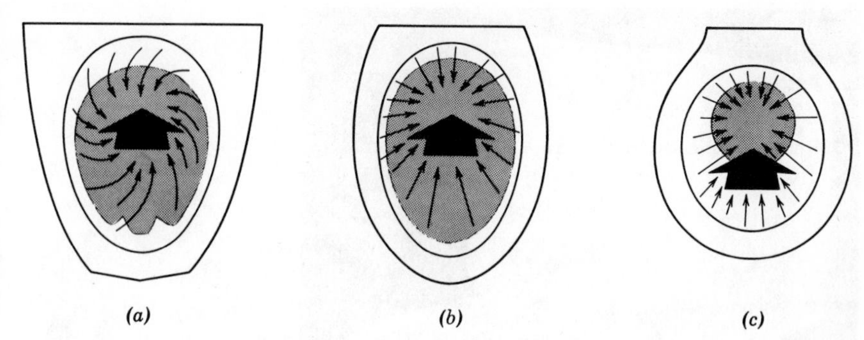

***Fig. 17.5** Toilet flushing actions.* (a) *Siphon vortex. Water enters through punchings around circumference of rim—centripetal action creates a vortex in center of bowl which draws contents down into well of bowl—this swirling action completely scours bowl—siphonic action causes expulsion of contents. Powerful, quiet, most sanitary toilet action.* (b) *Siphon jet. Water enters through rim and jets, filling trapway completely—siphonic action results in quick withdrawal of water from bowl preventing objectionable rise of water in bowl—location of jet makes toilet self-draining. Removal of visible water is all that is necessary to prevent freezing in an unheated structure during winter. Rapid and sanitary toilet action.* (c) *Washdown. As flushing action begins, water from rim fills bowl—trap at front of bowl acts as siphon and empties bowl—jet diverts portion of flushing water directly to trap to accelerate action. Suitable for economy installations. (American Standard)*

operation of the handle on the tank raises the tank ball from its seat at the bottom of the tank. The receding water draws the ball back into position, stopping the flow of water. The refill is controlled by a float valve. The tank has an overflow pipe to prevent accidents; also an after-fill tube that allows a stream of water to run into the bowl to reseal the trap against the entrance of sewer gas.

The off-the-floor, wall-hung toilet is popular because of the ease of cleaning around and beneath it (Fig. 17.6). The tank may be concealed within the wall but then an access panel for servicing must be provided; this panel can be part of the wall itself. The space within the wall must be several inches deeper than for conventional installation and, therefore, this type is adapted particularly for new construction when the greater depth can be provided in a home in the process of building.

A method of removing odors during the use of the toilet has been developed by the Plumbing Division of American Standard. A flushing button on the side of the tank is pulled out and bowl ventilating action begins within ten seconds. A vacuum is formed by the water siphoning

Fig. 17.6 Wall-hung toilet; easy to clean beneath. (Crane)

action and it draws the odors back through holes in the rim of the bowl. The air then passes through the discharge outlet of the toilet, on into the vent pipe and outdoors (Fig. 17.7).

Bidet

Although comparatively new in America, the bidet has been used for years in certain European countries, in Latin America, and the West Indies. The bidet, made of vitreous china, is a low-set basin, similar to a toilet, equipped with hot and cold water valves which direct a spray of water upward from inside the basin in order to rinse the body after use of the toilet (Fig. 17.8). Warm water (and mild soap) clean far better than paper, and physicians claim that the use of a bidet reduces rashes, irritations, and infections because it provides better cleanliness, better hygiene, and better health. Seated on the bidet, facing the water valves, the user finds them within comfortable reach. Water may also be retained in the basin for bathing all the lower parts of the body, even the feet if desired. Elderly people often find this fixture especially adapted to their needs, but even young children can learn to use it when they begin to acquire cleanliness habits.

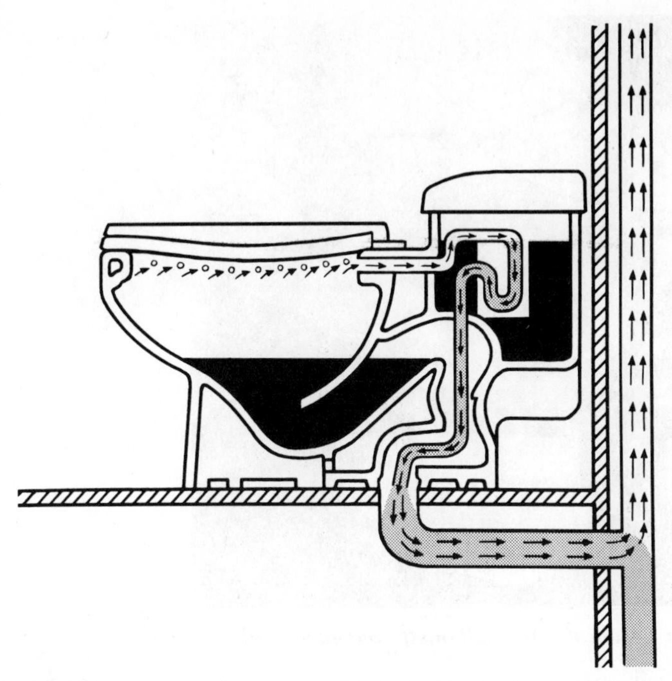

***Fig. 17.7** A toilet that ventilates itself. The ventilating unit draws air and odors through the holes in the rim of the bowl by the vacuum created by water passing through the unit. There are no moving parts. (American Standard)*

Fittings

Fittings are needed to supply the flow of water to and from the sink, toilet, and bathtub; without them we would not enjoy the benefits of running water. They include the many styles of faucets, the valves, shower heads, drains, strainers, and traps. The best fittings are of brass, usually plated with chromium. Brass offers maximum resistance to corrosion, years of trouble-free service, and wears and retains the chromium plating better than other metals do. One manufacturer supplies an antidrip unit that provides smooth water flow. Avoid dripping faucets. Even a slow drip wastes 15 gallons of water per day. Handles that turn the bathroom faucets on and off may be of crystal-clear acrylic plastic in a choice of colors: amber, white and charcoal. Acrylic plastic is noted for its long-wearing qualities. Other handles are of chromium-plated brass in a variety of shapes, easy to grasp and turn.

Fig. 17.8 A bidet beside the low silhouette one-piece toilet. (Kohler)

Safety Features

Trap. Certain other accessories are needed and certain precautionary measures must be observed if sanitary conditions are to prevail in the home. One such accessory is the trap. The pipe draining each plumbing fixture should be equipped with a trap. There are two types, the P-shaped and the S-shaped (Fig. 17.8). There is less likelihood of water siphoning off from the P-shaped trap and it is preferred. The main house trap is installed in the house drain, inside the foundation wall of the house. This trap, filled with water, prevents the entrance of noxious gases from the street sewer. These gases, notably carbon monoxide, methane, and hydrogen sulfide, are deleterious to health and may be flammable and could cause disastrous explosions.

Venting System. A venting system is necessary to protect the water seal in the traps. The system consists of an air pipeline that extends through the roof of the house, with a branch of the air pipe connected to the sewer side of each trap (Fig. 17.9). Toilets discharge into a soil pipe, which is about 4 inches in diameter. Sinks, lavatories, bathtubs, and clothes washers discharge into a smaller waste

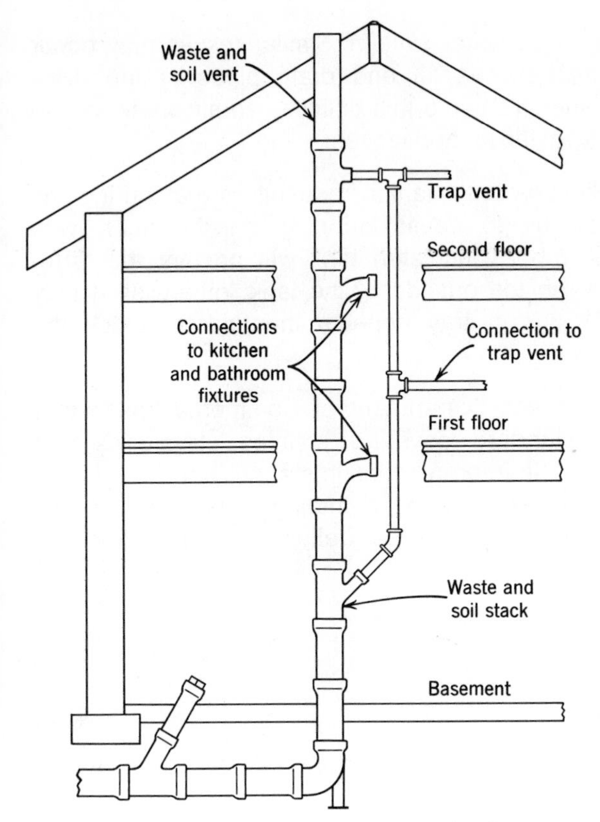

***Fig. 17.9** Main plumbing connections for home traps and vents.*

pipe. These pipes, usually of cast iron, empty into the house drain which is connected to the city sewer, or in rural areas to a septic tank. At the upper end, waste and soil pipes join above the highest installed plumbing fixture and extend above the house roof to become the venting system. The pipe above the roof should be wide enough in diameter to prevent any possibility of freezing over during winter months.

Drains. Floor drains are frequently present in the basement. During extended periods of little or no use, the water will tend to evaporate from the traps, allowing the entrance of sewer gas. To prevent this from occurring, water should be poured into the traps occasionally, perhaps once every month or two.

Air Break. It was noted that faucets should be placed far enough above tubs and bowls to prevent dirty water from passing into them and contaminating the house water system. Similar results may occur from cross connections between fill and drain pipes of the dishwasher and clothes washer. To avoid this difficulty an air break should be left in the fill pipes to these appliances.

Sweating. In humid climates toilet tanks frequently sweat during the summer months sufficiently to cause dripping on the floor, with subsequent damage. A rubber insulated tank will prevent the forming of such condensate on the outside of the tank; otherwise it may be necessary to install a drip tray beneath the tank to catch the water.

Standards. The United States Department of Commerce has established a Commercial Standard covering stainless lavatories and sinks for residential use. It informs consumers on how to identify fixtures that meet the standards by the plumbing fixture label, grade mark, or certificate. The bulletin gives definitions, measurements, and drawings with charts.

Terminology

Water. Essential for life. Three sources of home supply: kitchen sink, bathroom, laundry.

Water heater. Need to know tank size, input rating, recovery capacity. Location as close to point of use as possible. Size is monthly usage (Table 17.1) divided by 30. Recovery capacity is number of gallons of water that can be heated through 100°F in one hour.

Electric heater. Heating unit on outside of tank or inside immersed in the water.

Gas heater. Vertical flue up center of tank or external flue around tank wall between it and outside wall of heater.

BATHROOM FIXTURES

Sink. Round, oval, or rectangular in shape, has overflow outlet; faucets often connected to mixing spout, open end of which should be at least one inch higher than water level in basin. New faucets are usually single control or dial operated.

Tub. Rectangular or square in shape, slip-resistant bottom, niche for soap, grab bar, mixing faucet, rubber stopper, or pop-up waste.

Shower receptor. Part of tub or separate compartment; temperature regulator advisable.

Toilet. Of vitreous china, action by siphon jet, reverse trap, or wash down, siphon jet quieter. Tank, placed at various heights, has common housing with bowl. Off-the-floor bowl easy to clean beneath. Some bowls have ventilating action through holes in rim.

Bidget. Low-set basin with hot and cold water valves, directs spray to rinse body after use of toilet; cleanses better than paper.

Trap. P or S shaped, P preferred; it is filled with water to prevent entrance of noxious gases from sewer.

Venting system. Air pipeline extends through house roof; branches connected to sewer side of each trap. Protects water seal in traps.

Drain. Floor drain in basement; should keep water in drain traps to prevent entrance of sewer gases.

Standards. Commercial standard—informs consumer how to identify standard fixtures.

Experiments and Projects

1. Visit a plumbing firm. Arrange a demonstration and discussion on water heaters, bathroom fixtures, etc. If possible, visit the municipal water plant.
2. Determine the type of toilet installed in your bathroom. Note the flushing action. Examine the toilet tank. How much water is used for each flushing operation? Note the manner of refilling the tank and the after-fill. Compare this operation with that of other toilets if possible.
3. How is your shower activated?
4. Examine types and placement of faucets and traps in your home, or where you live.
5. Look at the roof to locate the vent pipe.
6. Examine the water heater. How is it heated? If it is gas heated, can you tell whether it has a central or external flue? Is it vented to the outdoors?
7. If your water is hard, how do you soften it for use in dishwashing and laundry?

CHAPTER 18

Indoor Climate Control

Almost twenty percent of the energy consumption of the United States is in residences. Of this amount 11.0% is used for heating and 0.7% for cooling. Since energy for heating and cooling is such a large part of the residential energy consumption, it is logical to seek a sizeable reduction of energy in house heating and cooling.

In order to conserve energy and at the same time maintain comfort conditions, it is important to know what the contributing factors are to comfortable heating in the winter and cooling in the summer and to know which of these can be changed or modified to reduce energy consumption, and the cost of operation, without discomfort.

INDOOR CLIMATE CONTROL

Indoor climate control includes a study of heating and cooling systems, ventilating systems, air-cleaning devices, and moisture control devices.

The elements that produce comfort conditions in the home are those which control the rate of heat loss from the body. In winter, the air around the body must be kept warm so that not too much heat escapes. In summer, air temperatures must be held down so heat can escape from the body. Other factors that influence the comfort of the occupants of a room are humidity and movement of air. Too much humidity causes excessive condensation; too little causes undersirable loss of moisture from mucous membranes of nose and throat, causing irritation, and loss of moisture from the skin, resulting in roughness and even cracking. Optimum conditions are usually obtained by a temperature of 75°F, 30% relative humidity, and air movement of 15 to 25 feet per minute. In the interest of energy conservation the President has recommended that the temperature be lowered to 68° in homes throughout the United States.

BASIC CONSIDERATIONS IN HEATING AND COOLING

Heat always moves from a warm area to a cooler one. In cold weather, heat flows out through the walls, windows, floors, and ceilings. In hot weather, it flows into the house. Heat that leaves the house is known as "heat loss." Heat added when it comes into the house is called "heat gain." Both are measured in British Thermal Units.

An estimate must be made of the quantity of heat movement when planning a heating or cooling system for a home. After estimating how much heat leaves or enters through the walls, windows, ceiling, and floor (heat loss by transmission), an extra amount must be added for heating or cooling the air that "breathes" through crevices (heat loss by infiltration). If the house is to be cooled, additional amounts are figured for heat that enters through the glass from the sunshine and the heat and moisture entering and generated inside the house by the members of the family.

Structural factors that determine the ease of maintaining rooms at a comfortable temperature are the amount of insulation, the window area, the orientation to the sun, the absence or presence of double or triple glass or storm sash, the number and size of cracks around the windows, doors, and foundation as well as where the heat is released into the house. The size of the system needed to heat or cool the house and the energy costs are dependent largely on the structural factors and the prevailing weather conditions.

WINTER AIR CONDITIONING

Calculation of the heat loss is the first and most important step in designing a heating system.

The next most significant factors in comfortable heating (given sufficient heat delivery, adequately controlled) **are where the heat is released into the house and how it is utilized after it gets there.** The heat loss of a structure is affected (1) by transmission of heat

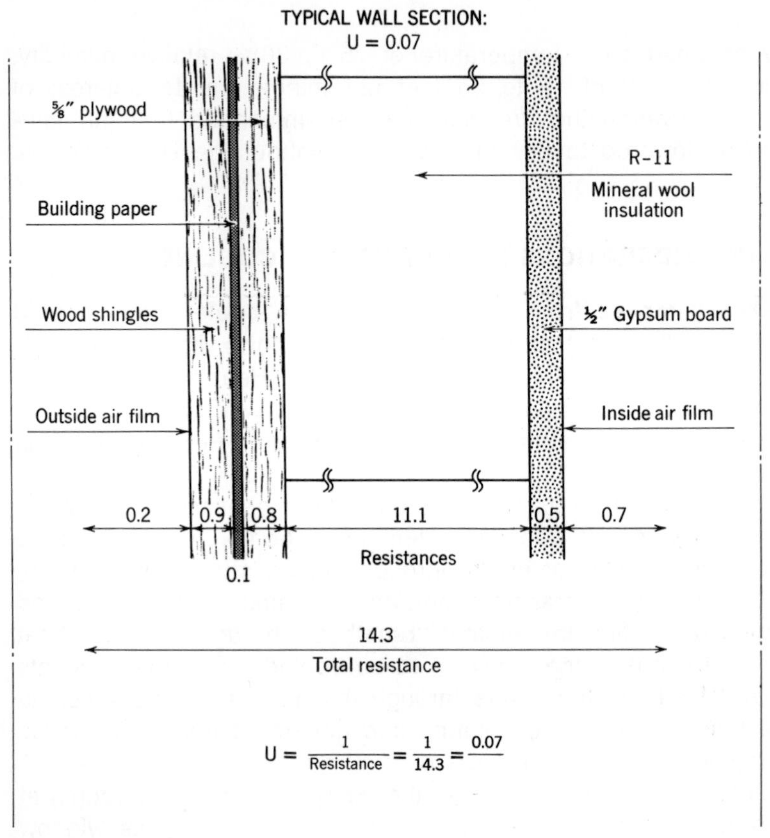

Fig. 18.1 Thermal resistances of materials in a wall section.

HEAT LOSS

through the structure, (2) by infiltration of air from outdoors into the building, and (3) by the design temperature difference.[1]

Heat Loss by Transmission

From 65 to 75% of heat lost in a building is by transmission. Heat flows by conduction through the building members and is given off on the outside of the building by convection and radiation. The amount of heat loss varies depending on the building materials used

[1]Design Temperature Difference is the difference between the indoor and outdoor design temperatures expressed in degrees Fahrenheit. The Indoor Design Temperature is the temperature desired for comfort indoors. It is usually 75°F. The Outside Design Temperature is arrived at by averaging minimum seasonal outdoor temperatures from records of the U.S. Weather Bureau (Fig. 18.2, Fig. 18.3).

in the outside walls, ceiling, windows and doors, and the total area exposed to the outdoor temperature. Each material in a building offers some resistance to the flow of heat (Fig. 18.1). Thermal resistances have been calculated for most building materials. The total resistance of a wall or roof is determined by adding the resistances of each of the components of the wall or roof (Fig. 18.1).

Thermal resistances have been calculated for most building materials. In examining the cross section of the wall shown in Figure 18.1, note the thermal resistances (the R values) of each of the components of the wall:

	R Value
Outside air film	0.2
Wood shingles	0.9
Building paper	0.1
5/8″ plywood	0.8
3 5/8″ fiberglass	11.1
1/2″ gypsum board	0.5
Inside air film	0.7
Total resistance units	14.3

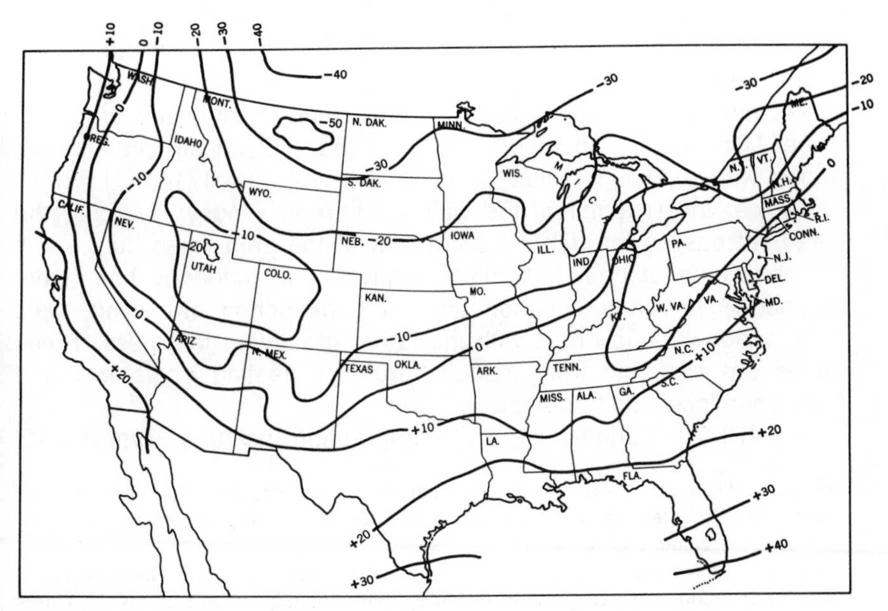

Fig. 18.2 Suggested outdoor temperatures for heating design.

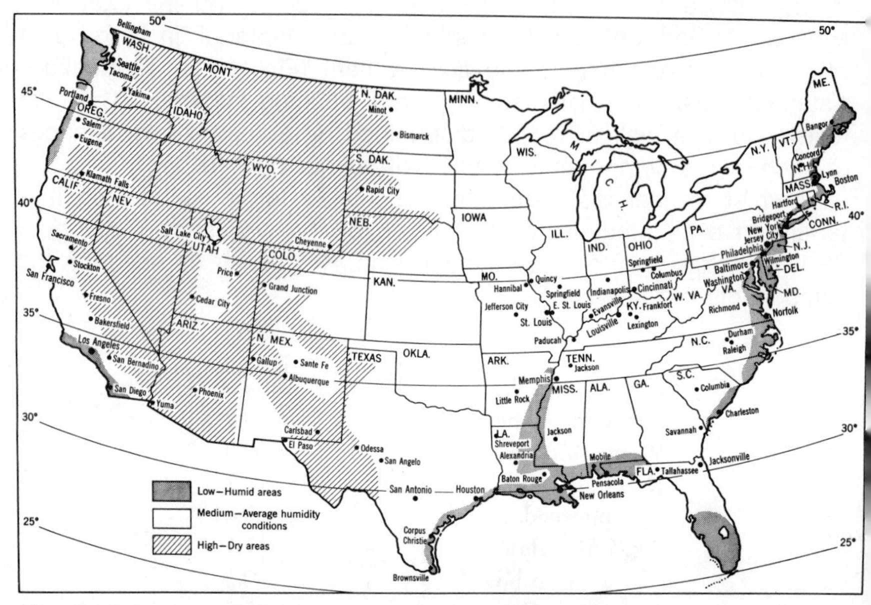

Fig. 18.3 Summer daily temperature range, United States.

From the thermal resistance of individual building materials or the combined resistances of materials in wall, roof, and floor constructions, a very useful figure can be calculated: the overall **coefficient of heat transmission.**[2] The reciprocal of the resistance is called the coefficient of heat transmission and is usually designated as the U value: $U = 1/R$. To find the U value of the wall construction shown above with a total R value of 14.3, $U = 1/R$, $U = 1/14.3$ $U = 0.07$. The lower the U factor of the wall, floor, roof, window, or door, the less heat passes through a square foot of the area in an hour.

Coefficients of heat transmission have been made by competent engineers[3] for each common type of construction of ceiling, wall, floor, windows, and doors with insulation of varying thickness. These tables are available from manufacturers of heating equipment and from your local utility company.

The ARI (Air-Conditioning and Refrigeration Institute) Form ALC-1[4]

[2]These coefficients are expressed in Btu per (hour) (square foot) (Fahrenheit degree difference in temperature between the air on the two sides) and are based on an outside wind velocity of 15 mph.

[3]Manual J, Environmental Systems Library; ASHRAE Handbook of Fundamentals.

[4]Air Conditioning and Refrigeration Institute, 1346 Connecticut Avenue, N.W., Washington, D.C.

is a much simplified method of calculating heat losses and heat gains to determine the size of equipment needed for heating and cooling.

HEAT LOSS BY INFILTRATION

The amount of heat loss through infiltration in a building is from 25 to 35%. How much air is infiltrated depends on how airtight the building may be. Infiltration is affected by the number of doors and windows and how well they fit, whether they are weather stripped as well as how many times the doors are opened and how long they remain open, the number and sizes of cracks in the structure, and the presence of dampers in fireplace chimneys and other openings. Wind direction and wind velocities also affect infiltration losses. More air leakage occurs when outside temperatures are low and when wind velocities are high.

TOTAL HEAT LOSS OF A RESIDENCE

In determining heat losses for a complete house, the transmission heat losses and the infiltration heat losses for each room in the house are calculated and totaled. Heat loss from ducts located in unheated spaces must be added to the heating load. (Duct heat loss multipliers have been developed and are tabulated for easy use.) Although they are interior heat sources (people and heat-generating equipment partially balance the heat loss of a building), these factors are not normally considered in heating calculations.

Heat loss in Btuh by transmission through building structure = Area in sq. ft. of each kind of wall, glass, floor, ceiling, or roof × Design temp. difference × Coefficient of heat transmission

Heat loss by infiltration = Volume × design temp. difference × infiltration factor in Btuh

The sum of the heat loss by transmission and the heat loss by infiltration and the duct heat losses is the figure that is used to determine the size of the heating plant that is required.

To Determine the Size of the Heating Plant Required

Furnaces are rated by the Btu **input** capacity, listed on the nameplate. Btuh input is used in calculating fuel consumption costs.

The Btu **output** depends on the effectiveness with which a unit utilizes the fuel supplied (efficiency). The output of a funace in Btuh is the total heating capacity. The Btu **output** is used to determine the correct furnace size. The furnace output must equal or exceed the total heat loss of the house—transmission and infiltration.

It is important that a heating system be properly sized for the house in which it is to be installed. A properly sized system should maintain a temperature of at least 72°F during the coldest weather to be expected in an area. Too small a system will not supply enough heat for comfort. Too large a system supplies too much heat for short periods and then cuts off resulting in wide temperature fluctuations.

THE HEAT DISTRIBUTION SYSTEM

Cold walls and uncomfortable drafts may be present even though the room thermostat registers 70°F, if the outside wall is not insulated, or the heat is not put into the room where the cold comes in—along outside walls and beneath the windows (Fig. 18.4). When the outside temperature is 0°F, 70° air touches the 59° wall, the air is cooled, contracts, and drops to the floor where it displaces warmer air, creating a continuous uncomfortable draft. More heat will not make the outside wall warmer but only results in producing stronger drafts.

Putting the heat into a room along the outside walls and beneath the windows is called perimeter heating. The heat blankets the inside surfaces of the exterior walls, thus preventing the room air from being chilled by contact with cold walls and windows.

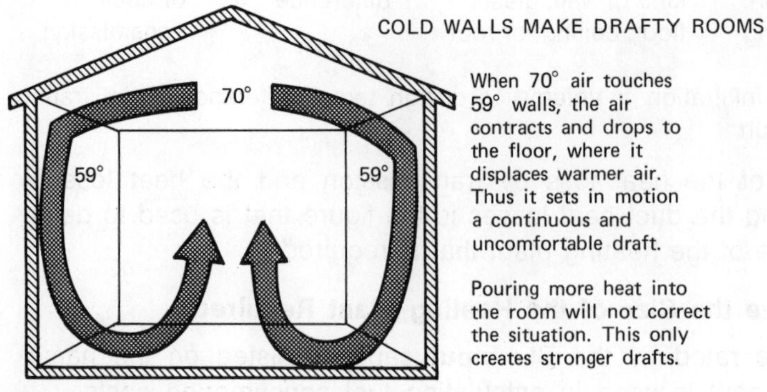

***Fig. 18.4** Cold walls and uncomfortable drafts may be present even though the room thermostat registers 70°F if the outside wall is not insulated, or the heat is not put into the room where the cold comes in.*

***Fig. 18.5** When heat is put into a room on an inside wall instead of along the outside wall and under the window, it accelerates the "down-slip" of cold air from the glass. The differences in temperature between the upper part of the room and the floor are greater, creating stronger drafts.*

When heat is put into a room on an inside wall instead of along the outside wall and under the window, it accelerates the "down-slip" of cold air from the glass (Fig. 18.5). The convection currents are working against you. The differences in temperature between the upper part of the room and the floor are greater, creating stronger drafts.

When the source of heat is located where the cold comes in, along the outside wall and under the window (Fig. 18.6), it moves air up to warm the glass and provides local radiant warmth. The differences in temperature between the upper and lower parts of the room are less, and there is less draft.

Floors must be warm. This means that heat must either be put into the floors or beneath them. This may mean that the basement or crawl space area should be heated.

The heat should be delivered in short but frequent intervals; or it should be delivered continuously at adjusted temperatures so there are no noticeable changes in room air temperature.

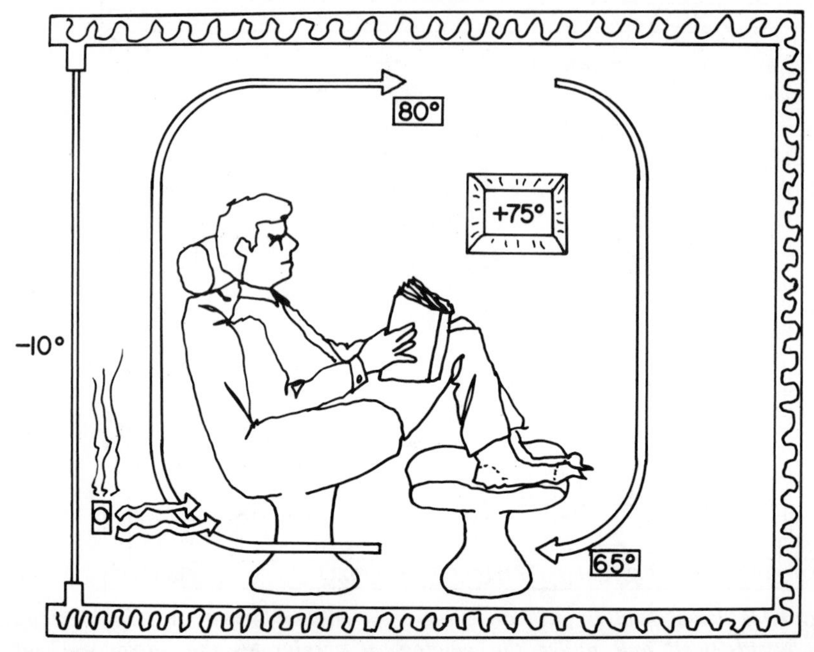

Fig. 18.6 When the source of heat is located along the outside wall and under a window where the cold comes in, it moves air up to warm the glass and provides local radiant warmth. The difference in temperature between the lower and upper parts of the room is less and there is less draft.

CONTROLS

Thermostats

The location of the thermostat is an important factor in achieving home comfort. If only one thermostat is used, it should be placed in the most lived-in room in the house. It should be mounted on the wall at from 2½ to 4 feet from the floor—at sitting height. Usually a good place is on an inside partition about 18 inches from an outside wall.

A thermostat should not be placed behind a door or on interior passages. These are protected areas and are the last to feel changes in outdoor weather. It should not be placed where it is exposed directly to air from opening doors, or on a wall exposed to direct sunshine. It should not be placed on a warm wall such as one with warm pipes or air ducts inside or one with the kitchen range or refrigerator on the other side or above radiators or warm air registers.

TV sets, lamps, and radios release heat and should not be placed near the thermostat.

In a home with a fireplace, a nearby thermostat will be affected by the extra heat and, therefore, the furnace will not come on. Other rooms in the house will become cold. The fire in the fireplace will draw huge volumes of air from the rest of the house and send them up the chimney so the other rooms will become even colder. The problem can be taken care of by placing the thermostat in another room, or a system with additional thermostats may be installed. Opening a window in the room in which the fireplace is located will help. The air for the fire will then come from the window rather than from the rest of the house.

Some houses, particularly split levels, have areas which have different heat demands because of exposures and heat loss. These can be heated more satisfactorily by installing more than one indoor thermostat and dividing the heating system into zones, each regulated by a thermostat.

Fuel savings can be realized by lowering the thermostat by 8 to 12° for 8 hours per night. Fuel savings can amount to ¾% for each degree Fahrenheit that the setting is lowered. Setting the thermostat back more than 8 to 12° does not save fuel because the furnace will be on longer to reheat the house in the morning. A thermostat should be set at the temperature desired and left there rather than changed up and down often during the day. Heat will not be delivered any faster by setting the temperature at the top of the scale.

Continuous Air Circulation

Continuous air circulation (C.A.C.) is a method of controlling the blower on a hot air system so that the blower runs most of the time in mild weather and practically all of the time when the outside temperature is freezing or colder. Constant circulation keeps the heat evenly distributed and maintains a constant delivery. Blower speeds are usually set as low as practicable. Constant operation of the blower will not wear out the motor, and the cost of running it will be more than offset by increased comfort and some fuel saving. C.A.C. is essential if air cleaning equipment is to operate at its highest efficiency.

MOISTURE CONTROL

Humidity is the amount of evaporated water in the air. Air can hold larger amounts of water at warm temperatures than at cold temperatures. The air in an average one-story house can hold about eight quarts of water at 70°F, less than four pints at 32°F, and only about a half pint at 0°.

Either too little or too much humidity in a home can be unhealthy as well as damaging to the house. If there is too little humidity, mucous membranes of the nose and throat dry out and their efficiency in repelling air-borne disease germs is reduced, hair and skin dry out, the evaporation rate of the body increases, and the air temperature must be correspondingly increased to keep comfortable; interior woodwork of the house will dry and shrink, and furniture becomes loosened in the joints; rugs and carpets "fuzz"; static electricity is greater; leather goods and bindings become dry and brittle.

If there is too much humidity, the rate of evaporation of the body is reduced resulting in the feeling of discomfort; moisture condenses on windows in cold weather and freezes and, when it thaws, may drip down to damage sills, moisture may enter the walls, collect, and freeze, possibly damaging the framing of the house; paint on the outside of the house may blister and peel; the interior woodwork will swell causing doors, windows, and drawers to stick; mold and mildew may form.

Most older homes suffer from inadequate humidity during the winter months, and some newer homes have excessive humidity. Many older homes are loosely constructed and have a high amount of air infiltration into the house. Very tightly constructed houses have excessive humidity because of moisture released by cooking, laundering, showers, etc.

Although the human body is comfortable with a relative humidity of from 20 to 60% at a temperature of 75°F, the degree of humidity that can be practically maintained indoors without objectionable condensation is dependent on the outside temperature, the type and amount of insulation, the type of walls and ceilings, the presence or absence of vapor barriers, the amount of moisture entering the building, the presence of storm sash or double glass on the windows, and the moisture production in the house.

When water vapor comes into contact with a cold surface, it condenses to form water, or if the temperature is below 32°F, to form frost. Table 18.1 indicates the percent of relative humidity that can be maintained within the home before condensation occurs on windows with single and double glass. With single glass, when the temperature is 0°F, moisture will condense on the windows when the relative humidity is 16%. This is drier than the Sahara desert where the average humidity is 22%. By using double glass at the windows, no condensation will take place when the outdoor temperature is zero until the relative humidity reaches 50%.

Table 18.1
Moisture Condensation on Windows[a]

Outside Air Temperature, °F	Percent Relative Humidity	
	Single Glass	Double Glass
−30	7	38
−20	8	40
−10	11	45
0	16	50
10	21	56
20	29	62
30	38	69
40	48	76

[a]Heavy drapes and closed blinds will cause condensation at higher temperatures.

One of the greatest hazards of too much moisture may not be easily seen. During the winter, in the heat transfer through walls and ceilings, air is cooled as it approaches the outer walls. If the air is cooled beyond its dew point, it begins to lose its water vapor, and condensation will occur. The condensed water may then saturate the structural elements and the insulation, causing deterioration and decreased efficiency.

Vapor Barriers

To prevent the moisture from inside the house from getting into the walls, vapor barriers should be installed in the walls. Vapor barrier protection is considered essential if the average temperature in January is 35° or less, regardless of the kind of house. Vapor barriers should be installed on the **inside** or **warm** side of all exposed walls and ceilings to prevent transmission and condensation within a wall (Fig. 18.7*a*).

Insulation can be purchased with or without an integral vapor barrier in the form of a lining of metallic foil or asphalt-impregnated paper on one side of the insulation. If no vapor barrier is attached to the blankets or batts, metallic foil or a sheet of polyethylene plastic should be installed at the time the insulation is put in (Fig. 18.2).

When insulation is blown into *existing walls,* it is impossible to install these types of vapor barriers. In these cases, vapor-barrier paint can be used on the *inside* walls of the house. Two or three coats of

Fig. 18.7 (a) *Fiberglas Friction Fit building insulation is used in side walls of new construction, generally in electrically heated homes and apartments. A polyethylene vapor barrier is then installed over the insulation.* (b) *Kraft Faced building insulation, widely used for home insulation, is also used in staggered stud wall construction for sound control.*

a good quality alkyd base, semigloss paint, preferably over a base coat of an aluminum paint, will offer fair vapor protection.

Vapor barriers safeguard insulation from moisture. Insulation is only effective when dry. Heat will pass quickly through wet insulation. Any vapor barrier damaged during installation should be replaced or repaired.

Humidifiers

During the winter months, it is usually necessary, in all but very tightly constructed houses, to add moisture. It is the infiltration into a building that creates the need for adding moisture. Every cubic foot of outside air that leaks into a house causes a similar cubic footage of air to leak out of the house at some other point. The air which leaks out has taken up moisture from the living processes in the house. The greater the infiltration, the greater the amount of water vapor must be added inside the house in order to bring relative humidity to a safe and comfortable level. More moisture is lost through exfiltration than gained from infiltration.

Moisture may be added to a house with a humidifier. Humidifiers discharge moisture into the house air. Humidifiers for use with air circulation heating systems may consist of a pan inside the furance filled with water, through an automatic valve, over which the warm air steam passes. Another type uses absorbent plates. The lower edges are immersed in a water pan, the upper surfaces are exposed to the circulating airstream inside the furance. For hydronic systems, water pans may be hung on the sides of radiators, or evaporating plate humidifiers can be within hot water convectors. All of these types with their variations have limited humidification capacities.

A "power" humidifier has greater capacity than the types already mentioned. One for warm air furnaces has an evaporating pan with an electric heating element in it. Another passes the air through a water spray. One introduces a water mist directly into the air stream circulating through the furnace.

Appliance-type humidifiers which can be moved from room to room are also available. Some of these discharge water mist into the room air. Some circulate air through a wet pad.

A mechanical humidifier should be equipped with a humidity controller that will shut it off when a preset humidity is reached.

To achieve the best performance from a portable humidifier, the homeowner should give special attention to where it is placed. It should not be placed against an outside wall since the temperature there is not typical of the temperature in the house as a whole. The cooler air near the outside wall will not absorb as much moisture as the warmer air surrounding the interior walls. A humidifier, in a multiple-level house, should be placed on the first floor near a warm inside wall facing the stairwell with at least 6 inches distance from the wall to allow for circulation.

Dehumidifiers

Heavily insulated and tightly constructed houses may need dehumidifiers during the winter months. Houses in damp climates may need dehumidifiers during the summer months.

A dehumidifier is designed to protect the house and its furnishings and not to make the family feel more comfortable. In a dehumidifier a fan draws vapor-laden air across a cold evaporator, causing the vapor to condense into droplets of water which are collected in a drip pan or drain hose.

An air conditioner also dehumidifies in essentially the same way. However, an air conditioner does something a dehumidifier does not do. It expels the heat it collects to the outdoors. The heat which the dehumidifier's refrigerating mechanism collects stays inside the room and the energy it uses is also released as heat. Its tendency is the raising of the room temperature, not the lowering of it.

Under A.H.A.M. (Association of Home Appliance Manufacturers) standards, dehumidifier capacity is expressed in terms of the number of pints of water a unit can remove in 24 hours of continuous operation at a room temperature of 80°F at 60% relative humidity. Most dehumidifiers collect from 24 to 48 pints of water per 24 hours. The moisture-removing ability of room air conditioners is expressed in pints removed in a single hour. Small units remove 2 pints per hour and the largest ones remove as much as 9 pints per hour.

A dehumidifier is useful in locations such as basements, attics, closets, or storage rooms. The space should be closed so that no more moisture-laden air can get into the area.

The simplest models run continuously as long as they are turned on. More expensive models are equipped with humidity and overflow controls. When the humidity reaches a desirable level, the dehumidifier will shut off. The overflow control is usually a float device on the collecting pan and shuts off the dehumidifier when the float is raised close to the top.

AIR CLEANING

Air Filters

Dirt comes into the house from outside air and by people entering from the outside. Dirt and lint are also produced in the house and get into the air. Grease particles and odors from cooking, ash particles and odors from tobacco smoke, hair and feathers from pets, lint from clothing and bedding, and bits of fiber from floor covering all contribute to airborne dirt. Removing this dirt from the air is accomplished to greater or less degree by the installation of air filters in the central heating or cooling system or in individual cooling units; or by the use of portable room air cleaners. Filters differ in their ability to remove dirt from the air. Table 18.2 shows the efficiency of different types of filters based on the National Bureau of Standards Dust Spot Test.

Table 18.2
Comparative Efficiences of Major Types of Domestic Air Filters

Filter Type	Cleaning Efficiency (Percent)
Disposable	5–8
Permanent	10–35
Electrostatic	
Self-charging	10–15
Charged media	40–50
Two-stage electronic	70–99

Disposable Filters. This is the usual type of filter installed with forced air furnaces and air conditioning systems. The particle collecting medium is usually made of Fiberglas pads, wire screen, or perforated metal plates covered with a sticky coating to help catch the dust. Large dirt particles are effectively stopped, but most small particles pass right through (Fig. 18.8).

Permanent Filters. These are similar in construction to disposable filters, but are built for cleaning and reuse. Some recommend coating with a special liquid after each cleaning to improve air filter effectiveness temporarily.

Electrostatic Self-Charging Filters. These filters are made of honeycombed or shredded polyethylene. As air circulates over the surface of the plastic, a static charge is set up which attracts dirt particles. Most are quite low in filtering efficiency, and the efficiency diminishes as the humidity in the air increases. Filters of this type can be cleaned easily and are reusable.

Charged-Media Type Electrostatic Filters. Such filters usually consist of a pleated fibrous pad on which an electrical circuit is printed. A high-voltage power pack, from 3000 to 6000 volts, charges the pad

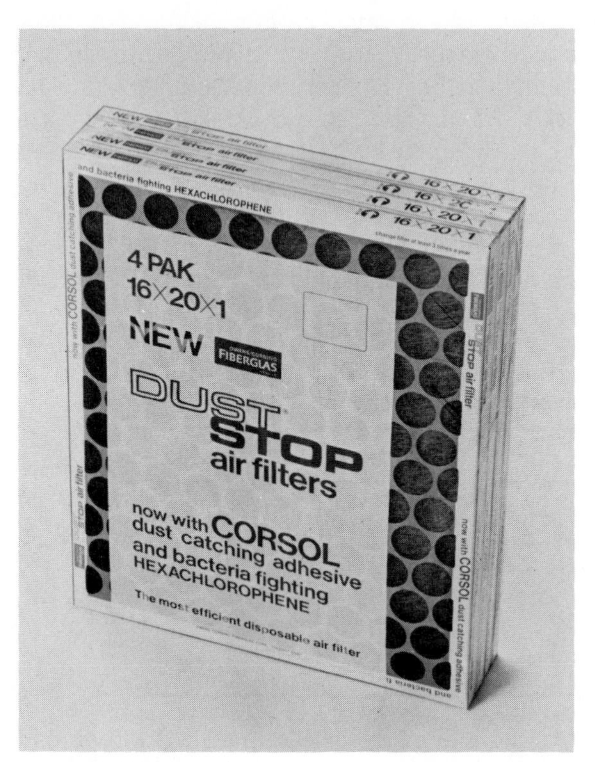

***Fig. 18.8** Fiberglas Dust-Stop air filters are treated with a special dust-catching adhesive, and protected with thin sheet-metal grilles, stapled to a fiberboard frame. When dust is removed from the air, the home is a cleaner, healthier place in which to live. (Owens-Corning Fiberglas)*

to attract and hold the dirt. As dust and dirt are collected, the resistance to the flow of air increases. It then is necessary to replace the pad. Annual replacement costs are quite high. The efficiency is decreased as the humidity is increased. These filters generally work well during the heating season, but lose much of their effectiveness in the cooling season when the relative humidity is high.

Electronic Air Cleaners. This type of filter contains two sections, an ionizing section and a collecting section. As the air comes into the ionizing section, the dirt particles passing through it become electrically charged by a series of wires strung between electrodes. In the collecting section of the filter, electrically charged metal plates attract the charged dirt particles like a magnet catches iron filings; and they remain on the plates until they are washed off. Some models

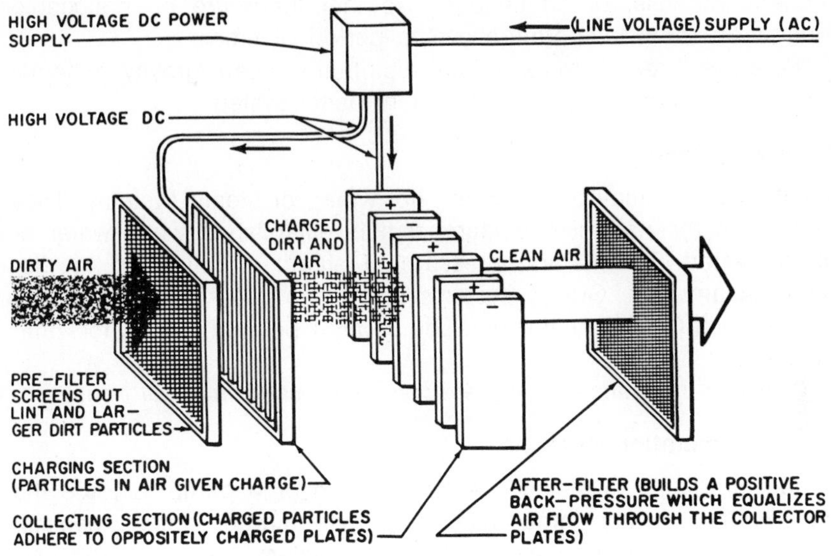

***Fig. 18.9** Two stage electronic filter. (Honeywell)*

offer automatic washing devices. The cleaning efficiency of electronic air cleaners is from 70 to 99%, depending on the velocity of the air (Fig. 18.9).

All types of air filters must be kept clean in order to allow the heating or cooling system to work efficiently. Dirty filters can reduce circulation and can, therefore, cut down heat output as much as 25%. Air cleaners installed in a central heating system function only when the blower is running. Most blowers run only when heat is called for. Therefore, to control dust, not only an efficient air cleaner is needed, but the forced-air circulation system must be installed so that the blower operates continuously.

HEATING SYSTEMS

Central Heating

In a central heating system, fuel is burned in one area, and the heat that is produced is distributed from this location to the living areas of the house by pipes or ducts.

A good heating system should provide sufficient quantities of heat and deliver it where it is needed to provide equal and even temperatures in all parts of the room. It should be delivered in short frequent intervals, or continuously so that there are no noticeable changes in room temperature over a period of time.

There are three systems of heat distribution used: gravity systems, forced circulation systems, and radiant panel systems.

Gravity Systems

Gravity systems may be warm air, hot water, or steam systems. They depend on the natural circulation that is established when water or air is heated. The natural pressure of steam is relied on to distribute steam heat. Gravity systems cannot be used in basementless houses because the heater must be located below the radiators and registers.

Relatively inexpensive, they are frequently used in small houses.

Forced Circulation Systems

Forced circulation systems use either a fan or a pump to force the heated air or water throughout the house. Larger homes can be evenly and comfortably heated with these systems.

Forced warm air systems can provide heating, ventilation, air circulation, air cleaning, humidification, and air cooling—probably the most versatile system available. These systems may use any one of three kinds of furnaces: conventional up flow, down flow, or horizontal; all of them use pipes or ducts to distribute warm air to the rooms and return cold air from them.

Forced hydronic systems have three variations: the series loop, one-pipe, and two-pipe systems. All provide excellent heating. With proper control, the temperature of the circulating water can be balanced with the heating needs of the house. Several types of radiators, convectors, or baseboard radiators can be used. Special attachments to radiators and convectors can be used to humidify.

Duct heaters provide a different type of forced air system. This method is a circulating air system with a central fan connected by ducts to air registers in each room. Each duct has a heater installed in it, each room has its own thermostat. Cooling can be added readily by placing cooling coils in the fan cabinet. Similarly, air cleaning and humidification can be added.

Radiant Panel Systems

With radiant panel heating, there is no visible evidence of a heating system. Hot water pipes, warm air conduits or electric heating cables are embedded within the floors or ceilings and occasionally in the walls. The warm panels give a combination of conduction and radiant heat; the room air is warmed by direct contact with the panels (conduction), and heat waves are radiated in all directions, warming every object they reach, including people.

Rugs and floor coverings slow down heat release from the floor panels. This makes it hard to heat all the rooms evenly if some have floor coverings and some do not. Also, carpets and rugs may be damaged by the "bottled up" heat. This problem is eliminated with ceiling panels. Proper controls are necessary to obtain good temperature regulation since the panels take a while to warm up or cool down. It is difficult to provide air circulation, air cooling, air cleaning, and humidification with these systems. Floor panel surfaces should not exceed 80° nor ceiling panels 120°F.

The Heat Pump

An electric heat pump heats and cools. It is a reversible refrigeration unit. In the winter, the heat is extracted from the outside air, ground or water, and is distributed through the house by a duct system that is especially designed for it. In the summer, heat is extracted from the air inside the house and discharged outside the house.

When outdoor temperatures are low, heat pumps using outdoor air or the ground as heat sources operate at reduced efficiencies. Although water is the best heat source, a cheap adequate supply of good quality water is not available in many areas.

Heat pumps generally switch from heating to cooling and from cooling to heating by reversing the direction of refrigerant flow and the roles of the condenser and evaporator in the system (Fig. 18.10).

Heat pumps are available as room units or central-system units. In warm climates a heat pump may furnish all of the heating and cooling necessary. In more severe climates, electric resistance heating is usually added to provide enough heat for the coldest months.

Area Heating

Area heaters are used in some homes instead of central heating, or area heaters may be used to supplement the heat from the central heating system. Some of the types available are:

The Room Heater. Early types of room heaters burned coal or wood. Now they can be obtained in automatic gas, oil and electric models, some equipped with thermostats. Noncirculating room heaters, heat by conduction and by radiation.

Circulating Room Heaters. These operate somewhat like a warm air furnace. Cool air is drawn off the floor into the bottom, discharged from the top. An electric circulating fan often is used to blow the air downward and into remote corners of the room (Fig. 18.11).

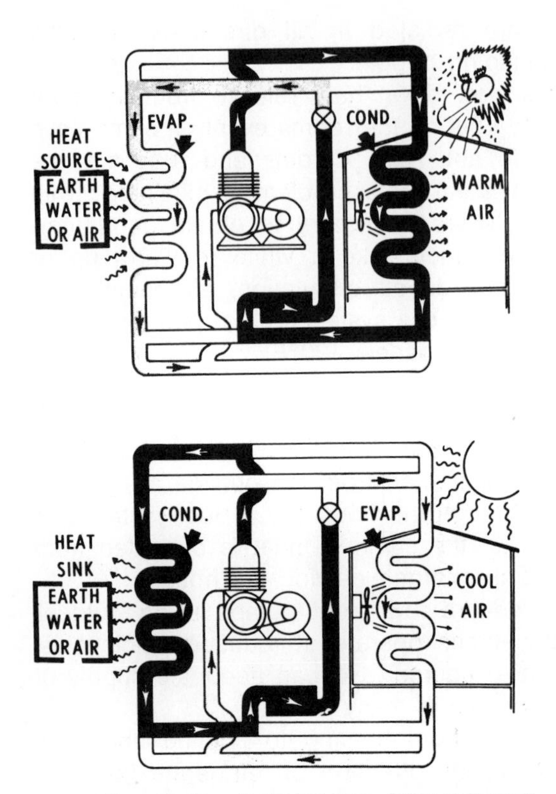

–Above: Heating cycle of a heat pump. Below: Cooling cycle.

Fig. 18.10 The heat pump is an application of the compression system of refrigeration. Air heated by the condenser or cooled by the evaporator is circulated to the rooms of the house. (Agriculture Information Bulletin 306, Agriculture Research Service, U.S.D.A.)

Floor Furnaces. These are essentially small oil, gas, or electric gravity warm air furnaces suspended from the floor joists in an open space under the floor. Heat is delivered through a large register flush in the floor. Cold air is returned through the outer edges of the same register. Control is either manual or by thermostat. Surface temperatures on the registers are frequently high enough to burn a person who might touch it with feet or hands. Several floor furnaces are required to do a satisfactory heating job in an ordinary house.

Wall Heaters. These are installed against a wall, partially or completely recessed. Air is drawn into the bottom by gravity, warmed, and then discharged at the top. Some have small circulating fans to take

***Fig. 18.11** Fan-forced baseboard portable heater. Five-position heat-selector dial. Automatic thermostat maintains temperature selected. Parabolic reflector and fan control direction of heat discharge. (Toastmaster)*

in room air at the top and discharge heated air from the bottom to warm the floor. Room thermostat control is available.

Baseboard Heaters. Portable electric baseboard heaters (Fig. 18.11) and oil-fired baseboard units which heat water, may be installed along the outside walls or under large windows for the benefits of perimeter heating. Each unit of this type is controlled by its own thermostat.

Fireplaces. Since almost 90% of the heat goes up the chimney—along with heated air from the house—this is the most inefficient way to heat a room. The steel air-circulation fireplace is an improvement since it has air passages on either side of the fireplace for gravity circulation of heated air.

A fireplace requires large quantities of air to burn wood or coal. A window in the room in which the fireplace is located should be partially open when the fireplace is in use to prevent the heated air from the rest of the house from being used for combustion of the wood. The damper should be tightly closed when there is no fire to prevent the heated indoor air from escaping through the chimney.

FUELS

Coal, oil, gas, and electricity all provide good heating provided the heating equipment is correctly installed. The major differences between them are cost and convenience.

Since all fuels at the present time are in short supply, and the predictions indicate that every effort must be made to reduce the consumption of fuel for the forseeable future, constant consideration of fuel conservation measures will be required.

Fuel conservation may be achieved by:

1. Setting the thermostats at the lowest acceptable level in the winter and highest acceptable level in the summer.
2. Drawing blinds and draperies in unoccupied rooms.
3. Keeping the damper closed when the fireplace is not in use.
4. Keeping filters in the air distribution systems clean.
5. Having the furnace cleaned and adjusted at least once a year.
6. The most important single element in reducing energy consumption for both heating and cooling is increased insulation of present buildings and increased minimum requirements for insulation for all new buildings. The National Bureau of Standards estimates that 40–50% savings could be realized through proper insulation and construction practices of both old and new structures.

INSULATION

Reducing Heat Losses by Transmission

Heating a home comfortably and economically begins with retarding the heat loss through the proper use of insulation. By slowing down the flow of heat through a structural section, the heat loss during a given time through that section can be reduced. At the same time, the amount of heat that must be added to equal the losses is reduced. Insulation adds to the comfort of the building occupants by eliminating cold floors and cold outside walls. Insulation also helps keep the heat out in the summer months. In both new houses and in replacement units, smaller heating equipment can be used with adequate insulation. Walls and ceilings will not require cleaning as frequently because dust and dirt are not attracted as readily to warm surfaces as to cold surfaces.

Insulating materials are available in various shapes and sizes. In selecting insulation, the material's adaptability to its use in the building, the cost, the ease of installation and its effectiveness must be considered.

Insulation of high quality and effectiveness should:

1. Have a high thermal efficiency.
2. Have high chemical and physical stability.
3. Not be subject to rot or decay.
4. Last the life of the building.
5. Be clean and sanitary.
6. Not attract or harbor insects and rodents.

7. Not be combustible.
8. Be moisture resistant. (A 1% increase in moisture has been shown to decrease the efficiency 5%.)
9. Batt and blanket types must be rugged enough to withstand handling. Fill insulators must be of proper density. Materials should not settle or change density.

Insulating materials are of three general types: fibrous, reflective, and glass and plastic foams.

Fibrous Types

Fibrous types of insulation may be rigid or flexible. Rigid insulation is usually produced in board or plank form while flexible insulation is found in blankets, batts, or loose fill.

Fibrous insulating materials are lightweight and porous. Their insulating value is provided by thousands of very small air pockets, each one of which helps to retard the flow of heat. Fibrous insulating materials include both rock wool and glass fiber. The effectiveness of fibrous insulation is affected by the thickness and density of the fibers, moisture and mean temperature (Fig. 2.3 and Fig. 18.7).

Reflective Types

Reflective types of insulation are usually a metal foil or foil surfaced material. They operate on the principle that polished metallic surfaces will reflect infrared or radiant heat rays (Fig. 18.12). To be effective, the foil must be exposed to an air space since it affects the portion of heat which travels across the air space by radiation. The number of reflecting surfaces, not the thickness of the material, determines its insulating value. Each reflecting surface is equivalent to ½ inch of insulating board. Several manufacturers of fibrous type insulation are now incorporating reflective surfaces in their batts and blankets. Metal foil is an excellent vapor barrier, and, if installed facing an air space, it supplements the regular insulating value of fiber type insulation. Some authorities state that reflective insulation is more effective in preventing downward flow of heat. In other words, it is more effective in floors over unheated areas.

Foam Types

Polyurethane and polystyrene foam have been used for insulation of residences in increasing amounts during recent years. In addition to the higher R values per square inch and their ease of application, foams provide an excellent vapor barrier. The Federal Trade Commission has now charged that foamed polyurethane and all forms of polystyrene and its copolymers may constitute serious fire hazards. The

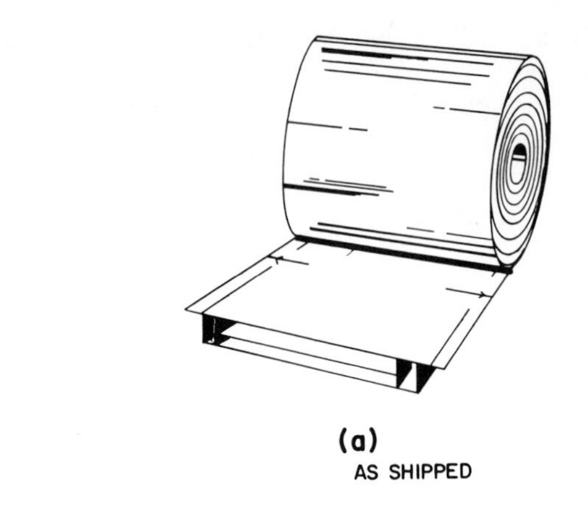

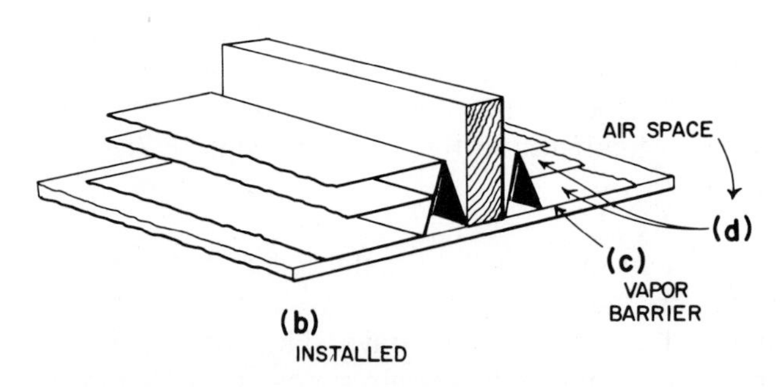

REFLECTIVE INSULATION

***Fig. 18.12** Reflective insulation.*

manufacturers, according to FTC have misrepresented these combustible plastics when they claimed them to be "non-burning" and "self-extinguishing". Urethane foam will burn and give off noxious gases. For these reasons, many local building codes now forbid their use.

The All Weather Comfort Standard developed by the building industry recommends the following minimum R values for electrically heated and for air conditioned homes:

Ceilings	R-19 to R-24
Outside walls	R-11 to R-13
Floors over unheated spaces	R-13
Basement walls	R-11

A comparison of the R values of some insulating materials is shown in Table 18.3. All R values are for a one-inch thickness of material, except for vertical air space. R value for vertical air space is for air space from ¾ to 4 inches thick. The total value for an air space is not proportional to its thickness.

Table 18.3
Insulating Values of Various Materials

Material	R per Inch Thickness
BATT OR BLANKET	
Wood or cellulose fiber with paper backing and facing	4.00
Mineral wool (rock, slag, or glass)	3.80–2.80
Pulp or paper	3.70
Wood fiber (redwood, hemlock, fir)	3.33
LOOSE FILL	
Mineral wool (rock, slag, or glass)	3.80–2.80
Pulp or paper	3.70
Cork (dry)	3.57
Wood fiber (redwood, hemlock, fir)	3.60–3.33
Sawdust or shavings (dry)	2.22
Vermiculite, expanded	2.20–2.08
BOARD OR RIGID	
Expanded urethane, foamed in place, sprayed, or preformed	8.33–5.88
Polystyrene foam, extruded or expanded	4.50–3.85
Glass fiberboard	4.34–4.00
Polystyrene molded from beads	4.17–3.45
Corkboard	3.70
Wood fiberboard, laminated sheathing	2.90–2.00
Foam or cellular glass	2.50
Homosote	2.38
AIR	
Air space, vertical, tight construction	0.90
Vertical wall air surfaces	
Indoor, still air	0.68
Outdoor, 15 mph wind	0.17

FHA (Federal Housing Administration) Standards

The new FHA Standards call for ceiling insulation with a value of R-11 to R-19 depending on local climate. All floors and exterior walls must receive insulation with a value of R-7 to R-11. It is estimated that only 25% of the houses built under the old requirements received as much wall insulation that all FHA houses will now be getting.

Where Insulation Should be Placed

Insulation should be placed in the outside walls and in the attic floor. In a flat-roofed home, the roof should be insulated to a maximum as a barrier against direct heat fom the sun. Overhangs should be insulated. Porch ceilings as well as porch walls should be insulated if the house extends over the porch. Foundation walls need insulation, particularly from the grade line up. Foundation wall insulation for houses erected on concrete slab is especially important. Insulation should be inserted between the footing wall and the edge of the slab and should extend downward into the ground at least 18 inches or bend back under the slab for the same distance. When the cement slab is not surrounded by a footing wall, the insulation should be applied directly to the exposed edge of the slab and bent back under it.

Floor over crawl spaces should be insulated except when the crawl space is used as a heating plenum.

When there is living space above a garage, the ceiling and wall separating the garage and house should be insulated.

Insulation is not needed beneath the floor if the house has a basement except in houses heated electrically where the basements are not heated continuously.

Insulation of Pipes and Ducts

Water pipes in outside walls or in unheated portions of the house should be insulated to protect them from freezing. Condensation and dripping from cold water pipes can be eliminated by insulating the pipes with "antisweat" pipe covering.

Loss of heat from steam and hot water pipes can be prevented by insulation. This is especially important in long runs of pipe. Valves and fittings should also be covered with material molded to shape.

Bright tin of bare pipes of warm air furnaces act as reflective insulation and should not be covered. Application of one layer of asbestos paper **increases** the heat loss and, therefore, should be avoided.

Double Glazing

To cut down on heat loss in glass areas, double or triple glass should be used.

The most common form of double glazing consists of storm windows and doors. Storm sashes are the only insulating-windows which help to reduce the infiltration of cold air, dirt, and soot occurring around the sash of operating windows. Because storm sash, even when used, are not ordinarily in place the year around, it is recommended that all operating windows be weatherstripped. When windows are weatherstripped, heat loss is about the same for all types of double-glazed windows.

To reduce moisture condensation between the glass when storm sash is installed on the inside, the inside sash must fit more tightly than when installed on the outside. Storm sash installed on the inside should be weatherstripped. Storm sash installed on the outside should fit rather loosely and the inner window sash should be weatherstripped. A glass storm window cuts the heat loss through a regular window by more than half, if the storm window is an inch or more away from the regular glass.

Double- or triple-glazed windows consist of two or three panes of glass separated by an air space with the edges of the glass sealed at the factory. Sealed double glass can be installed in an operating sash or can be set in a frame to form a fixed window. Sealed double or triple glass reduces maintenance problems. They are not changed with changing seasons as are storm sash, and there are only two glass surfaces to be washed instead of four. It is easier also to provide ventilation by opening the windows whereas with storm sash this cannot be done unless special provision is made in the storm sash for a portion to open when desired.

A double-glazed window with ¼-inch air space will reduce heat transmission to about half what it would be if a single glass was used.

Draperies

In cold weather, when the window glass is chilled, a closed drapery will reduce the radiative heat loss of people sitting near a window, enough to improve their thermal comfort appreciably. It is true, however, that most draperies are not fitted tightly to the window, and room air can and dies move freely by convection into and out of the space behind the drapery. For this reason, closing a drapery at night provides only a minor reduction of the winter heat losses of a house—considerably less than the reduction obtained 24 hours a day by using storm windows or double glazing. However, even with these in use, there will be more comfort for people near windows if draperies are closed in cold weather.

Reducing Heat Loss by Infiltration

From ¼ to ⅓ of heat losses in a home are by infiltration. Reduction of infiltration losses can be accomplished by cutting down air infiltration around windows and doors and their frames by means of weatherstripping, caulking, and storm sash. Storm sash and double and triple glass also cut down the heat loss through the glass itself.

Weatherstrips

Weatherstrips can be obtained for most types of windows: double hung, casement, sliding, and awning. These can be applied to windows of existing houses as well as to new windows. The space around the sash is filled with strips of rubber, spring metal, or other material. Both wood and metal windows are widely available with weatherstrips applied at the factory. It is usually less expensive to buy such windows for a new house than it is to get the unweatherstripped windows and have a mechanic add the strips at the site.

Doors should also be weatherstripped. With doors, the weatherstrips have to be done when the door is hung in its frame.

Caulking

Air and considerable moisture can infiltrate through cracks between window and door frames and the walls into which they are set. Durable, waterproof caulking compounds should be used to fill up the cracks between the window and door frames and the walls, especially at the top and bottom, but also, especially in masonry houses, at the sides as well.

Caulking should also be done around such places as chimneys, eaves, and dormers. Caulking should be done with a caulking gun forcing the compound into the open cracks.

SUMMER AIR CONDITIONING

Just as during the winter months, a comfortable summer indoor climate requires control of temperature and humidity, air circulation, and air cleaning. Attic fans, room fans, or evaporative coolers are frequently used to reduce the temperature and increase air circulation in a house but do little to reduce the humidity. Since humidity is as much a source of discomfort in summer as temperature in most parts of the country, the only satisfactory systems are those which control both temperature and humidity. Considerable medical evidence exists pointing to the benefits of summer air conditioning to those suffering from heart ailments and to the elderly. Studies made by one laboratory have shown that physical tasks performed at 100°F require 50% more energy than the same tasks performed at 75°F. Cooling and dehumidifying the air reduce fatigue and result in longer periods of work at greater efficiency.

Cooling a house is more difficult than heating a house. Cooking, housecleaning, bathing, operating appliances, living, and breathing add both heat and moisture to the air in the house. These help the heating system but increase the capacity needed for the cooling system.

Heat Gains

Keeping the heat out of the house is the most important consideration in achieving summer comfort. If the sun's rays can be kept off the roof, the window areas, and the walls, and the hot air kept from penetrating the house, the indoor temperature will not rise as much. Shading the roof and walls under an umbrella of decidous trees will best protect the roof and wall areas from the sun. Where this is not possible, other measures can be taken. A smooth white roof will reflect 70% of the sun's heat compared to 50% by a dark roof. Clean white asbestos shingles reflect about 55%. White asphalt shingles, which have a coarser surface, reflect 25%. Marble chips used on a flat roof will reflect 30 to 50% of the sun's heat; the smaller the chips, the higher the reflectivity. A white roof loses its reflectance as it becomes dirty and, therefore, cannot be counted on for sustained cooling.

Heavier ceiling insulation is recommended for summer than for winter air conditioning because the roof is thermally more important in summer and because cooling is more expensive than heating.

A good air wash through the attic is also recommended. In a poorly vented attic, heat will build up during the day and seep into the house overnight. An exhaust fan may be installed in one gable end, forcing hot air out while replacement air is pulled in through a louvered vent at the opposite gable or up through screened openings in the underside of roof overhangs. A thermostat may be installed which starts the fan when the air temperature reaches 100°F, bringing in outside air so that the temperature in the attic would seldom be higher than 100°. Without a thermostatically controlled fan the temperature often is 130°F.

The heat gain through the walls is next in importance to the roof Direct sun can heat up a dark wall to 150°F. A good layer of insulation will substantially reduce the heat flow through the walls. (Fig. 18.7).

Glass transmits 35 times as much heat from the sun and 10 times as much heat from the outside air as an equal area of insulated wall.

Double glazing will stop only about 10% more sun heat than a single pane of glass. However, double glazing will reduce the amount of heat from outside air 50% over single glass. This could be very important in a house with large glass areas. A heat reflecting glass is now available which can save more in heat load than any other glass.

Light-colored venetian blinds of aluminum or white plastic are fairly efficient shading devices for glass area. If mounted on the outside with a slat setting of 45°, they are 70% effective in reducing the load from the sun. If mounted on the outside and fully closed, they may be about 50% effective. Vertical inside blinds and draperies usually cost more and are less efficient shading devices. Louver-type insect shades when used on the south are 80 to 90% effective in reducing heat load. On east and west windows they are 25 to 75% effective. Overhangs on south windows, if properly sized, are the most effective shading devices. East and west walls cannot be protected from the sun's rays by overhangs because of the low angle of the sun in the early morning and later afternoon. Canvas awnings are from 65 to 75% effective in reducing heat load on south, west, and east windows. Free-standing walls or fences combined with roof overhangs are very good against low sun. Engineers recommend that on a hot day, inside shades, blinds, or draperies be kept fully closed over large glass areas whether exposed to direct sun or not because reflected glare from the sun can add a substantial heat load, even through a north window.

Heat created by cooking, the clothes dryer, and the shower and baths can best be taken care of with exhaust fans. The 400 Btu per hour given off by each occupant, as well as the heat from the lights and electrical equipment in use, are sources of indoor heat which must be taken into consideration in estimating the heat gain in a house.

Moisture

It requires 1000 Btu of cooling capacity to remove every pound of water. Therefore, it is important to keep the moisture load in the house to a minimum. Moisture comes from the outdoors and is also generated indoors by cooking, bathing, and clothes drying. Use of vapor barriers in the walls and ceiling stop high-pressure outside vapor from infiltrating into the house. Vapor barriers over crawl spaces, under slab and around foundation walls keep moisture from entering in these areas.

The biggest single source of indoor moisture can be an unvented automatic clothes dryer. A clothes dryer giving up 10 pounds of moisture an hour will constitute an extra load of 10,000 Btu. Thus it is important that an automatic clothes dryer be vented to the outside. Clothes hung on lines in the basement or in the bathroom to dry can also add quantities of moisture to the house.

Cooking discharges moisture into the house. Heat should be reduced in cooking as soon as the boiling point is reached and covers should be used on pans to reduce the amount of steam generated. An exhaust fan vented to the outside should be installed over the range. To keep from drawing too much cool air from the rest of the house when the fan is on, a nearby kitchen window should be opened a few inches for replacement of air.

Keeping the bathroom door closed and a window open while bathing and showering will keep some moisture from the rest of the house. An exhaust fan vented to the outside is recommended. Shower curtain should be nonabsorbent to prevent evaporation into the air later.

Calculation of Heat Gains

A heat gain calculation, sometimes called a cooling load estimate, for each house is essential to determine accurately the capacity of the equipment. Two identical houses in the same neighborhood will have different cooling capacity requirements because of their orientation to the sun. One house might face north with only its narrow sides and small windows exposed to the hot east and west sun. The other house may have the long side facing into hot east or west sun and require larger cooling equipment. A house without trees must handle about twice the heat load as the same house with trees shading the roof and east and west walls. A house in New York needs as big a cooling unit as the same house would need in Texas because the high temperatures reached are the same. The cooling season in Texas, however, is longer than in New York so the system would run several times as long in Texas.

In calculating heat gain, in addition to transmission and infiltration gains, gains due to the number of people occupying the space and a latent heat load factor (moisture load) are included. For any specific outside design temperature, the heat gain is affected by the daily temperature range. In some areas of the country the outside temperature may drop 15°F at night (low), whereas in some of the dry areas it may drop as much as 30°F (high). Heat gain estimates are computed for the sunny hours. (Heat loss transmission calculations are made for night conditions when no relief from the sun is possible.)

Until recently there have been almost as many systems for estimating cooling equipment loads as there are societies and manufacturers interested in the subject. There is now a simplified, industry-developed, uniform system that produces reliable results.

Cooling Equipment

The cooling machinery is essentially the same as that in a household refrigerator but is larger in size. It can use electricity, gas, or oil, and a compression or an absorption type system (see **Refrigerators**). The compressor in a central system is frequently installed outdoors. The condenser may be air cooled or water cooled. The air-cooled condenser is less efficient than one which is water cooled and usually requires a larger air conditioner. However, the air-cooled condenser usually has a lower service and maintenance cost. An evaporative condenser uses a fan to cool hot refrigerant gas like the air-cooled condenser but also has water sprayed continuously over its coil to increase its cooling efficiency. The water is recirculated within the unit and needs to be replaced only occasionally.

Flexible connections between the cooling equipment and the duct system will reduce the noise caused by fans and high pressure. Fiberglas ducts are said to be less noisy than metal ducts. Indoor equipment should be installed as far from the bedroom as possible and surrounded with acoustical material. Outdoor equipment should be as remote from the house as feasible and not where it will disturb the neighbors. An acoustical fence and shrubbery will help reduce the noise level.

The Air Conditioning and Refrigerating Institute (ARI) has established a standard for testing, rating, and certifying central air conditioning equipment in terms of Btu. Equipment meeting the standards (ARI 210–66) bears the ARI Seal of Certification. ARI also has a similar standard for heat pump equipment (ARI 240–67).

Proper size is important. The rated heat-removal capacity of the unit must closely match the estimated heat gain of the house. An oversized unit will give extra cooling under unusual conditions—such as when there are a number of guests in the house on a hot afternoon. However, under normal conditions it will be off so much of the time that it will do a poor job of dehumidifying, which it does only when the system is on. An undersized unit will do a good job of dehumidifying but may not give satisfactory cooling in very hot weather.

Cooling Systems

Central air conditioners are designed to cool and dehumidify a whole house. If the cooling system is tied into the house heating system, the cost of installation is usually lower than the cost of installing two systems—one for heating and one for cooling. When building a new house, it is wise to make provision for both heating and cooling, even though all of the cooling equipment may not be installed immediately.

Ducted Circulating Air Systems. It is possible to use the same ducts and registers for both heating and cooling. Cooling often requires more circulating air as well as larger ducts and registers than are needed for heating. When building a new house these can be sized for both heating and cooling. In an existing house with a forced warm air duct system, modifications will frequently have to be made.

In cooling, a more comfortable indoor climate is achieved by delivering cool air high on the wall. Perimeter registers may provide excellent cooling if the air is delivered fast enough to flow at least 4 feet upward toward the ceiling. In older houses with registers on inside walls, the registers should be equipped with adjustable vanes to deliver cold air directed toward the ceiling during the cooling season and hot air toward the floor during the heating season.

Return air registers should be located at the ceiling if cool air is delivered from the floor and at the floor if cool air comes from the top of the room.

Central Hydronic Systems. A combined heating and cooling hydronic system can be made an all-year system by using a special heating-cooling convector instead of conventional radiators. Each has a small fan and a filter. The convectors are connected to the pipes which carry hot water in the winter. In the summer, cold water is circulated from a water chiller. Each convector is controlled by a separate thermostat. This is a rather expensive installation.

In a house with a hydronic heating system, a supplementary cooling system can be installed with ducts and cooling equipment in the attic or basement of the house, or it may be cooled with individual cooling units for each room. The latter are usually more expensive in installation and operation than a single unit to cool the house.

The Heat Pump

The heat pump is a piece of equipment which performs the functions of an air conditioner in summer and which also heats the house on cold days (Fig. 18.10).

Room Air Conditioners

Room air conditioners eliminate duct work, are easy to install, and allow individual room temperature control. They are less efficient than the central units, and if a house requires a number of individual units, the total cost may be higher than that of a central unit.

The capacity of the unit needed is based on the heat gain of the area to be cooled just as in a central system. A much simplified form for estimating the cooling load has been developed by the Association of Home Appliance Manufacturers (AHAM). Local dealers should have the form. AHAM has also formulated a standard (AHAM-CN-1) which covers the capacity rating in Btu and also some performance characteristics. Equipment meeting these standards carry the AHAM seal. When the cooling load estimate in Btu has been determined, this is matched to the capacity rating of the equipment shown on the nameplate.

Room units require from ⅓ to 1½ horsepower. Some operate on 120 volts and some on 240 volts. It is recommended that even the smaller units be on a separate circuit.

Room units may be window or wall types. The window unit is mounted in a partially opened window with its major portion projecting outdoors. The wall unit is permanently framed and mounted in an outside wall. Neither is attractive from the outdoors, but the interior portions of many of these units are handsomely designed. Each type is available in models that heat as well as cool.

How long a unit will last depends chiefly on its resistance to corrosion. A heavy gauge "bonderized"[5] outside metal cabinet and cooling coils treated against corrosion will increase the life of the unit.

A high quality thermostat which turns a unit on and off to maintain a preset temperature is a feature on most models. An air outlet which directs the cold supply of air over the heads of the occupants of a room provides a more comfortable room environment.

Filters are of two types: those which are discarded and replaced when they become dirt clogged and those which can be washed and reused. Either type should be located so they can easily be removed or replaced.

When the entire cabinet is lined with insulation, top, bottom, and sides, to protect the cabinet projecting outside from the heat and rain, more efficient operation will result. This part of the cabinet will reflect a high proportion of the sun's rays if painted white.

The National Bureau of Standards has initiated a voluntary labeling program for major household appliances and equipment to effect energy conservation. The NBS is proposing that manufacturers place on the label the energy efficiency and energy consumption values

[5]A special base treatment which retards rust.

based on a standard test procedure. It's goal is to encourage consumers to purchase appliances that are more efficient than other similar appliances.

Energy efficiency ratios (EER = *Btu/hr*) have been determined for room air conditioners. The range of EER Watts for room air conditioners is from 4.7 to 12.2. The higher the EER the greater the efficiency. Greater efficiency means that the appliance will perform its function with less energy than will less efficient products.

Room Fans

Room fans increase the air movement, which increases moisture evaporation. This cools the air.

The most effective air movement is obtained by placing ventilation openings on opposite sides of the space to be cooled at sitting height. To prevent short circuiting of air in window fan installations, adjoining windows must be closed and the door to the room leading to the rest of the house left open.

Fans can be operated at night when the outdoor air is cooler and can bring in large quantities of cool air. The indoor temperature may be reduced to within one or two degrees above the outdoor temperature.

Fans are rated by the amount of air they move in cubic feet per minute (c.f.m.). However, there are several different rating tests to determine fan capacity.

To determine the size of the fan needed:

1. Find the volume of the area to be cooled (length × width × height).
2. Divide the volume by 1½ air changes an hour to get the minimum c.f.m. requirement. (If only one air change an hour is desired, the c.f.m. requirement is the same as the volume.)
3. Select a fan with a c.f.m. rating larger than the requirement to allow for differences in test procedure and efficiency.

A consumer purchasing a 20 inch or larger fan should be sure that it is accompanied by a label showing its air-moving capacity as a *ventilator* or a *circulator.* Different test methods are used to establish the rating on these two types of service. The capacity of a fan as a ventilator may only be from ⅕ to ½ its capacity as a circulator.

A window fan is easily installed but is sometimes noisy in operation. if the outside air is humid at night, the humid air will be brought into the house along with dust and pollen.

Attic Fans. A 36- to 42-inch fan with ½ horsepower motor will provide 60 air changes per hour for the average three-bedroom house.

An attic fan can also remove accumulated attic heat during the day. When used during the day the attic should be closed off from the rest of the house to prevent the fan from drawing hot air into the house.

Evaporative Coolers

Evaporative coolers circulate large amounts of air through the house. The outdoor air is cooled by forcing the air through a waterspray or through continuously wet fiber pads. Since the air enters the house at 100% relative humidity, this system is only feasible in very dry climates. Twenty to 40 house-air changes per hour are necessary. Five to 10 gallons of water per hour are needed to cool the average size home.

Living with Air Conditioning

The family with an air conditioned home will better achieve the full benefits of their unit if they follow the recommendations of specialists in the field.[6]

1. Set the thermostat at the room temperature you want. Most people prefer an indoor temperature of 76 or 78°.
2. Put the system into operation at the start of the hot weather season. Do not stop it and open the windows on a mild day or a cool evening. The unit will lose control of the indoor heat and humidity and will have to work much harder to regain control after it is started again. Dust and dirt come in, too. The cooling unit and filter can remove these annoyances while operating if the doors and windows are closed.
3. Keep the shades and draperies closed on all windows and doors not protected from the sun on the outside and on windows or doors where the sun's heat is reflected from a drive, patio, or unshaded porch.
4. Use a kitchen ventilating fan while cooking. Close the doors to the rest of the house. Open a kitchen window a few inches to replace the air removed by the fan. The range, with all burners going, releases more heat than a 2- or 3-ton cooler could remove if it did nothing else.
5. Do cleaning jobs that require water—mopping, washing windows—during the coolest part of a warm day. These jobs release moisture faster than the cooler can remove it on a warm day.
6. Do not dry laundry in the bathroom or in the basement. Vent an automatic dryer to the outdoors. Iron on a cool day or evening when the system has enough unused capacity to remove the heat and moisture that are generated.
7. Keep the bathroom door closed when bathing. Then open the window slightly after bathing and ventilate the moisture-laden air to the outdoors with a fan. Use a plastic or other nonabsorbent shower curtain.

[6]Honeywell, **How to Get the Most From Your Heating or Cooling Dollar.**

Terminology

A comfortable indoor environment. Provided by establishing air conditions which permit the body to reject excess heat at a satisfactory rate through regulation of temperature (air and radiant), air movement, relative humidity, and air cleaning.

Heat loss. The amount of heat which leaves the house by transmission and infiltration in cold weather.

Heat gain. The amount of heat and moisture entering from outside the house and generated inside the house during the summer months.

R values of insulating materials. Indicate the ability of the material to resist the passage of heat.

Coefficient of heat transmission. The reciprocal of the resistance and is usually designated as the U value: $U = 1/R$. Tables of U values have been compiled for common types of ceiling, wall, floor, window, and door construction.

Vapor barriers. Prevent moisture from inside the house from getting into the walls (and into the insulation) and condensing within the walls during cold weather.

Outdoor design temperature. Has been determined from U.S. Weather Bureau records for every major city in the United States and Canada.

Continuous air circulation (C.A.C.). A method of controlling the blower in forced warm air heating systems to keep heat evenly distributed and maintain a constant delivery.

Heat pump. A reversible refrigeration unit. It both heats and cools. In the summer, heat is extracted from the air inside the house and discharged outside the house. In the winter, the heat is extracted from the outside air, ground, or water and is distributed through the house by a duct system.

Experiments and Projects

1. Ask a builder to take you to a home that is under construction. Observe the type and amounts of insulation on walls, ceilings, floors, crawl spaces, and overhangs. What is the size of louvers under the roof? What type of heating system is being installed? Is summer air conditioning being installed or provided for? In what

compass direction does the living room face? Dining room? Kitchen? Bedroom? Do trees shade any walls or roof of the house?

2. Place a thermometer in a living room (1) on the floor, (2) at 2½ feet from the floor (sitting height), (3) at 4–5 feet from the floor (usual thermostat location), (4) near the ceiling. Note differences in temperature. At which of these levels should the thermostat be located for most comfort to persons in the living room?
3. Observe where the thermostat is located in several homes of your acquaintances. Does the location of each one meet the criteria of heating engineers outlined in this text?
4. On a day in which the temperature is below freezing outdoors, stand in a heated room near a window which is not equipped with storm sash or double glass and on which the sun is not shining. Why do you feel colder near the window than in another part of the room? Draw the drapery or shade completely over the window. Stand in the same position as previously. Explain why you feel warmer.
5. What type of heating system do you have in your home? Where are the hot air registers located in the living room? in the dining room? in the bedrooms? in the family room? Is each located where convection currents are working for you?
6. What are some tips you could give to friends who are going to build a new home, which will include summer air conditioning as well as a heating plant, so that they may attain maximum comfort and minimum expenditure for equipment and operation of a heating and air conditioning system?
7. From Figure 18.2 find the outside design temperature for your home. If you plan to maintain an indoor temperature of 70°F, what is the design temperature difference which would be used in calculating heat loss?
8. Obtain a "Residential Heat Gain and Heat Loss Estimate" form from your local utility or heating contractor. Calculate the heat loss for the living room in your home or the living room on which you can obtain information on the type of construction.
9. If the living room in problem 8 did not have 3 inches of insulation in the walls and ceiling, and did not have storm sash or double glass windows, calculate the difference in heat loss if these were added.

CHAPTER 19

Maximizing Satisfaction in Work in the Home

Achievement and efficiency are valued symbols of success for all kinds of workers in this country. Thus the person, almost always the woman, who manages homemaking responsibilities successfully and keeps housework under control tends to gain feelings of satisfaction, security, and pride in her position whether she is a full-time homemaker or also gainfully employed. These feelings tend to be shared by family members and help children to develop wholesome attitudes toward homemaking responsibilities. A sound reason for any woman to become an effective home manager, in the words of the late Lillian Gilbreth, is to make homemaking "as interesting and satisfying as it is important." The overall purpose of effectiveness in performing the work of the home is to maximize the satisfaction and the well-being of the homemaker and the family as well as to enable them to share in the life of the community. In the United States, the acquisition of labor-saving appliances has become an important aid in achieving these goals.

The affluence of this country has enabled the majority of families to obtain a large number and variety of household appliances designed to facilitate the performance of every household task. But does the ownership of a so-called "houseful" of labor-saving equipment assure an efficient homemaker and a satisfied family? The answer tends to vary among families although study indicates that only through effective management can the homemaker and the family gain the greatest possible benefits from their household equipment and other resources.

CONTRIBUTION OF EQUIPMENT TO ATTAINING FAMILY GOALS

Since needs and wants are satisfied and goals attained through the management of human and material resources, a first step in the management of household equipment is to identify the goals of the family and its members and to understand how equipment can contribute to the attainment of these goals through the functions it can perform.

Generally household equipment is designed to serve one or more of the following functions:

1. Regulate the indoor environment for the comfort and health of individuals concerned. This refers to temperature, ventilation, humidity, and purity of air and provision of light, sanitation, and safety as well as adding attractiveness to the house. This kind of environment can help promote an individual's efficiency in work and enjoyment of rest and leisure.
2. Conserve the time and energy of workers in the home through making the work easier and/or faster. Performing household tasks with less time and energy frees the homemaker and other family members to pursue their personal interests. This is illustrated by the automatic washer that has reduced family laundering time from a day or more weekly to a few hours especially when combined with other laundry aids and the use of easy-care fabrics.
3. Improve the quality of the output—produce a better product and/or do a better job. For example, the use of a standard recipe with accurate measuring and functional mixing appliances combined with a standard-sized baking pan made of suitable material and design and used in a thermostatically controlled oven helps to assure a quality cake. Similarly a contemporary vacuum sweeper can get the house cleaner with less effort than the formerly used brooms, dust mops, and feather dusters.
4. Increase the satisfaction of the worker through increased and improved achievement with less time and effort. Pride of ownership may also add pleasure.

DETERMINATION OF NEED FOR HOUSEHOLD EQUIPMENT

A second factor closely related to the first one in the effective management of household equipment is to determine the goals of family members as well as to determine the actual need for any appliance. Need is related to the kind and frequency of the activities carried on in the home and the effort required in their performance in order to achieve satisfactory results. Generally the work required for different homemaking activities such as providing food and clothing for the family, house cleaning, and laundering is directly related to the number of family members, and sometimes also outsiders, and includes their age, sex, health, and life-style. In addition, the desirability of some equipment such as air-treatment facilities is also affected by the type, size and quality of dwelling, its location and site, permanence of residence, and whether residence is owned or rented.

Because few individuals and families have adequate resources including money to fulfill all of their needs and wants, they must weigh and compare the cost-benefit ratio of each possibility for satisfying different needs and wants in order to decide what will be of greatest value to them now and in the future. Deciding whether to purchase a certain piece of equipment may mean competition for the available income with other kinds of equipment and with other goods and services. Usually such a choice results in decisions to satisfy some wants and at the same time to postpone or forego some other wants.

Many homemakers are fascinated, for example, by the time-saving possibilities in food preparation with the use of a microwave oven. As a result they face decisions: Should they invest in a microwave oven when they own a serviceable range or will the microwave oven be the best replacement for the present outmoded or worn-out range? Answering questions as the following can aid in arriving at a reasonable decision. Does the family prepare many of its meals at home and serve the kind of foods best cooked in the microwave oven? What is the quality of such foods? How important is it to the homemaker to save time in cooking and what will she do with the free time? Of what significance is the possible saving of fuel through shorter cooking time and the fact that there is less heat in the kitchen than with the use of the regular gas or electric range? Would the use of pressure sauce pans and/or more ready-prepared foods be adequate in speeding up food preparation? Can essential space be provided for the convenient and safe operation of the microwave oven? Will or can the persons who use the oven adjust their food-preparation practices to the speed of this appliance? With the slower range the homemaker could set the table, prepare the salad, feed the baby, and attend to a variety of tasks while foods were cooking and baking. With the microwave oven she must be more alert to exact timing to get all foods ready at the same time in optimum condition. Will the microwave oven be adequate or must other cooking or baking equipment be used? Are reliable servicing facilities available locally to keep the oven operating safely and effectively?

Will using the money for the microwave oven serve the family better than using it for another purpose? What other needs and wants may need to be sacrificed for the purchase of the microwave oven?

Other similar equipment as well as other questions are pertinent. When deciding whether to purchase a washer and/or a dryer the adequacy of the water and the fuel supply is crucial. In addition the family may find it beneficial to compare the cost, convenience, and results of using a commercial laundry—both the regular and the self-service types—with ownership or leasing of laundry equipment. To figure the cost of laundering at home the purchase price and depreciation, upkeep, and repair or leasing costs plus the cost of space, plumbing, fuel connections, operation including water and water heating, and laundry aids are usually included. The worker's time and transportation are also involved. So may be the heat, humidity, and noise of home laundering. With regard to the dryer the thrifty homemaker, especially when she lives in a low humidity, low-smog area, may choose line drying and like the open-air exercise. Setting the basket with wet laundry on a cart and using a clothespin apron or line bag will minimize the physical effort of carrying and stooping.

The questions and factors that are significant in purchasing equipment as well as other consumer goods and services have to be answered cooperatively by family members on the basis of their value system, desired life-style, and their situation including finances.

The personalized aspect of need-determination is further illustrated with certain personal care items. For example, what is the effect on a person's well being of different vibrators as individual pieces of equipment and as included in chairs and beds? Are automatic massagers beneficial to health and for reducing weight and size? Only a physician can help the person answer these questions. Whether any healthy person needs an automatic fingernail buffer or automatic shoe polishing equipment seems to depend on the pleasure he gets from the use of gadgets instead of his own muscles and his available money and concern for fuel conservation.

Another factor of need-determination requires coordination of all household equipment owned to avoid unnecessary duplication. When is more than one appliance needed to perform the same purpose? What small electric kitchen appliances are needed for cooking and baking when the range includes such special features as automatic oven and broiler controls, rotisserie, thermostatically controlled surface units, timer, clock, lights and the continuous or the self-cleaning oven? Why do some homemakers pay for such features but seldom or never use them? How do you decide which features are really needed? May a single person or couple, who use extensively ready-prepared and ready-to-serve food as well as eat out frequently, substitute well-chosen, small electric appliances for a range?

Since research shows that many homemakers own a variety of small household utensils and tools including small electric appliances which they seldom or never use, the beginning family may want to analyze its own situation carefully to decide what equipment it really wants before starting its purchasing or gift lists. For instance, what foods will they eat which will be better when prepared with an electric mixer or blender? How many speeds will they actually use for either of these appliances? Does the serving of waffles once or twice yearly make a waffle iron a good investment? Or should the family owning a waffle iron make an effort to use it more frequently? When may dual-purpose equipment be serviceable and reduce the number of appliances needed? Unused equipment means a waste of money and may clutter up storage space to retard efficiency.

One more decision in the determination of need is when to replace currently owned equipment with newer models How do you decide whether the advantages of the new model will be worth its cost or even mean a saving through lower operation and/or maintenance costs?

Often the decision about new models of household equipment as well as of other consumer goods involves psychological need. When one recalls that the pioneer families produced the majority of goods and services they used with almost no labor-saving appliances, one may question the real need for some of today's appliances and furnishings when the family also has easy access to a multitude of ready-to-use and easy-care products and commercial services. Probably the answer is that in this country the "good life" has become the "easy life" with little physical effort, one lived in a comfortable environment with freedom and leisure, surrounded by conveniences, and promoted by the desire to keep up with, or better surpass, the Joneses. Thus yesterday's luxuries become today's necessities and a piece of equipment with little practical value may be worth its cost because of the psychological satisfactions it provides.

Finally, the determination of need in any endeavor such as the occupation of homemaking is related to the characteristics and significance of the occupation. What are the responsibilities and contributions of the homemaker that may justify large expenditures for equipment?

The Homemaker's Job—Work in the Home

Although the roles of family members are changing so that some responsibilities between husband and wife are neither clearly defined nor separated, the great majority of women retain the major responsibility for the management and the work in the home with only limited assistance from husband, children, and paid outside help. At the same time some people question the significance of the homemaking job and whether it should remain the wife's major responsibility. Regardless of the opinions about homemaking it remains a unique occupation. Even though it is characterized by a great diversity of activities and demands knowledge and variety of skills, any person may assume the job without any preparation. Much of the work is coordinated with the biological needs of the family and the customs of society, but for some aspects the homemaker is free to choose and organize. She may indulge in some leisure during working hours but her responsibility may extend over a 24-hour day when she is the mother of young children. She tends to have longer working hours than members of other vocations. Collectively the more than 50 million homemakers of this nation probably use more time, money, and some other resources than even the largest corporation. Therefore when certain resources are to be used and conserved more wisely it may be well to include homemakers.

Because the homemaker is not paid (hopefully she works for love), the monetary value of her work is seldom recognized. Nevertheless in 1965 the Economic Research Department of Chase Manhattan Bank conducted a survey and concluded that the variety of tasks performed by the homemaker is worth $8285.68 a year at the regular pay rate of the labor market. In April, 1973 the editors of **Changing Times** in an article, entitled "What's a Housewife Worth?," categorized her job as housework, child care, and keeper of the nest. They estimated the value of her work at $235.40 weekly for 99.6 hours of work, of which 44.5 hours were devoted to child care.

In the October, 1973 **Journal of Home Economics,** in an article, "Household Work Time," the author reported that research showed the employed homemaker spent an average of 4 to 8 hours daily and the nonemployed 5 to 12 hours at housework. This does not include all the custodial care of children such as their sleeping time. Still the employed homemaker worked a 66 to 77 hour week at household duties and additional job. The amount of time the woman spent in housework varied with the number of children, the age of the youngest child or age of wife in a childless family, and her employment of 15 or more hours weekly.

The estimated money earned for doing the housework ranged from $3900 annually by the nonemployed young, childless wife to $9000 for the nonemployed mother of seven to nine children with the youngest child under one year of age for an 84-hour work week.

The foregoing monetary figures gain in significance when compared with the fact that in 1970 "the medium income for all women who were year-round full time workers was $5440." Only for the professional, technical, and kindred category did income for women reach $8000.

Still the homemaker's monetary contribution is probably not her greatest value to the family. Her responsibilities as "keeper of the nest" are more difficult to replace by hired help and have the greatest impact on family members. These seem to be well described by Margaret Mead in the November, 1963 issue of ***Redbook*** in the following words:

> Through the ages, human beings have remained human because there were women whose duty it was to provide continuity in their lives—to be there when they went to sleep and when they woke up, to ease pain, to sympathize with failure and rejoice with success, to listen to tales of broken hearts, to soothe and support and sustain and stimulate husbands and sons as they faced the vicissitudes of a hard outside world. Throughout history children have needed mothers, men have needed wives. The young, the sick, the old, the unhappy and the triumphantly victorious have needed special individuals to share with them and to care for them.

The fact the homemaker's job, especially as a mother, is not an easy one, is clearly expressed by Saperstein as follows: "It probably takes more endurance, more patience, more intelligence, more healthy emotion to raise a decent happy human being than to be an atomic physicist, a politician or a psychiatrist." Of course this responsibility tends to succeed best when cooperatively shared with the husband.

Although equipment, ready-to-use and easy-care products, and commercial services have reduced the physical work and energy expenditures of the homemaker, intellectual requirements including decision making and managerial responsibilities, especially for the gainfully employed homemaker, are increasing. Possibly in the future computers may assist the homemaker and the family with the multitude of managerial decisions they face.

On the basis of the complexity and the significance of the homemaker's responsibilities may it be logical to conclude that when equipment is managed to free her from some time-consuming, repetitive housework for humanistic, intellectual, and spiritual endeavors in the home and community and possibly also for employment of value to society, equipment is a wise investment?

SELECTION OF HOUSEHOLD EQUIPMENT

After a decision has been made to procure a certain household appliance to satisfy a certain need or want, more managerial decisions arise such as which one to select from the many different brands and models on the market, whether to purchase or lease, to buy now or wait for a sale, to pay cash or use credit—and which form of credit—and from whom to obtain

Gaining some understandinf of the operating principles and different materials used and their likely performance lays a foundation for an intelligent selection of any appliance. It aids the prospective consumer in understanding the advertising, specification sheets, fact tags, warranties, and instructions for use and care provided by the manufacturer. As manufacturers make special efforts to provide adequate, useful, easy-to-understand information, a careful study of this information becomes more rewarding in comparing and selecting a piece of equipment. When such study is supplemented by observation of demonstrations and consulting the reports of a consumer testing agency a satisfying choice should result.

Usually purchasing an appliance from a local, well established merchant with a reputation for integrity and the facilities to service this equipment is preferable to purchasing from door-to-door sales people or from an out-of-town dealer with no nearby establishment.

Whether to wait to accumulate cash or for a sale before purchase is often a psychological decision. At times even the use of credit may be desirable for health or financial reasons. This can be illustrated by obtaining a freezer when foods are plentiful and comparatively inexpensive, or a sewing machine when the homemaker produces clothing and/or household furnishings at a lower cost or better quality than buying ready-made ones. She may even earn money by sewing for others.

Even though the leasing of household appliances is not widely practiced, it may prove advantageous under special circumstances as for people who move frequently or those who want to be freed from unkeep and repair.

In conclusion equipment is well selected when it performs effectively and safely the purposes for which it was chosen with thrifty utilization of energy at a total cost in harmony with the benefits it provides and the family financial situation.

LOCATION AND INSTALLATION OF HOUSEHOLD EQUIPMENT

Another aspect of household equipment management is concerned with the location and installation of large equipment and the space for storage and operation of all small appliances. Effective placement organizes the equipment in the major activity areas with the furnishings, storage facilities, work surfaces, and other appliances, materials, and supplies with which it is used. The arrangement is effective when it provides space for the number of individuals who use it at one time with postural positions essential in its use. At the same time it enables the individual to work with healthful body movements, a minimum retracing of steps and handling of materials, and without safety hazards. Here following the manufacturer's instructions for installation is important. The information for kitchen planning presented in chapter 11 can be applied to the design of other activity areas such as those for sewing, bathing and dressing, sleep, and play.

Storage in the Home

A necessity in every activity center for both work and play is adequate, appropriate storage for everything needed such as small equipment, furnishings, materials, and supplies. Effective storage is conveniently located and designed to facilitate getting and putting away items through being visible and accessible. It also conserves items stored and is flexible for changing needs.

Open storage is convenient but may be unsightly or fail to protect articles. Drawers, pull-out shelves, and sliding racks provide accessibility but may be difficult to operate when heavily loaded. Although preferences, work habits, available space, and items stored determine storage needs, the following guidelines are widely applicable in the home:

1. Design space to fit objects stored. Flexibility through adjustable shelves, partitions, and movable accessories increases usefulness.
2. Store the most frequently used items at the most easily accessible spaces. This means between wrist and shoulder level for easy reaching.
3. Locate items where they are most frequently used first. For instance, baking pans are more conventiently stored at the preparation center than at the oven.
4. Stack or store only identical items more than one deep. To move other items before getting the desired one is impractical. Subshelves, files, racks, revolving shelves, and hooks can expand storage space to avoid awkward stacking.
5. Place items so they can be picked up ready to use. A well-located knife rack illustrates this idea. The prepositioning of stored items reduces handling and in this case avoids cut fingers.
6. Group items that are regularly used together in activity centers. Small items which are used in more than one center may be duplicated. For instance, large kitchen spoons are needed in the mixing area and at

the range. Some cleaning supplies and minor pieces of equipment may be needed in kitchen, bathroom, and laundry. Duplication costs may be more desirable than constantly moving them around.

7. Label all containers in which contents are not visible. Store all packaged items with labels toward the front.
8. Use containers such as canisters which can be opened and closed with one hand.
9. Identify contents of storage places such as cabinets and drawers which are used by more than one person. Lists posted on the inside of cabinet or closet doors, and in drawers help every user to find what he wants and to replace items where they can be quickly found the next time. For young children, who cannot read, pictures or simple drawings such as an outline of their foot for hosiery, may encourage them to develop habits of orderliness in an interesting way and help them to feel that they are a member of the team especially when they help prepare the identifications for their clothing and toy storage within their reach.
10. Provide for safety in storage. Heed warning labels and be alert to items dangerous for children. Store them so that children can not get them. Also explain to children why certain items and spaces are "off-limits" to them as soon as they are old enough to understand. Poisonous products must be clearly labeled and never stored in food containers. Warning signs with regard to temperature and storage need to be observed. Avoid storage of flammable products as gasoline in the house, garage, or automobile.
11. Use partitions in drawers. Also drawers too deep for items stored may be made more useful by addition of flat, lightweight boxes. When the stored items are categorized and organized in these boxes, a box may be lifted to obtain the desired article without disarranging the rest of the contents.
12. Keep heavy or bulky items that are frequently used such as mixer and blender at waist level, preferably where they can be used without lifting and carrying. Keeping them on the kitchen cabinet or on an easily raised shelf in the mixing area conserves effort and tends to increase their use. Often appliances are not used because of the difficulty of getting them and putting them away.

USE AND CARE OF HOUSEHOLD EQUIPMENT

Of major importance in gaining the greatest possible benefits from household equipment is its correct use and care. This is best achieved through following the manufacturer's instructions which should be filed for ready accessibility. Misuse and improper care can result in safety hazards, poor performance, inferior product, high repair costs, and, for automatic appliances, high energy consumption, and sometimes excessive noise and pollution.

Closely related to the proper space and use and care of household equipment for efficient operation is the environment in which the equipment is operated and where the homemaker works.

ENVIRONMENT FOR WORK

The homemaker's physical and emotional well-being, her feelings about the work, and her productivity on the job are influenced by the environment in her work place. Even though more research is needed to identify optimum working conditions, the following factors appear to be significant in the effective performance of work including the use of equipment in the home:

1. Adequate, appropriate, conveniently located, properly furnished, well-arranged space.
2. A place that is attractive enough for the esthetic satisfaction of individuals who use it. The overall decor including color and design is suitable for the purpose of the space and pleasing to the most frequent users. Generally bright colors and bold design are suitable for activity areas such as playroom and kitchen, whereas subdued color and restful design are appropriate for quiet areas as bedroom. Warm colors tend to enhance rooms with limited exposure to sun but also cause area to appear smaller. In contrast, cool colors are desirable for areas with much exposure to sun and give a feeling of spaciousness. Mirrors and windows can minimize feelings of claustrophobia when working in a small space.
3. Orderliness for safety and to avoid feelings of confusion, frustration, and futility. Appropriate storage can contribute to orderliness.
4. Air treatment to provide the temperature, humidity, ventilation, and air, free from dust, pollen, obnoxious odors, and other pollutants to provide air that is conducive to good health, comfort, and achievement.
5. Lighting that is right for the eyes and appropriate for the activity or job.
6. A minimum of loud, harsh, irritating noise. Although individuals differ in their reactions to various types and intensities of sound measured in decibels, studies show that long-time exposure to noise unpleasant for the individual contributes to mental and emotional discomfort and fatigue and tends to reduce achievement. Over a period of time loud noise such as some contemporary music and industrial processes can injure hearing.

As man-made noise is increasing, noise conditioning is recommended for homes as well as public and industrial places. Although not all unpleasant noise can be eliminated from the home both outdoor and indoor noise can be modified. Outdoor noises such as automobiles, trucks, airplanes, construction and other sources can largely be kept out of the house by heavy insulation throughout the building and the use of storm windows and doors. Building a berm and planting trees and shrubbery between house and street also helps. In fact, trees and shrubbery around the house are useful in muffling sound and reducing some air pollutants.

Indoor noise can be reduced through the use of acoustical ceiling materials, carpeting, resilient floor coverings, draperies, wall hangings, and upholstery—especially when made of textured fabrics. Quality equipment kept in good working order and correctly installed helps to minimize operating noise. Industry is also striving to reduce operating noise in equipment. Placing resilient mats under some equipment such as mixers and blenders decreases noise.

Leaky or dripping faucets, creaking floors, and squeaking doors deserve immediate repair. Personal habits of the worker can be quiet or noisy. Even the kinds of shoes worn influence the sounds of activity. Moreover individuals may decide whether the right kind and volume of music may add to the pleasure of an activity and help to reduce feelings of fatigue thereby increase achievement but reduce accidents and errors.

7. Provision for safety in the home. Follow suggestions for the selection and use of electric and gas equipment presented in previous chapters. Keep all in good working condition. Use only a sturdy stepstool or ladder to reach high places (adjust curtains, get an item from shelf, or replace light bulb). Long-handled tongs are also useful in getting some items from high shelves. Use skid-resistant floor treatment and waxes. Avoid the use of rugs and mats that slip. Especially for older people convenient bars to grasp for getting in and out of bath tubs and nonslip bottoms of tubs help to prevent falling. So do hand rails which on stairs and steps are essential, and helpful in hallways. Flame-resistant clothing for children and old people, particularly for sleepwear, is important. Open-flame unvented gas heaters are dangerous. If these must be used, keep a window slightly opened and place the heaters where the flame can not blow out. Gas ranges should not be placed in drafts nor where curtains can blow across burners. Keep the kind of fire extinguisher appropriate for possible fires in the home in a convenient place in good working condition. Be sure that every family capable of handling them knows how to operate them. Usually the local fire department can provide

help in selecting an effective fire extinguisher and in prescribing safety procedures in case of a fire in the home.[1]

8. Clothing that is appropriate for the worker and the job. Clothes that are attractive enough for the worker to feel good about her appearance, that are comfortable, nonrestrictive of body movement, right for the temperature, and safe for the worker add to her comfort and possible achievement. Dangling parts that can ignite on gas burners or overturn hot utensils and long, full skirts that may trip the worker are undesirable. Shoes that are comfortable and easy to keep on are an asset.

9. Provision of a file-and-business center. A place to keep legal and financial papers, equipment instructions and guarantees, and other important papers, as well as a place to write and keep records is a time and energy-saving convenience. Its location depends on the dwelling and available space plus preferences of the family.

10. Meeting needs of physically handicapped homemaker or other family members. The facilities needed and the adjustments necessary in the dwelling depend on the person and the type and severity of the handicap. Research and resulting publications give many useful suggestions for equipment, clothing, and adapting the house to the needs of the physically handicapped so that they can be more nearly independent and participate in family activities.

WORK PROCEDURES IN THE HOME

Some research has revealed that homemakers in general have not reduced the time devoted to home responsibilities as much as could be expected with the use of labor- and time-saving equipment and products and commercial services. Apparently the time saved by a time saving convenience is quickly absorbed by other chores. Reasons for this seem unclear. Perhaps following effective work procedures would be helpful. These may well begin with proper handling of the body.

HEALTHFUL BODY MECHANICS

The human body may be viewed as being composed of three weights: the head, chest, and pelvis. These are supported by the legs and feet. In healthful posture the body weights are aligned without strain by

[1]It is recommended that families with children practice procedures to increase their safety in case of fire and hazardous weather, such as tornadoes.

ligaments and muscles which sustain the bone structure. When any body segment is out of alignment, extra effort, sometimes known as "static work," is required. This is illustrated by the worker who bends his back or hangs his head at a too-low work surface and the one who must raise his shoulders at a too-high surface. Both require extra physical effort through muscle contraction to hold the body in an unnatural position. These contracted muscles also tend to retard blood circulation. Such muscular effort can be more tiring than the work performed. Thus the physical cost of even light work can be high and cause fatigue when the natural body structure is ignored. For this reason, the proper handling of the body, popularly known as "body mechanics" (the scientific study of body motions), is important in all movements such as standing, walking, sitting, lying, climbing, pulling, pushing, lifting, stooping, and stretching. The following guidelines are widely recommended to maximize achievement and to minimize fatigue:

1. Include some physical activity in the daily routine to help maintain optimum health and normal weight. Some housekeeping tasks both indoors and outdoors in yard and garden can provide wholesome exercise when carried on with desirable body motions. Some examples are hoeing, pushing sweeper or lawnmower, mopping, taking the baby or the dog for a walk, and bedmaking.

2. Keep all parts of the body aligned. For this the equipment and the work surface must suit the stature of the worker and the task. These two facts—that the arm can exert its greatest strength when bent at the elbow so that the forearm is almost perpendicular to the upper arm and that the arm can move freely at shoulder and elbow in normal position—help to determine the comfortable height of work surface. For both standing and sitting at work, the work surface needs to be a few inches lower than the elbow for the majority of household tasks. Writing seems to be one exception when work surface seems comfortable at elbow height. The creative homemaker and her family can adjust a number of work surfaces to meet the needs of the individuals who use them most often. In addition to the regular dough board, other pull-out boards can be added to the kitchen cabinet to provide the right height for different tasks while standing or sitting. Sometimes it is possible to place a lightweight plywood board over a drawer that has been pulled out or use the board as a lap board over the arms of a chair to obtain a comfortable height work surface.[2]

[2]A short person, such as a child, would find a broad, sturdy, nontipping footstool useful for a variety of activities at work surfaces too high for comfortable use, such as the washbasin or kitchen cabinet and range.

3. Use natural, relaxed, rhythmic motions for all body movements. Avoid tense, jerky ones.

4. Follow a tempo suited for the worker and the task. Hurrying accelerates fatigue.

5. Lift, hold, and carry loads close to the body. Even an empty hand tires quickly when extended.

6. Let the strongest muscles, which are best suited for the task, bear most of the burden. Leg muscles are stronger than the back. Therefore bend the hips and knees, not the back when stooping. Proper stooping is important because the stress from bending the back on the intervertebral discs can cause back injury. Lifting, especially a heavy object, combined with bending the back greatly increases this stress or strain. Avoid placing heavy objects on the floor or other low surface when they must be lifted later as much as possible. The frequent lifting of a baby, often from low surfaces, contributes to the fatigue of a mother of small children.

To minimize back strain, squat and use leg muscles when lifting an object from the floor. When pushing a heavy or hard to move object, crouch with bent legs and push at center of gravity of the object while rising out of the crouching position. Placing the feet apart gives a broader base for lifting and pushing.

7. Whenever feasible use both hands at the same time rather than letting one remain idle. While dusting, washing, wiping an uncluttered surface as windows, walls, or table top the use of both hands, in rhythmic tempo, can speed up the work. In cleaning a surface with various objects, one hand may move the objects while the other one wipes or handles vacuum-cleaner wand.

8. Distribute a heavy load being carried between both hands. If this is impossible, bend the knees, not the back, while holding the load close to the body.

9. Use gravity when possible. Cutting food on a cutting board is easier, faster, and safer than cutting it while holding it in the hand.

10. Take advantage of momentum in such tasks as dusting, washing windows, sweeping, mopping, and hoeing. Continuous circular motions are easier and use less energy than short, jerky ones as well as sudden starts, stops, and sharp right-angle turns.

11. Select equipment that will not contribute to "static work" of the user. The built-in oven and range top can provide for proper posture of the most frequent user when installed at the right height for her body structure. So can the long-handled dust pan, broom, mop, and hoe. Grass and pruning shears that are operated with both hands tend to be less tiring then those operated with the fingers of one hand.

Although the finger movements take less energy than the arm movements the fingers tire more readily.

Seemingly more research is needed in biomechanics or human engineering to design equipment that conforms to the structure of the body rather then causing the body to conform to the equipment in the home as well as in industry and business. Even improperly designed scissors and pliers can cause the user to twist wrist and finger joints out of their normal position and make their use tiring.

Equipment that requires reaching across a broad surface to grasp and lift an object up or down may be uncomfortable epecially if the object is heavy or the worker must stretch to reach it.

12. Be aware of the fatigue of standing or keeping the body erect. Standing takes more energy and puts more pressure on joints, tendons, and ligaments then sitting or reclining. Fatigue is less when moving about than standing still. Resilient floor covering is more comfortable than hard flooring such as cement. Especially for homemakers with such health problems as heart ailments and varicose veins, sitting in order to work when possible and reclining during rest periods in order to ease the work of the heart are desirable.

13. Determine when sitting at work offers any advantages. The old exhortation "Sit while doing housework" is not necessarily the best policy. Although sitting may be desirable for health reasons, it may not reduce energy requirements. In order for sitting at work to offer advantages to the average homemaker these conditions must prevail:

(a) The task permits the worker to remain seated for some time. Getting up and down takes more energy than to remain standing.

(b) The worker, the chair, and the work surface are in right proportion to each other and right for the task. This means the surface clears the thighs and permits all parts of the body to remain in their natural alignment. Usually the work surface should be about three inches below the elbow although this may vary with the task.

(c) The work space provides adequate knee room; to push knees to the side is uncomfortable and tiring.

(d) The chair supports the body. It has a firm, broad seat, a flexible back support of the right height to fit the small part of the back, and permits feet to be flat on the floor or on an adequate foot rest. The hips and knees are bent at a comfortable angle with no pressure under knees to interfere with circulation. Recommended for sitting at work is a secretary's chair that is adjustable in seat height from floor, in the depth of seat from back rest, and the height of the back support.

Such a chair would permit comfortable sitting for such tasks as machine and hand sewing, writing, reading, and the preparation of vegetables when used with pull-out shelf or lap board.

(e) Equipment, materials, and supplies essential for the task are centered in front of worker to avoid excessive stretching. A chair with proper size swivel casters adds further convenience to sitting at work so that ironing at an adjustable board is also feasible.

14. Remember, it is more important to use body movements in harmony with body structure than to reduce the number of body movements. Many people need more, not less, exercise. Food not used for energy tends to become body fat.

WORKING EFFICIENCY

If life is to rise above mere existence and become human, warm, vibrant, and stimulating one must not become too busy "for the embrace, swift though it may occasionally be, or too hurried for the easy chuckle or the hearty laugh; too rushed for the gentle word or the sincere compliment; too hurried to take time to hear what others are saying; too pressed for the thoughtful note or simple gift. . . ."[3]

Although efficiency should not be an ultimate goal for the homemaker, in housework it is an important means for realizing the foregoing ideals.

Although there are no exact rules for maximizing achievement and satisfaction in housework and minimizing fatigue, the following concepts seem widely applicable for working efficiency:

1. Keeping in mind values held and goals to attain in deciding what household activities are worth doing, which could or should be eliminated, and which can be simplified.

2. Analyzing personal attitudes toward work in the home. Attitudes influence performance and contribute to fatigue—an unpleasant feeling of tiredness, lethargy, and disinterest. Strong negative emotions such as intense dislike for one's situation, work, and/or associates, lack of self-confidence, boredom, frustration, anxiety, and hatred as well as being overly conscientious, and striving toward impossible goals, when sustained over a period of time, contribute to debilitating tensions, fatigue, and even illness without any physical exertion. Obviously, both satisfaction and work efficiency become impossible unless such tensions are relieved through self-understanding, elimination

[3]Bernice Moore, "Time, Tensions, and Mental Health, Time Man's Constant—Its Use, His Variable," **Journal of Home Economics,** 49, December, 1957, pp. 758–763.

of offending factors, addition of pleasant interests, learning to accept the situation, and/or using professional help.

3. Practicing good health habits. Poor health can cause pathologic fatigue which often precedes and usually accompanies illness.

4. Developing the knowledge and skills essential for the successful management and for efficiency in the performance of housework.

5. Planning ahead and getting organized. Prepare a mental or written list of tasks in order of their urgency and importance with a flexible schedule for getting them done. Decide when, where, with what, with whom, and how each task is to be achieved. Assemble all necessary facilities before beginning the task. For example, well-planned shopping can reduce the number of trips especially for forgotten items. Reasonable timing provides for possible delays and interruptions. The importance of putting people first in home management will often interfere with work efficiency but should not upset the homemaker.

6. Following work patterns that include rest, change of pace (and sometimes also new interests), and recreative leisure before being satiated with the job and becoming overly fatigued. Sustained physical exertion generally results in physical fatigue which can reduce achievement and increase accidents but which can be relieved through rest and sleep.

7. Streamlining the three steps of getting ready, doing, and clearing away that are inherent in almost all tasks. This is illustrated by grouping tasks that need the same facilities as cleaning the whole house before putting away cleaning equipment and materials, letting the dishes remain in the drainer or dishwasher between some meals, or leaving out sewing so it can be picked up quickly at odd moments.

8. Using trays and carts to transport a number of items from one place to another to reduce steps as in setting and clearing the table. Carts make transport of bulky, large items easier but do not necessarily conserve more energy than carrying items.

9. Carrying out simple time-and-motion studies on repetitive household tasks that also seem excessively time consuming and/or fatiguing in order to discover ways of alleviating distasteful situations. A family member or friend may observe the homemaker as she carries out a task such as dishwashing in her usual way by hand or with dishwasher. Recording the steps or trips made plus the arm and body motions and sometimes time used makes evaluation and determination of possible improvements easier. Such a study with experimentation to develop better ways of doing the work can add interest and challenge to otherwise routine tasks. Figures 19.1 and 19.2 show results of changing arrangements after a time and motion study.

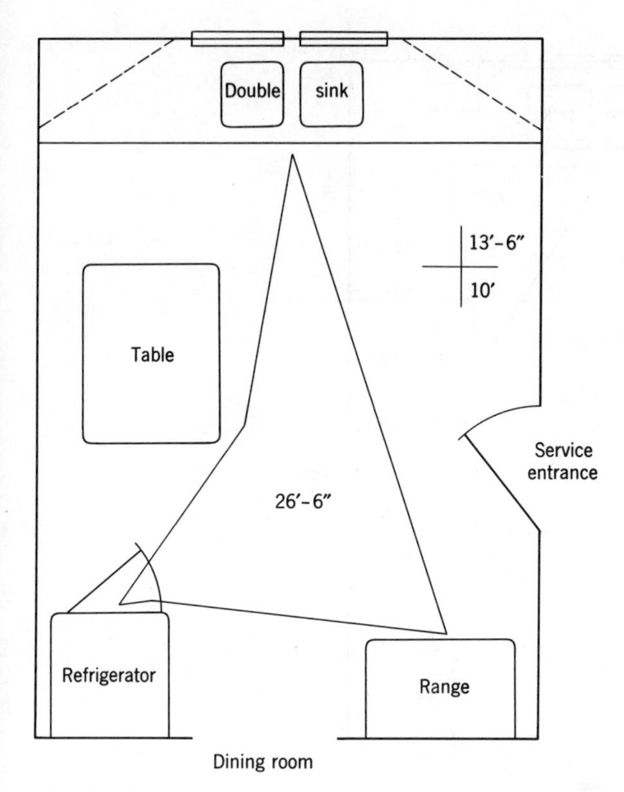

***Fig. 19.1** This old-fashioned kitchen requires many unnecessary steps. Wall cabinet space is almost nonexistent. No counter areas are near the refrigerator and range, and family members passing between the back door and dining room will tend to interfere with the homemaker's work at the range.*

10. Reducing the need for work. This is often possible through the increased use of labor-saving conveniences as the appropriate equipment, ready-to-use and easy-care products, commercial services, changing a desired product—for instance, instead of baking an apple pie, bake or stew or eat the apples raw—and reducing the frequency of the job. The last suggestion is illustrated by reducing the amount of housecleaning necessary by keeping more dirt out through the use of air conditioning with closed windows, foot scrapers, and mats at outside doors and storm windows and doors. It is further helped when everyone learns to put away properly what he has used and clean up his own mess. Moreover, properly regulated cooking and baking temperatures used with appropriate utensils reduces possibilities of foods scorching and boiling over whereas soil defying fabrics decrease need for cleaning. For instance stainless steel needs less polishing

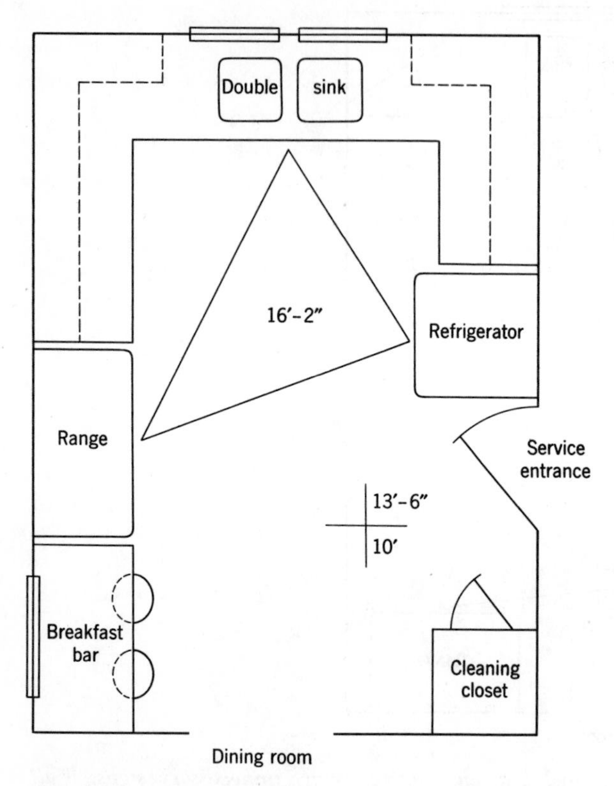

***Fig. 19.2** A study of tasks carried on in the kitchen plus a bit of imagination can relocate refrigerator and range, greatly increase cabinet space, and noticeably reduce the size of the work triangle.*

than silver, and wash-and-wear fabrics need little or no ironing. A reasonable number of decorative objects in the home decreases time for dusting. Moreover, hand-washed dishes tend to be more sanitary when not dried with a dish towel although glassware and silverware may lack the desired sparkle. Which methods are used is a family choice.

11. Lowering standards through intelligent neglect particularly during periods of stress as at the time of a new baby, sick family member, or heavy outside responsibilities. Dishes need not be washed after every meal nor linens or clothes laundered after each use. Some lower standards also reduce use of the decreasing resources of water and fuel.

12. Realizing that the greatest weight a person carries is his own body. This explains why climbing stairs (lifting the body) is a high

energy use activity and why normal weight is an advantage. Moreover, any homemaker susceptible to fatigue and health problems will find a house on one level desirable. Since energy requirements in walking and climbing stairs are increased through carrying a load, as child or vacuum cleaner, homemakers in multilevel houses will find planning to minimize trips up and down and duplicating some equipment and materials used on both floors rewarding.

13. Sharing work in the home with other family members especially by the employed homemaker. Such sharing may provide some opportunity for "togetherness," teaching children a sense of responsibility and how to do the work, and helping other family members to realize what is involved in work in the home.

14. Recognizing particularly tiring activities such as long periods of physical and mental effort especially with poor posture, restricted movement, continuous postural adjustments, and/or an uncomfortable environment, heavy physical exertion, unaccustomed action or work, lack of knowledge or ability to perform a task, work that requires extreme alertness and close attention, and working under pressure, as well as emotional stress even when not related to the work. Understanding of these factors can help the person to avoid many of these situations and to handle them when necessary.

15. Freeing the mind through use of automatic equipment and controls enables the homemaker to carry on several tasks at one time and to dovetail others. With an automatic washing machine and thermostatically controlled oven and iron, a homemaker may be washing the family laundry, baking an oven meal, ironing and looking after the children, or chatting with someone at the same time without undue strain, provided these procedures have been planned.

16. Striving to maintain desirable relationships within the family and with outsiders can not only enrich life but also contribute to greater achievement in work. Friction among individuals may cause fatiguing tensions while lack of cooperation among members of a group tends to delay action. Any off-the-job stress affects on-the-job performance and satisfaction.

17. Practicing safety at work. To prevent accidents a safe environment must be accompanied by safe working procedures such as immediately wiping up spills, keeping small items such as toys in certain out-of-traffic areas, not using stairs for storage, avoiding use of shoes and other clothing that may cause stumbling, not letting cords of electric appliances dangle where crawling youngsters can grasp them and pull appliances down on themselves, not permitting handles of cooking utensils to extend over edge of range so utensils can easily be

knocked down, disconnecting electric appliances when working on them and not handling them with wet hands, and using cutting boards for slicing and chopping. Thinking before acting will help the person avoid these and other hazards.

18. Striving to conserve resources and to minimize pollution. Although this is not a guideline for working efficiency, it can greatly influence work in the home and the well being of the family. The following suggestions tend to conserve resources and reduce pollution.

(a) Avoid overheating, overcooling, and overlighting the house as well as heating, cooling, and lighting any unused areas.

(b) Use adequate insulation throughout the building with storm windows and doors. This reduces fuel consumption for both heating and cooling so that the money saved can soon pay for the cost of installation.

(c) Operate such appliances as the dishwasher, the dryer only with full loads but not overloaded.

(d) Keep all equipment in good working condition. One dripping faucet can waste gallons of water daily, and a freezer that needs defrosting uses excessive electricity. Use automatic appliances wisely—only when they are really an asset.

(e) Encourage merchants to use only essential wrappings for all merchandise. Carrying a shopping bag decreases use of containers. Containers that are readily disposable or can be reused or recycled are preferable. Some plastics cause a special disposal problem.

(f) Avoid discarding products that can be used or recycled. Encourage the community to collect items such as cans, glass containers, paper, and metals for future use. The homemaker and the family can stretch their imagination and creative ability to use many items generally thrown away. Some can be utilized in arts and crafts. The disposal of equipment no longer used presents a still-unsolved problem.

(g) Reduce unnecessary automobile trips. Many times walking would be possible and preferable for health of both adults and children. In some situations, car pools, use of public transportation, or bicycles would be desirable. Of course a car of reasonable size and power in good condition and wisely driven also reduces fuel consumption and pollution.

19. Providing for the physically handicapped homemaker or other family member. Although time and energy conserving conveniences are similar for the normal and the handicapped homemaker, some adjustments in the dwelling and the equipment, and probably special

tools and clothing, are needed for the handicapped person. Since it is desirable for any handicapped person to be as independent as possible and to participate in family activities, as well as for the homemaker to assume as many homemaking responsibilities as she can manage, it is recommended that the family study research findings to learn how to meet the needs of the particular type of disability of the family member. A variety of publications and other aids are available.

Achieving satisfaction in managing the work in the home is a lifelong process shared by wife, husband, and children and supported by a favorable public opinion toward homemaking. Success, which probably requires a combination of mental, physical and spiritual effort, can be rewarding for the entire family and beneficial to society.

Experiments and Projects

1. Identify basic concepts and develop major generalizations that are important for the effective management of household equipment.
2. Become familiar with types of research in household equipment. Select a study related to the management of equipment. Write an abstract including how findings may be applied in the home. Report briefly to class.
3. In cooperation with another class member carry out a time and motion study of some repetitive household task such as food preparation, laundering, dishwashing, house cleaning. Explore and present possible ways for handling such tasks especially when they are disliked, too time consuming, tiring, difficult, and/or unfamiliar in order to minimize boredom, frustration, pressures, and fatigue and to increase satisfaction with work in the home. Different students can select different tasks. In demonstrating improvements they can explain and show (a) how to select and to maximize use of equipment, (b) treat the environment, (c) organize the work area, (d) handle the body, (e) use effective work methods, and (f) cope with negative attitudes, lack of knowledge, and ability.
4. Become familiar with different sources of information and help for the selection, use, and care of household equipment available to the majority of homemakers. Include books, bulletins, magazines, materials by manufacturers including instruction and guarantees, demonstrations, and places that would most likely handle problems with the performance and servicing of appliances. Evaluate with regard to reliability, adequacy, worth to homemaker, and clarity. Ar-

range a display of these materials which also illustrates their positive and negative aspects.

5. Investigate the needs of homemakers and other family members with the most common physical disabilities to learn possible adjustments of space, environment, equipment, and work methods. Also needed are special facilities for them to be as independent as possible and to carry on some activities including work in the home.
6. Demonstrate good body mechanics for different activities and how some household tasks may provide wholesome exercise. Illustrate some practical ideas for adjusting home facilities to provide for healthful use of body.
7. Select a personal problem related to household equipment management and develop a possible solution. Evaluate progress made.

CHAPTER 20
Metrication

The United States is the only technically advanced country in the world which is not now using, or, is not committed to begin using, the International Metric System of measurement. For this reason it appears that metrication for the United States of America is inevitable, and the sooner the conversion is begun, the less difficult and the less costly it will be.

Many U.S. industries have adopted the metric system. Both General Motors and Ford are manufacturing automobiles in metric measurement. The National Education Association recommended that the metric system be taught starting with 1972–1973 year. The schools in the state of California will adopt the metric system in 1976.

The International Metric System, Système International d'Unités (S.I.), has been adopted at the international standardization level. It is a simplified and coordinated system of measuring made possible by the Flemish mathematician Simon Stevin's invention of the decimal point in 1585. Stevin realized that if he moved his dot to the right of any given number and filled in the space with a zero, he had multiplied that number by 10; moving it to the left he divided by 10.

In any measuring system, the original difficulty encountered is arriving at a basic unit of measurement. Our present system was originally based on the length of barley grains, royal thumbs, and the like. Thomas Jefferson proposed a system of measurement based on the swing of a simple pendulum. In 1790 the French Academy of Sciences settled on a basic unit of measure equaling one ten-millionth of the earth's quarter meridian stretching from the North Pole to the equator and passing through Paris. This unit was called the meter.

The basic measure for weight (or mass) was tied to this meter in that the mass of one cubic centimeter of water under given conditions was established as one gram. For volume, the liter was pegged as the space occupied by one kilogram of water at a certain temperature. Since then there have been refinements and additions, mainly because scientists needed measurements for such things as electric current, the ampere; time, the second; and temperature, degree Kelvin, which in common use is translated to degree Celsius (formerly called Centigrade); and luminous intensity, the candela. These are the six base units. All other units, such as those for speed and volume, are derived from the base unit. Area is measured in square meters, speed in meters per second, density in kilograms per cubic meter. The newton, the unit of force, is a relationship involving meters, kilograms, and seconds. The pascal, the unit of pressure, is defined as one newton per square meter.

Since the metric system is based on the decimal system, multiples and submultiples of any given unit are related by powers of 10. For example: meter (m), kilometer (1000 m), millimeter (0.001 m). Latin prefixes are used to denote values that are smaller than the base unit. Thus a decigram is one-tenth of a gram; a dekagram is 10 grams. Greek prefixes are used for values greater than the base unit. (Table 20.1).

Table 20.1

Prefixes for Quantities Smaller than the Base Unit		Prefixes for Quantities Larger than the Base Unit	
Deci (one-tenth)	0.1	Deka (ten)	10
Centi (one-hundredth)	0.01	Hecto (one hundred)	100
Milli (one-thousandth)	0.001	Kilo (one thousand)	1000
Micro (one-millionth)	0.000001	Mega (one million)	1,000,000

If the metric system is made mandatory in the United States, many thoughtful decisions will be needed in the areas of household equipment and food preparation. All of our current tape measures, yardsticks, rulers, thermometers, thermostats, bathroom scales, and road maps—not to mention the odometers on cars—will be obsolete. This means new cookbooks, new measuring devices, new meat and candy thermometers, new calorie charts, and a new Celsius scale on the oven. Two types of conversion are possible:

1. **Soft conversion.** Metric measures could be substituted for nonmetric ones now in use without any change in standard dimensions. A 30-inch range would be a 0.762-meter range.

2. **Hard conversion.** Actual sizes of household equipment would change by basing sizes on metric-based modules. A 30-inch range would be a 0.800-meter range, which is about 31.5 inches.

A decision on which type of conversion should be made should give consideration to any changes in modules adopted by the building industry and the machine tool industry. A natural preference for easy numbers would probably lead to the adoption of modules of whole numbers or simple fractions that are easy to remember and easy to use in calculations—in other words, to a hard conversion. To avoid complete chaos, however, the new modules adopted must be compatible with those in related areas and with those presently in use. Many people in the United States who have purchased foreign cars have experienced the inconvenience of limited service available because the garage mechanic did not have tools to fit the engine parts of foreign cars and service facilities properly equipped to service foreign cars were not located where service was needed. To avoid a similar situation in servicing all types of equipment during a transition period from our customary units of measurement to the metric units, stocks of replacement parts and servicing tools to fit appliances for each system of measurement will have to be maintained.

Temperature conversion from one system to another when baking should not present a difficult problem for the homemaker. Our present oven temperature controls calibrated on the Fahrenheit scale are commonly in 25-degree intervals. Food specialists[1] assure us that temperatures in Celsius could be rounded for oven temperatures without detriment to the product (Table 20.2).

For all practical purposes the conversion of oven temperatures from Fahrenheit to Celsisus could be accomplished by dividing the Fahrenheit temperature by 2.

Measurements of both dry and liquid ingredients in food preparation in the United States is by volume in cups and spoonfuls. The capacity of a one-cup measure is one-fourth quart, our tablespoon is one-sixteenth of a cup, and our teaspoon is one-third of a tablespoon. The question arises: Should we continue to use the measuring equipment we now have but express capacities in metric units, or should changes in capacities be made in measuring equipment which are consistent with metric units? In other words, should our cup measure retain its present capacity of ¼ of a quart or 236.6 milliliters, or should it be ¼ of a liter or 250 milliliters?

[1]Dr. Fern Hunt.

Table 20.2

Oven Baking Temperatures

°F	°C	Possible Rounding Temperatures °C
250	121	120
275	135	135
300	149	150
325	163	165
350	177	175
375	190	190
400	205	205
425	218	220
450	232	230
475	246	245

[1]Dr. Fern Hunt.

If it were decided to change to a standard cup capacity of 250 milliliters, should the subdivisions be in tenths (25 milliliter graduations), should subdivisions continue to be in halves, quarters, and thirds, or should the subdivisions be in tenths, quarters, and thirds?

Should the ratio of tablespoons to cups be retained if a 250-milliliter standard cup is adopted? If the 16:1 ratio was retained, use of old favorite recipes could be continued with new measuring devices, but yield from the recipe would be increased slightly.

In most European countries, only liquid ingredients are measured by volume. Dry or solid ingredients are weighed. In converting to a new system of measurement, should we also convert the measurement of dry ingredients from volume to weight? Measurements could be more accurate by weight than by volume if women used their scales or balances correctly. There is a question, however, whether we need measurements that accurate in the kinds of cooking we do. Changing from volume measurements of dry ingredients to weights of dry ingredients might be a major source of difficulty to most American homemakers. A kitchen scale would have to be substituted for a measuring cup. To become accustomed to the variations in weight of the same volume of different materials could be troublesome, for example:

1 cup of butter = 224 grams
1 cup of sugar = 196 grams
1 cup of cake flour = 98 grams

At the present time in the United States, the capacity of saucepans, saucepots, and kettles is indicated in quarts when level full of water. If the capacity is designated in liters, should the actual size of the saucepan be changed to an even number of liters at the same time? For example, our present one-quart saucepan would hold 0.946 liters or 946 milliliters. Should it be changed in size to hold exactly one liter? The difference in capacity is not critical, and the change could be made when present tools are replaced.

The sizes of baking utensils and of skillets is now indicated in inches. In a soft conversion, a 10-inch skillet would become a 25.40-centimeter skillet; an 8-inch × 1½-inch cake pan would become a 20.32 × 3.81-centimeter cake pan. In the case of the skillet, there would probably be no reason to change until new tools were needed and then the new tools could be those which would produce a 25- or 26-centimeter skillet. In the case of the cake pan, any change in size should be dependent upon the decision made on the standard measuring cup size that is adopted. If a 250-ml cup and a tablespoon of 1/16 of a 250-ml cup were adopted, resulting in an increased volume of the recipe, the actual pan size might have to be increased. This could be rounded to the nearest even number of centimeters to adequately accommodate the volume.

Temperature conversions from Fahrenheit to Celsius are familiar to most students and practitioners in the scientific areas. However, even these people are not accustomed to the use of Celsius measurements in other applications. If the Weather Bureau forecast a high temperature for the day of 25°C, most people would experience some uncertainty on whether they should don a warm or light coat (Fig. 20.1).

In the metric system, the unit of mass is the kilogram, and the unit of weight is the newton. Up until now, we have "weighed" an object to obtain its "weight" (which was wrong in any case, as we were in fact determining its mass and not its weight). By definition, the mass of any object is the amount of matter that it contains; its weight the force with which the earth or any other heavenly body attracts it. The mass of any object never varies, but its weight may vary from one place to another. An object with a mass of 1 kilogram weighs exactly 9.80665 newtons. If such an object is transported to the moon, it will weigh only 2 newtons on the moon's surface and on its way to the moon it passes through a point where it becomes completely weightless. But under all these conditions its mass remains constant at 1 kg. The homemaker buying butter is interested in the quantity of the butter and not in the force with which the earth attracts it—in other words, she is interested in the mass of the butter not in its weight. Weight,

All You Will Need to Know About Metric
(For Your Everyday Life)

10

Metric is based on Decimal system

The metric system is simple to learn. For use in your everyday life you will need to know only ten units. You will also need to get used to a few new temperatures. Of course, there are other units which most persons will not need to learn. There are even some metric units with which you are already familiar: those for time and electricity are the same as you use now.

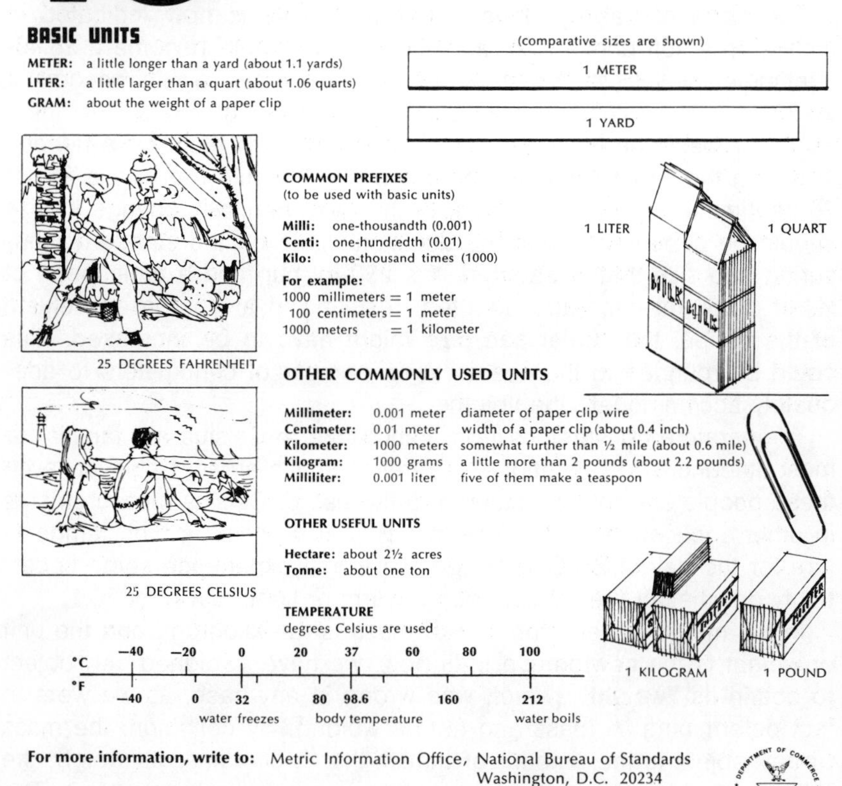

BASIC UNITS

METER: a little longer than a yard (about 1.1 yards)
LITER: a little larger than a quart (about 1.06 quarts)
GRAM: about the weight of a paper clip

(comparative sizes are shown)

1 METER

1 YARD

25 DEGREES FAHRENHEIT

25 DEGREES CELSIUS

COMMON PREFIXES
(to be used with basic units)

Milli: one-thousandth (0.001)
Centi: one-hundredth (0.01)
Kilo: one-thousand times (1000)

For example:
1000 millimeters = 1 meter
100 centimeters = 1 meter
1000 meters = 1 kilometer

1 LITER

1 QUART

OTHER COMMONLY USED UNITS

Millimeter:	0.001 meter	diameter of paper clip wire
Centimeter:	0.01 meter	width of a paper clip (about 0.4 inch)
Kilometer:	1000 meters	somewhat further than ½ mile (about 0.6 mile)
Kilogram:	1000 grams	a little more than 2 pounds (about 2.2 pounds)
Milliliter:	0.001 liter	five of them make a teaspoon

OTHER USEFUL UNITS

Hectare: about 2½ acres
Tonne: about one ton

TEMPERATURE
degrees Celsius are used

°C: −40, −20, 0, 20, 37, 60, 80, 100
°F: −40, 0, 32, 80, 98.6, 160, 212
water freezes (0 °C / 32 °F); body temperature (37 °C / 98.6 °F); water boils (100 °C / 212 °F)

1 KILOGRAM

1 POUND

For more information, write to: Metric Information Office, National Bureau of Standards
Washington, D.C. 20234

Fig. 20.1 All you will need to know about metric. (National Bureau of Standards)

on the other hand, is mainly of interest in the design of structures and in the natural sciences. The standard objects that are placed on a balance for measuring mass should be termed "masses" rather than "weights." What we now refer to as a set of weights should be a set of masses. Instead of "net weight 1 kg,"—"net mass 1 kg." Instead of "My weight is 60 kg,"—"My body mass is 60 kg."

In other countries which have metricated, it has been established that conversion factors from our present system to the metric should not be memorized. The continual conversion of metric to nonmetric units can complicate the changeover. An easier and more effective system is to become familiar with the actual sizes and quantities of the new units.

HOW TO START THINKING METRIC

Symbols

Every metric unit has one, and only one, correct symbol. The symbols were chosen with great care on the international level in order that no confusion about the symbols for units could ever arise. They remain the same in all languages in the world. Symbols are never followed by a period:

15 kg and **not** 15 kg.

No addition such as "s" should be made to the symbols to indicate plurality:

12 cm and **not** 12 cms

(The symbol s is for second.) A space is always left between a figure and the subsequent symbols:

15 km and **not** 15km

The Method of Writing Numbers

In order to facilitate the reading of numbers consisting of many digits, such numbers are divided into groups of three, starting from the decimal sign and working to the left and to the right. The groups are separated by a space and no longer by a comma or a point or any other means: 13 500. This method is necessary because the comma is now used as a decimal sign, in accordance with the practice followed in established metric countries: 0,036 41.

Sequences, Coordinates, and Other Sets

Since confusion may arise if the comma is used to separate elements of sequences, coordinates, and so on, if these elements already contain the comma as a decimal sign, it is recommended that the semicolon should be used consistently to separate elements. This is also in line with the usage in metric countries.

Examples

Sequences,
Decimals: 1,5; 2,0; 2,5; 3,0; . . .
Whole numbers: 1; 2; 4; 8; 16; . . .
Coordinates,
Decimals: (1,75; 2,80)
Whole numbers: (2; 4)

Where the Comma is Not Used

When writing numbers the comma is the **decimal** sign and it therefore only belongs with **decimal** methods of writing. In all other methods it is recommended that existing usage be retained.

Examples

Indication of time such as 3:45 P.M.
Numbering of paragraphs such as 2.13 or 2.12.2

The Method of Writing Units

The names of all units are in lower-case letters when written out in full, for example, kilometer, gram, liter. The symbols for most units are also in lower-case letters: km, g, l. Only in cases where the name of a unit is derived from the name of a person should the symbol be either a capital letter or commence with a capital letter if the symbol consists of more than one letter:

	gram	symbol g
but	joule	symbol J

UNITS FOR EVERYDAY USE

The base unit of length is the meter. The meter is somewhat longer than a yard.

Length and Distance

Meter (m)	Replaces yard and foot. The dimensions of a room will be expressed in meters: 4 m × 5 m. Dress materials will be sold by the meter and parts of a meter.
Kilometer (km) 1000 m	Replaces the mile. Road maps will indicate distances in kilometers and odometers will be graduated in kilometers.
Centimeter (cm) 1/100 m	Replaces foot and inch. Height will be measured in centimeters. A woman of average height will be 163 cm tall.
Millimeter (mm) 1/1000	Replaces the inch. The millimeter will be used in engineering drawings and the lengths and diameters of screws, thickness of wire, sheet metal, etc. will be measured in millimeters.

Area

The main unit of area is the square meter. This represents the area of a square of which the dimensions are 1 m × 1 m.

Square meter (1 m^2)	Replaces square yard and square foot. The floor area of a room or house will be given in square meters.
Square centimeters (cm^2) $\frac{1}{10\ 000}$ m^2	Replaces square inch for small areas.
Hectare (ha) 10 000 m^2	Replaces acre. The areas of farms will be indicated in hectares. Surveying and mapping, which includes the sizes of properties, property transactions, and the registration of deeds will be conducted in metric units. Original deeds of transfer now in customary units could be converted to metric when transfer takes place.

Volume, Capacity

The main unit is the cubic meter and represents the volume of a cube of which each side is 1 m in length.

Cubic meter (m^3)	Replaces cubic yard and cubic foot.
Cubic decimeter (dm^3) $\frac{1}{10}$ m^3 or Liter	Replaces gallon and pint. A special name for the cubic decimeter is the liter, and it is used to express the volume of liquids as well as the capacity of such appliances as refrigerators.
1 kiloliter (kl) (same as 1 m^3) 1000 1	Replaces gallon and pint. Water bills will be calculated in kiloliters consumed by the household.
1 milliliter (ml) $\frac{1}{1000}$ l	Replaces the pint and fluid ounce. The quantity of cooking oil, soft drinks, etc. will be given in milliliters on the container.
1 megaliter (Ml) 1 000 000 l	Replaces gallon for large volumes of water. The water content of dams will be measured in megaliters.

Mass

The base unit of mass is the kilogram. The kilogram is about 10% more than 2 pounds. Half a kilogram of butter (500 grams) will, therefore, probably replace the existing pound of butter.

Kilogram (kg) 1 000 g	Replaces 2 pounds and over. Meat, flour, sugar, fresh vegetables, etc. will probably be sold by the kilogram. Body mass will be determined in kilograms.
Gram (g) 1/1 000 kg	Replaces ounces and pounds. Capacity of canned juices, vegetables, and fruits will be in grams.
Megagram (Mg 1 000 000 g or Metric ton (t)	Replaces the existing ton for big loads.

Velocity, Speed

Velocity is expressed in kilometers per hour.

Kilometers per hour (km/h)	Replaces miles per hour. The odometer of all new cars will be graduated in km/h.

Force, Weight

Newton (N) kg.m/s^2	Replaces pound-force.

Pressure

Pascal (Pa) (N/m^2) or Bar (bar)	Replaces pound-force per square inch. The bar may be used for pressure in pressure cookers as well as for tire pressures.
Millibar (mbar) 1/1 000 bar	Used in airplanes to determine altitude.

Temperature

Celsius (°C)	Replaces Fahrenheit. Atmospheric and body temperature will be expressed in °C. Oven temperatures for baking purposes will be given in °C.

Energy, Work, Quantity of Heat

The basic unit of energy is the joule. As in the case of the calorie, the joule is too small for everyday use, and the kilojoule will be used commonly.

Joule (J) N.m	Replaces the calorie.
Kilojoule (1 000 J)	Replaces the kilocalorie. Used mainly in dietetics.
Joule per kilogram kelvin J/kg K	Specific heat. Replaces Calories per gm °C or BTU per pound °F.
Watt per meter kelvin W/mK	Replaces Calories/cm^3/°C/second to express thermal conductivity.

Electrical Units

The electrical units used in our customary system are part of the International Metric System, and most of the symbols are the same.

Coulomb (C)	Electric charge.
Ampere (A)	C/s.
Watt (W) J/s.	Power.
Volt (V) W/A	Voltage, electromotive force.
Ohm (Ω) V/A	Electrical resistance.
Farad (F) A.s/V	Electric capacitance.
Henry (H) Vs/A	Inductance.
Siemens (S) A/V	Conductance.

Light

The basic unit in both our customary units and the International Metric System is the candela.

Candela (cd)	The basic unit.
Candela per square meter cd/m^2	Luminance.
Lumen (lm) cd.sr (sr = steradian = solid angle)	Luminous flux.
Lux (lx) lm/m^2	Illumination.

THINKING METRIC

A useful visual comparison of customary and metric units has been prepared by the National Bureau of Standards (Fig. 20.1).

APPROXIMATE CONVERSION FACTORS[2]

Try to use the conversion factors as seldom as possible, rather make an attempt to think metric. Should you, however, need a conversion

[2]The conversion factors are approximate and should not be used where a high degree of accuracy is required.

factor (e.g., for a recipe), it should be used once only. The metric quantity can then be written next to the quantity in our customary units and after that should be used always. A continual conversion of units will complicate the changeover considerably.

Length, Distance

Metric to Nonmetric	Nonmetric to Metric
1 km = 0,6 mile	1 mile = 1,6 km
8 km = 5 miles	5 miles = 8 km
1 m = 1,1 yd	1 yd = 0,9 m
or 3,3 ft	1 ft =0,3 m
or 39,4 in.	1 in. = 25,4 mm
1 mm = 0,04 in. (= 1/25 in.)	

Area

Metric to Nonmetric	Nonmetric to Metric
1 ha = 2,5 acres	1 acre = 0.4 ha
1 m² = 1,2 sq yd	1 sq yd = 0,8 m²
= 10,8 sq ft	1 sq in. = 6,5 cm²
1 cm² = 0,2 sq in.	1 sq in. = 6,5 cm²

Volume, Capacity

Metric to Nonmetric	Nonmetric to Metric
1 m³ = 1,3 cu yd or 35.3 cu ft	1 cu yd = 0,8 m³
1 cm³ = 0,06 cu in.	1 cu in. = 16,4 cm³
1 l = 0,2 gallon = 1,75 pt, 1¾ pt	1 gallon = 4,5 l
4 l = 7 pt	1 pt = 567 ml or 7 pt = 4 l
1 ml = 0,04 fluid oz	1 fluid oz = 28,4 ml

Mass

Metric to Nonmetric	Nonmetric to Metric
1 metric ton = 2 205 lb = 1,1 tons	1 ton = 0,9 metric ton
1 kg = 2,2 lb	1 lb = 0,45 kg = 454 g
1 g = 0,04 oz	1 oz = 28,3 g

Velocity, Speed

Metric to Nonmetric	Nonmetric to Metric
1 km/h = 0,6 m.p.h.	1 m.p.h. = 1,6 km/h
8 km/h = 5 m.p.h.	5 m.p.h. = 8 km/h

Pressure

Metric to Nonmetric	Nonmetric to Metric
1 pascal, or bar = 14.5 pound force per sq in.	1 pound force per sq in = 1 pascal or 1 bar

Temperature

10°C = 50°F. For every increase or decrease of 5 degrees of the Celsius scale, 9 degrees are added to or subtracted from the Fahrenheit scale.

Number of degrees Celsius	= (Number of degrees Fahrenheit − 32) × 5⁄9
Number of degrees Fahrenheit	= (Number of degrees Celsius × 9⁄5) + 32

Energy

The joule (J) replaces the calorie (cal), and kilojoule (kJ) replaces the kilocalorie.

Metric to Nonmetric	Nonmetric to Metric
1 J = 0,24 cal	1 Cal = 4,18 kJ
1 kJ = 0,24 Cal	1 cal = 4,18 J

Force, Weight

Metric to Nonmetric	Nonmetric to Metric
1 N (newton) = 0,225 pound-force	1 pound-force = 4,45 N

Work

Metric to Nonmetric	Nonmetric to Metric
1 kW (kilowatt) = 1,34 horsepower	1 horsepower = 0,75 kW

Terminology

The International Metric System, the Système International d'Unités, known internationally as "S.I." is a decimal system in which changes of measurement from one S.I. unit into another is made by multiplying or dividing by a power of 10. The system is also what is known as "coherent." This means that quantities expressed in S.I. units used in calculations end up with a quantity of S.I. units: If a car goes 60 meters in 5 seconds, its average speed is 60 divided by 5 meters per second and the answer is in the S.I. unit of speed or velocity. By way of contrast, the conventional system of weights and measures used in the United States is by no means as convenient: a car going 60 yards in 5 seconds would not usually be described as going at 60 divided by 5 yards per second—we would prefer to give the speed in miles per hour or possibly feet per second.

The International System of Units (S.I.) is built up from three kinds of units: base units, derived units, and supplementary units.

There are seven base units:

Physical Quantity	Unit		Symbol	Equivalent Conv. Unit
Length	Meter		m	39.4 in., 1.1 yd
Mass	Kilogram		kg	2.2 lb
Time	Second		s	Same
Electric current	Ampere		A	Same
Temperature	Celsius		C	32°F = 0°C
Luminous intensity	Candela		cd	Same
Amount of substance	Mole		mol	Contains as many elementary entities as there are atoms in 0.012 kilogram of carbon 12
Some Derived Units		Derivation		
Area	Square meter		m^2	1/2 sq yd
Volume	Cubic meter		m^3	1.3 cu yd
Speed	Meter per second		m/s	3.3 ft/sec
Force	Newton	Kg m/s^2	N	0.225 pound force
Work, energy, quantity of heat	Joule	N/m	J	0.24 cal
Power	Watt	J/s	W	Same
Electrical potential, EMF	Volt	W/A	V	Same
Electrical resistance	Ohm	V/A	Ω	Same
Frequency	Hertz	1/s	Hz	Same
Luminous flux	Lumen	cd sr	Im	Same

Supplementary Units

Physical Quantity	Unit	Symbol
Plane angle	Radian	rad
Solid angle	Steradian	sr

Experiments and Projects

1. About 11% of U.S. manufacturing companies already use the metric system to some degree. Give five examples of products already sold marked in metric quantities.
2. Make a note of your measurements in metric units: height in meters, chest/bust in centimeters, waist in centimeters, and mass in kilograms.
3. Construct a cube, 10 millimeters on each side. What is the length of each side in meters? In centimeters? What is the area of the cube in mm^2, in m^2, in cm^2? What is the volume in liters, m^3, dm^3, and cm^3?
4. What is normal human body temperature in degrees Celsius? What is the outdoor temperature in degrees Celsius? What is the temperature of the room in degrees Celsius?
5. Set the oven control now calibrated in degrees Fahrenheit to the temperature equivalent of the following baking temperatures: 120°C, 165°C, 175°C, 200°C, 220°C.
6. Measure the difference in volume in liters of the following recipe using, first, the customary measuring cups, tablespoons, teaspoons, etc. and then using a 250-ml cup, 15-ml tablespoon, and 5-ml teaspoon.

Pancakes

Beat together	1 egg
	1¼ cups milk
Add	1¼ cups sifted flour
	1 tsp sugar
	1½ tsp baking powder
	½ tsp salt
	2 T salad oil

Beat together until ingredients are well blended.

7. In what metric units would you express the following: miles per hour, Btu, energy (work), force, pressure, mass, electric current, electrical potential, power, frequency, electric resistance, luminous intensity?

Magazines

American Dyestuff Reporter
American Fabrics
American Home
Appliance Manufacturer
Better Homes and Gardens
Changing Times
Chemical Engineering
Consumers' Bulletin
Consumer Reports
Detergent Age
Family Circle
Family Safety
Farm Journal
Good Housekeeping
Home Furnishings Daily
Home Furnishing Review
House and Garden
House Beautiful
Housewares Review
Illuminating Engineering
Journal of Home Economics
Ladies Home Journal
Light
Light and Lighting
McCall's
Merchandising Week
Modern Textiles
Practical Forecast for Home Economics

Redbook
Seventeen
Sunset Magazine
Soap and Chemical Specialties

Manufacturers

Admiral Corporation, 3800 Cortland St., Chicago 60047
Aluminum Goods Manufacturing Company, Manitowoc, Wis. 54220
Amana Refrigeration, Inc., Amana, Iowa 52203
American Radiator and Standard Sanitary Corp., 40 West 40th St., New York 10018
Armstrong Cork Company, 1010 Concord St., Lancaster, Pa. 17604
Blackstone Industries, Inc., 2485 South Broad St., Trenton, N.J. 08610
Calgon Corp., 1965 Calgon Center, Pittsburgh, Pa. 15219
Caloric Corporation, Topton, Pa. 19562
Club Aluminum Products Company, 825 26th St., LaGrange Park, Ill. 60528
Colgate-Palmolive Co., 302 Park Ave., New York 10022
Corning Glass Works, Corning, N. Y. 14830
Crane Company, 300 Park Ave., New York 10022
Culligan, Inc., 1 Culligan Parkway, Northbrook, Ill. 60062
Dominion Electric Company, Mansfield, Ohio 44902
Dow Chemical Company, Midland, Mich. 48640
E. I. Du Pont de Nemours & Co., 1007 Market St., Wilmington, Del. 19898
Ecko Products, 1949 N. Cicero Ave., Chicago 60639
Foley Manufacturing Company, 3300 N.E. Fifth, Minneapolis 55413
Formica Corporation, 4632 Spring Grove Ave., Cincinnati 45232
Frigidaire Division, General Motors Corporation, 300 Taylor St., Dayton, Ohio 45401
General Electric Company, Consumers Institute, Appliance Park, Louisville, Ky. 40225
General Electric Company, Lamp Division, Nela Park, Cleveland 44114
General Electric Company, Portable Appliance Dept., 1285 Boston Ave., Bridgeport, Conn. 06602

Hamilton Beach, Division of *Scovill Manufacturing Company*, Racine, Wis. 53401
Hamilton Manufacturing Company, Two Rivers, Wis. 54241
Harper-Wyman Co., 930 N. York Rd., Hinsdale, Ill. 60521
Hobart Manufacturing Company, KitchenAid Division, Troy, Ohio 45373
Honeywell, Inc., 2701 Fourth Ave., South, Minneapolis 55408
Hoover Company, North Canton, Ohio 44720
Hotpoint Home Economics Institute, 5600 West Taylor, Chicago 60644
International Nickel Co., Inc., 1 New York Plaza, New York 10004
S. C. Johnson and Son, Inc., Racine, Wis. 53401
Kelvinator, American Motors Corp., 14250 Plymouth Rd., Detroit 48232
Lever Bros., Co., 390 Park Ave., New York 10022
Lewyt Corporation, 43–22 Queens, Long Island City, N. Y. 11101
Magic Chef, Inc., 740 King Edward Ave. S.E., Cleveland, Tenn. 37311
Maytag Co. (The), 512 N. Fourth Ave. West, Newton, Iowa 50208
Mirro Aluminum Co., Manitowoc, Wis. 54220
National Presto Industries, Inc., Eau Claire, Wis. 54701
Nesco, Inc., 947 W. St. Paul Ave., Milwaukee, Wis. 55116
Norge Sales Corp., Merchandise Mart Plaza, Chicago 60654
John Oster Manufacturing Co., 5055 Lydell Ave., Milwaukee 53217
Owens-Corning Fiberglas Corp., Fiberglas Tower, Toledo, Ohio 43601
Parker Rust Proof Company, 3605 Perkins Ave., Cleveland, Ohio 44114
Philco Corporation, Tioga and C Sts., Philadelphia 19134
Proctor & Gamble Co., Ivorydale Technical Center, Cincinnati 45217 or P.O. Box 599 45201
Procter and Schwartz Corp., 700 W. Tabor Rd., Philadelphia 19120
Proctor-Silex Corp., 333 W. 65th St., Chicago 60638
Rival Manufacturing Co., 36 and Bennington, Kansas City, Mo. 64129
Robertshaw Controls Company, 1701 Byrd Ave., Richmond, Va. 23226
Ronson Corp., 1 Ronson Rd., Woodbridge, N.J. 07095
Geo. D. Roper Corporation, 1905 W. Court St., Kankakee, Ill. 60701
St. Charles Manufacturing Company, St. Charles, Ill. 60174
Speed Queen, Division of *McGraw-Edison Company*, Ripon, Wis. 54971
Sunbeam Corporation, 3600 Roosevelt Rd., Chicago 60650
Tappan Company, 250 Wayne, Mansfield, Ohio 44902
Toastmaster Division, *McGraw-Edison Co.*, Elgin, Ill. 60120
U.S. Steel Corp., 600 Grant, Pittsburgh, Pa. 15230
Warwick Electronics, Inc., 7300 N. Lehigh, Chicago 60648
WasteKing Corp., 3300 E. 50th St., Los Angeles 90058
Wear-Ever Aluminum, Inc., New Kensington, Pa. 15068
West Bend Company, West Bend, Wis. 53095

Westinghouse Electric Corporation, Mansfield, Ohio 44902
Westinghouse Electric Corporation, Lamp Division, Bloomfield, N.J. 07003
Whirlpool Corporation, Benton Harbor, Mich. 49022
Wiremold Company, Hartford, Conn. 06110
Youngstown Kitchens, Division of *American-Standard*, Warren, Ohio 44483

Manufacturers publish information at frequent intervals. Teachers should request new materials at least once a year. (Also contact local utilities—gas, electricity, and water.)

Associations

American Association of Textile Chemists and Colorists, P.O. Box 12215, Research Triangle Park, N.C. 27709

American Gas Association, Inc., 1515 Wilson Blvd., Arlington, Va. 22209

American Gas Association, Inc., Laboratories, 8501 East Pleasant Valley Road, Cleveland, Ohio 44131

American Heart Association, 44 East 23rd, New York 10010

American Home Economics Association (AHEA), 1600 Twentieth St. N.W., Washington 20009

Home Economics in Business (HEIB)

American Institute of Architects, 1735 New York Ave. N.W., Washington 20006

American Iron and Steel Institute, 150 East 42nd St., New York 10017

American National Standards Institute (ANSI), 1430 Broadway, New York 10018

American Society of Heating, Refrigerating, and Air-Conditioning Engineers, United Engineering Center, 345 East 47th St., New York 10017

Association Home Appliance Manufacturers (AHAM), 20 N. Wacker Drive, Chicago 60606

Bakelite Company, 30 East 42nd St., New York 10017

Better Light Better Sight Bureau, 750 Third Ave., New York 10017

Consumers Union, Mt. Vernon, N.Y. 10550

Detergents Inc., 1147 Chesapeake Ave., Columbus, Ohio 43212

Edison Electric Institute, 750 Third Ave., New York 10017

Electrical Womens Round Table (EWRT), Ruth Whiting, Executive Secretary-Treasurer, 5630-D Coach and Four Drive, West Dayton, Ohio 45440

Electronic Industries Association, 2001 I St., N.W., Washington 20006

Equitable Life Assurance Society of the U.S., 393 Seventh Ave., New York 10001

Federal Trade Commission, Washington 20580
Gas Appliance Manufacturers Association (GAMA), 60 East 42nd St., New York 10017
Household Finance Corporation, Money Management Institute, Prudential Plaza, Chicago 60601
Illuminating Engineering Society, 345 East 47th, New York 10017
Industry Committee on Interior Wiring Design, Room 1650, 750 Third Ave., New York 10017
Institute of Electrical and Electronics Engineers, Inc. (IEEE), 345 East 47th, New York 10017
National Better Business Bureau, Inc., 230 Park Ave., New York 10017 (also local offices)
National Bureau of Standards, Washington 20234
National Electrical Manufacturers Association (NEMA), 155 East 44th St., New York 10017
National Housewares Manufacturers Association, 1130 Merchandise Mart, Chicago 60654
Porcelain Enamel Institute, Inc., 1900 L St. N.W., Washington 20036
Soaps and Detergent Association, 475 Park Ave. South, New York 10006
Society of the Plastics Industry, Inc. (The), 250 Park Ave., New York 10017
Small Homes Council, Mumford House, Univ. of Illinois, Urbana, Ill. 61503
Superintendent of Documents, U.S. Government Printing Office, Washington 20250
(Write to this address for most of the USDA publications. Send cash or Post office money order; do not send stamps.)
Underwriters' Laboratories, Inc., 207 East Ohio St., Chicago 60611
Vacuum Cleaner Manufacturers' Association, 1615 Collamer St., Cleveland 44110
Water Conditioning Foundation, 1780 Maple St., Northfield, Ill. 60093

These associations either publish informational material or can suggest reliable sources.

Common Abbreviations

AGA. American Gas Association
AHAM. Association Home Appliance Manufacturers
AHEA. American Home Economics Association
ANSI. American National Standards Institute
ASHRAE. American Society of Heating, Refrigerating, and Air-conditioning Engineers
BLBS. Better Light Better Sight Bureau

EEI. Edison Electric Institute
EWRT. Electrical Womens Round Table
GAMA. Gas Appliance Manufacturers Association
HEIB. Home Economics in Business
IEEE. Institute of Electrical and Electronic Engineers
IES. Illuminating Engineering Society
MACP. Major Appliances Consumer Action Panel
NBS. National Bureau of Standards
NEMA. National Electrical Manufacturers Association
UL. Underwriters' Laboratories

Index